SEARCH METHODOLOGIES

SEARCH METHODOLOGIES
Introductory Tutorials in Optimization and
Decision Support Techniques

Edited by

EDMUND K. BURKE
GRAHAM KENDALL

 Springer

Edmund K. Burke
University of Nottingham
United Kingdom

Graham Kendall
University of Nottingham
United Kingdom

Library of Congress Control Number: 2005051623

ISBN-10: 0-387-23460-8   ISBN-13: 978-0387-23460-1
ISBN-10: 0-387-28356-0 (e-book)   ISBN-13: 978-0387-28356-2 (e-book)

Printed in the United States of America.   Printed on acid-free paper.

9 8 7 6 5 4 3 2

springer.com

1004778414

Contents

Foreword

This is not so much a foreword as a testimonial. As I embarked on the pleasant journey of reading through the chapters of this book, I became convinced that this is one of the best sources of introductory material on the search methodologies topic to be found. The book's subtitle, "Introductory Tutorials in Optimization and Decision Support Techniques", aptly describes its aim, and the editors and contributors to this volume have achieved this aim with remarkable success.

The chapters in this book are exemplary in giving useful guidelines for implementing the methods and frameworks described. They are tutorials in the way tutorials ought to be designed. I found no chapter that did not contain interesting and relevant information, and found several chapters that can only be qualified as delightful.

Those of us who have devoted a substantial portion of our energies to the study and elaboration of search methodologies often wish we had a simple formula for passing along the core notions to newcomers to the area. (I must confess, by the way, that I qualify as a relative newcomer to some of the areas in this volume.) While simplicity, like beauty, to some degree lies in the eyes of the beholder, and no universal or magical formula for achieving it exists, this book comes much closer to reaching such a goal than I would have previously considered possible. It will occupy a privileged position on my list of recommended reading for students and colleagues who want to get a taste for the basics of search methodologies, and who have an interest in equipping themselves to probe more deeply.

If good books may be likened to ladders that help us ascend to higher rungs of knowledge and understanding, we all know there are nevertheless many books written in technical areas that seem to be more like stumbling blocks, or at best broken stepping stools that deposit us in isolated nooks offering no clear access to continued means of ascent. Not so the present book. Its chapters lead to ideal sites for continued exploration, and offer compelling motivation for further pursuit of its ideas and frameworks. If my reckoning is not completely amiss, those who read this volume will find abundant reasons for shar-

ing my conviction that we owe its editors and authors a true debt of gratitude
for putting this work together.

Fred Glover, Professor
Leeds School of Business, University of Colorado
Boulder, CO, USA

Preface

We first had the idea for this book over three years ago. It grew out of a one day workshop entitled, *Introductory Tutorials in Search, Optimization and Decision Support Methodologies (INTROS)*, which was held in Nottingham in August 2003. The aim of the workshop was to deliver basic introductions to a broad spectrum of search methodologies from across disciplinary boundaries. It was supported by the UK Engineering and Physical Sciences Research Council (EPSRC) and the London Mathematical Society (LMS) and was attended by over one hundred delegates from all over the world. We were very fortunate to have eleven of the world's leading scientists in search methodologies presenting a range of stimulating and highly informative tutorials. All of the INTROS presenters have contributed to this volume and we have enhanced the content by inviting additional, specifically targeted, complementary chapters. We are pleased to be able to present such a comprehensive, multidisciplinary collection of tutorials in this crucially important research area.

We would like to take this opportunity to thank the many people who have contributed towards the preparation of this book. We owe a great debt of gratitude to the authors of the chapters. As one would expect from such a distinguished group of scientists, they have prepared their excellent contributions in a thoroughly reliable and professional manner. Without them, of course, the book would not have been possible. We are extremely grateful to our copy editor, Piers Maddox who excelled himself in bringing together, into one coherent structure, the various documents that we sent to him. We are also very grateful to Gary Folven, Carolyn Ford and their staff at Springer who have provided us with invaluable advice and support during every step of the way. We would like to offer our gratitude to Fred Glover for writing the foreword for the book. His warm praise is particularly pleasing. A special thank you should go to Emma-Jayne Dann and Alison Payne for all the administrative support they have given us, both in the preparation of this volume and in the organization of the INTROS workshop that underpinned it. We are also very thankful to EPSRC and LMS for the financial support they gave us to hold this workshop. Finally, we offer a special thank you to the INTROS delegates for their enthusiasm and their encouragement.

We hope you enjoy reading this volume as much as we have enjoyed putting it together. We are already planning a second edition and, if you have any comments which can help us improve the book, please do not hesitate to contact us. We would welcome your advice.

EDMUND K. BURKE AND GRAHAM KENDALL
ekb@cs.nott.ac.uk and gxk@cs.nott.ac.uk
June 2005

Chapter 1

INTRODUCTION

Edmund K. Burke, Graham Kendall

Automated Scheduling, Optimisation and Planning Research Group
The School of Computer Science and IT, University of Nottingham, UK

1.1 INTER-DISCIPLINARY DECISION SUPPORT: MOTIVATION

The investigation of search and optimization technologies underpins the development of decision support systems in a wide variety of applications across industry, commerce, science and government. There is a significant level of diversity among optimization and computational search applications. This can be evidenced by noting that a very small selection of such applications includes transport scheduling, bioinformatics optimization, personnel rostering, medical decision support and timetabling. More examples of relevant applications can be seen in Pardalos and Resende (2002), Leung (2004) and Dell'Amico et al. (1997). The exploration of decision support methodologies is a crucially important research area. The potential impact of more effective and more efficient decision support methodologies is enormous and can be illustrated by considering just a few of the potential benefits: more efficient production scheduling can lead to significant financial savings; higher quality personnel rosters lead to a more contented workforce; more efficient healthcare scheduling will lead to faster treatment (which could save lives); more effective cutting/packing systems can reduce waste; better delivery schedules can reduce fuel emissions.

This research area has received significant attention from the scientific community across many different academic disciplines. Indeed, a quick look at any selection of key papers which have impacted upon search, optimization and decision support will demonstrate that the authors have been based in a number of different *departments* including Computer Science, Mathematics, Engineering, Business, Management, and others. It is clearly the case that the investigation and development of decision support methodologies is inherently multi-disciplinary. It lies firmly at the interface of Operational Research and Artificial Intelligence (among other disciplines). However, not only is the underlying methodology inherently inter-disciplinary but the broad range of application areas also cuts across many disciplines and industries. We firmly believe that scientific progress in this crucially important area will be made far more effectively and far more quickly by adopting a broad and inclusive multi-disciplinary approach to the international scientific agenda in this field. The way forward is inter-disciplinary.

This observation provides one of the key motivations for this book. The book is aimed primarily at first-year postgraduate students and final-year undergraduate students. However, we have also aimed it at practitioners and at the experienced researcher who wants a brief introduction to the broad range of decision support methodologies that is available in the scientific literature. In our experience, the key texts for these methodologies lie across a variety of volumes. This reflects the broad range of disciplines that are represented here. We wanted to bring together a series of entry-level tutorials, written by world-leading scientists from across the disciplinary range, in one single volume.

1.2   THE STRUCTURE OF THE BOOK

This book was originally motivated by the thought of being able to give first year Ph.D. students a single volume that would give them a basic introduction to the various search and optimization techniques that they might need to use during their research. In this respect the book can be read in a sequential manner. However, each chapter also stands alone and so the book can be dipped into when you come across a technique which you are not familiar with, or just need to find some general references on a particular topic.

If you want to read the book all the way through, we hope that the way we have ordered the chapters makes sense. We start by introducing (in Chapters 2 and 3) some classical search and optimization techniques which, although not always suitable (particularly when your problem has a very large search space), are still important to have in your "tool box" of methodologies. Many of the other chapters introduce various search and optimization techniques, some of which have been used for over 30 years (e.g. genetic algorithms, Chapter 4) and some which are relatively new (e.g artificial immune systems, Chapter 13). Some of the chapters consider some of the more theoretical aspects of search and optimization. The chapter by Darrell Whitley and John Paul Watson, for example, introduces *Complexity Theory and the No Free Lunch Theorem* (Chapter 11) whilst Colin Reeves considers *Fitness Landscapes* in Chapter 19.

One element of every chapter is a section called *Tricks of the Trade*. We recognize that it is sometimes difficult to know where to start when you first come across a new problem. Which technique or methodology is the most appropriate? This is a *very* difficult question to answer and forms the basis of much research in the area. It often requires experiments with a range of these techniques. *Tricks of the Trade* is designed to give you some guidelines on how you should get started and what you should do, if you run into problems. Although tricks of the trade is towards the end of each chapter, we believe that it could be one of the first sections you read.

We have also included *Sources of Additional Information* in every chapter. These sections are designed as pointers to useful books, web pages, etc, which might be where you turn to next, once you have read the chapter in this book.

As this book is aimed primarily at the *beginner* (first-year Ph.D. student, final-year undergraduate, practitioner/researcher learning a new technique, etc) we thought it might be useful to explain some basic concepts which many books just assume that the reader already knows. Indeed, our own Ph.D. students and final-year undergraduates often make this complaint. We realize that the following list is not complete. Nor can it ever be, as we are not aiming to write a comprehensive encyclopedia. If you feel that any important terms are missing, please let the editors (authors of this introduction) know and we will consider including them in future editions. All of these concepts are, purposely, explained in an informal way so that we can get the basic ideas across to the reader. More formal definitions can be found elsewhere (see the *Sources of Additional Information* and *References*), including in the chapters of this book.

1.3   BASIC CONCEPTS AND UNDERLYING ISSUES

In this section we will go through a number of basic terms and issues and offer a simple description or explanation. In the spirit of attempting to explain these concepts to beginners, we will restrict the formal presentation of these concepts as much as possible. Instead, we will attempt to explain the basic ideas which underpin the terminology and the (often mathematical) formulations. Many of these terms are described and discussed throughout the book (see the index).

Artificial intelligence   Artificial Intelligence is a broad term which can be thought of as covering the goal of developing computer systems which can solve problems which are usually associated with requiring human level intelligence. There are a number of different definitions of the term and there has been a significant amount of debate about it. However, the philosophical arguments about what is or is not Artificial Intelligence do not fall within the remit of this book. The interested reader is directed to the following (small) sample of general AI books: Negnevitsky (2005), Russell and Norvig (2003), Callan (2003), Luger (2002), MacCarthy (1996), Cawsey (1998), Rich and Knight (1991) and Nilsson (1998).

Operational research (Operations research)   These two terms are completely interchangeable and are often abbreviated to OR. Different countries tend to use one or other of the terms but there is no significant difference. The field was established in the 1930s and early 1940s as scientists in Britain became involved in the *operational* activities of Britain's radar stations. After the war, the field expanded into applications within industry, commerce and government and spread throughout the world. Gass and Harris, in the preface to their excellent Encyclopedia of Operations Research and Management Science (Gass and Harris, 2001), present several definitions. However, as with Artificial Intelligence (above), we are not really concerned with the intricacies of different definitions in this book. The first definition they give says

> Operations Research is the application of the methods of science to complex problems arising in the direction and management of large systems of men, machines, materials and money in industry, business, government and defense.

This presents a reasonable summary of what the term means. For more discussion, and a range of definitions, on the topic, see Bronson and Naadimuthu (1997), Carter and Price (2001), Hillier and Lieberman (2005), Taha (2002), Urry (1991) and Winston (2004). For an excellent and fascinating early history of the field see Kirby (2003).

Management science   This term is sometimes abbreviated to MS and it can, to all intents and purposes, be interchanged with OR. Definitions can be found in Gass and Harris (2001). However, they sum up the use of these terms nicely in their preface when they say

> Together, OR and MS may be thought of as the science of operational processes, decision making and management.

Feasible and infeasible solutions   The idea of feasible and infeasible solutions is intuitive but let us consider the specific problem of cutting and packing, so that we have a concrete example which we can relate to. This problem arises in many industries: for example, in the textile industry where pieces for garments have to be cut from rolls of material, in the newspaper industry where the various text and pictures have to be laid out on the page and in the metal industry where metal shapes have to be cut from larger pieces of metal—see Dowsland and Dowsland (1992) for a more detailed review of this area. Of course, all these industries are different but let us consider a generic problem where we have to place a number of pieces onto a larger piece so that we can cut out the smaller pieces. Given this generic problem a feasible solution can be thought of as all the shapes being placed onto the larger sheet so that none of them overlap and all the pieces lie within the confines of the larger sheet. If some of the pieces overlap each other or do not fit onto the larger sheet, then the solution is infeasible. Of course, the problem definition is important when considering whether or not a given solution is feasible. For example, we could relax the constraint that says that *all* of the shapes have to be placed on the larger sheet, as our problem might state that we are trying to cut out as many of the smaller shapes as possible, but it is not imperative that we include all the smaller pieces. A feasible solution is often defined as one that satisfies the *hard constraints* (see below).

Hard constraints For any given problem, there are usually constraints (conditions) that *have* to be satisfied. These are often called *hard constraints*. To continue with the cutting and packing example from above, the condition that no pieces can overlap is an example of a hard constraint. To take a new example, if we consider a nurse rostering problem, then an example of a hard constraint is the condition that no nurse can be allocated to two different shifts at the same time. If we violate a hard constraint, it leads to an *infeasible* solution. More information about research on nurse rostering problems can be seen in Burke et al. (2004).

Soft constraints and evaluation functions A *soft constraint* is a condition that we would like to satisfy but which is not absolutely essential. As an example, from nurse rostering again, we may have a soft constraint that says that we would like nurses to be able to express preferences about which shifts they would like to work. However, if this constraint is not fully met, a solution is still feasible. It just means that another solution which does meet the condition (i.e. more nurses have their personal working preferences met) would be of higher quality. Of course, there could be many competing soft constraints, which may provide a trade off in the *evaluation function* (measure of the quality of the solution which is also sometimes known as the *objective, fitness* or *penalty* function), as the improvement of one soft constraint may cause other soft constraint(s) to become worse. This is a situation where a multi-objective approach might be applicable (see Chapter 10).

Many problems have an evaluation function represented by a sum of each of the penalty values obtained for not satisfying each of the various constraints. Some problems simply ignore the hard constraints in the evaluation function and just disregard infeasible solutions. Another approach is to set a penalty value for the hard constraints but to set it very high so that any solution which violates the hard constraints is given a very high evaluation. A further possibility is to have dynamic penalties so that, at the start of the search, the hard constraints are given relatively low penalty values, so that the infeasible search space is explored. As the search progresses, the hard constraint penalty values are gradually raised so that the search eventually only searches the feasible regions of the search space.

Deterministic search   This term refers to a search method or algorithm which always returns the same answer, given exactly the same input and starting conditions. Several of the methods presented in this book are not deterministic i.e. there is an element of randomness in the approach so that different runs on exactly the same starting conditions can produce different solutions. Note, however, that the term *"non-deterministic"* can mean something more than simply not being deterministic. See Chapter 11 for an explanation.

Optimization   Within the context of this book, optimization can be thought of as the process of attempting to find the best possible solution amongst all those available. Therefore, the task of optimization is to model your problem in terms of some evaluation function (which represents the quality of a given solution) and then employ a search algorithm to minimize (or maximize, depending on the problem) that objective function. Most of the chapters in this book are describing methodologies which are aiming to optimize some function. However, most of the problems are so large that it is impossible to guarantee that the solution obtained is the optimal one. The term optimization can lead to confusion because it is sometimes also used to describe a process which returns the guaranteed optimal solution (which is, of course, subtly different from the process which just aims to find the best solution possible).

Local and global optimum   Figure 1.1 illustrates the difference between a local and global optimum. A local optimum is a point in the search space where all neighboring solutions are worse than the current solution. In Figure 1.1, there are four local optima. A global optimum is a point in the search space where *all* other points in the search space are worse than (or equal to) the current one.

Exhaustive search   By carrying out an exhaustive search, you search every possible solution and return the optimal (best) one. For small problems, this is an acceptable strategy, but as problems become larger it becomes impossible to carry out an exhaustive search. The types of problem that often occur in real world search and optimization problems tend to grow very large very quickly. We will illustrate this by considering a very well known problem: the traveling salesman problem (often referred to as TSP). This can be thought of as the problem of attempting to minimize the distance taken by a traveling salesman who has to visit a certain number of cities exactly once and return home. See Johnson and McGeoch (1997) or Lawler et al. (1990) for more details about the TSP. With a very small number of cities, the number of possible solutions is relatively small and a computer method can easily exhaustively check all possibilities (the search space) and return the best one. For example, the problem with five cities has a search space of size 12. So all 12 possibilities can be very easily checked. However, for a 50-city problem (10 times the number of cities), the number of solutions rises to about 10^{60}. Michalewicz and Fogel, in their excellent book on modern heuristics, consider exactly this 50 city problem. They say,

> There are only 1,000,000,000,000,000,000,000 [10^{20}] liters of water on the planet so a 50-city TSP has an unimaginably large search space. Literally, it's so large that as human, we simply can't conceive of sets with this many elements.

> (Michalewicz and Fogel, 2004)

Therefore, for large problems (and large does not have to be that large), an exhaustive search is simply not an option. However, even if it is a possibility (i.e. the search space is small enough to allow us to carry out an exhaustive search) we must know how to systematically navigate the search space. This is not always possible.

Complexity   This term refers to the study of how difficult search and optimization problems are to solve. It is covered in Chapter 11.

Order (Big O notation)   This term and an associated notation is used at various places in this book and so we define it here. Suppose we have two functions $f(x)$ and $g(x)$ where x is, of course, a variable. We say that $g(x)$ is of the order of $f(x)$ written $g(x) = O(f(x))$ if, for some constant value K, $g(x) \leq Kf(x)$ for all values of x which are greater than K. This notation is often used when discussing the time complexity of search algorithms. In a certain sense, $f(x)$ *bounds* $g(x)$ once the values of x get beyond the value of K.

Heuristics   When faced with the kind of problem discussed in the exhaustive search section above, we have to accept that we need to develop an approach to obtain high-quality solutions—but optimality cannot be guaranteed (without *checking out* all the possibilities). Such an approach is called a heuristic. The following two definitions provide good descriptions.

> A heuristic technique (or simply heuristic) is a method which seeks good (i.e. near-optimal) solutions at a reasonable computation cost without being able to guarantee optimality, and possibly not feasibility. Unfortunately, it may not even be possible to state how close to optimality a particular heuristic solution is.

> (Reeves, 1996)

> A "rule of thumb" based on domain knowledge from a particular application, that gives guidance in the solution of a problem.... Heuristics may thus be very valuable most of the time but their results or performance cannot be guaranteed.

> (Oxford Dictionary of Computing, 1996)

There are many heuristic methods available to us. Some examples are simulated annealing (Chapter 7), genetic algorithms (Chapter 4), genetic programming (Chapter 5) and tabu search (Chapter 6). The term "approximate" is sometimes used in connection with heuristic methods but it is important not to confuse with approximation methods (see Chapter 18).

Constructive heuristics   Constructive heuristics refer to the process of building an initial solution from scratch. Take university examination timetabling as an example (Burke and Petrovic, 2002; Petrovic and Burke, 2004; Schaerf, 1999). One way to generate a solution is to start with an empty timetable and gradually schedule examinations until they are all timetabled. The order in which the examinations are placed onto the timetable is often important. Examinations which are more difficult to schedule (as determined by a heuristic measure of difficulty) are scheduled first in the hope that the *easier* examinations can *fit around* the difficult ones.

Constructive heuristics are usually thought of as being fast as they are often a single-pass approach.

Local search heuristics   Local search can be thought of as a heuristic mechanism where we consider *neighbors* of the current solution as potential replacements. If we accept a new solution from this neighborhood, then we *move* to that solution and then consider its neighbors (see hill climbing (below) for some initial discussion of this point). What we mean by *neighbor* is dependent upon the problem solving situation that we are confronted with. Some of the techniques presented in this book can be described as local search methods. For example, see simulated annealing (Chapter 7) and tabu search (Chapter 6). Hill climbing is also a local search method (see below). For more information about local search see Aarts and Lenstra (1997). Note the difference between a constructive heuristic which builds a solution from scratch and a local search heuristic which moves from one solution to another. It is often the case that a constructive heuristic is used to generate a solution which is employed as the starting point for local search.

Hill climbing   Hill climbing is probably the most basic local search algorithm. It is easy to understand and implement but suffers from getting stuck at a local optimum (see below). In the following discussion, we will assume we are trying to maximize a certain value. Of course, minimizing a certain value is just an analogous problem, but then we would be *descending* rather than *climbing*.

The idea behind hill climbing is to take the current solution and generate a neighbor solution (see local search) and move to that solution only if it has a higher value of the evaluation function (see above). The algorithm terminates when we cannot find a better-quality solution. The problem with hill climbing is that it can easily get stuck in a local optimum (see above). Consider Figure 1.1.

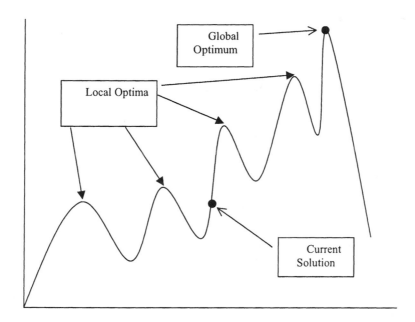

Figure 1.1. Hill Climbing getting stuck in a local optimum and the concept of local and global optima.

If the current solution is the one shown in Figure 1.1, then hill climbing will only be able to find one of the local optima shown (the one directly above it in this case). At that point, there will be no other better solutions in its neighborhood and the algorithm will terminate.

Both simulated annealing (Chapter 7) and tabu search (Chapter 6) are variations of hill climbing but they incorporate a mechanism to help the search escape from local optima.

Metaheuristics This term refers to a certain class of heuristic methods. Fred Glover first used it and he defines it (Glover, 1997) as follows:

> A meta-heuristic refers to a master strategy that guides and modifies other heuristics to produce solutions beyond those that are normally generated in a quest for local optimality. The heuristics guided by such a meta-strategy may be high level procedures or may embody nothing more than a description of available moves for transforming one solution into another, together with an associated evaluation rule.

Osman and Kelly (1996) offer the following definition:

> A meta-heuristic is an iterative generation process which guides a subordinate heuristic . . . .

The study and development of metaheuristics has become an extremely important area of research into search methodologies. In common usage, in the literature, the term tends to be used to refer to the broad collection of relatively *sophisticated* heuristic methods that include simulated annealing, tabu search, genetic algorithms, ant colony methods and others (all of which are discussed in detail in this book). The term is employed sometimes with and sometimes without the hyphen in the literature. It is also sometimes interchanged with the term "modern heuristics" (see Rayward-Smith et al. (1996). For more information about metaheuristics, see Glover and Kochenberger (2003), Osman and Kelly (1996), Voss et al. (1999), Ribeiro and Hansen (2002) and Resende and de Sousa (2004).

Evolutionary methods   Evolutionary methods can be thought of as representing a subset of the metaheuristic approaches and are typified by the fact that they maintain a *population* of candidate solutions and these solutions compete for survival. Such approaches are inspired by evolution in nature.

Some of the methods in this book are evolutionary. Chapter 4 (Genetic Algorithms) represents perhaps the best known evolutionary approach but there are many others including genetic programming (Chapter 5) and ant algorithms (Chapter 14).

Hyper-heuristics   Hyper-heuristics can be confused with metaheuristics but the distinction between the two terms is quite clear. Hyper-heuristics are simply methods which search through a search space of heuristics (or search methods). They can be defined as *heuristics to choose heuristics*. Most implementations of metaheuristics explore a search space of solutions to a given problem but they can be (and sometimes are) employed as hyper-heuristics. The term hyper-heuristic only tells you that we are operating on a search space of heuristics. It tells you nothing else. We may be employing a metaheuristic to do this search and we may not. The actual search space being explored may include metaheuristics and it may not (but very little work has actually been done which includes metaheuristics among the search space being addressed). Chapter 17 describes hyper-heuristics in more detail and readers are also referred to Burke et al. (2003).

1.4   SOURCES OF ADDITIONAL INFORMATION

This section provides a list of journals (in alphabetical order) across a range of disciplines that regularly publish papers upon aspects of decision support methodologies. This list is certainly not exhaustive. However, it provides a starting point for the new researcher and that is the sole purpose of presenting it here. We have purposefully not provided URL links to the journals as many will change after going to press, but an internet search for the journal title will quickly locate the home page.

- ACM Journal of Experimental Algorithmics
- Annals Of Operations Research
- Applied Artificial Intelligence
- Applied Intelligence
- Artificial Intelligence
- Artificial Life
- Computational Intelligence
- Computer Journal
- Computers & Industrial Engineering
- Computers & Operations Research
- Decision Support Systems
- Engineering Optimization
- European Journal Of Information Systems
- European Journal Of Operational Research
- Evolutionary Computation
- Fuzzy Sets And Systems
- Genetic Programming and Evolvable Machines
- IEEE Transactions On Computers
- IEEE Transactions On Evolutionary Computation
- IEEE Transactions On Fuzzy Systems
- IEEE Transactions On Neural Networks
- IEEE Transactions On Systems Man And Cybernetics Part A—Systems And Humans
- IEEE Transactions On Systems Man And Cybernetics Part B—Cybernetics
- IEEE Transactions On Systems Man And Cybernetics Part C—Applications And Review
- IIE Transactions
- INFORMS Journal On Computing

- Interfaces
- International Journal Of Systems Science
- International Transactions On Operational Research
- Journal Of Artificial Intelligence Research
- Journal Of Global Optimization
- Journal Of Heuristics
- Journal Of Optimization Theory And Applications
- Journal of Scheduling
- Journal Of The ACM
- Journal Of The Operational Research Society
- Knowledge-Based Systems
- Machine Learning
- Management Science
- Mathematical Programming: Series A and B
- Mathematics of Operations Research
- Neural Computation
- Neural Computing & Applications
- Neural Networks
- Neurocomputing
- Omega - International Journal of Management Science
- Operations Research
- Operations Research Letters
- OR Spectrum
- SIAM Journal on Computing
- SIAM Journal On Optimization

The following list of references just includes those volumes and papers which give an overview of search and optimization methodologies and some well studied search/optimization problems. More detailed bibliographies and sources of additional information are given at the end of each chapter throughout the book.

References

Aarts, E. and Lenstra, J. K. (eds), 1997, *Local Search in Combinatorial Optimization*, Wiley, New York.

Bronson, R. and Naadimuthu, G, 1997, Operations Research, Schaum's Outlines, 2nd edn, McGraw-Hill, New York.

Burke, E. K., De Causmaecker, P., Vanden Berghe, G. and Van Landeghem, R., 2004, The state of the art of nurse rostering, *J. Scheduling* 7:441–499.

Burke, E. K., Kendall, G., Newall, J. P., Hart, E., Ross, P. and Schulenburg, S., 2003, Hyper-heuristics: An emerging direction in modern search technology, in: *Handbook of Metaheuristics*, F. Glover and G. Kochenberger, eds, Chapter 16, pp. 457–474.

Burke, E. K., Petrovic, S., 2002, Recent research directions in automated timetabling, *Eur. J. Oper. Res.* **140**:266–280.

Carter, M. W., and Price, C. C., 2001, *Operations Research: A Practical Introduction*, CRC Press, Boca Raton, FL.

Callan R., 2003, *Artificial Intelligence*, Palgrave Macmillan, London.

Cawsey A., 1998, *The Essence of Artificial Intelligence*, Prentice-Hall, Englewood Cliffs, NJ.

Dell'Amico, M., Maffioli, F. and Martello, S. (eds), 1997, *Annotated Bibliographies in Combinatorial Optimization*, Wiley, New York.

Dowsland, K. A. and Dowsland, W. B., 1992, Packing problems, *Eur. J. Oper. Res.* **56**:2–14.

Gass, S. I. and Harris, C. M., 2001, *Encyclopaedia of Operations Research and Management Science*, Kluwer, Dordrecht.

Glover, F. and Kochenberger, G. (eds), 2003, *Handbook of Metaheuristics*, Kluwer, Dordrecht.

Glover, F. and Laguna, M., 1997, *Tabu Search*, Kluwer, Dordrecht.

Hillier F. S. and Liberman G. J., 2005, *Introduction to Operations Research*, McGraw-Hill, New York (8th edn).

Johnson, D. S. and McGeoch, L. A., 1997, The travelling salesman problem: A case study, in: E. Aarts and J. K. Lenstra, eds, *Local Search in Combinatorial Optimization*, Wiley, New York, pp. 215–310.

Kirby, M. W., 2003, *Operational Research in War and Peace: The British Experience from the 1930s to 1970*, Imperial College Press, London.

Lawler, E. L., Lenstra, J. K., Rinnooy Kan, A. H. G. and Shmoys, D. B. (eds), 1985, *The Travelling Salesman Problem: A Guided Tour of Combinatorial Optimization*, Wiley, New York (reprinted with subject index 1990).

Leung, J. Y-T. (ed.), 2004, *Handbook of Scheduling*, Chapman and Hall/CRC Press, Boca Raton, FL.

Luger G. F. A., 2002, *Artificial Intelligence: Structures and Strategies for Complex Problem Solving*, Addison-Wesley, Reading, MA, 4th edn.

McCarthy, J., 1996, *Defending AI Research: A Collection of Essays and Reviews*, CSLI Publications, Stanford, CA.

Michaelwicz, Z. and Fogel D. B., 2004, *How to Solve It: Modern Heuristics*, Springer, Berlin, 2nd edn.

Negnevitsky M., 2005, *Artificial Intelligence: A Guide to Intelligent Systems*, Addison-Wesley, Reading, MA, 2nd edn.

Nilsson, N., 1998, *Artificial Intelligence: A New Synthesis*, Morgan Kaufmann, San Mateo, CA.

Oxford Dictionary of Computing, 1997, Oxford University Press, Oxford, 4th edn.

Osman, I. H. and Kelly J. P. (eds), 1996, *Metaheuristics: Theory and Applications*, Kluwer, Dordrecht.

Pardalos, P. M. and Resende, M. G. C. (eds), 2002, *Handbook of Applied Optimization*, Oxford University Press, Oxford.

Petrovic, S. and Burke, E. K., 2004, University timetabling, Chapter 45 of J. Y-T. Leung, ed., *Handbook of Scheduling*, Chapman and Hall/CRC Press, Boca Raton, FL.

Rayward-Smith V. J., Osman I. H., Reeves C. R. and Smith G. D., 1996, *Modern Heuristic Search Methods*, Wiley, New York.

Reeves, C. R., 1996, Modern heuristic techniques, in: V. J. Rayward-Smith, I. H. Osman, C. R. Reeves and G. D. Smith, *Modern Heuristic Search Methods*, Wiley, New York, pp. 1–25.

Resende, M. G. C. and de Sousa, J. P. (eds), 2004, *Metaheuristics: Computer Decision Making*, Kluwer, Dordrecht.

Ribeiro, C. C. and Hansen, P. (eds), 2002, *Essays and Surveys in Metaheuristics*, Kluwer, Dordrecht.

Rich E. and Knight K., 1991, *Artificial Intelligence*, McGraw-Hill, New York, 2nd edn.

Russell, S. and Norvig, P., 2003, *Artificial Intelligence: A Modern Approach*, Prentice-Hall, Englewood Cliffs, NJ.

Schaerf, A., 1999, A survey of automated timetabling, *Artif. Intell. Rev.* **13**:87–127.

Taha H. A., 2002, *Operations Research: An Introduction*, Prentice-Hall, Englewood Cliffs, NJ, 7th edn.

Urry S., 1991, *An Introduction to Operational Research: The Best of Everything*, Longmans, London.

Voss, S, Martello, S., Osman, I. H. and Roucairol, C. (eds), 1999, *Metaheuristics: Advances and Trends in Local Search Paradigms for Optimization*, Kluwer, Dordrecht.

Winston, W. L., 2004, *Operations Research: Applications and Algorithms*, Duxbury Press, Philadelphia, PA, 4th edn.

Chapter 2

CLASSICAL TECHNIQUES

Kathryn A. Dowsland
Gower Optimal Algorithms Ltd
Swansea, UK
and
The School of Computer Science and IT
University of Nottingham, UK

2.1 INTRODUCTION

The purpose of this chapter is to provide an introduction to three classical search techniques, branch and bound, dynamic programming and network flow programming, all of which have a well established record in the solution of both classical and practical problems. All three have their origins in, or prior to, the 1950s and were the result of a surge in interest in the use of mathematical techniques for the solution of practical problems. The timing was in part due to developments in Operations Research in World War II, but was also spurred on by increasing competition in the industrial sector and the promise of readily accessible computing power in the foreseeable future. A fourth technique belonging to this class, that of Integer Programming, is covered in Chapter 3. Given their age, it is not surprising that they no longer generate the same level of excitement as the more modern approaches covered elsewhere in this volume, and as a result they are frequently overlooked. This effect is reinforced as many texts such as this omit them—presumably because they have already been covered by a range of sources aimed at a wide variety of different abilities and backgrounds. In this volume we provide an introduction to these well-established classics alongside their more modern counterparts. Although they have shortcomings, many of which the more recent approaches were designed to address, they still have a role to play both as stand-alone techniques and as important ingredients in hybridized solution methods.

The chapter is meant for beginners and it is possible to understand and use the techniques covered without any prerequisite knowledge. However, some of the examples in the chapter are based on problems in graph theory. In all cases, the problems and specialist terms are defined in full, but a basic knowledge of graph theory terminology such as that provided in Balakrishnan (1997) would be useful. Some of the examples also belong to a class of problems known as *linear programs* (LPs) and some of the discussion in the section on network flows makes use of the relationship between network flow problems and linear programming problems. Although knowledge of LPs is not necessary to understand the algorithms or examples as these are all couched in purely combinatorial terms we start the chapter with a brief introduction to linear programming. Further details can be found in Anderson et al. (1997).

The chapter is organized as follows. The overview of LPs is followed by three sections introducing branch and bound, dynamic programming and network flow programming. In each case an introductory description is followed by two or more examples of their use in solving different problems, including a worked numerical example in each case. Each section ends with a brief discussion of more advanced issues. Section 2.6 looks at some problems that frequently occur as sub-problems in the solution of more complex problems and suggests algorithms based on the techniques covered in Sections 2.3–2.5 for their solution. Section 2.7 takes a brief look at potential future applications and Section 2.8 provides some hints and tips on how to get started with each of the techniques. Additional sources of information not covered in the references are given at the end of the chapter.

2.2   LINEAR PROGRAMMING

2.2.1   Introduction

This section provides a brief overview of those aspects of LPs that are relevant to the remainder of this chapter. We start by outlining the basic features of an LP model and then go on to look at an important concept of such models—that of duality. We do not go into any detail with regard to solution algorithms for two reasons. Firstly, they are not necessary in order to understand the material presented in the remainder of the chapter. Secondly, LP packages and solution code are available from a wide variety of sources so that it is no longer necessary for a potential user to develop their own solution code.

2.2.2   The Linear Programming Form

A linear programming problem is an optimization problem in which both the objective (i.e. the expression that is to be optimized) and the constraints on the solution can be expressed as a series of linear expressions in the decision variables. If the problem has n variables then the constraints define a set of hyper-planes in n-dimensional space. These are the boundaries of an n-dimensional region that defines the set of feasible solutions to the problem and is known as the feasible region. The following example illustrates the form of a simple linear programming problem.

A Simple Example   A clothing manufacturer makes three different styles of T-shirt. Style 1 requires 7.5 minutes of cutting time, 12 minutes of sewing time, 3 minutes of packaging time and sells at a profit of £3. Style 2 requires 8 minutes of cutting time, 9 minutes of sewing time, 4 minutes of packaging time and makes £5 profit. Style 3 requires 4 minutes of cutting time, 8 minutes of sewing time and 2 minutes of packaging time and makes £4 profit. The company wants to determine production quantities of each style for the coming month. They have an order for 1000 T-shirts of style 1 that must be met, and have a total of 10 000 minutes available for cutting, 18 000 minutes for sewing and 9000 minutes available for packaging. Assuming that they will be able sell as many T-shirts as they produce in any of the styles, how many of each should they make in order to maximize their profit?

We can formulate the problem mathematically as follows. First we define three decision variables x_1, x_2 and x_3 representing the number of T-shirts manufactured in styles 1, 2 and 3 respectively. Then the whole problem can be written as

$$\text{maximize } 3x_1 + 5x_2 + 4x_3 \tag{2.1}$$
$$\text{subject to } 7.5x_1 + 8x_2 + 4x_3 \leq 10000 \tag{2.2}$$
$$12x_1 + 9x_2 + 8x_3 \leq 18000 \tag{2.3}$$
$$3x_1 + 4x_2 + 2x_3 \leq 9000 \tag{2.4}$$
$$x_1 \geq 1000 \tag{2.5}$$
$$x_1, x_2, x_3 \geq 0 \tag{2.6}$$

Expression (2.1) defines the profit. This is what we need to maximize and is known as the *objective function*. The remaining expressions are the *constraints*. Constraints (2.2)–(2.4) ensure that the hours required for cutting, sewing and packaging do not exceed those available. Constraint (2.5) ensures that at least 1000 T-shirts of style 1 are produced and constraint (2.6) stipulates that all the decision variables must be non-negative. Note that all the expressions are linear in the decision variables, and we therefore have a linear programming formulation of the problem.

The General LP Format   In general a linear program is any problem of the form

$$\text{max(or min)} \qquad \sum_{i=1}^{n} c_i x_i$$

$$\text{such that} \qquad \sum_{i=1}^{n} a_{1i} x_i \sim b_1 \qquad\qquad (2.7)$$

$$\vdots$$

$$\sum_{i=1}^{n} a_{mi} x_i \sim b_m$$

where $\sim$ is one of $\geq$, $=$ or $\leq$.

The important point about a linear programming model is that the feasible region is a convex space and the objective function is a convex function. Optimization theory therefore tells us that as long as the variables can take on any real non-negative values (possibly bounded above) then the optimal solution can be found at an extreme point of the feasible region. It is also possible to derive conditions that tell us whether or not a given extreme point is optimal. Standard LP solution approaches based on these observations can solve problems involving many thousands of variables in reasonable time. As we will see later in this chapter there are special cases where these techniques work even when the variables are constrained to take on integer or binary values. The general case where the variables are constrained to be integer, known as integer programming, is more difficult and is covered in Chapter 3.

Although it makes sense when formulating LPs to use the flexibility of the formulation above in allowing either a maximization or minimization objective and any combination of inequalities and equalities for the constraints, much linear programming theory (and therefore the solution approaches based on the theory) assume that the problem has been converted to a standard form in which the objective is expressed in terms of a maximization problem, all the constraints apart from the non-negativity conditions are expressed as equalities, all right-hand side values $b_1, \ldots, b_m$ are non-negative and all decision variables are non-negative. A general formulation can be converted into this form by the following steps.

1 If the problem is a minimization problem the signs of the objective coefficients $c_1, \ldots, c_n$ are changed.

2 Any variable not constrained to be non-negative is written as the difference between two non-negative variables.

3 Any constraint with a negative right-hand side is multiplied by -1.

4 Any constraint which is an inequality is converted to an equality by the introduction of a new variable, (known as a slack variable) to the left-hand side. The variable is added in the case of a $\leq$ constraint and subtracted in the case of a $\geq$ constraint. The formulation is often written in matrix form:

$$
\begin{aligned}
\max \quad & CX \\
\text{s.t.} \quad & AX = b \\
& X \geq 0
\end{aligned} \tag{2.8}
$$

where $C = (c_1, \ldots, c_n)$, $b = (b_1, \ldots, b_m)^\mathrm{T}$, $X = (x_1, \ldots, x_n)^\mathrm{T}$ and $A = (a_{ij})_{m \times n}$.

2.2.3 Duality

An important concept in linear programming is that of *duality*. For a maximization problem in which all the constraints are $\leq$ constraints and all variables are non-negative, i.e. a problem of the form

$$
\begin{aligned}
\max \quad & CX \\
\text{s.t.} \quad & AX \leq b \\
& X \geq 0
\end{aligned} \tag{2.9}
$$

the dual is defined as

$$
\begin{aligned}
\min \quad & b^\mathrm{T} Y \\
\text{s.t.} \quad & A^\mathrm{T} Y \geq C^\mathrm{T} \\
& Y \geq 0
\end{aligned} \tag{2.10}
$$

The original problem is known as the *primal*. Note that there is a dual variable y_i associated with each constraint in the primal problem and a dual constraint associated with each variable x_i in the primal. (The dual of a primal with a more general formulation can be derived by using rules similar to those used to convert a problem into standard form to convert the primal into the above form, with equality constraints being replaced by the equivalent two inequalities.)

The dual has the important property that the value of the objective in the optimal solution to the primal problem is the same as the optimal solution for the dual problem. Moreover, the *theorem of complementary slackness* states that if both dual and primal have been converted to standard form, with $s_1, \ldots, s_m$ the slack variables in the primal problem and $e_1, \ldots, e_n$ the slack variables in the dual, then if $X = (x_1, \ldots, x_n)$ is feasible for the dual and $Y = (y_1, \ldots, y_m)$ is feasible for the primal, X and Y are optimal solutions to the primal and dual respectively if and only if $s_i y_i = 0, \forall i = 1, m$ and $e_j x_j = 0, \forall j = 1, n$. This relationship is an important feature in the specialist solution algorithm for the minimum cost network flow algorithm presented later in this chapter and underpins other LP-based solution approaches.

2.2.4   Solution Techniques

Solution approaches for linear programming fall into two categories. Simplex-type methods search the extreme points of the feasible region of the primal or the dual problem until the optimality conditions are satisfied. The technique dates from the seminal work of Dantzig (1951). Since then the basic technique has been refined in a number of ways to improve the overall efficiency of the search and to improve its operation on problems with specific characteristics. Variants of the simplex approach are available in a number of specialist software packages, as a feature of some spreadsheet packages, and as freeware from various web-based sources.

Although they perform well in practice, simplex-based procedures suffer from the disadvantage that they have poor worst-case time-performance guarantees. This deficiency led to the investigation of interior point methods, so called because they search a path of solutions through the interior of the feasible region in such a way as to arrive at an optimal point when they hit the boundary. Practical interior point methods can be traced back to the work of Kamarkar (1984), although Khachiyan's (1979) ellipsoid method was the first LP algorithm with a polynomial time guarantee. Recent interior point methods have proved efficient, in particular for large LPs, and there has also been some success with hybridizations of interior point and simplex approaches. While the choice of solution approach may be important for very large or difficult problems, for most purposes standard available code based on simplex-type approaches should suffice.

2.3   BRANCH AND BOUND

2.3.1   Introduction

When faced with the problem of finding an optimum over a finite set of alternatives an obvious approach is to enumerate all the alternatives and then select the best. However, for anything other than the smallest problems such an approach will be computationally infeasible. The rationale behind the branch and bound algorithm is to reduce the number of alternatives that need to be considered by repeatedly partitioning the problem into a set of smaller sub-problems and using local information in the form of bounds to eliminate those that can be shown to be sub-optimal. The simplest branch and bound implementations are those based on a constructive approach in which partial solutions are built

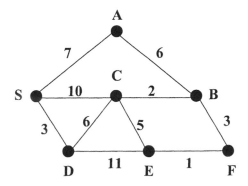

Figure 2.1. Shortest path network.

up one element at a time. We therefore start by introducing the technique in this context before going on to show how it can be extended to its more general form.

Assume that we have a problem whose solutions consist of finite vectors of the form $(x_1, x_2, \ldots, x_k)$ where k may vary from solution to solution. Those combinations of values that form feasible vectors are determined by the problem constraints. The set of all possible solutions can be determined by taking each feasible value for x_1, then for each x_1 considering all compatible values for x_2, then for each partial solution $(x_1, x_2, \ldots)$ considering all compatible x_3 etc. This process can be represented as a tree in which the branches at level i correspond to the choices for x_i given the choices already made for $x_1, \ldots, x_{i-1}$, and the nodes at level i correspond to the partial solutions of the first i elements. This is illustrated with reference to Figures 2.1 and 2.2. Figure 2.2 is the tree enumerating all simple routes (i.e. routes without loops) from S to F in the network shown in Figure 2.1. The branches at level 1 represent all the possibilities for leaving S and the branches at lower levels represent all the possibilities for extending the partial solution represented by the previous branches by one further link. The node at the top of the tree is sometimes referred to as the *root*, and the relationship between a given node and one immediately below is sometimes referred to as *parent/child* or *father/son*. All nodes that can be reached from the current node by traveling down the tree are referred to as *descendants* and nodes without any descendants are *terminal nodes*.

If we wish to search the tree in an ordered fashion we need to define a set of rules determining the order in which the branches are to be explored. This is known as the *branching strategy*. The two simplest strategies are known as *depth-first search* and *breadth-first search*. Depth-first search (also known as branch and backtrack) moves straight down a sequence of branches until

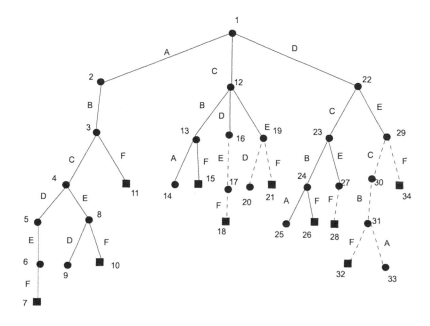

Figure 2.2. Tree enumerating all simple routes.

a terminal node is reached before backtracking up to the nearest junction. If there are any unexplored branches below this junction it will select one of these, again continuing downwards until a terminal node is reached. If all children of the current node have been explored, the search backtracks to the previous node and continues from there. In contrast, breadth-first search enumerates all the branches at one level before moving on to the next level. Depth-first search is likely to find a feasible solution early and it does not have the vast storage requirements of breadth-first search. However, breadth-first search allows comparisons to be made across the tree, facilitating removal of dominated sub-solutions. In Figure 2.2 the nodes are numbered in depth-first search order using a strategy of ordering the branches at each level from left to right. For a breadth-first search the search order would be 1, 2, 12, 22, 3, 13, 16, 19, 23, 29, 4, 11, etc.

As mentioned above, although it is possible to enumerate the whole tree in a small example like this, the size of the tree grows explosively with problem size, and for most real-life problems complete enumeration is not possible. For example, complete enumeration of all feasible allocations of n tasks to n machines would result in a tree with $n!$ terminal nodes (i.e. $\sim 2.4 \times 10^{18}$ terminal nodes for 20 machines). The branch and bound process allows us to prove that some partial solutions cannot lead to optimal solutions and to cut them and their descendants from the search tree—a process known as *pruning*. This is achieved through the use of upper and lower bounds satisfying *lower_bound* $\leq$ $z \leq$ *upper_bound*, where z is the optimal solution obtained over all descendants of the current node. For a minimization problem the upper bound is a quantity UB such that we know we can do at least as well as UB. This is usually the best feasible solution found so far. The lower bound at node i, LB_i is an optimistic estimate of the best solution that can be obtained by completing the partial solution at node i, i.e. we know that in exploring below node i we cannot do better than LB_i. Nodes where $LB_i \geq UB$ need not be explored further and we say that they are *fathomed*. When all nodes are fathomed the upper bound is the optimal solution. Note that for a maximization problem the roles of the upper and lower bounds are reversed.

2.3.2 Branch and Bound Based on Partial Solutions

The success of a branch and bound implementation for a large problem depends on the number of branches that can be pruned successfully. This is largely dependent on the quality of the bounds, but the branching strategy can also have a significant effect on the number of nodes that need to be explored. The basic branch and bound approach and the influence of bound quality and branching strategy will be illustrated with reference to our shortest path example.

Example 1. Finding the Shortest Path   We start by illustrating the use of simple bounds to prune the tree using the branching strategy defined by the node numbering. Our upper bound will be the cost of the shortest path found to date. Thus at the start we have $UB = \infty$. For the local lower bound at each node in the search tree we will simply use the sum of the costs on the links traveled so far. Table 2.1 gives details of the order in which the nodes are visited and the upper and lower bounds at each node.

The first seven branches correspond to the construction of the first path SABCDEF and the lower bound column gives the cost incurred by the partial solution at each stage. Note that when we reach node 7 we have our first complete path and so the upper bound is reduced to the cost of the path, i.e. 33. We now know that the optimal solution cannot be worse than 33. From node 7 we backtrack via nodes 6 and 5 to the junction at node 4 before continuing down the tree. Each time a cheaper path is completed the upper bound is reduced but the local lower bounds remain below the upper bound until we reach node 16. Here the cost of the partial solution SCD = 16 and the upper bound is 15. Thus this node is fathomed, and we backtrack immediately to node 12. Similarly, nodes 19, 27 and 29 are fathomed by the bounds and the branches below them can be pruned. When the search is complete the cost of the optimal solution is $UB = 14$ and is given by the path to node 26, SDCBF. All the

Table 2.1. Using simple bounds to prune the tree.

	Node	LB	UB		Node	LB	UB		Node	LB	UB
1	1	0	∞	16	4	15	21	31	19	15	15
2	2	7	∞	17	3	13	21	32	12	10	15
3	3	13	∞	18	11	16	16	33	1	0	15
4	4	15	∞	19	3	13	16	34	22	3	15
5	5	21	∞	20	2	7	16	35	23	9	15
6	6	32	∞	21	1	0	16	36	24	11	15
7	7	33	33	22	12	10	16	37	25	17	15
8	6	32	33	23	13	12	16	38	24	11	15
9	5	21	33	24	14	18	16	39	26	14	14
10	4	15	33	25	13	12	16	40	24	11	14
11	8	20	33	26	15	15	15	41	23	9	14
12	9	31	33	27	13	12	15	42	27	14	14
13	8	20	33	28	12	10	15	43	23	9	14
14	10	21	21	29	16	16	15	44	22	3	14
15	8	20	21	30	12	10	15	45	29	14	14

branches represented by the dotted lines in Figure 2.2 have been pruned and the search has been reduced from 33 to 23 branches.

This is already a significant reduction but we can do better by strengthening the bounds. The bound is based on the cost to date and does not make use of any estimate of the possible future cost. This can easily be incorporated in two ways. First we know that we must leave the current vertex along a link to an unvisited vertex. Thus we will incur an additional cost of at least as much as the cost of the cheapest such link. Similarly, we must eventually enter F. Thus we must incur an additional cost at least as great as the cheapest link into F from the current vertex or a previously unvisited vertex. However, we cannot simply add both these quantities to the original bound as, at vertices adjacent to F, this will incur double counting. This highlights the need for caution when combining different bounds into a more powerful bound. We can now define our new lower bound LB_i as follows.

Let $(x_1, \ldots, x_j)$ be the current partial solution. Define L_1 = cost of path $(x_1, \ldots, x_j)$, L_2 = cheapest link from x_j to an unvisited vertex and L_3 = cheapest link into F from any vertex other than those in path $(x_1, \ldots, x_{j-1})$. Then $LB_i = L_1 + L_2 + L_3$ if x_j is not adjacent to F, $LB_i = L_1 + \max(L_2, L_3)$ otherwise. Table 2.2 gives details of the search using this new bound.

Note how the lower bound is now higher at each node and a good estimate of the cost of each path is obtained before the path is completed. Lower bounds of ∞ are recorded whenever the branches from a particular node are exhausted. There is more work incurred in calculating this bound, not only because the

Table 2.2. Search using improved bounds.

	Node	LB (L_1, L_2, L_3)	UB		Node	LB (L_1, L_2, L_3)	UB		Node	LB (L_1, L_2, L_3)	UB
1	1	4(0, 3, 1)	∞	13	8	21(20, 1, 1)	33	26	15	15	15
2	2	14(7, 6, 1)	∞	14	10	21	21	27	13	∞	15
3	3	15(13, 2, 1)	∞	15	8	∞	21	28	12	16(10, 5, 1)	15
4	4	21(15, 5, 1)	∞	16	4	∞	21	29	1	4(0, 3, 1)	15
5	5	33(21, 11, 1)	∞	17	3	16(13, 3, 1)	21	30	22	10(3, 6, 1)	15
6	6	33(32, 1, 1)	∞	18	11	16	16	31	23	12(9, 2, 1)	15
7	7	33	33	19	3	∞	16	32	24	14(11, 3, 1)	15
8	6	∞	33	20	2	∞	16	33	25	∞	15
9	5	∞	33	21	1	4(0, 3, 1)	16	34	24	14(11, 3, 1)	15
10	4	21(15, 5, 1)	33	22	12	13(10, 2, 1)	16	35	26	14	14
11	8	21(20, 1, 1)	33	23	13	15(12, 3, 1)	16	36	24	17(11, 6, 1)	14
12	9	∞	33	24	14	∞	16	37	23	15(9, 5, 1)	14
				25	13	15(12, 3, 1)	16	38	22	15(3, 11, 1)	14

Table 2.3. Search using improved branching strategy.

	Node	LB (L_1, L_2, L_3)	UB		Node	LB (L_1, L_2, L_3)	UB		Node	LB (L_1, L_2, L_3)	UB
1	1	4(0, 3, 1)	∞	6	24	17(11, 6, 1)	14	11	1	11(0, 10, 1)	14
2	22	10(3, 6, 1)	∞	7	23	15(9, 5, 1)	14	12	12	13(10, 2, 1)	14
3	23	12(9, 2, 1)	∞	8	22	15(3, 11, 1)	14	13	13	15(121, 3, 1)	14
4	24	14(11, 3, 1)	∞	9	1	8(0, 7, 1)	14	14	12	17(10, 6, 1)	14
5	26	14	14	10	2	14(7, 6, 1)	14	15			

actual calculation is more complex, but also because the bound at a given node
may change when returning to the node in a backtracking step. This strengthens the bound and reduces the total number of branches searched to 19.

Finally, we consider the branching strategy. So far we have used simple
depth first search taking the branches at each node in the given order. In general, the efficiency of the search will be increased if the upper bound can be
reduced sooner, or if the levels of the tree can be organized so as to incur high
lower bounds closer to the root of the tree. Here we apply a strategy that will
favor the former, and bias the search towards finding shorter routes first. This
is achieved by exploring the branches at each node in increasing cost order instead of from left to right. Table 2.3 shows the result of the search using this
strategy.

Now the search starts by selecting the D branch from the root node and quickly finds the optimal solution of 14. This results in early pruning of the nodes from the other two branches and the whole search is completed in seven branches.

This example has illustrated how the size of the search tree is dependent on both the quality of the bounds and the branching strategy. However, it should be noted that this is not the most efficient way of solving the shortest path problem and better approaches are suggested in Section 2.6. Nevertheless, since early papers on the technique in the early 1960s, it has proved successful in solving a wide range of both classical and practical problems. Examples include algorithms for a range of graph-theoretic problems, for example node coloring (Brown, 1972; Zykov, 1949), clique and independent set problems (Bron and Kerbosch, 1973), location problems (Erlenkotter, 1978; Jarivnen et al., 1972) and the traveling salesman problem (Little et al., 1963; Held and Karp, 1970; Balas and Christofides, 1981), and for several classical combinatorial optimization problems such as knapsack problems (Martello and Toth, 1981, 1990), set covering and set partitioning (Garfinkel and Nemhauser, 1969) and generalized assignment problems (Ross and Soland, 1975). We use one of these, Brown's graph coloring algorithm, to consolidate the ideas of the last section.

Example 2. Brown's Algorithm for Graph Coloring   An example of a problem that has been tackled using a variety of branch and bound approaches is that of graph coloring. The graph coloring problem is that of minimizing the number of colors needed to color the vertices of a graph so no two adjacent vertices are given the same color. The graph coloring problem is an interesting example as it is the underlying model for many timetabling and scheduling problems. Brown's 1972 algorithm is an example of a branch and bound approach based on partial solutions. The algorithm is a straightforward application of the branch and bound process using simple bounds. As with the shortest path problem, its efficiency can be improved by applying some intelligence to the branching strategy. However, the strength of the algorithm lies in its backtracking strategy that essentially prunes away branches by backtracking up several levels at a time where appropriate. The algorithm can be summarized as follows.

The branches in the tree correspond to the decision to color a given vertex in a given color. In the simplest version, the vertices are pre-ordered and the branches at level i correspond to choosing a color for the ith vertex. The colors are also ordered and the branches at each node are searched in color order. The upper bound is given by the number of colors in the best solution to date and the lower bound on each partial solution is given by the number of colors used up to that point. If the upper bound is equal to Q, then when the search backtracks to a vertex v_i for which there are no un-explored branches corresponding to colors below Q in the ordering, that node is obviously bounded and a further backtracking step must be executed. Rather than simply backtracking to the previous node, the search recognizes the fact that in order to progress it is necessary for an alternative color to become free for v_i. This will only be achieved if a neighbor of v_i is uncolored. Therefore, the search backtracks up the tree until a neighbor of v_i is encountered before attempting a forward branch. If further backtracking, say at vertex v_j, is required before v_i has been successfully re-colored then re-coloring a neighbor v_j may also achieve the desired result so v_j's neighbors are added to those of v_i in defining a potential backtracking node. In order to manage this backtracking strategy in a complex search, those vertices that are neighbors of backtracking vertices are stored in a queue in reverse order and branching takes place from the first vertex in the queue. A formal definition of the algorithm is given below. The list of identifiers, J, is the set of nodes which trigger the bounding condition and $Queue$ is an ordered list of the neighbors of elements in J.

Step 0. Define orderings
Order the vertices $1, 2, \ldots, n$ and colors $c_1, c_2, \ldots$.
$\Gamma^-(i)$ denotes neighbors of vertex i which precede i in the ordering.

Step 1. Find initial coloring.
Color the vertices in order using the lowest indexed feasible color.

Step 2. Store new solution.
Let q be the number of colors used. Set the upper bound equal to q and store the current q-coloring. Set list of identifiers, $J = \emptyset$.

Step 3. Backtrack.
3.1 Find first vertex corresponding to local LB = q.
Let i^* be the first vertex colored q.
3.2 Update list of backtracking vertices and corresponding queue of neighbors.
Remove all $j < i^*$ from J.
Set $J = J \cup \{i^*\}$.
Set $Queue = \bigcup_{j \in J} \Gamma^-(j)$ in reverse order.
3.3 Backtrack to level defined by first vertex on the queue.
Let i' be the first vertex on the queue. Let q' be its color.
If $i' = k$ and vertices $1, 2, \ldots, k$ are colored $c_1, c_2, \ldots, c_k$ then STOP.
Stored solution is optimal.
Otherwise uncolor all $i \geq i'$.

Step 4. Branch.
4.1 Recolor i'.
Color i' in the first feasible color $\{q' + 1, q' + 2, \ldots, q - 1\}$.

If no feasible color set $i^* = i'$ and go to 3.2.

4.2 Recolor remaining vertices.

Attempt to color vertices $i = i' + 1, n$ in colors 1 to $q - 1$ using first feasible color.

If vertex i requires color q then set $i^* = i$ and go to 3.2.

Otherwise go to step 2.

Note that Steps 1 and 4.2 amalgamate several forward branches into one step, building on the partial coloring until the bounding condition is reached. Similarly, in Step 3.3 several backtracking steps are amalgamated. Note also that the lower bounds are not stored explicitly as the first node with a local lower bound of q will always correspond to the point where color q was used for the first time.

The algorithm is illustrated with reference to the graph in Figure 2.3 using the given ordering of the vertices and colors in alphabetical order:

Step 1. Initial coloring = $1A, 2B, 3A, 4C, 5A, 6D, 7B, 8D$; $q = 4$.

Step 2. $J = \emptyset$.

Step 3. $i^* = 6$, $J = \{6\}$. *Queue* = $\{4, 2, 1\}$. Backtrack to node 4. $q' = C$.
   Partial coloring = $1A, 2B, 3A$.

Step 4. Vertex 4 already colored in $q - 1$; $i^* = 4$; go to 3.2.

Step 3. $J = \{6, 4\}$. *Queue* = $\{3, 2, 1\}$. Backtrack to node 3. $q' = A$.
   Partial coloring= $1A, 2B$.

Step 4. Color 3 in color C and continue coloring $1A, 2B, 3C, 4A, 5B, 6C$,
   $7A, 8C$; $q = 3$.

Step 2. $J = \emptyset$.

Step 3. $i^* = 3$; $J = \{3\}$; *Queue* = $\{2\}$. Stopping condition reached and
   solution with $q = 3$ is an optimal solution in three colors.

As with the shortest path implementation, the efficiency of the search can be improved by an intelligent ordering of the vertices to encourage good solutions to be found more quickly. The simplest improvement is to pre-order the vertices in decreasing degree order. However, there is no reason why the ordering of the vertices should remain the same throughout the tree, and greater efficiency gains can be obtained using some form of dynamic ordering such as selecting the vertex with the largest number of colors already used on its neighbors to color next—a strategy known as DSATUR. It is also worth considering the stopping condition in terms of the vertex ordering. The condition is valid because backtracking beyond the point where vertices 1 to k are colored in colors 1 to k will simply result in equivalent colorings with a different permutation of colors. This condition will be achieved more quickly if the or-

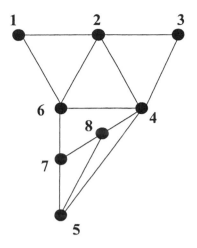

Figure 2.3. Coloring example.

dering starts with a large clique. Thus a good strategy is to find a large clique and place these vertices in a fixed order at the start and then to use a dynamic ordering for the remaining vertices.

2.3.3   A Generalization

So far the discussion has focused on search trees based on building up partial solutions. Such approaches have proved popular for a variety of problems. However, they are just a special case of a more general strategy in which the branches correspond to adding constraints to the problem. In the case of a partial solution, the constraints take the form $x_i = a_i$. The more general format underpins the branch and bound strategy for integer programming and will be treated in detail in Chapter 3. Therefore we will not go into detail here. Instead we will briefly illustrate the approach with an alternative tree search approach to the graph coloring problem.

Zykov's Algorithm for Graph Coloring   Consider any two non-adjacent vertices v_i and v_j. In any solution to the graph coloring problem there are two possibilities. Either they will be allocated the same color or they will be allocated different colors. The optimal solution subject to them being in the same color is the optimal coloring in a graph with v_i and v_j merged into a single vertex, while the optimal solution in the latter case is the optimal coloring in the original graph with edge (v_i, v_j) added. Obviously the better of these two solutions is optimal with respect to the original problem. We also observe that if we continue a sequence of adding edges and/or merging vertices in an arbitrary graph then we will eventually be left with a complete graph (i.e. a

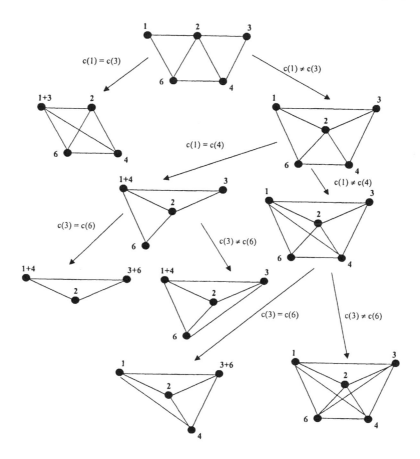

Figure 2.4. Zykov's search tree for optimal coloring: $c(i)$ denotes color of vertex i.

graph in which every vertex is adjacent to every other). A complete graph with n vertices obviously requires n colors. These observations form the basis of Zykov's algorithm (1949) in which there are two branches at each level of the tree corresponding to the decision as to whether two non-adjacent vertices will be allocated the same or different colors. The two child nodes represent the coloring problems in the two suitably modified graphs and the terminal nodes will all be complete graphs. The smallest complete graph defines the optimal coloring.

This is illustrated in Figure 2.4, which shows the search tree that results from the problem of finding the optimal coloring of the sub-graph defined by vertices 1, 2, 3, 4 and 6 of the graph in Figure 2.3.

The left-hand branch at each level constrains two non-adjacent vertices to be the same color and the child node is obtained by taking the graph at the parent node and merging the two vertices. The right-hand branch constrains the same two vertices to be different colors and the child is formed by adding an edge between the two relevant vertices in the parent graph. Branching continues until the resulting child is a complete graph. Here the terminal nodes reading from left to right are compete graphs of size 4, 3, 4, 4 and 5 respectively. The optimal solution is given by the complete graph on three vertices in which vertices 1 and 4 are allocated to one color, 3 and 6 to a second color and vertex 2 to the third.

A suitable upper bound is again the best solution found so far. A lower bound on the optimal coloring in each sub-graph can be defined by the largest clique it contains (a clique is a set of vertices such that each vertex in the set is adjacent to every other). Finding the largest clique is itself a difficult problem but a heuristic can be used to get a good estimate. In Figure 2.4, using a depth first search and exploring the $c(i) = c(j)$ branch first at each node, we might recognize that the parent of the node representing the optimal solution contains two cliques of size 3. Similarly, its parent contains cliques of size 3. Thus we can backtrack straight up to the root node saving a total of four branches. Branching strategies can be designed to produce dense sub-graphs quickly, thereby increasing the chances of finding large cliques early in the tree. More recent versions of the algorithm make use of theoretical results that state that certain classes of graph, known as perfect graphs, are easy to color. Branching strategies are designed to produce graphs belonging to one of the classes of perfect graph as early in the tree as possible. These can then be colored optimally, thus saving all the branches required in order to reduce them to complete graphs. See Golumbic (1980) for a wide ranging treatment of perfect graphs and associated algorithms.

2.3.4 Other Issues

Bounds The most important feature of a branch and bound algorithm is probably the quality of the bounds, and it is usually worth putting considerable effort into ensuring that these are as tight as possible. In the case of lower bounds this is often achieved by exploiting as much information about the problem as possible. For example, Dowsland (1987) used a clique model to solve a class of packing problems known as the pallet loading problem. She combined the bounds in a published maximum clique algorithm with bounds derived from geometric aspects of the physical problem and showed that the percentage of problems solved within a given time frame rose from 75% to 95%. In some cases a problem may become easy to solve if some of the constraints are removed. This is a process known as relaxation and the solution to the relaxed problem will always provide a valid bound on the solution to the original problem. This approach was used by Christofides and Whitlock (1977) in their solution to a guillotine cutting problem in which a large stock-sheet of material must be cut into a set of smaller rectangular pieces, using a sequence of guillotine cuts, so as to maximize the value of the cut pieces. In their version of the problem the demand for pieces of each dimension was constrained, whereas the unconstrained version of the problem is relatively easy to solve. Their solution uses a tree search in which each branch represents a cut, and the nodes define the set of rectangles in a partial solution. Bounds are obtained by solving the unconstrained problems for each of the sub-rectangles in the partial solution. Although such relaxation bounds can be quite effective, there is often a gap between them and the solution to the constrained problem. This gap can sometimes be reduced by incorporating a suitable penalty for the violated constraints into the objective function. This is the basis of Lagrangian relaxation that has proved a popular bound for a variety of branch and bound algorithms. For example, Beasley (1985) uses the approach for a non-guillotine version of the cutting stock problem. In Lagrangian relaxation an iterative approach is used to set parameters that will increase the tightness of the bound. Such an approach is obviously time-consuming but for moderately sized problems in a wide variety of application areas the computational effort is well worth the number of branches it saves. Details of Lagrangian relaxation can be found in Fisher (1985).

The search efficiency is also influenced by the upper bound. As we have seen, pruning is more effective when good solutions are found early. Earlier pruning may result if a heuristic is used before the start of the search to find a good solution that can be used as an upper bound at the outset. Similarly, using a heuristic to complete a partial solution and provide a local upper bound should also help in fathoming nodes without branching all the way down to terminal nodes.

It is also worth noting here that bounding conditions based on information other than numeric upper and lower bounds may be useful in avoiding infeasible solutions or solutions that are simply permutations or sub-sets of solutions already found. An example of this approach is the maximal clique algorithm of Bron and Kerbosch (1973) which includes bounding conditions based on relationships between the branches already explored and those yet to be visited from a given node.

Branching   There are also issues concerning branching strategies that have not been discussed. We have assumed that the search is always carried out in a depth first manner. An alternative that can be successful if the bounds are able to give a good estimate of the quality of solutions below each branch is to use the best-first strategy, in which nodes at different levels across the breadth of tree may be selected to be evaluated next according to an appropriate definition of "best". In our examples the ordering of branches was geared towards finding good solutions as early as possible. An alternative strategy is to select branches that will encourage bounding conditions to be satisfied sooner rather than later. This approach is taken by Bron and Kerbosch in their branch and bound algorithm for finding cliques in a graph. At each node the next branch is chosen so as to encourage the bounding condition to occur as early as possible. Comparisons between this version and a version in which the branches are selected in the natural order show that the advantage, in terms of computation time, of using the more complex strategy increases rapidly with problem size.

Miscellaneous   Although good bounds and branching strategies have resulted in many successful branch and bound algorithms, it should be noted that the size of the search tree will tend to grow exponentially with the problem size. It is therefore important to make sure that the underlying tree is as small as possible. For example, thought should be given to avoiding branches that lead to solutions that are symmetric to each other if at all possible. It may also be possible to apply some form of problem reduction in a pre-processing phase. For example, Garfinkel and Nemhauser (1969) outline a set of reductions for both set covering and set partitioning problems. It should also be noted that such a strategy may provide further reductions when applied to the sub-problems produced within the search itself.

Finally, it is worth noting that in many implementations, the optimal solution is found quickly and most of the search time is spent in proving that this is in fact the optimal solution. Thus, if there is insufficient time to complete the search the best solution to date can be taken as a heuristic solution to the problem. A useful trick that will guarantee a heuristic solution within $\alpha \times 100\%$ of the optimum is to replace the upper bound with $UB(1 - \alpha)$. The tighter bound should enable the search to be completed more quickly.

2.4   DYNAMIC PROGRAMMING

2.4.1   Introduction

Like branch and bound, dynamic programming (DP) is a procedure that solves optimization problems by breaking them down into simpler problems. It solves the problem in stages, dealing with all options at a particular stage before moving on to the next. In this sense it can often be represented as a breadth first search. However, unlike the levels of the tree in branch and bound which partition the problem by adding constraints, the stages in DP are linked by a recursive relationship. The name dynamic programming derives from its popularity in solving problems that require decisions to be made over a sequence of time periods. Even when this is not the case the name dynamic programming is still widely used, but the term *multistage programming* is sometimes used as an alternative.

The basis of DP is Bellman's *principle of optimality* (Bellman, 1957) which states that "the sub-policy of an optimal policy is itself optimal with regard to start and end states". As an illustration, consider the shortest route problem. If we are told that in Figure 2.1 the optimal route from S to F goes via E then we can be sure that that part of the route from S to E is the optimal route between S and E, and that part from E to F is the optimal route from E to F. In other words, each sub-path of the optimal path is itself the shortest path between its start and end points. Any DP implementation has four main ingredients. These are *stages*, *states*, *decisions* and *policies*. At each stage, for each feasible state we make a decision as to how to achieve the next stage. The decisions are then combined into sub-policies that are themselves combined into an overall optimal policy. DP is a very general technique that has been applied at varying levels of complexity. These have been classified into four levels: deterministic, stochastic, adaptive and residual. Our treatment here will be limited to deterministic problems.

The design of a DP algorithm for a particular problem involves three tasks: the definition of the stages and states, the derivation of a simple formula for the cost/value of the starting stage/state(s) and the derivation of a recursive relationship for all states at stage k in terms of previous stages and states.

The definition of the stages and states will obviously depend on the problem being tackled, but there are some definitions that are common. Stages are frequently defined in terms of time periods from the start or end of the planning horizon, or in terms of an expanding subset of variables that may be included at each stage. Common definitions of states are the amount of produce in stock or yet to be produced, the size or capacity of an entity such as a stock-sheet, container, factory, budget, etc, or the destination already reached in a routing problem.

2.4.2   Developing a DP Model

Forward Recursion and the Unbounded Knapsack Problem   We will firstly illustrate the concepts of DP by reference to a classical optimization problem—the unbounded knapsack problem. The problem can be stated as follows. Given a container of capacity b and a set of n items of size w_i and value v_i for $i = 1, \ldots, n$, such that the number of each item available is unbounded, maximize the value of items that can be packed into the container without exceeding the capacity.

The problem can be formulated as follows:

$$\text{max} \quad \sum_{i=1}^{n} v_i x_i \tag{2.11}$$

$$\text{s.t.} \quad \sum_{i=1}^{n} w_i x_i \leq b \tag{2.12}$$

$$x_i \geq 0 \text{ and integer}$$

where x_i = the number of copies of item i in the solution.

We can formulate this problem as a DP as follows. Define $F_k(S)$ to be the maximum value for a container of capacity S using items of sizes 1 to k. Here the items available represent the stages and the capacity available, the states. $F_1(S)$ is the value that can be obtained if the only pieces available are those of type 1. This is given by

$$F_1(S) = \text{int} \left(\frac{S}{w_1} \right) v_1 \tag{2.13}$$

All that remains is to define a recursive relationship for $F_k(S)$ in terms of previous stages and states. This is achieved as follows. The optimal solution either makes use of at least one item of type k, or it does not contain any items of type k. In the latter case, $F_k(S) = F_{k-1}(S)$. In the former case, one copy of item k takes up w_k units of capacity and adds v_k units of value. Bellman's principle tells us that the remaining $S - w_k$ units of capacity must be packed optimally. This packing may contain further items of type k and is given by $F_k(S - w_k)$. Thus we have for $k > 1$,

$$\begin{aligned} F_k(S) &= \text{max}\{F_{k-1}(S), F_k(S - w_k) + v_k\} \text{ for } S \geq w_k \\ F_k(S) &= F_{k-1}(S) \text{ otherwise} \end{aligned} \tag{2.14}$$

The solution to the overall problem is given by $F_n(b)$. We will illustrate the procedure with the following example.

Let $n = 3$, $b = 19$, $w_1, w_2, w_3 = 3, 5$ and 7 respectively and $v_1, v_2, v_3 = 4, 7$ and 10 respectively. The values of $F_k(S)$ for $k = 1, 3$ and $S = 0, 19$ are given in Table 2.4. The values in column $k = 1$ are first calculated using (2.13). Then the subsequent columns are calculated in order starting from the top and working down using (2.14). For $S \geq w_k$ the appropriate value is obtained by comparing the value in row S in the previous column with the sum of v_k and the value in the current column w_k rows up. The value of 27 in row 19, column 3 tells us that the optimal value is 27. In order to determine how this solution is achieved we need to work backwards. We need to find out whether this value came from $F_2(19)$ or $F_3(19 - 7) + 10$. The latter option is the correct one.

Table 2.4. Unbounded knapsack calculations.

S/k	1	2	3	S/k	1	2	3
0	0	0	0	10	12	14	14
1	0	0	0	11	12	15	15
2	0	0	0	12	16	16	17
3	4	4	4	13	16	18	18
4	4	4	4	14	16	19	20
5	4	7	7	15	20	21	21
6	8	8	8	16	20	22	22
7	8	8	10	17	20	23	24
8	8	11	11	18	24	25	25
9	12	12	12	19	24	26	27

We therefore record one item of type 3 and check the source of the value 17 in $F_3(12)$. This is either $F_2(12)$ or $F_3(5) + 10$. Once again the latter option is correct. We record a second item of type 3 and move to $F_3(5)$. $F_3(5) = F_2(5)$ so we check the source of the value of $F_2(5) = F_2(0) + 7$. Thus we record an item of type 2. As we have reached the top row of the table corresponding to capacity = 0 the solution is completed and is given by $x_1 = 0$, $x_2 = 1$, $x_3 = 2$.

Backward Recursion and a Production Planning Problem   In the above example the recursion worked in a forward direction with the stages corresponding to an increasing subset of variables. Our second example is taken from the field of production planning and is typical of many multi-period problems in that it is solved using backwards recursion, i.e. by working backwards from the end of the planning period. The problem can be summarized as follows.

Production of a single product is to be planned over a fixed horizon of n time periods. At the start there are S_0 units in stock and no stock is required at the end. Each time period t_i has a known demand d_i which *must* be met. Up to Q units can be produced in any one time period. The cost of making q units is given by $c(q)$ and economies of scale mean that $c(q)$ is not linear in q. Surplus units can be stored from one time period to the next at a warehousing cost of w per unit. There are natural definitions of the stages and states for this problem in terms of the time periods and stock levels. However, there is no simple formula for deciding what should be done in time period 1. Instead, we reorder the time periods in reverse order and relate stage k to period $(n - k)$. If we start the last period with S units in stock, then we must meet the demand d_n exactly as we are to finish without any surplus. Thus we must produce $d_n - S$

units. The formula for the optimal policy at the starting stage is therefore

$$F_0(S) = c(d_n - S) \tag{2.15}$$

We now need to define a recursive formula for $F_k(S)$ in terms of previous stages. If we start period $n - k$ with S units in stock and make q units we end with $S + q - d_{n-k}$ in stock. This will incur a warehousing cost and will define the starting stock for the next time period. The optimal policy from this point on has already been calculated as $F_{k-1}(S+q-d_{n-k})$. Thus the recursive formula is given by

$$F_k(S) = \min_{d_{n-k}-S \le q \le Q} \{c(q) + w(S + q - d_{n-k}) + F_{k-1}(S + q - d_{n-k})\} \tag{2.16}$$

The lower limit on q ensures that production is sufficient to meet demand.

We also need to define the set of states that need to examined at each stage k. This can be limited in three ways. First, there is no point in having more stock than can be sold in the remaining time periods. Second, it is not possible to have more stock than could be produced up to that time period less that already sold. Third, it is not feasible to have a stock level that will not allow demand in future periods to be met. Thus, for period $n - k$ we have $\text{MIN}_k \le S \le \min\{\text{MAX1}_k, \text{MAX2}_k\}$, where

$$\text{MAX1}_k = \sum_{i=n-k}^{n} d_i \qquad \text{MAX2}_k = S_0 + \sum_{i=1}^{n-k-1} (Q - d_i)$$

and

$$\text{MIN}_k = \max \left\{ 0, \max_{n-k \le j \le n} \sum_{i=n-k}^{j} (d_i - Q) \right\}$$

Once again we illustrate the formulation with a concrete example. Let $n = 4$, $Q = 5$, $S_0 = 1$, $w = \pounds 2000$ and production costs and demands be as given in Table 2.5.

Working in units of $\pounds 1000$ the calculations for the stages are then

Stage 0.
   $\text{MAX1}_0 = 2$, $\text{MAX2}_0 = 6$, $\text{MIN}_0 = 0$.
   $F_0(0) = c(2) = 13$, $F_0(1) = c(1) = 7$, $F_0(2) = c(0) = 0$.
Stage 1.
   $\text{MAX1}_1 = 3$, $\text{MAX2}_1 = 2$, $\text{MIN}_1 = 0$.
   $F_1(0) = \min\{c(1)+0 \cdot w + F_0(0), c(2)+1 \cdot w + F_0(1), c(3)+2 \cdot w + F_0(2)\}$
      $= \min \{7 + 0 + 13, 13 + 2 + 7, \underline{16 + 4 + 0}\} = 20$.
   $F_1(1) = \min \{0 + 0 + 13, 7 + 2 + 7, 13 + 4 + 0\} = 13$.
   $F_1(2) = \min \{\underline{0 + 2 + 7}, 7 + 4 + 0\} = 9$.

Table 2.5. Production costs and demands.

	Production costs					
Units	0	1	2	3	4	5
Cost (£1000s)	0	7	13	16	20	24

	Demands			
Period	1	2	3	4
Demand	3	6	1	2

Stage 2.

$MAX1_2 = 9$, $MAX2_1 = 3$, $MIN_1 = 1$.

$F_2(1) = \min \{24 + 0 + 20\} = 44$.

$F_2(2) = \min \{20 + 0 + 20, \underline{24 + 2 + 13}\} = 39$.

$F_2(3) = \min \{16 + 0 + 20, \underline{20 + 2 + 13}, 24 + 4 + 9\} = 35$.

Stage 3.

We do not need to calculate limits on S as we know that starting stock = 1.

$F_3(1) = \min \{\underline{16 + 2 + 44}, 20 + 4 + 39, 24 + 6 + 35\} = 62$.

Note that in many cases the full range of values for q from $S - d_{n-k}$ to Q have not been included in the minimization as they would lead to overstocking or under stocking. For example, in calculating $F_1(0)$ we do not consider any value of q above 3 as this would give more than two units at the start of the last time period. Similarly, we do consider q less than 3 in $F_3(1)$ as we need at least one unit in stock at the start of time period 2.

As with the knapsack problem, the calculations give the cost of the optimal solution but we need to work backwards in order to derive the optimal solution. Starting with $F_3(1)$ we note that the minimum value resulted from manufacturing three units, which leaves one unit in stock once the demand for three units has been met. Thus, the policy from time period 2 onwards is given by $F_2(1)$. This is optimized by producing five units, leaving 0 in stock. We therefore move to $F_1(0)$ which is optimized in two way—producing 1 and leaving 0 in stock or producing 3 and leaving 2 in stock. This implies that there are two optimal solutions. The former is completed using $F_0(0)$ and the latter using $F_0(2)$. The two solutions are summarized in Table 2.6.

2.4.3   Other Issues

One of the main criticisms of a dynamic programming approach is that the number of sub-problems that need to be solved is dependent not only on the

Table 2.6. The two optimal solutions.

| | | Production plan 1 | | | |
| | | | | | Cost (£) |
Period	Starting stock	Make	Sell	Closing stock	production + warehousing
1	1	3	3	3	18 000
2	1	5	6	0	24 000
3	0	3	1	2	20 000
4	2	0	2	0	0
Total					62 000

| | | Production plan 2 | | | |
| | | | | | Cost (£) |
Period	Starting stock	Make	Sell	Closing stock	production + warehousing
1	1	3	3	3	18 000
2	1	5	6	0	24 000
3	0	1	1	0	7 000
4	0	2	2	0	13 000
Total					62 000

stages but also on the states. While the number of stages is usually related to the size of the problem in the traditional sense (i.e. is a function of the number of variables) the number of states are frequently related to the size of the constants in the problem. For example, in the knapsack problem the number of states depends on the capacity of the container, while the number of states for the production planning problem are essentially bounded by a function of the maximum production capacity, Q. For real-life problems such quantities may be extremely large. This is often exacerbated by the fact that the states may be multi-dimensional. For example, in the standard DP formulation for two-dimensional cutting problems the states are defined by rectangles of dimension $X \times Y$. Our two examples were also relatively simple in that the recursive relationship relied only on the solutions at the previous stage. Many DP formulations require recursive relationships that use all previous stages, thus necessitating the results of all previous calculations to be stored, resulting in pressure on available memory in addition to long computation times. It is therefore important that some thought is given to reducing the number of states. Findlay et al. (1989) use a model similar to our production planning example to plan daily production in an oil field so as to meet a quarterly production target. The states at each stage are given by the amount still to be produced before the end of the quarter. Thus in their basic model the number of states is given by the number of days in the quarter multiplied by the total quarterly target. However, there are upper and lower bounds on daily production and by using these to produce three bounds on the feasible states at each stage, the total size of the search space can be reduced to less than half its original size.

Although the need to calculate and store all sub-solutions is often seen as a drawback of DP it can also be viewed as an advantage, as there is no need to carry out a whole new set of calculations if circumstances change. For example, in the production planning example, if for some reason we only managed to make two units instead of three in the third period, we could adopt the optimal policy from that point on simply by selecting the policy given by $F_1(1)$ instead of $F_1(2)$. This flexibility is cited as one of the reasons for the choice of DP as a solution technique by Findlay et al. (1989), as oil production is regularly affected by problems that may cause a shortfall in production on a particular day. Another example of the usefulness of being able to access the solutions to all sub-problems without additional computational effort arises in the solution of two-dimensional cutting problems. The bounds used by Christofides and Whitlock (1977) in their branch and bound algorithm cited in the previous section are calculated using a DP approach. The bound at the root node requires the solution to the unconstrained guillotine cutting problem in a rectangle of dimensions $X \times Y$ and the bounds at the other nodes require solutions to the same problem in smaller rectangles. These are precisely the problems solved in the various stages. Therefore, once the bound at the root node has been calculated, bounds for all the other nodes are available without further computation.

It is also worth emphasizing that DP is a very general approach. While this can be regarded as one of its strengths it can also be a drawback in that there are few rules to guide a beginner in its use for a completely new problem. In many cases it is relatively easy to define the stages of an implementation but it is more difficult to find a suitable definition for the states. Although there are examples of DP being used to solve a variety of problems, the vast majority still lie in the areas of multi-period planning, routing and knapsack-type problems, where it is relatively easy to adapt existing approaches. We have already mentioned the production planning problem tackled by Findlay et al. (1989). Other examples are a multi-period model for cricketing strategy (Clarke and Norman, 1999), a model for optimizing the route taken in orienteering (Hayes and Norman, 1984), and a multiple choice knapsack model for pollution control (Bouzaher et al., 1990).

2.5 NETWORK FLOW PROGRAMMING

2.5.1 Introduction

Network flow programming deals with the solution of problems that can be modeled in terms of the flow of a commodity through a network. At first glance it appears that such models might be very limited in their application, perhaps encompassing areas such as the flow of current in electrical networks, the flow of fluids in pipeline networks, information flow in communications networks and traffic flow in road or rail networks. However, their scope is far wider. They not only encompass a wide range of graph and network problems that appear to have little to do with flows, such as shortest path, spanning tree, matching and location problems, but also model a wide range of other problems ranging from scheduling and allocation problems to the analysis of medical x-rays. Network flow problems can be categorized as integer programming problems with a special structure. For the basic network flow models that deal with homogeneous flows this structure impacts on the solution process in two ways. First, the constraint matrix of the LP formulation has the property that it is *totally unimodular*. This implies that any solution at the extreme points of the feasible region will be integer valued. From a practical point of view this means that as long as all the constants in a problem are integer valued then solution via the simplex method will also be integer valued. Thus integer programs that have the special structure of a network flow problem can be solved without recourse to any of the specialist IP techniques described in Chapter 3. However, the structure of the problem also means that it can be solved directly by combinatorial algorithms that are simpler to implement than the full simplex algorithm. The inspiration for, and the verification of, the optimality of these procedures is rooted in the underlying LP theory. We will start by looking at the maximum flow problem, the simplest but least flexible of the network flow formulations, in order to introduce the basic concepts and building blocks that will be used in the solution of a more flexible model, the minimum cost flow problem.

2.5.2 The Maximum Flow Problem

Introduction The maximum flow problem is that of maximizing the amount of flow that can travel from source S to sink T in a network with capacities or upper bounds on the flow through each of its arcs. The problem can be stated as follows:

Let S = source, T = sink, x_{ij} = flow in arc (i, j),

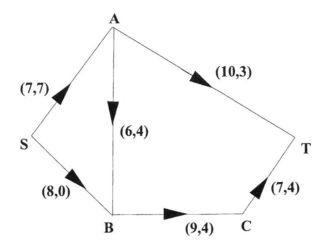

Figure 2.5. Maximum flow principle.

V = total flow from S to T, u_{ij} = upper bound on arc (i, j),

max V

$$\text{s.t.} \sum_{\substack{j \\ (i,j) \in A}} x_{ij} - \sum_{\substack{k \\ (k,i) \in A}} x_{ki} \begin{cases} = V \text{ if } i = S \\ = -V \text{ if } i = T \\ 0 \text{ for all other } i \end{cases} \qquad (2.17)$$

$$x_{ij} \leq u_{ij} \text{ for all } (i, j) \in A \qquad (2.18)$$

$$x_{ij} \geq 0$$

Constraints (2.17) ensure that the total flow arriving at the sink and the flow leaving the source are both equal to the objective, V, while at all other nodes the amount of flow entering the node is equal to that leaving.

As stated above, the problem could be solved by applying the simplex algorithm to the above formulation. However, a simpler and more intuitive algorithm is the Ford–Fulkerson labeling algorithm (Ford and Fulkerson, 1956) that is based on the idea of flow augmenting chains. Given a feasible flow in a network (i.e. a flow satisfying constraints (2.17) and (2.18)), a *flow augmenting* chain from S to T is a chain of arcs (in any orientation) such that the flow can be increased in forward arcs and decreased in backward arcs. This concept will be illustrated with reference to Figure 2.5.

The two-part labels on the arcs represent the capacity and the flow respectively. A total of seven units of flow travel from S to T. We can increase this flow along a chain of arcs in two ways. We could take chain {(S,B), (B,C), (C,T)} made up entirely of forward arcs and increase the flow by three units. The limit of the increase is three, as any larger increase would violate the ca-

CLASSICAL TECHNIQUES   47

pacity of arc (C,T). Alternatively we could take the chain {(S,B), (A,B), (A,T)}, in which arc (A,B) is a backward arc as it is oriented in the opposite direction to the chain. Using this chain we can increase the flow by four units by increasing the flow in arcs (S,B) and (C,T) and decreasing the flow in arc (A,B). This will have the effect of diverting four units of flow from (A,B) to (A,T), thus allowing an additional four units to arrive at B from S. The limit of four units derives from the fact that this will reduce the flow in (A,B) to 0. Ford and Fulkerson proved the result that a flow is optimal if and only if it has no flow augmenting chain. This suggests that the maximum flow problem can be solved by repeatedly finding a flow augmenting chain and augmenting the flows in the chain, until no flow augmenting chain exists. The Ford–Fulkerson labeling algorithm provides a mechanism for doing this while guaranteeing to identify a flow augmenting chain if it exists. It builds up one or more partial chains by successively labeling nodes with a two-part label (p_i, b_i) where p_i defines the predecessor of node i in the chain and b_i is an upper bound on the capacity of the chain up to node i. The objective is either to reach node T in which case a flow augmenting chain has been found, or to terminate without reaching T, in which case no flow augmenting chain exists.

The Ford–Fulkerson Labeling Algorithm   For maximum flow from source S to sink T.

Notation: x_{ij} is the current flow in arc (i, j), u_{ij} is the capacity of arc (i, j).
Step 1. Find an initial feasible flow (all flows = 0 will do). *Find a flow augmenting chain as follows.*
Step 2. Label $S(-, \infty)$ and set all other nodes as unlabeled.
Step 3. Select a forward arc (i, j) from a labeled to an unlabeled node such that $u_{ij} - x_{ij} > 0$, or select a backward arc (j, i) from an unlabeled node to a labeled such that $x_{ij} > 0$. If no such arc exists STOP *current flow is maximal.*
Step 4. If forward arc label $j(i, \min\{b_i, u_{ij} - x_{ij}\}$. If backward arc label $i(-j, \min\{b_j, x_{ij}\}$.
Step 5. If T not labeled go to step 3. Otherwise adjust flow as follows.
Step 6. Trace path back from T using labels p_i to determine preceding node. Increase flows in forward arcs on the path by b_T and decrease flows in backward arcs on the path by b_T.
Step 7. Go to step 2.

We illustrate the algorithm with reference to Figure 2.5. We start with the given flow.

Labeling: S$(-, \infty)$, B(S, min$(\infty, 8 - 0) = 8$), C(B, min$(8, 9 - 4) = 5$), T(C, min$(5, 7 - 4) = 3$), $b_T = 3$.

Thus we can augment the flow by three units. Using the labels p_i and working back from p_T we get chain S, B, C, T. All arcs are forward so the flow is increased by three units in each to give (S,A) seven units, (S,B) three units, (A,B) four units, (A,T) three units, (B,C) seven units, (C,T) seven units.

Attempt to find a flow augmenting chain given updated flows.
Labeling: S $(-, \infty)$, B(S, 5), C(B, 2), A$(-$B, 4), T(A, 4).

Note that in this small example we can see that labeling C will lead to a dead end. However, we have labeled it here to show that in the general case all labeled nodes need not appear in the final augmenting chain. The chain is given by SBAT where the link from B to A is backward so that flow is decreased by four units on this link and increased on all others. This gives: (S,A) seven units, (S,B) seven units, (A,B) zero units, (A,T) seven units, (B,C) seven units, (C,T) seven units.

Attempt to find a flow augmenting chain given updated flows.
Labeling: $S(-, \infty)$, $B(S, 1)$, $C(B, 1)$.

No further nodes are available for labeling. Thus the current flow of 14 units is optimal.

2.5.3   Minimum Cost Flow Problem

Introduction   Having introduced the basic concepts via the maximum flow problem we now move on to the more flexible model of the minimum cost flow in a closed network. This problem consists of a cyclic network, i.e. a network without a source and sink, and has upper and lower bounds on the capacities of the arcs as well as a cost associated with each arc. The objective is to find a feasible flow in the network such that the total cost is minimized. The LP formulation to the problem is as follows.

Let x_{ij} be the flow in arc (i, j), u_{ij} the upper bound on arc (i, j), l_{ij} the lower bound on arc (i, j) and c_{ij} the cost:

$$\min \sum_{(i,j)\in A} c_{ij} x_{ij}$$

$$\text{s.t.} \sum_{(i,j)\in A} x_{ij} - \sum_{(k,i)\in A} x_{ki} = 0 \quad \forall i \tag{2.19}$$

$$x_{ij} \le u_{ij} \quad \forall(i, j) \in A \tag{2.20}$$

$$x_{ij} \ge l_{ij} \quad \forall(i, j) \in A \tag{2.21}$$

$$x_{ij} \ge 0 \quad \forall(i, j) \in A$$

Constraint (2.19) ensures that the flow into each node equals that flowing out, while constraints (2.20) and (2.21) are the upper and lower bound constraints respectively. As with the maximum flow problem, this problem can be solved using the simplex method, but there are also a number of specialist solution techniques. Here we introduce one of these—the out-of-kilter algorithm.

The Out-of-Kilter Algorithm   The derivation of the out-of-kilter algorithm (Minty, 1960; Fulkerson, 1961) is based on linear programming theory, but the algorithm can be implemented without any prior LP knowledge. We associate a real number $\pi(i)$ with each node i. These numbers are sometimes called *node potentials* and are simply the dual variables associated with the flow balancing constraints (2.19). By using the optimality conditions for linear programming to determine equations for the dual variables associated with constraints (2.20) and (2.21) in terms of c_{ij}, $\pi(i)$ and $\pi(j)$, and using the complementary slackness conditions, it can be shown that the solution is optimal if and only if the following kilter conditions are satisfied for each arc (i, j):

- if $x_{ij} = l_{ij}$ then $c_{ij} + \pi(i) - \pi(j) > 0$

- if $l_{ij} < x_{ij} < u_{ij}$ then $c_{ij} + \pi(i) - \pi(j) = 0$

- if $x_{ij} = u_{ij}$ then $c_{ij} + \pi(i) - \pi(j) < 0$

For a given arc we can represent these conditions diagrammatically by a two-dimensional plot in which x_{ij} is plotted on the x-axis and $c_{ij} + \pi(i) - \pi(j)$ is plotted on the y-axis. The set of all points satisfying the kilter conditions form two vertical lines defined by $x = l_{ij}$ for $y \ge 0$ and $x = u_{ij}$ for $y \le 0$, connected by a horizontal line segment from $(l_{ij}, 0)$ to $(u_{ij}, 0)$. This is called the *kilter line* and diagram is a *kilter diagram*. Figure 2.7 shows the six kilter diagrams relating to the arcs of the network in Figure 2.6.

The bold lines are the kilter lines. The markers on the diagrams are plots of $(x_{ij}, c_{ij} + \pi(i) - \pi(j))$ for different flows and potentials. When the marker

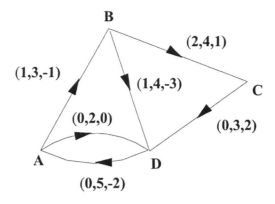

Figure 2.6. Minimum cost flow problem: the three-part-labels are (l_{ij}, u_{ij}, c_{ij}).

lies on the kilter line the corresponding arc is said to be *in kilter*, if not it is said to be *out of kilter*. We will refer to this figure again when we illustrate the out-of-kilter algorithm.

The out-of-kilter algorithm works with solutions that satisfy the flow balance constraints (2.19), but may violate the upper and lower bounds. By changing the flows or the potentials it gradually moves each arc into kilter without moving any other arc away from the kilter line in the process. It can be stated as follows.

Out-of-kilter algorithm for min-cost flow in a closed network. Find an initial flow satisfying the flow balance equations and a set of node potentials $\pi(i)$ $\forall i$. Let $y(i, j) = c_{ij} + \pi(i) - \pi(j)$ (note that all flows and potentials = 0 will do).

If $y(i, j) > 0$, $x_{\min}(i, j) = \min(x_{ij}, l_{ij})$, $x_{\max}(i, j) = \max(x_{ij}, l_{ij})$.
If $y(i, j) = 0$, $x_{\min}(i, j) = \min(x_{ij}, l_{ij})$, $x_{\max}(i, j) = \max(x_{ij}, u_{ij})$.
If $y(i, j) < 0$, $x_{\min}(i, j) = \min(x_{ij}, u_{ij})$, $x_{\max}(i, j) = \max(x_{ij}, u_{ij})$.
While any arcs out of kilter and procedure is successful do.
Attempt to update flows.
Select and out-of-kilter arc (p, q).
If $x(p, q) > x_{\min}(p, q)$ then set $s = p, t = q, v = x_{pq} - x_{\min}(p, q)$.
If $x(p, q) < x_{max}(p,q)$ then set $s = q, t = p, v = x_{\max}(p, q) - x_{pq}$.

Attempt to find a flow augmenting chain from s to t to carry up to v additional units of flow without using (p, q) and without exceeding $x_{\max}(i, j)$ in forward arcs or falling below $x_{\min}(i, j)$ in backward arcs. [Note: this can be achieved by starting the labeling algorithm with $s(-, v)$ and respecting $x_{\max}$ and $x_{\min}$ when adjusting the flows.]

If successful increase flow in the chain and increase/decrease flow in (p, q) by b_t. Otherwise *attempt to update node potentials as follows.*

Let L be the set of all arcs (i, j) labeled at one end and not the other such that $l_{ij} \leq x_{ij} \leq u_{ij}$.

For those arcs in L labeled at i: if $y(i, j) > 0$ set $\delta_{ij} = y(i, j)$, otherwise set $\delta_{ij} = \infty$.

For those arcs in L labeled at j: if $y(i,j) < 0$ set $\delta_{ij} = -y(i,j)$, otherwise set $\delta_{ij} = \infty$.

Set $\delta = \min \{\delta_{ij} : (i, j) \in L\}$.

If $\delta = 0$ then stop—no feasible flow. Otherwise set $\pi(k) = \pi(k) + \delta$ for all unlabeled nodes and update $y(i, j)$ for all arcs labeled at one end and not the other.

Repeat.

When the algorithm terminates either all the arcs are in kilter and the current flows are optimal or no feasible solution exists.

We illustrate the algorithm with reference to the network in Figure 2.6 and the kilter diagrams in Figure 2.7. Note that $x_{\min}$ and $x_{\max}$ are simply the minimum and maximum flow values that ensure that an arc never crosses or moves further away from the kilter line in a horizontal direction and δ serves the same purpose for moves in a vertical direction.

Initialization: We start with all flows and potentials equal to zero. Thus $c_{ij} + \pi(i) - \pi(j)$ is simply the arc cost for all arcs. This situation is given by the solid circles in Figure 2.7. Note that arcs (A, D) and (C, D) are already in kilter, but arc (D, A) is not, even though its flow lies within its lower and upper bounds.

Iteration 1: We select out-of-kilter arc (A, B). We would like to increase the flow in this arc by three units. Thus $v = 3, s = B, t = A$.

Labeling: $B(-, 3), D(B, 3), A(D, 3)$. Labeling has been successful. Therefore we increase flow in chain $\{(B, D), (D, A)\}$ and in arc (A, B) by three units and update $x_{\min}$ and $x_{\max} \forall$ updated arcs. This is shown by the solid squares in Figure 2.7. Note that arc (A, B) is now in kilter.

Iteration 2: Select out-of-kilter arc $(B, C), s = C, t = B, v = 2$.

Labeling: $C(-, 2)$; no further labeling possible as we cannot increase flow in (C, D) without moving away from the kilter line. $L = \{(C, D)\}$ and as (C, D) is labeled at C, so $\delta_{CD} = 2$. Note that this is the maximum distance this arc can move down without leaving the kilter line.

Unlabeled nodes are A, B, D. Increase potentials on these nodes by two to give $\pi(A) = 2, \pi(B) = 2, \pi(C) = 0, \pi(D) = 2$. This will change the $y(i, j)$ values for arcs (B, C) and (C, D) as shown by the solid triangles. Note that for (C, D) this also changes $x_{\max}$.

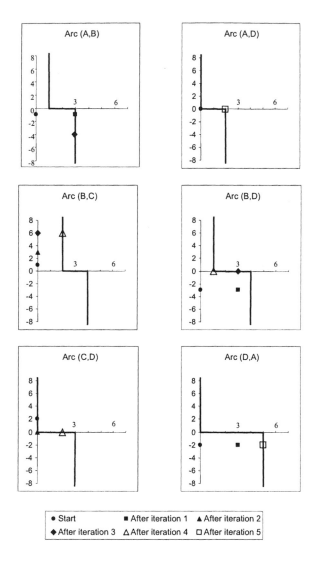

Figure 2.7. Kilter diagrams for Figure 2.6 problem.

Iteration 3: We try again with arc (B, C), $s = C$, $t = B$, $v = 2$.

Labeling: $C(-, 2)$, $D(C, 2)$, $A(D, 2)$; no further labeling is possible as flow in (A, B) is at full capacity and decreasing flow in (B, D) will move away from the kilter line. $L = \{(A, B), (B, D)\}$. $\delta_{AB} = \infty$, $\delta_{BD} = 3$, $\delta = 3$.

Unlabeled node B. Increase $\pi(B)$ by three, giving $\pi(A) = 2$, $\pi(B) = 5$, $\pi(C) = 0$, $\pi(D) = 2$ and changing positions of arcs (A, B), (B, C) and (B, D) as given by the solid diamonds. Arc (B, D) is now in kilter and has a new value for $x_{\min}$.

Iteration 4: We try again with arc (B, C), $s = C$, $t = B$, $v = 2$.

Labeling: $C(-, 2)$, $D(C, 2)$, $B(-D, 2)$. t is labeled.

Adjust flows in chain $\{(C, D), (B, D)\}$ and increase flow in (B, C) by two units as shown by the unfilled triangles. Arc (B, C) is now in kilter.

Iteration 5: Select only remaining out of kilter arc (D, A), $s = A$, $t = D$, $v = 2$.

Labeling: $A(-, 2)$, $D(A, 2)$. t is labeled. Increase flow in (D, A) and (A, D) by two units as shown by the unfilled squares.

All arcs are in kilter, therefore the solution is optimal. Flows are as given by the last update for each arc, i.e. $x_{AB} = 3$, $x_{AD} = 2$, $x_{BC} = 2$, $x_{BD} = 1$, $x_{CD} = 2$ and $x_{DA} = 5$, with a total cost of -13.

2.5.4 Other Issues

The previous sections introduced two relatively simple algorithms for the maximum flow and minimum cost flow problems. The out-of-kilter algorithm also has the advantage that it is easy to find an initial solution as the upper and lower bounds do not need to be satisfied. However, these are not necessarily the most efficient algorithms in each case. For example, it is possible to design pathological cases of the max-flow problem for which the number of iterations made by the Ford–Fulkerson algorithm is only bounded by the capacity on the arcs. There are also inherent inefficiencies in the algorithms in that subsequent iterations may need to recalculate labels already calculated in previous iterations. A more efficient algorithm is the network simplex algorithm which can be found in Ahuja et al. (1993). The algorithm is so called because it can be shown that the iterations used to improve an initial solution correspond exactly to those computed by the simplex method, but the special structure of the problem means that the algorithm can be expressed purely in network terms without recourse to the simplex tableau. For large problems the additional efficiencies of the network simplex algorithm may pay off, but for small to moderate problems the out-of-kilter algorithm should be fast enough, and has the advantage that it is easy to implement from scratch and that code is available from a number of sources.

As already mentioned, the efficiency of network flow solution algorithms means that it is worthwhile attempting to model any new problem as a network flow problem. A wide range of problems can be modeled using the two formulations already given. However, the scope of network flow approaches is even wider when we consider problems that can be modeled using the dual formulations of network flow problems. Examples of this type of model include Mamer and Smith's (1982) approach to an infield repair kit problem and Johnson's (1968) open cast mine planning problem. Even when the network model is not flexible enough to incorporate all the constraints in a problem it may still provide an effective optimization tool. For example, Glover et al. (1982) model the problem of determining the allocation of space on aircraft to different fare structures as a min-cost flow problem. Their model does not incorporate all the constraints needed to model the problem and thus infeasible solutions may be returned. If this occurs the solution is excluded from the model and the solution process reiterated until a feasible solution results. The authors comment that this was far more efficient than attempting to include all such constraints directly into an IP model.

We have limited our focus to problems in which flows are homogeneous and there is no gain or leakage of flow along an arc. Problems in which multiple commodities share the arc capacities and problems where flow is not preserved along an arc have also been widely studied. Unlike the simple problems covered here, both these problems are NP-complete (see Chapter 11). Algorithms for solving such problems are beyond the scope of this chapter but can be found in many network flow texts. Ahuja et al. (1993) give a comprehensive treatment of network flows including models, algorithms and practical applications.

2.6   SOME USEFUL MODELS

The previous sections have outlined three general approaches to optimization problems. In this section we focus on two classes of problem that frequently occur as sub-problems in the solution of larger or more complex problems over a wide range of application areas. In each case solution approaches based on the methods covered in the previous sections are described.

2.6.1   Shortest Path Problems: Dynamic Programming Approaches

The first class of common sub-problems are those involving shortest paths. As observed in Section 2.3, the branch and bound shortest path algorithm presented there is not the most efficient way of tackling such problems. In this section more efficient approaches based on dynamic programming are presented. We start by considering the problem of finding the shortest path from a source vertex, s, to all other vertices in a graph.

Bellman's Shortest Path Algorithm   This problem can be solved by Bellman's shortest path algorithm in which the stages are defined by the number of links allowed in the path and the states are defined by the vertices. The formulae for the starting state and the recursive relationship can be defined as follows:

$$F_0(v) = 0 \text{ if } v = s, \quad F_0(v) = \infty \text{ otherwise}$$

$$F_k(v) = \min \left\{ F_{k-1}(v), \min_{w \in Q_{k-1},(w,v) \in E} F_{k-1}(w) + c_{vw} \right\}$$

where Q_k is the set of vertices whose values were updated at stage k and E is the set of links in the network. Bellman's shortest path algorithm can then be stated as follows:

Set $F_0(v) \forall v \in V$
While $Q_k \neq \emptyset$ and $k \leq n$ do
    Calculate $F_k(v) \forall v \in V$ and determine Q_k
End while
If $Q_k = \emptyset$ then $F_k(v)$ defines length of shortest path from s to $v \forall v$
If $k = n$ and $Q_k \neq \emptyset$ then network contains negative cost circuits and shortest paths cannot be defined.

If paths from every vertex to every other are required then rather than execute Bellman's algorithm n times it is more efficient to use Floyd's shortest path algorithm (Floyd, 1962).

Floyd's Shortest Path Algorithm This is also a DP-type approach but in this case $F_k(i,j)$ represents the shortest path between i and j allowing only vertices 1 to k as intermediate points. The initial states are given by $F_0(i,j) = C_{ij}$, where C_{ij} is the cost of link (i,j), and the recursive relationship by $F_k(i,j) = \min\{F_{k-1}(i,j), F_{k-1}(i,k) + F_{k-1}(k,j)\}$. As $F_k(i,i) = 0 \, \forall k$, unless the network contains a negative cost circuit neither $F_k(i,k)$ or $F_k(k,j)$ will be updated during iteration k. Therefore, the subscript k is usually dropped in practice and matrix $F(i,j)$ is overwritten at each iteration. The algorithm can then be stated as follows.

Step 1.
    $k = 0$, $F(i, j) = c(i, j) \forall (i, j)$.
Step 2.
    $\forall k = 1, n$
    $\forall i = 1, n$ st $i \neq k$ and $F(i, k) \neq \infty$
    $\forall j = 1, n$ st $j \neq k$ and $F(k, j) \neq \infty$
    $F(i, j) = \min \{F(i, j), F(i, k) + F(k, j)\}$
    if $F(i, i) < 0$ for any i STOP (*negative cost circuit detected*)
    end loops

It should also be noted here that if all costs are non-negative then an algorithm due to Dijkstra (1959) is a more efficient than Bellman's algorithm. Dijkstra's algorithm can be found in most basic texts covering graph and network algorithms (e.g. Ahuja et al., 1993).

2.6.2   Transportation Assignment and Transhipment: Network Flow Approaches

In this section we consider a second commonly occurring class of sub-problems—the family of transportation-type problems, including the assignment and transhipment problems. All three problems can be modeled as minimum cost flow problems. Once the appropriate model has been derived, solutions can be obtained using the out-of-kilter algorithm or any other minimum cost flow algorithm. The focus of this section is therefore on defining appropriate models.

The Transportation Problem   The transportation problem is that of determining the amount of product to be supplied from each of a given set of supply points to each of a given set of demand points, given upper bounds on availability at each supplier, known demands at each demand point, and transportation costs per unit supplied by supplier i to demand point j. The problem can be formulated as follows:

$$\min \sum_{i=1}^{n} \sum_{j=1}^{m} c_{ij} x_{ij}$$

$$\text{s.t.} \sum_{i=1}^{n} x_{ij} \geq d_j \qquad\qquad (2.22)$$

$$\sum_{j=1}^{m} x_{ij} \leq s_i$$

We can model this problem as a minimum cost flow network by defining nodes, arcs, lower and upper bounds and costs as follows.

Nodes:

- A dummy source node S.

- A dummy sink node T.

- One node for each supplier, $i = 1, n$.

- One node for each demand point j, $j = 1, m$.

Arcs:

- An arc from each supplier node i to each demand node j with lower bound $= 0$, upper bound $= s_i$ and cost $= c_{ij}$. Note that if all $c_{ij} \geq 0$ then the upper bound can be replaced by $\min \{s_i, d_j\}$.

- An arc from S to each supplier node with lower bound $= 0$, upper bound $= s_i$ and cost $= 0$.

- An arc from each demand node, j, to T with lower bound $= d_j$ upper bound $= M$, where M is some suitably large number and cost $= 0$. Note if all $c_{ij} \geq 0$ then the upper bound can be replaced by d_j.

- An arc from T to S with lower bound $= 0$, upper bound $= \sum_{i=1}^{n} s_i$, and cost $= 0$.

The minimum cost flow in the network defined above will give the optimal solution to the transportation problem and the flows in the arcs (i, j) define the value of the variables x_{ij}.

Many management science and operational research texts cover the stepping stone algorithm or MODI method for the transportation problem. It is interesting to note that the rules used by the MODI method for determining those x_{ij} that should be considered for increase are precisely the kilter conditions that define those arcs that lie to the left of the kilter line in the network flow model, and that the stepping stone process for finding a suitable path to update the solution corresponds to finding a flow augmenting chain. Many texts also suggest a heuristic method known as Vogel's approximation method, or VAM, to obtain an initial solution. This approach can be taken to find an initial flow in the above model, thus reducing the number of iterations required when compared with a starting solution of 0.

The Assignment and Transhipment Problems   Several relatives of the transportation problem are regularly encountered as sub-problems in more complex models. Here we consider the assignment and transhipment problems.

The assignment problem is that of assigning tasks to resources. It is assumed that there are n tasks and n resources and a cost c_{ij} associated with the assignment of task i to resource j. The objective is to assign each task to exactly one resource and each resource to exactly one task such that the total cost of the assignments is minimized. This problem can be regarded as a transportation problem in which $n = m$ and $s_i = d_j = 1$, $\forall i$ and $\forall j$. Thus any assignment problem can be modeled using the transportation model above with the appropriate values for s_i and d_j.

The transhipment problem is a transportation problem in which goods may travel from supplier to demand point via any or none of a set of intermediate points—known as transhipment points. These points may also be supply points or demand points in their own right. The network flow model for the transportation problem can easily be adapted to the transhipment model by adding new nodes for the transhipment points and adding arcs from each supply point and to each demand point with appropriate bounds and transhipment costs. If a transhipment point is also a supply point or demand point then it is also connected to S or T respectively.

2.6.3   Other Useful Models

The above classes of problem are two of the most frequently occurring sub-problems that can be solved using the methods covered in this chapter. Two other classes of problem that deserve special mention are the binary and bounded knapsack problems and matching problems in bipartite graphs.

We have already presented a dynamic programming approach for the unbounded knapsack problem in Section 2.4. The binary version of the problem occurs when at most one piece of each type is available. An effective solution approach to this problem is a tree search in which there are two branches from each node, one corresponding to fixing the value of a variable, x_i at 0 and the other to fixing the same variable at 1. Upper and lower bounds are easily obtained, simply by sorting the variables in v_i/w_i order and adding each to the knapsack in turn, until the capacity is exceeded. In its basic form the algorithm is an implementation of the standard integer programming branch and bound algorithm as discussed in Chapter 3. However, the bounds can be improved by adding problem specific information. Martello and Toth (1990) give details of several variants of the method and suggest a series of increasingly powerful bounds. They also show how the method can be extended to the more general case in which the variables are not binary, but are restricted by upper bounds.

Matching problems also occur widely. A matching in a graph is a set of edges no two of which have a vertex in common. Maximum matching problems in arbitrary graphs can be solved using the blossom algorithm due to Edmonds, and described in Ahuja et al. (1993). However, if the graph is bipartite, i.e. the vertices can be partitioned into two subsets such that there are no edges between any two vertices in the same subset, then the problem can be solved more simply using network-flow-type algorithms. Such models are useful in practice, as bipartite graphs often appear in allocation or scheduling problems.

2.7   AREAS FOR FUTURE APPLICATION

In this section we outline some potential areas for the future application of the methods described in this chapter. All three approaches have been used extensively in the solution of a broad range of problems for several decades. The increase in computer power available to individuals and organizations since their introduction has led to a continuously expanding range of problems that can be solved to optimality within a feasible amount of time. At the same time, theoretical advances in potential application areas have lead to improved bounds, once again increasing the scope of branch and bound approaches. There are currently many researchers active in each of the areas. Thus it is likely that we will continue to see new theoretical developments as well as new branch and bound or DP implementations for old problems and the application of classical approaches to new practical problems. However, there is also considerable potential for combining these techniques with some of the more modern approaches covered elsewhere in this book. Indeed there is considerable evidence that this is already happening. Such integration can be at one of three levels: pre- or post-processing, true hybridization, and cross-fertilization of ideas. We take a brief look at each of these in turn.

2.7.1   Pre- and Post-processing

The simplest form of integration is to use a classical approach as part of a staged solution to a problem. Tree search approaches are often used to enumerate all the variables required for the optimization phase of the problem. For example, in crew scheduling problems the set of feasible tours of duty satisfying all the necessary constraints are enumerated first and then these are used as input to an optimization algorithm, or in the case of timetabling and scheduling problems the allocation of events to rooms may be carried out in a post-processing phase once the schedule has been determined. In other cases pre-processing may be used to reduce the size of the solution space. For example, Dowsland and Thompson (2000) solve a nurse scheduling problem using tabu search. Before calling the tabu search routine they use a tree search approach to solve a modified knapsack problem that enables them to determine the precise number of additional nurses required to cover the weekly demand on a ward. This allows them to minimize the size of the solution space and to simplify the evaluation function in the tabu search part of the solution process.

2.7.2   True Hybrids

A greater degree of integration is provided by true hybrid techniques approaches in which a classical approach is embedded into a modern search tool or vice versa. Many researchers have already tried this. There are several instances of the integration of branch and bound with genetic algorithms. For example, Cotta et al. (1995) use a tree search to find the best child from a set of possibilities given by a loosely defined crossover, while Nagar et al. (1995) use a genetic algorithm (see Chapter 4) to search out a promising set of ranges on the values of a series of variables and then use a tree search to find the optimal solution within the range. Classical techniques have also been embedded into neighborhood search approaches such as simulated annealing (see Chapter 7), tabu search (see Chapter 6) and variable depth search. For example, there are many problems in which the variables can be partitioned into two sets, A and B, such that if the values of A are fixed the problem of optimizing B reduces to a network flow problem. The size of the solution space can be reduced to cover only the variables in A, with each solution being completed by solving for the variables in B. Network-flow-type problems are especially amenable to this sort of role, as neighborhood moves will typically involve changing a cost or bound, or adding or deleting an arc. All of these changes can be accommodated in the out-of-kilter algorithm using the previous solution to initialize the solution process for its neighbor(s), thus minimizing the computational effort in resolving each new sub-problem. Examples of this type of strategy include Hindi et al. (2003) who solve transhipment sub-problems to complete solutions in their variable depth search approach to lot sizing problem, and Dowsland and Thompson (2000) who use a network flow problem to allocate nurses on days to the morning or afternoon shifts. This reduces the number of variables in the search space for each nurse by up to a factor of 32.

As discussed in other chapters, many neighborhood searches can be improved by extending the size of the neighborhood. One way of doing this is to introduce chains of moves. Rather than select a chain at random or enumerate all chains in the neighborhood, it makes sense to find optimal or improving chains directly using some form of shortest path algorithm. For example, Dowsland (1998) uses a mixture of tree search and Floyd's algorithm to find improving chains of moves for a nurse scheduling problem. Other forms of compound moves may not involve chains. In such cases other techniques can be used to find optimal moves. For example, Potts and van de Velde (1995) use dynamic programming to search for good moves in neighborhoods made up of multiple swaps for the traveling salesman problem. Gutin (1999) also considers the TSP but utilizes a bipartite matching problem in order to determine the best option from a neighborhood defined by removing and reinserting k vertices in the tour. Other researchers have worked with this neighborhood for different definitions of k.

2.7.3   Cross-fertilization

As well as true hybrid techniques such as those outlined above there are also examples of cross-fertilization of ideas in which an ingredient from a classical approach has been embedded into a more modern method. One example is the incorporation of bounds into a heuristic search in order to avoid or leave non-promising areas of the solution space. Hindi et al. (2003) use this strategy in their variable depth search approach to avoid wasting time in solving transhipment problems when it can be shown that the resulting solution cannot result in an improvement, while Dowsland (1998) uses bounds in combination with a tabu list to avoid wasting time in areas of the search space that cannot lead to a better solution than the best found so far. Similar approaches have been taken with genetic algorithms by Tamura et al. (1994) and Dowsland et al. (2004) who use mutation and crossover respectively to destroy partial solutions that exceed a tree-search-type bound. A second example of cross-fertilization is in the use of ejection chains, a concept suggested by Glover and Laguna (1997) that is a generalization of alternating chains—a term used for the way in which flow is updated in certain classes of network flow problems.

2.8 TRICKS OF THE TRADE

2.8.1 Introduction

For newcomers to the field, the prospect of applying any of the above techniques to a given problem can appear daunting. This section suggests a few tips on overcoming this feeling and getting started on a basic implementation, and then going on to identify possible areas for algorithm improvement. We start with a few general observations that apply to all three techniques and follow this with more specialist advice for each of the techniques in turn.

1 *Get a basic understanding of the technique and how it might be applied to a given problem.* This involves reading suitable books and articles. Due to the relatively long history of the techniques covered in this chapter there is a wealth of introductory material available. Although some of the seminal material and/or early texts in each field provide valuable insights and technical detail they often use specialist terminology and can be difficult to understand. Therefore they are probably best left until some practical experience has been gained, and the best place to start would be one of the more up-to-date texts given in the references and sources of additional information at the end of this chapter. These will provide a thorough background but may not include examples that are closely related to the reader's own problem. It is therefore desirable to supplement these sources with journal articles related to the relevant application area.

2 *Don't reinvent the wheel.* There are many published articles describing implementations of these techniques to classical combinatorial optimization problems. If your problem can be formulated as, or is closely related to, one of these problems then it is likely that an effective implementation has already been published, and in many cases suitable code may also be readily available. Sometimes the relationship to a classical problem is obvious from the problem definition, but this is not always the case. It is therefore well worth considering different ways of modeling

a problem, e.g. using graph-theoretic models or trying different LP-type formulations and comparing these with well-known classical problems.

3 *Don't be too pessimistic.* Although the complexity of most branch and bound and dynamic programming algorithms is likely to be exponential (or at best pseudo-polynomial) they can still be effective tools for solving moderately sized problems to optimality, and may well compete with the more modern methods described later in this book when used as heuristics. While there may be many good practical reasons for selecting a heuristic rather than an exact approach to a problem, many real-life NP-hard problems (see Chapter 11) have proved amenable to solution by both branch and bound and dynamic programming. However, this may require a little ingenuity on the part of the algorithm designer. If the initial implementation seems too slow or memory intensive it is well worth spending some time and effort trying to make improvements. The situation with regard to network flow programming is somewhat different, in that the simple models covered in detail in this chapter can all be solved to guaranteed optimality in polynomial time and are the obvious first-choice approaches for problems that have the required structure.

2.8.2   Tips for Branch and Bound

The issues arising in developing and implementing a branch-and-bound approach can be broadly divided into three categories.

1 *Representations.* For most problems there are several different potential tree search representations and search strategies each of which will impact differently on solution time and quality. In order to appreciate this it is worth reading articles that use different representations for the same problem. Before starting to code any implementation think about what you have read in relation to your own problem. You should also think about whether you want to apply a depth-first search or if there may be advantages in a more memory intensive breadth or best-first search.

2 *Coding.* Although it is relatively easy for a beginner to construct a tree search on paper it can be difficult to translate this into computer code. We therefore recommend finding code or detailed pseudo-code for a similar tree structure to the one you are planning on and using it as a template for your own program structure. It is also often helpful to code the three operations of branching, backtracking and checking the bounding conditions separately. It is also a good idea to start with very simple bounds and a branching strategy based on a natural ordering of the vertices and test on a small problem, building up more sophisticated bounds and branching strategies as necessary.

3 *Performance.* Once the basic code is working the following pointers will help in getting the best out of an implementation.

Analyse the time taken to calculate the bounds and their effectiveness in cutting branches and consider how the trade-off influences overall computation time. While computationally expensive bounds or branching strategies are likely to increase solution times for small problems, they may well come into their own as problem size grows.

In the same way as it is important to test heuristics for solution quality on problem instances that closely match those on which they are to be used in terms of both problem size and characteristics, so it is important to ensure that a branch and bound approach will converge within a realistic time-scale on typical problem instances.

Remember that branch and bound can be used as a heuristic approach, either by stopping after a given time or by strengthening the bound to search for solutions that improve on the best so far by at least $\alpha\%$. In this case it is important to consider whether you want to bias the search towards finding good solutions early or precipitating the bounding conditions earlier.

2.8.3   Tips for Dynamic Programming

Due to its generality, getting started with dynamic programming can be difficult but the following pointers may be of assistance.

1 *Representation.* When faced with a new problem simply deciding on the definition of stages and states that form the basis of a correct dynamic programming formulation can be difficult, and access to a successful formulation to a similar problem can be invaluable. As a starting point it is worth considering if the problem can be classified as a multi-period optimization problem, routing problem, or knapsack-type problem. If so, then there are a wealth of examples in standard texts or journal articles that should help.

Remember that the objective is to produce something that is considerably more efficient than complete enumeration. Therefore, it is important to ensure that the calculations relating to the first stage are trivial, and that there is a relatively simple recursive relationship that can be used to move from one stage to the next. If there is no apparent solution when defining the stages in a forwards direction then consider the possibility of backward recursion.

2 *Performance.* DP can be expensive both in terms of computational time and storage requirements and the main cause of this is the number of

states. Therefore, once a basic structure has been determined it is worthwhile considering ways of cutting down on the number of states that need to be evaluated. It is also worthwhile ensuring that only those stages/states that may be required for future reference be stored.

If the environment requires the solution of several similar problems, e.g. if the DP is being used to calculate bounds or optimize large neighborhoods, consider whether or not the different problems could all be regarded as sub-problems of one large problem, thereby necessitating just one DP to be solved as a pre-processing stage. As with branch and bound, it is important to ensure that the time and memory available are sufficient for typical problem instances.

2.8.4   Tips for Network Flow Programming

The problems facing a beginner with network flow programming are different in that there are standard solution algorithms available off-the-peg, so that the only skill involved is that of modeling. Because these algorithms are readily available and operate in polynomial time then it is certainly worth considering whether any new problem might be amenable to modeling in this way. Some tips for recognizing such problems are given below.

If the problem involves physical flows through physical networks then the model is usually obvious. However, remember that network flow models simply define an abstract structure that can applied to a variety of other problems. Typical pointers to possible network flow models are allocation problems (where flow from i to j represents the fact that i is allocated to j), sequencing problems (where flow from i to j represents the fact that i is a predecessor of j), and problems involving the selection of cells in matrices (where flow from i to j represents the selection of cell (i,j)). However, this list is by no means exhaustive.

Other tips for modeling problems as network flows are to remember that network links usually represent important problem variables and that even if the given problem does not have an obvious network flow structure the dual problem might. In addition, although typical practical problems will be too large and complex to draw the whole network it is worthwhile making a simplified sketch of potential nodes and links. If the problem has complexities such as hierarchies of resources, multiple pricing structures, etc., this may be facilitated by simplifying the problem first and then trying to expand it without compromising the network flow structure.

2.9   CONCLUSIONS

Although the classical techniques described in this chapter were developed to meet the challenges of optimization within the context of the computer technology of the 1950s, they are still applicable today. The rapid rate of increase of computing power per unit cost in the intervening years has obviously meant a vast increase in the size of problems that can be tackled. However, this is not the sole reason for the increase. Research into ways of improving efficiency has been continuous both on a problem-specific and generic basis—for example all the techniques lend themselves well to parallelization. Nevertheless, they do have drawbacks. The scope of network flow programming is limited to problems with a given structure, while the more general methods of branch and bound and DP may require vast amounts of computer resource. DP solutions to new problems are often difficult to develop whereas it may not be easy to find good bounds for a branch and bound approach to messy practical problems. Hence the need for the more recent techniques described elsewhere in this volume.

However, it should be noted that these techniques have not eclipsed the classical approaches, and there are many problems for which one of the techniques described here is still the best approach. Where this is not the case and a more modern approach is appropriate it is still possible that some form of hybrid techniques may enhance performance by using the strengths of one approach to minimize the weaknesses of another.

SOURCES OF ADDITIONAL INFORMATION

Basic material on DP and network flows can be found in most undergraduate texts in management science and OR. Their treatment of branch and bound tends to be limited to integer programming. The generic version of branch and bound covered in this chapter can be found in most texts on combinatorial algorithms/optimization, or in subject oriented texts (e.g. algorithmic graph theory, knapsack problems). Below are a small sample of relevant sources of information:

- http://www.informs.org/Resources/ (INFORMS OR/MS Resource Collection: links to B&B, DP and network flow sources).

- http://www.ms.ic.ac.uk/jeb/or/contents.html (Dr J. Beasley, Imperial College).

- http://web.mit.edu/~jorlin/www/ (Professor J. Orlin, MIT).

- http://www.math.ilstu.edu/~sennott/ (Sennott, 1998).

- Bather, J., 2000, *Decision Theory: An Introduction to Dynamic Programming and Sequential Decisions,* Wiley, New York.

- Bellman, R., 2003, *Dynamic Programming,* Dover, New York.

- Christofides, N., 1975, *Graph Theory—An Algorithmic Approach,* Academic, New York.

- Hu, T. C., 1981, *Combinatorial Algorithms,* Addison-Wesley, Reading, MA.
- Lawler, E. L., Lenstra, J. K., Rinnooy Kan, A. H. G., and Shmoys, D. B., 1985, *The Travelling Salesman Problem,* Wiley, New York.
- Nemhauser, G. L., and Wolsey, L. A., 1988, *Integer and Combinatorial Optimisation,* Wiley, New York.
- Papadimitriou, C. H., 1982, *Combinatorial Optimisation: Algorithms and Complexity,* Prentice-Hall, Englewood Cliffs, NJ.
- Reingold, E. M., Nievergelt, J., and Deo, N., 1977, *Combinatorial Algorithms,* Prentice-Hall, Englewood Cliffs, NJ.
- Sennott, L. I., 1998, *Stochastic Dynamic Programming,* Wiley, New York.
- Taha, H. A., 2002, *Operations Research,* Prentice-Hall, Englewood Cliffs, NJ.

References

Ahuja, R. K., Magnanti, T. L. and Orlin, J. B., 1993, *Network Flows: Theory, Algorithms and Applications,* Prentice-Hall, Englewood Cliffs, NJ.

Anderson, D. R., Sweeney, D. J. and Williams, T. A., 1997, *Introduction to Management Science: Quantitative Approaches to Decision Making,* West Publishing, Minneapolis, MN.

Balakrishnan, V. K., 1997, *Schaum's Outline of Graph Theory,* McGraw-Hill, New York.

Balas, E. and Christofides, N., 1981, A restricted Lagrangian approach to the travelling salesman problem, *Math. Program.* **21**:19–46.

Beasley, J. E., 1985, An exact two-dimensional non-guillotine cutting tree-search procedure, *Oper. Res.* **33**:49–64.

Bellman, R., 1957, *Dynamic Programming,* Princeton University Press, Princeton, NJ.

Bouzaher, A., Braden, J. B. and Johnson, G. V., 1990, A dynamic programming approach to a class of non-point source pollution control problems, *Manage. Sci.* **36**:1–15.

Bron, C. and Kerbosch, J., 1973, Finding all cliques of an un-directed graph— alg 457. *Commun. ACM* **16**:575–577.

Brown, J. R., 1972, Chromatic scheduling and the chromatic number problem, *Manage. Sci.* **19**:456–463.

Christofides, N. and Whitlock, C., 1977, An algorithm for two-dimensional cutting problems, *Oper. Res.* **25**:30–44.

Clarke, S. R. and Norman, J. M., 1999, To run or not? Some dynamic programming models in cricket, *J. Oper. Res. Soc.* **50**:536–545.

Cotta, C., Aldana, J. F., Nebro, A. J. and Troya, J. M., 1995, Hybridising genetic algorithms with branch and bound techniques for the resolution of the TSP, *Proc. Int. Conf. on Artificial Neural Networks and Genetic Algorithms,* C. C. Poras et al., ed., pp. 277–280.

Dantzig, G. B., 1951, Maximization of a linear function of variables subject to linear inequalities, *Activity Analysis of Production and Allocation,* T. C. Koopmans, ed., Wiley, New York.

Dijkstra, E. W., 1959, A note on two problems in connection with graphs, *Numer. Math.* **1**:269.

Dowsland, K. A., 1987, An exact algorithm for the pallet loading problems, *Eur. J. Oper. Res.* **31**:78–84.

Dowsland, K. A., 1998, Nurse scheduling with tabu search and strategic oscillation, *Eur. J. Oper. Res.* **106**:393–407.

Dowsland, K. A. and Thompson, J. M., 2000, Solving a nurse scheduling problem with knapsacks, networks and tabu search, *J. Oper. Res. Soc.* **51**:825–833.

Dowsland, K. A., Herbert, E. A. and Kendall, G., 2004, Using tree search bounds to enhance a genetic algorithm approach to two rectangle packing problems, *Eur. J. Oper. Res.,* in press.

Erlenkotter, D., 1978, A dual-based procedure for uncapacitated facility location. *Oper. Res.* **26**:992–1009.

Findlay, P. L., Kobbacy, K. A. H. and Goodman, D. J., 1989, Optimisation of the daily production rates for an offshore oil field, *J. Oper. Res. Soc.* **40**:1079–1088.

Fisher, M. L., 1985, An applications oriented guide to Lagrangian relaxation, *Interfaces* **15**:10–21.

Floyd, R. W., 1962, Algorithm 97—shortest path, *Commun. ACM* **5**:345.

Ford, L. R. and Fulkerson, D. R., 1956, Maximal flow through a network, *Can. J. Math.* **18**:399–404.

Fulkerson, D. R., 1961, An out-of-kilter method for minimal cost flow problems, *SIAM J. Appl. Math.* **9**:18–27.

Garfinkel, R. S. and Nemhauser, G. L., 1969, The set partitioning problem: set covering with equality constraints, *Oper. Res.* **17**:848–856.

Glover, F., Glover, R., Lorenzo, J. and McMillan, C., 1982, The passenger mix problem in the scheduled airlines, *Interfaces* **12**:73–79.

Glover, F. and Laguna, M., 1997, *Tabu Search,* Kluwer, Dordrecht.

Golumbic, M. C., 1980, *Algorithmic Graph Theory and Perfect Graphs,* Academic, New York.

Gutin, G. M., 1999, Exponential neighbourhood local search for the travelling salesman problem, *Comput. OR* **26**:313–320.

Hayes, M. and Norman, J. M., 1984, Dynamic programming in orienteering—route choice and the siting of controls, *J. Oper. Res. Soc.* **35**:791–796.

Held, M. and Karp, R. M., 1970, The travelling salesman problem and minimum spanning trees, *Oper. Res.* **18**:1138–1162.

Hindi, K. S., Fleszar, K. and Charalambous, C., 2003, An effective heuristic for the CLSP with setup times, *J. Oper. Res. Soc.* **54**:490–498.

Jarvinen P, Rajala, J. and Sinervo, H., 1972, A branch and bound algorithm for seeking the p-median, *Oper. Res.* **20**:173.

Johnson, T. B., 1968, Optimum pit mine production scheduling, *Technical Report,* University of California, Berkeley, CA.

Kamarkar N. K., 1984, A new polynomial-time algorithm for linear programming, *Combinatorica* **4**:373–395.

Khachiyan L. G., 1979, A polynomial algorithm in linear programming, *Dokl. Akad. Nauk SSSR* **244**:1093–1096 (in Russian) (English transl.: 1979, *Sov. Math. Dokl.* **20**:191–194).

Little, J. D. C., Murty, K. G., Sweeney, D. W. and Karel, C., 1963, An algorithm for the travelling salesman problem, *Oper. Res.* **11**:972–989.

Martello, S. and Toth, P., 1981, A branch and bound algorithm for the zero-one multiple knapsack problem, *Discr. Appl. Math.* **3**:275–288.

Martello, S. and Toth, P., 1990, *Knapsack Problems: Algorithms and Computer Implementations,* Wiley, New York.

Mamer J. W. and Smith, S. A., 1982, Optimising field repair kits based on job completion rate, *Manage. Sci.* **28**:1328–1334.

Minty, G. J., 1960, Monotone networks, *Proc. R. Soc.* A **257**:194–212.

Nagar, A., Heragu, S. S. and Haddock, J., 1995, A meta-heuristic algorithm for a bi-criteria scheduling problem, *Ann. OR* **63**:397–414.

Potts, C. N. and van de Velde, S. L., 1995, Dynasearch—iterative local improvement by dynamic programming: Part 1. The TSP, *Technical Report,* University of Twente, Netherlands.

Ross, G. T. and Soland, R. M., 1975, A branch and bound algorithm for the generalised assignment problem, *Math. Program.* **8**:91–103.

Tamura, H., Hirahara, A., Hatono, I. and Umano, M., 1994, An approximate solution method for combinatorial optimisation—hybrid approach of genetic algorithm and Lagrangian relaxation method, *Trans. Soc. Instrum. Control Eng.* **130**:329–336.

Zykov, A. A., 1949, On some properties of linear complexes, *Math Sb.* **24**:163–188.

Chapter 3

INTEGER PROGRAMMING

Robert Bosch
Oberlin College
Oberlin OH, USA

Michael Trick
Carnegie Mellon University
Pittsburgh PA, USA

3.1 INTRODUCTION

Over the last 20 years, the combination of faster computers, more reliable data, and improved algorithms has resulted in the near-routine solution of many integer programs of practical interest. Integer programming models are used in a wide variety of applications, including scheduling, resource assignment, planning, supply chain design, auction design, and many, many others. In this tutorial, we outline some of the major themes involved in creating and solving integer programming models.

The foundation of much of analytical decision making is linear programming. In a linear program, there are *variables*, *constraints*, and an *objective function*. The variables, or decisions, take on numerical values. Constraints are used to limit the values to a feasible region. These constraints must be linear in the decision variables. The objective function then defines which particular assignment of feasible values to the variables is optimal: it is the one that maximizes (or minimizes, depending on the type of the objective) the objective function. The objective function must also be linear in the variables. See Chapter 2 for more details about Linear Programming.

Linear programs can model many problems of practical interest, and modern linear programming optimization codes can find optimal solutions to problems with hundreds of thousands of constraints and variables. It is this combination of modeling strength and solvability that makes linear programming so important.

Integer programming adds additional constraints to linear programming. An integer program begins with a linear program, and adds the requirement that some or all of the variables take on integer values. This seemingly innocuous change greatly increases the number of problems that can be modeled, but also makes the models more difficult to solve. In fact, one frustrating aspect of integer programming is that two seemingly similar formulations for the same problem can lead to radically different computational experience: one formulation may quickly lead to optimal solutions, while the other may take an excessively long time to solve.

There are many keys to successfully developing and solving integer programming models. We consider the following aspects:

- be creative in formulations,

- find integer programming formulations with a strong relaxation,

- avoid symmetry,

- consider formulations with many constraints,

- consider formulations with many variables,

- modify branch-and-bound search parameters.

To fix ideas, we will introduce a particular integer programming model, and show how the main integer programming algorithm, branch-and-bound, operates on that model. We will then use this model to illustrate the key ideas to successful integer programming.

3.1.1   Facility Location

We consider a facility location problem. A chemical company owns four factories that manufacture a certain chemical in raw form. The company would like to get in the business of refining the chemical. It is interested in building refining facilities, and it has identified three possible sites. Table 3.1 contains variable costs, fixed costs, and weekly capacities for the three possible refining facility sites, and weekly production amounts for each factory. The variable costs are in dollars per week and include transportation costs. The fixed costs are in dollars per year. The production amounts and capacities are in tons per week.

The decision maker who faces this problem must answer two very different types of questions: questions that require numerical answers (for example, how many tons of chemical should factory i send to the site-j refining facility each week?) and questions that require yes–no answers (for example, should the site-j facility be constructed?). While we can easily model the first type of question by using continuous decision variables (by letting x_{ij} equal the

Table 3.1. Facility location problem.

		Site			
		1	2	3	Production
Variable cost	factory 1	25	20	15	1000
	factory 2	15	25	20	1000
	factory 3	20	15	25	500
	factory 4	25	15	15	500
Fixed cost		500 000	500 000	500 000	
Capacity		1 500	1 500	1 500	

number of tons of chemical sent from factory i to site j each week), we *cannot*
do this with the second. We need to use integer variables. If we let y_j equal 1
if the site-j refining facility is constructed and 0 if it is not, we quickly arrive
at an IP formulation of the problem:

minimize

$$52 \cdot 25x_{11} + 52 \cdot 20x_{12} + 52 \cdot 15x_{13}$$
$$+ 52 \cdot 15x_{21} + 52 \cdot 25x_{22} + 52 \cdot 20x_{23}$$
$$+ 52 \cdot 20x_{31} + 52 \cdot 15x_{32} + 52 \cdot 25x_{33}$$
$$+ 52 \cdot 25x_{41} + 52 \cdot 15x_{42} + 52 \cdot 15x_{43}$$
$$+ 500\,000y_1 + 500\,000y_2 + 500\,000y_3$$

subject to

$$x_{11} + x_{12} + x_{13} = 1000$$
$$x_{21} + x_{22} + x_{23} = 1000$$
$$x_{31} + x_{32} + x_{33} = 500$$
$$x_{41} + x_{42} + x_{43} = 500$$

$$x_{11} + x_{21} + x_{31} + x_{41} \leq 1500y_1$$
$$x_{12} + x_{22} + x_{32} + x_{42} \leq 1500y_2$$
$$x_{13} + x_{23} + x_{33} + x_{43} \leq 1500y_3$$
$$x_{ij} \geq 0 \quad \text{for all } i \text{ and } j$$
$$y_j \in \{0, 1\} \quad \text{for all } j$$

The objective is to minimize the yearly cost, the sum of the variable costs
(which are measured in dollars per week) and the fixed costs (which are measured in dollars per year). The first set of constraints ensures that each factory's
weekly chemical production is sent somewhere for refining. Since factory 1
produces 1000 tons of chemical per week, factory 1 must ship a total of 1000
tons of chemical to the various refining facilities each week. The second set
of constraints guarantees two things: (1) if a facility is open, it will operate
at or below its capacity, and (2) if a facility is not open, it will not operate
at all. If the site-1 facility is open ($y_1 = 1$) then the factories can send it up
to $1500y_1 = 1500 \cdot 1 = 1500$ tons of chemical per week. If it is not open

$(y_1 = 0)$, then the factories can send it up to $1500y_1 = 1500 \cdot 0 = 0$ tons per week.

This introductory example demonstrates the need for integer variables. It also shows that with integer variables, one can model simple logical requirements (if a facility is open, it can refine up to a certain amount of chemical; if not, it cannot do any refining at all). It turns out that with integer variables, one can model a whole host of logical requirements. One can also model fixed costs, sequencing and scheduling requirements, and many other problem aspects.

3.1.2   Solving the Facility Location IP

Given an integer program (IP), there is an associated linear program (LR) called the *linear relaxation*. It is formed by dropping (relaxing) the integrality restrictions. Since (LR) is less constrained than (IP), the following are immediate:

- If (IP) is a minimization problem, the optimal objective value of (LR) is less than or equal to the optimal objective value of (IP).

- If (IP) is a maximization problem, the optimal objective value of (LR) is greater than or equal to the optimal objective value of (IP).

- If (LR) is infeasible, then so is (IP).

- If all the variables in an optimal solution of (LR) are integer-valued, then that solution is optimal for (IP) too.

- If the objective function coefficients are integer-valued, then for minimization problems, the optimal objective value of (IP) is greater than or equal to the ceiling of the optimal objective value of (LR). For maximization problems, the optimal objective value of (IP) is less than or equal to the floor of the optimal objective value of (LR).

In summary, solving (LR) can be quite useful: it provides a bound on the optimal value of (IP), and may (if we are lucky) give an optimal solution to (IP).

For the remainder of this section, we will let (IP) stand for the Facility Location integer program and (LR) for its linear programming relaxation. When we solve (LR), we obtain

Objective		
x_{11} x_{12} x_{13}		
x_{21} x_{22} x_{23}		
x_{31} x_{32} x_{33}		
x_{41} x_{42} x_{43}		
y_1 y_2 y_3		

	3340 000	
·	·	1000
1000	·	·
·	500	·
·	500	·
$\frac{2}{3}$	$\frac{2}{3}$	$\frac{2}{3}$

This solution has factory 1 send all 1000 tons of its chemical to site 3, factory 2 send all 1000 tons of its chemical to site 1, factory 3 send all 500 tons to site 2, and factory 4 send all 500 tons to site 2. It constructs two-thirds of a refining facility at each site. Although it costs only 3340 000 dollars per year, it cannot be implemented; all three of its integer variables take on fractional values.

It is tempting to try to produce a feasible solution by rounding. Here, if we round y_1, y_2, and y_3 from 2/3 to 1, we get lucky (this is certainly not always the case!) and get an integer feasible solution. Although we can state that this is a good solution—its objective value of 3840 000 is within 15% of the objective value of (LR) and hence within 15% of optimal—we cannot be sure that it is optimal.

So how can we find an optimal solution to (IP)? Examining the optimal solution to (LR), we see that y_1, y_2, and y_3 are fractional. We want to force y_1, y_2, and y_3 to be integer valued. We start by *branching* on y_1, creating two new integer programming problems. In one, we add the constraint $y_1 = 0$. In the other, we will add the constraint $y_1 = 1$. Note that any optimal solution to (IP) must be feasible for one of the two subproblems.

After we solve the linear programming relaxations of the two subproblems, we can display what we know in a tree, as shown in Figure 3.1.

Note that the optimal solution to the left subproblem's LP relaxation is integer valued. It is therefore an optimal solution to the left subproblem. Since there is no point in doing anything more with the left subproblem, we mark it with an "×" and focus our attention on the right subproblem.

Both y_2 and y_3 are fractional in the optimal solution to the right subproblem's LP relaxation. We want to force both variables to be integer valued. Although we could branch on either variable, we will branch on y_2. That is, we will create two more subproblems, one with $y_2 = 0$ and the other with $y_2 = 1$. After we solve the LP relaxations, we can update our tree, as in Figure 3.2.

Note that we can immediately "× out" the left subproblem; the optimal solution to its LP relaxation is integer valued. In addition, by employing a *bounding* argument, we can also × out the right subproblem. The argument goes like this: Since the objective value of its LP relaxation ($3636 666\frac{2}{3}$) is greater than the objective value of our newly found integer feasible solution

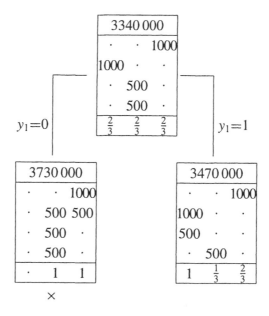

Figure 3.1. Intermediate branch and bound tree.

(3470 000), the optimal value of the right subproblem must be higher than (worse than) the objective value of our newly found integer feasible solution. So there is no point in expending any more effort on the right subproblem.

Since there are no active subproblems (subproblems that require branching), we are done. We have found an optimal solution to (IP). The optimal solution has factories 2 and 3 use the site-1 refining facility and factories 1 and 4 use the site-3 facility. The site-1 and site-3 facilities are constructed. The site-2 facility is not. The optimal solution costs 3470 000 dollars per year, 370 000 dollars per year less than the solution obtained by rounding the solution to (LR).

This method is called *branch and bound*, and is the most common method for finding solutions to integer programming formulations.

3.1.3   Difficulties with Integer Programs

While we were able to get the optimal solution to the example integer program relatively quickly, it is not always the case that branch and bound quickly solves integer programs. In particular, it is possible that the *bounding* aspects of branch and bound are not invoked, and the branch and bound algorithm can then generate a huge number of subproblems. In the worst case, a problem with n binary variables (variables that have to take on the value 0 or 1) can have 2^n subproblems. This exponential growth is inherent in any algorithm for integer programming, unless $P = NP$ (see Chapter 11 for more details), due to the range of problems that can be formulated within integer programming.

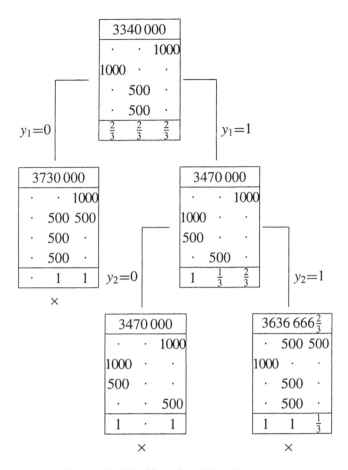

Figure 3.2. Final branch and bound tree.

Despite the possibility of extreme computation time, there are a number of techniques that have been developed to increase the likelihood of finding optimal solutions quickly. After we discuss creativity in formulations, we will discuss some of these techniques.

3.2 BE CREATIVE IN FORMULATIONS

At first, it may seem that integer programming does not offer much over linear programming: both require linear objectives and constraints, and both have numerical variables. Can requiring some of the variables to take on integer values significantly expand the capability of the models? Absolutely: integer programming models go far beyond the power of linear programming models. The key is the creative use of integrality to model a wide range of common structures in models. Here we outline some of the major uses of integer variables.

3.2.1   Integer Quantities

The most obvious use of integer variables is when an integer quantity is required. For instance, in a production model involving television sets, an integral number of television sets might be required. Or, in a personnel assignment problem, an integer number of workers might be assigned to a shift.

This use of integer variables is the most obvious, and the most over-used. For many applications, the added "accuracy" in requiring integer variables is far outweighed by the greater difficulty in finding the optimal solution. For instance, in the production example, if the number of televisions produced is in the hundreds (say the fractional optimal solution is 202.7) then having a plan with the rounded off value (203 in this example) is likely to be appropriate in practice. The uncertainty of the data almost certainly means that no production plan is accurate to four figures! Similarly, if the personnel assignment problem is for a large enterprise over a year, and the linear programming model suggests 154.5 people are required, it is probably not worthwhile to invoke an integer programming model in order to handle the fractional parts.

However, there are times when integer quantities are required. A production system that can produce either two or three aircraft carriers and a personnel assignment problem for small teams of five or six people are examples. In these cases, the addition of the integrality constraint can mean the difference between useful models and irrelevant models.

3.2.2   Binary Decisions

Perhaps the most used type of integer variable is the *binary variable*: an integer variable restricted to take on the values 0 or 1. We will see a number of uses of these variables. Our first example is in modeling binary decisions.

Many practical decisions can be seen as "yes" or "no" decisions: Should we construct a chemical refining facility in site j (as in the introduction)? Should we invest in project B? Should we start producing new product Y? For many of these decisions, a binary integer programming model is appropriate. In such a model, each decision is modeled with a binary variable: setting the variable equal to 1 corresponds to making the "yes" decision, while setting it to 0 corresponds to going with the "no" decision. Constraints are then formed to correspond to the effects of the decision.

As an example, suppose we need to choose among projects A, B, C, and D. Each project has a capital requirement ($1 million, $2.5 million, $4 million, and $5 million respectively) and an expected return (say, $3 million, $6 million,

$13 million, and $16 million). If we have $7 million to invest, which projects should we take on in order to maximize our expected return?

We can formulate this problem with binary variables x_A, x_B, x_C, and x_D representing the decision to take on the corresponding project. The effect of taking on a project is to use up some of the funds we have available to invest. Therefore, we have a constraint:

$$x_A + 2.5x_B + 4x_C + 5x_D \leq 7$$

Our objective is to maximize the expected profit:

$$\text{Maximize } 3x_1 + 6x_2 + 13x_3 + 15x_4$$

In this case, binary variables let us make the yes–no decision on whether to invest in each fund, with a constraint ensuring that our overall decisions are consistent with our budget. Without integer variables, the solution to our model would have fractional parts of projects, which may not be in keeping with the needs of the model.

3.2.3 Fixed Charge Requirements

In many production applications, the cost of producing x of an item is roughly linear except for the special case of producing no items. In that case, there are additional savings since no equipment or other items need be procured for the production. This leads to a *fixed charge* structure. The cost for producing x of an item is

- 0, if $x = 0$

- $c_1 + c_2x$, if $x > 0$ for constants c_1, c_2

This type of cost structure is impossible to embed in a linear program. With integer programming, however, we can introduce a new binary variable y. The value $y = 1$ is interpreted as having non-zero production, while $y = 0$ means no production. The objective function for these variables then becomes

$$c_1y + c_2x$$

which is appropriately linear in the variables. It is necessary, however, to add constraints that link the x and y variables. Otherwise, the solution might be $y = 0$ and $x = 10$, which we do not want. If there is an upper bound M on how large x can be (perhaps derived from other constraints), then the constraint

$$x \leq My$$

correctly links the two variables. If $y = 0$ then x must equal 0; if $y = 1$ then x can take on any value. Technically, it is possible to have the values $x = 0$ and

$y = 1$ with this formulation, but as long as this is modeling a fixed cost (rather than a fixed profit), this will not be an optimal (cost minimizing) solution.

This use of "M" values is common in integer programming, and the result is called a "Big-M model". Big-M models are often difficult to solve, for reasons we will see.

We saw this fixed-charge modeling approach in our initial facility location example. There, the y variables corresponded to opening a refining facility (incurring a fixed cost). The x variables correspond to assigning a factory to the refining facility, and there was an upper bound on the volume of raw material a refinery could handle.

3.2.4 Logical Constraints

Binary variables can also be used to model complicated logical constraints, a capability not available in linear programming. In a facility location problem with binary variables y_1, y_2, y_3, y_4, and y_5 corresponding to the decisions to open warehouses at locations 1, 2, 3, 4 and 5 respectively, complicated relationships between the warehouses can be modeled with linear functions of the y variables. Here are a few examples:

- At most one of locations 1 and 2 can be opened: $y_1 + y_2 \leq 1$.

- Location 3 can only be opened if location 1 is $y_3 \leq y_1$.

- Location 4 cannot be opened if locations 2 or 3 are such that $y_4 + y_2 \leq 1$ or $y_4 + y_3 \leq 1$.

- If location 1 is open, either locations 2 or 5 must be $y_2 + y_5 \geq y_1$.

Much more complicated logical constraints can be formulated with the addition of new binary variables. Consider a constraint of the form: $3x_1 + 4x_2 \leq 10$ OR $4x_1 + 2x_2 \geq 12$. As written, this is not a linear constraint. However, if we let M be the largest either $|3x_1 + 4x_2|$ or $|4x_1 + 2x_2|$ can be, then we can define a new binary variable z which is 1 only if the first constraint is satisfied and 0 only if the second constraint is satisfied. Then we get the constraints

$$3x_1 + 4x_2 \leq 10 + (M - 10)(1 - z)$$
$$4x_1 + 2x_2 \geq 12 - (M + 12)z$$

When $z = 1$, we obtain

$$3x_1 + 4x_2 \leq 10$$
$$4x_1 + 2x_2 \geq -M$$

When $z = 0$, we obtain

$$3x_1 + 4x_2 \leq M$$
$$4x_1 + 2x_2 \geq 12$$

This correctly models the original nonlinear constraint.

As we can see, logical requirements often lead to Big-M-type formulations.

3.2.5   Sequencing Problems

Many problems in sequencing and scheduling require the modeling of the order in which items appear in the sequence. For instance, suppose we have a model in which there are items, where each item i has a processing time on a machine p_i. If the machine can only handle one item at a time and we let t_i be a (continuous) variable representing the start time of item i on the machine, then we can ensure that items i and j are not on the machine at the same time with the constraints

$$t_j \geq t_i + p_i \text{ IF } t_j \geq t_i$$
$$t_i \geq t_j + p_j \text{ IF } t_j < t_i$$

This can be handled with a new binary variable y_{ij} which is 1 if $t_i \leq t_j$ and 0 otherwise. This gives the constraints

$$t_j \geq t_i + p_i - M(1 - y)$$
$$t_i \geq t_j + p_j - My$$

for sufficiently large M. If y is 1 then the second constraint is automatically satisfied (so only the first is relevant) while the reverse happens for $y = 0$.

3.3   FIND FORMULATIONS WITH STRONG RELAXATIONS

As the previous section made clear, integer programming formulations can be used for many problems of practical interest. In fact, for many problems, there are many alternative integer programming formulations. Finding a "good" formulation is key to the successful use of integer programming. The definition of a good formulation is primarily computational: a good formulation is one for which branch and bound (or another integer programming algorithm) will find and prove the optimal solution quickly. Despite this empirical aspect of the definition, there are some guidelines to help in the search for good formulations. The key to success is to find formulations whose linear relaxation is not too different from the underlying integer program.

We saw in our first example that solving linear relaxations was key to the basic integer programming algorithm. If the solution to the initial linear relaxation is integer, then no branching need be done and integer programming is no harder than linear programming. Unfortunately, finding formulations with this property is very hard to do. But some formulations can be better than other formulations in this regard.

Let us modify our facility location problem by requiring that every factory be assigned to exactly one refinery (incidentally, the optimal solution to our original formulation happened to meet this requirement). Now, instead of having x_{ij} be the tons sent from factory i to refinery j, we define x_{ij} to be 1 if factory i is serviced by refinery j. Our formulation becomes

Minimize
$$1000 \cdot 52 \cdot 25x_{11} + 1000 \cdot 52 \cdot 20x_{12} + 1000 \cdot 52 \cdot 15x_{13}$$
$$+ \ 1000 \cdot 52 \cdot 15x_{21} + 1000 \cdot 52 \cdot 25x_{22} + 1000 \cdot 52 \cdot 20x_{23}$$
$$+ \ 500 \cdot 52 \cdot 20x_{31} + 500 \cdot 52 \cdot 15x_{32} + 500 \cdot 52 \cdot 25x_{33}$$
$$+ \ 500 \cdot 52 \cdot 25x_{41} + 500 \cdot 52 \cdot 15x_{42} + 500 \cdot 52 \cdot 15x_{43}$$
$$+ \ 500\,000y_1 + 500\,000y_2 + 500\,000y_3$$

Subject to
$$x_{11} + x_{12} + x_{13} = 1$$
$$x_{21} + x_{22} + x_{23} = 1$$
$$x_{31} + x_{32} + x_{33} = 1$$
$$x_{41} + x_{42} + x_{43} = 1$$
$$1000x_{11} + 1000x_{21} + 500x_{31} + 500x_{41} \leq 1500y_1$$
$$1000x_{12} + 1000x_{22} + 500x_{32} + 500x_{42} \leq 1500y_2$$
$$1000x_{13} + 1000x_{23} + 500x_{33} + 500x_{43} \leq 1500y_3$$
$$x_{ij} \in \{0, 1\} \quad \text{for all } i \text{ and } j$$
$$y_j \in \{0, 1\} \quad \text{for all } j.$$

Let us call this formulation the *base formulation*. This is a correct formulation to our problem. There are alternative formulations, however. Suppose we add to the base formulation the set of constraints

$$x_{ij} \leq y_j \quad \text{for all } i \text{ and } j$$

Call the resulting formulation the *expanded formulation*. Note that it too is an appropriate formulation for our problem. At the simplest level, it appears that we have simply made the formulation larger: there are more constraints so the linear programs solved within branch-and-bound will likely take longer to solve. Is there any advantage to the expanded formulation?

The key is to look at non-integer solutions to linear relaxations of the two formulations: we know the two formulations have the same integer solutions (since they are formulations of the same problem), but they can differ in non-integer solutions. Consider the solution $x_{13} = 1, x_{21} = 1, x_{32} = 1, x_{42} = 1, y_1 = 2/3, y_2 = 2/3, y_3 = 2/3$. This solution is feasible to the linear relaxation of the base formulation but is not feasible to the linear relaxation of the expanded formulation. If the branch-and-bound algorithm works on the base formulation, it may have to consider this solution; with the expanded formulation, this solution can never be examined. If there are fewer fractional solutions to explore (technically, fractional extreme point solutions), branch and bound will typically terminate more quickly.

Since we have added constraints to get the expanded formulation, there is no non-integer solution to the linear relaxation of the expanded formulation that is not also feasible for the linear relaxation of the base formulation. We say that the expanded formulation is *tighter* than the base formulation.

In general, tighter formulations are to be preferred for integer programming formulations even if the resulting formulations are larger. Of course, there are exceptions: if the size of the formulation is much larger, the gain from the tighter formulation may not be sufficient to offset the increased linear programming times. Such cases are definitely the exception, however: almost invariably, tighter formulations are better formulations. For this particular instance, the Expanded Formulation happens to provide an integer solution without branching.

There has been a tremendous amount of work done on finding tighter formulations for different integer programming models. For many types of problems, classes of constraints (or *cuts*) to be added are known. These constraints can be added in one of two ways: they can be included in the original formulation or they can be added as needed to remove fractional values. The latter case leads to a *branch and cut* approach, which is the subject of Section 3.6.

A cut relative to a formulation has to satisfy two properties: first, every feasible integer solution must also satisfy the cut; second, some fractional solution that is feasible to the linear relaxation of the formulation must not satisfy the cut. For instance, consider the single constraint

$$3x_1 + 5x_2 + 8x_3 + 10x_4 \leq 16$$

where the x_i are binary variables. Then the constraint $x_3 + x_4 \leq 1$ is a cut (every integer solution satisfies it and, for instance $x = (0, 0, .5, 1)$ does not) but $x_1 + x_2 + x_3 + x_4 \leq 4$ is not a cut (no fractional solutions removed) nor is $x_1 + x_2 + x_3 \leq 2$ (which incorrectly removes $x = (1, 1, 1, 0)$).

Given a formulation, finding cuts to add to it to strengthen the formulation is not a routine task. It can take deep understanding, and a bit of luck, to find improving constraints.

One generally useful approach is called the Chvatal (or Gomory–Chvatal) procedure. Here is how the procedure works for "$\leq$" constraints where all the variables are non-negative integers:

1 Take one or more constraints, multiply each by a non-negative constant (the constant can be different for different constraints). Add the resulting constraints into a single constraint.

2 Round down each coefficient on the left-hand side of the constraint.

3 Round down the right-hand side of the constraint.

The result is a constraint that does not cut off any feasible integer solutions. It may be a cut if the effect of rounding down the right-hand side of the constraint is more than the effect of rounding down the coefficients.

This is best seen through an example. Taking the constraint above, let us take the two constraints

$$3x_1 + 5x_2 + 8x_3 + 10x_4 \le 16 \qquad x_3 \le 1$$

If we multiply each constraint by 1/9 and add them we obtain

$$3/9x_1 + 5/9x_2 + 9/9x_3 + 10/9x_4 \le 17/9$$

Now, round down the left-hand coefficients (this is valid since the x variables are non-negative and it is a "$\le$" constraint):

$$x_3 + x_4 \le 17/9$$

Finally, round down the right-hand side (this is valid since the x variables are integer) to obtain

$$x_3 + x_4 \le 1$$

which turns out to be a cut. Notice that the three steps have differing effects on feasibility. The first step, since it is just taking a linear combination of constraints, neither adds nor removes feasible values; the second step weakens the constraint, and may add additional fractional values; the third step strengthens the constraint, ideally removing fractional values.

This approach is particularly useful when the constants are chosen so that no rounding down is done in the second step. For instance, consider the following set of constraints (where the x_i are binary variables):

$$x_1 + x_2 \le 1 \qquad x_2 + x_3 \le 1 \qquad x_1 + x_3 \le 1$$

These types of constraints often appear in formulations where there are lists of mutually exclusive variables. Here, we can multiply each constraint by 1/2 and add them to obtain

$$x_1 + x_2 + x_3 \le 3/2$$

There is no rounding down on the left-hand side, so we can move on to rounding down the right-hand side to obtain

$$x_1 + x_2 + x_3 \le 1$$

which, for instance, cuts off the solution $x = (1/2, 1/2, 1/2)$.

In cases where no rounding down is needed on the left-hand side but there is rounding down on the right-hand side, the result has to be a cut (relative to the included constraints). Conversely, if no rounding down is done on the right-hand side, the result cannot be a cut.

In the formulation section, we mentioned that "Big-M" formulations often lead to poor formulations. This is because the linear relaxation of such a formulation often allows for many fractional values. For instance, consider the constraint (all variables are binary)

$$x_1 + x_2 + x_3 \leq 1000y$$

Such constraints often occur in facility location and related problems. This constraint correctly models a requirement that the x variables can be 1 only if y is also 1, but does so in a very weak way. Even if the x values of the linear relaxation are integer, y can take on a very small value (instead of the required 1). Here, even for $x = (1, 1, 1)$, y need only be 3/1000 to make the constraint feasible. This typically leads to very bad branch-and-bound trees: the linear relaxation gives little guidance as to the "true" values of the variables.

The following constraint would be better:

$$x_1 + x_2 + x_3 \leq 3y$$

which forces y to take on larger values. This is the concept of making the M in Big-M as small as possible. Better still would be the three constraints

$$x_1 \leq y \qquad x_2 \leq y \qquad x_3 \leq y$$

which force y to be integer as soon as the x values are.

Finding improved formulations is a key concept to the successful use of integer programming. Such formulations typically revolve around the strength of the linear relaxation: does the relaxation well-represent the underlying integer program? Finding classes of cuts can improve formulations. Finding such classes can be difficult, but without good formulations, integer programming models are unlikely to be successful except for very small instances.

3.4   AVOID SYMMETRY

Symmetry often causes integer programming models to fail. Branch-and-bound can become an extremely inefficient algorithm when the model being solved displays many symmetries.

Consider again our facility location model. Suppose instead of having just one refinery at a site, we were permitted to have up to three refineries at a site. We could modify our model by having variables y_j, z_j and w_j for each site (representing the three refineries). In this formulation, the cost and other coefficients for y_j are the same as for z_j and w_j. The formulation is straightforward, but branch and bound does very poorly on the result.

The reason for this is symmetry: for every solution in the branch-and-bound tree with a given y, z, and w, there is an equivalent solution with z taking on y's values, w taking on z's and y taking on w. This greatly increases the number of solutions that the branch-and-bound algorithm must consider in order to find and prove the optimality of a solution.

It is very important to remove as many symmetries in a formulation as possible. Depending on the problem and the symmetry, this removal can be done by adding constraints, fixing variables, or modifying the formulation.

For our facility location problem, the easiest thing to do is to add the constraints

$$y_j \geq z_j \geq w_j \quad \text{for all } j$$

Now, at a refinery site, z_j can be non-zero only if y_j is non-zero, and w_j is non-zero only if both y_j and z_j are. This partially breaks the symmetry of this formulation, though other symmetries (particularly in the x variables) remain.

This formulation can be modified in another way by redefining the variables. Instead of using binary variables, let y_j be the number of refineries put in location j. This removes all of the symmetries at the cost of a weaker linear relaxation (since some of the strengthenings we have explored require binary variables).

Finally, to illustrate the use of variable fixing, consider the problem of coloring a graph with K colors: we are given a graph with node set V and edge set E and wish to determine if we can assign a value $v(i)$ to each node i such that $v(i) \in \{1, \ldots, K\}$ and $v(i) \neq v(j)$ for all $(i, j) \in E$.

We can formulate this problem as an integer programming by defining a binary variable x_{ik} to be 1 if i is given color k and 0 otherwise. This leads to the constraints

$$\sum_k x_{ik} = 1 \text{ for all } i \text{ (every node gets a color)}$$

$$x_{ik} + x_{jk} = 1 \text{ for all } k, (i, j) \in E \text{ (no adjacent get the same)}$$

$$x_{ik} \in \{0, 1\} \text{ for all } i, k$$

The graph coloring problem is equivalent to determining if the above set of constraints is feasible. This can be done by using branch-and-bound with an arbitrary objective value.

Unfortunately, this formulation is highly symmetric. For any coloring of graph, there is an equivalent coloring that arises by permuting the coloring (that is, permuting the set $\{1, \ldots, k\}$ in this formulation). This makes branch and bound very ineffective for this formulation. Note also that the formulation is very weak, since setting $x_{ik} = 1/k$ for all i, k is a feasible solution to the linear relaxation no matter what E is.

We can strengthen this formulation by breaking the symmetry through variable fixing. Consider a clique (set of mutually adjacent vertices) of the graph. Each member of the clique has to get a different color. We can break the symmetry by finding a large (ideally maximum sized) clique in the graph and setting the colors of the clique arbitrarily, but fixed. So if the clique has size k_c, we would assign the colors $1, \ldots, k_c$ to members of the clique (adding in constraints forcing the corresponding x values to be 1). This greatly reduces the symmetry, since now only permutations among the colors $k_c + 1, \ldots, K$ are valid. This also removes the $x_{ik} = 1/k$ solution from consideration.

3.5 CONSIDER FORMULATIONS WITH MANY CONSTRAINTS

Given the importance of the strength of the linear relaxation, the search for improved formulations often leads to sets of constraints that are too large to include in the formulation. For example, consider a single constraint with non-negative coefficients:

$$a_1 x_1 + a_2 x_2 + a_3 x_3 + \cdots + a_n x_n \leq b$$

where the x_i are binary variables. Consider a subset S of the variables such that $\sum_{i \in S} a_i > b$. The constraint

$$\sum_{i \in S} x_i \leq |S| - 1$$

is valid (it is not violated by any feasible integer solution) and cuts off fractional solutions as long as S is minimal. These constraints are called *cover constraints*. We would then like to include this set of constraints in our formulation.

Unfortunately, the number of such constraints can be very large. In general, it is exponential in n, making it impractical to include the constraints in the formulation. But the relaxation is much tighter with the constraints.

To handle this problem, we can choose to generate only those constraints that are needed. In our search for an optimal integer solution, many of the constraints are not needed. If we can generate the constraints as we need them, we can get the strength of the improved relaxation without the huge number of constraints.

Suppose our instance is

$$
\begin{aligned}
\text{Maximize} \quad & 9x_1 + 14x_2 + 20x_3 + 32x_4 \\
\text{Subject to} \quad & 3x_1 + 5x_2 + 8x_3 + 10x_4 \leq 16 \\
& x_i \in \{0, 1\}
\end{aligned}
$$

The optimal solution to the linear relaxation is $x^* = (1, 0.6, 0, 1)$ with objective 49.4. Now consider the set $S = (x_1, x_2, x_4)$. The constraint

$$x_1 + x_2 + x_4 \leq 2$$

is a cut that x^* violates. If we add that constraint to our problem, we get a tighter formulation. Solving this model gives solution $x = (1, 0, 0.375, 1)$ and objective 48.5. The constraint

$$x_3 + x_4 \leq 1$$

is a valid cover constraint that cuts off this solution. Adding this constraint and solving gives solution $x = (0, 1, 0, 1)$ with objective 46. This is the optimal solution to the original integer program, which we have found only by generating cover inequalities.

In this case, the cover inequalities were easy to see, but this process can be formalized. A reasonable heuristic for identifying violated cover inequalities would be to order the variables by decreasing $a_i x_i^*$ then add the variables to the cover S until $\sum_{i \in S} a_i > b$. This heuristic is not guaranteed to find violated cover inequalities (for that, a knapsack optimization problem can be formulated and solved) but even this simple heuristic can create much stronger formulations without adding too many constraints.

This idea is formalized in the *branch-and-cut* approach to integer programming. In this approach, a formulation has two parts: the *explicit constraints* (denoted $Ax \leq b$) and the *implicit constraints* ($A'x \leq b'$). Denote the objective function as Maximize cx. Here we will assume that all x are integral variables, but this can be easily generalized.

Step 1. Solve the linear program Maximize cx subject to $Ax \leq b$ to get optimal relaxation solution x^*.
Step 2. If x^* integer, then stop. x^* is optimal.
Step 3. Try to find a constraint $a'x \leq b'$ from the implicit constraints such that $a'x^* > b$. If found, add $a'x \leq b$ to the $Ax \leq b$ constraint set and go to step 1. Otherwise, do branch-and-bound on the current formulation.

In order to create a branch-and-cut model, there are two aspects: the definition of the implicit constraints, and the definition of the approach in Step 3 to find violated inequalities. The problem in Step 3 is referred to as the *separation problem* and is at the heart of the approach. For many sets of constraints, no good separation algorithm is known. Note, however, that the separation problem might be solved heuristically: it may miss opportunities for separation and therefore invoke branch-and-bound too often. Even in this case, it often happens that the improved formulations are sufficiently tight to greatly decrease the time needed for branch-and-bound.

This basic algorithm can be improved by carrying out cut generation within the branch and bound tree. It may be that by fixing variables, different constraints become violated and those can be added to the subproblems.

3.6 CONSIDER FORMULATIONS WITH MANY VARIABLES

Just as improved formulations can result from adding many constraints, adding many variables can lead to very good formulations. Let us begin with our graph coloring example. Recall that we are given a graph with vertices V and edges E and want to assign a value $v(i)$ to each node i such that $v(i) \neq v(j)$ for all $(i, j) \in E$. Our objective is to use the minimum number of different values (before, we had a fixed number of colors to use: in this section we will use the optimization version rather than the feasibility version of this problem).

Previously, we described a model using binary variables x_{ik} denoting whether node i gets color k or note. As an alternative model, let us concentrate on the set of nodes that gets the same color. Such a set must be an *independent set* (a set of mutually non-adjacent nodes) of the graph. Suppose we listed all independent sets of the graph: $S_1, S_2, \ldots, S_m$. Then we can define binary variables $y_1, y_2, \ldots, y_m$ with the interpretation that $y_j = 1$ means that independent set S_j is part of the coloring, and $y_j = 0$ means that independent set S_j is not part of the coloring. Now our formulation becomes

$$\text{Minimize} \sum_j y_j$$
$$\text{Subject to} \sum_{j : i \in S_j} y_j = 1 \text{ for all } i \in V$$
$$y_j \in \{0, 1\} \text{ for all } j \in \{1, \ldots, m\}$$

The constraint states that every node must be in some independent set of the coloring.

This formulation is a much better formulation that our x_{ik} formulation. This formulation does not have the symmetry problems of the previous formulation and results in a much tighter linear relaxation. Unfortunately, the formulation is impractical for most graphs because the number of independent sets is exponential in the number of nodes, leading to an impossibly large formulation.

Just as we could handle an exponential number of constraints by generating them as needed, we can also handle an exponential number of variables by *variable generation*: the creation of variables only as they are needed. In order to understand how to do this, we will have to understand some key concepts from linear programming.

Consider a linear program, where the variables are indexed by j and the constraints indexed by i:

$$\text{Maximize } \sum_j c_j x_j$$

$$\text{Subject to } \sum_j a_{ij} x_{ij} \leq b_i \text{ for all } i$$

$$x_j \geq 0 \text{ for all } j$$

When this linear program is solved, the result is the optimal solution x^*. In addition, however, there is a value called the *dual value*, denoted π_i, associated with each constraint. This value gives the marginal change in the objective value as the right-hand side for the corresponding constraint is changed. So if the right-hand side of constraint i changes to $b_i + \Delta$, then the objective will change by $\pi_i \Delta$ (there are some technical details ignored here involving how large Δ can be for this to be a valid calculation: since we are only concerned with marginal calculations, we can ignore these details).

Now, suppose there is a new variable x_{n+1}, not included in the original formulation. Suppose it could be added to the formulation with corresponding objective coefficient c_{n+1} and coefficients $a_{i,n+1}$. Would adding the variable to the formulation result in an improved formulation? The answer is certainly "no" in the case when

$$c_{n+1} < \sum_i a_{i,n+1} \pi_i$$

In this case, the value gained from the objective is insufficient to offset the cost charged marginally by the effect on the constraints. We need $c_{n+1} - \sum_i a_{i,n+1} \pi_i > 0$ in order to possibly improve on our solution.

This leads to the idea of variable generation. Suppose you have a formulation with a huge number of variables. Rather than solve this huge formulation, begin with a smaller number of variables. Solve the linear relaxation and get dual values π. Using π, determine if there is one (or more) variables whose inclusion might improve the solution. If not, then the linear relaxation is solved. Otherwise, add one or more such variables to the formulation and repeat.

Once the linear relaxation is solved, if the solution is integer, then it is optimal. Otherwise, branch and bound is invoked, with the variable generation continuing in the subproblems.

Key to this approach is the algorithm for generating the variables. For a huge number of variables it is not enough to check all of them: that would be too time consuming. Instead, some sort of optimization problem must be defined whose solution is an improving variable. We illustrate this for our graph coloring problem.

Suppose we begin with a limited set of independent sets and solve our relaxation over them. This leads to a dual value π_i for each node. For any other independent set S, if $\sum_{i \in S} \pi_i > 1$, then S corresponds to an improving variable. We can write this problem using binary variables z_i corresponding to whether i is in S or not:

$$\text{Maximize } \sum_i \pi_i z_i$$

$$\text{Subject to } z_i + z_j \leq 1 \text{ for all } (i, j) \in E$$

$$z_i \in \{0, 1\} \text{ for all } i$$

This problem is called the *maximum weighted independent set* (MWIS) problem, and, while the problem is formally hard, effective methods have been found for solving it for problems of reasonable size.

This gives a variable generation approach to graph coloring: begin with a small number of independent sets, then solve the MWIS problem, adding in independent sets until no independent set improves the current solution. If the variables are integer, then we have the optimal coloring. Otherwise we need to branch.

Branching in this approach needs special care. We need to branch in such a way that our subproblem is not affected by our branching. Here, if we simply branch on the y_j variables (so have one branch with $y_j = 1$ and another with $y_j = 0$), we end up not being able to use the MWIS model as a subproblem. In the case where $y_j = 0$ we need to find an improving set, except that S_j does not count as improving. This means we need to find the second most improving set. As more branching goes on, we may need to find the third most improving, the fourth most improving, and so on. To handle this, specialized branching routines are needed (involving identifying nodes that, on one side of the branch, must be the same color and, on the other side of the branch, cannot be the same color).

Variable generation together with appropriate branching rules and variable generation at the subproblems is a method known as *branch and price*. This approach has been very successful in attacking a variety of very difficult problems over the last few years.

To summarize, models with a huge number of variables can provide very tight formulations. To handle such models, it is necessary to have a variable generation routine to find improving variables, and it may be necessary to modify the branching method in order to keep the subproblems consistent with that routine. Unlike constraint generation approaches, heuristic variable generation routines are not enough to ensure optimality: at some point it is necessary to prove conclusively that the right variables are included. Furthermore, these variable generation routines must be applied at each node in the branch-and-bound tree if that node is to be crossed out from further analysis.

3.7   MODIFY BRANCH-AND-BOUND PARAMETERS

Integer programs are solved with computer programs. There are a number of computer programs available to solve integer programs. These range from basic spreadsheet-oriented systems to open-source research codes to sophisticated commercial applications. To a greater or lesser extent, each of these codes offers parameters and choices that can have a significant affect on the solvability of integer programming models. For most of these parameters, the only way to determine the best choice for a particular model is experimentation: any choice that is uniformly dominated by another choice would not be included in the software.

Here are some common, key choices and parameters, along with some comments on each.

3.7.1   Description of Problem

The first issue to be handled is to determine how to describe the integer program to the optimization routine(s). Integer programs can be described as spreadsheets, computer programs, matrix descriptors, and higher-level languages. Each has advantages and disadvantages with regards to such issues as ease-of-use, solution power, flexibility and so on. For instance, implementing a branch-and-price approach is difficult if the underlying solver is a spreadsheet program. Using "callable libraries" that give access to the underlying optimization routines can be very powerful, but can be time-consuming to develop.

Overall, the interface to the software will be defined by the software. It is generally useful to be able to access the software in multiple ways (callable libraries, high level languages, command line interfaces) in order to have full flexibility in solving.

3.7.2   Linear Programming Solver

Integer programming relies heavily on the underlying linear programming solver. Thousands or tens of thousands of linear programs might be solved in the course of branch-and-bound. Clearly a faster linear programming code can result in faster integer programming solutions. Some possibilities that might be offered are primal simplex, dual simplex, or various interior point methods. The choice of solver depends on the problem size and structure (for instance, interior point methods are often best for very large, block-structured models) and can differ for the initial linear relaxation (when the solution must be found "from scratch") and subproblem linear relaxations (when the algorithm can use previous solutions as a starting basis). The choice of algorithm can also be affected by whether constraint and/or variable generation are being used.

3.7.3   Choice of Branching Variable

In our description of branch-and-bound, we allowed branching on any fractional variable. When there are multiple fractional variables, the choice of variable can have a big effect on the computation time. As a general guideline, more "important" variables should be branched on first. In a facility location problem, the decisions on opening a facility are generally more important than the assignment of a customer to that facility, so those would be better choices for branching when a choice must be made.

3.7.4   Choice of Subproblem to Solve

Once multiple subproblems have been generated, it is necessary to choose which subproblem to solve next. Typical choices are depth-first search, breadth-first search, or best-bound search. Depth-first search continues fixing variables for a single problem until integrality or infeasibility results. This can lead quickly to an integer solution, but the solution might not be very good. Best-bound search works with subproblems whose linear relaxation is as large (for maximization) as possible, with the idea that subproblems with good linear relaxations may have good integer solutions.

3.7.5   Direction of Branching

When a subproblem and a branching variable have been chosen, there are multiple subproblems created corresponding to the values the variable can take on. The ordering of the values can affect how quickly good solutions can be found. Some choices here are a fixed ordering or the use of estimates of the resulting linear relaxation value. With fixed ordering, it is generally good to first try the more restrictive of the choices (if there is a difference).

3.7.6   Tolerances

It is important to note that while integer programming problems are primarily combinatorial, the branch-and-bound approach uses numerical linear programming algorithms. These methods require a number of parameters giving allowable tolerances. For instance, if $x_j = 0.998$ should x_j be treated as the value 1 or should the algorithm branch on x_j? While it is tempting to give overly big values (to allow for faster convergence) or small values (to be "more accurate"), either extreme can lead to problems. While for many problems, the default values from a quality code are sufficient, these values can be the source of difficulties for some problems.

3.8   TRICKS OF THE TRADE

After reading this tutorial, all of which is about "tricks of the trade", it is easy to throw one's hands up and give up on integer programming! There are so many choices, so many pitfalls, and so much chance that the combinatorial explosion will make solving problems impossible. Despite this complexity, integer programming is used routinely to solve problems of practical interest. There are a few key steps to make your integer programming implementation go well.

- Use state-of-the-art software. It is tempting to use software because it is easy, or available, or cheap. For integer programming, however, not having the most current software embedding the latest techniques can doom your project to failure. Not all such software is commercial. The COIN-OR project is an open-source effort to create high-quality optimization codes.

- Use a modeling language. A modeling language, such as OPL, Mosel, AMPL, or other language can greatly reduce development time, and allows for easy experimentation of alternatives. Callable libraries can give more power to the user, but should be reserved for "final implementations", once the model and solution approached are known.

- If an integer programming model does not solve in a reasonable amount of time, look at the formulation first, not the solution parameters. The default settings of current software are generally pretty good. The problem with most integer programming formulations is the formulation, not the choice of branching rule, for example.

- Solve some small instances and look at the solutions to the linear relaxations. Often constraints to add to improve a formulation are quite obvious from a few small examples.

- Decide whether you need "optimal" solutions. If you are consistently getting within 0.1% of optimal, without proving optimality, perhaps you should declare success and go with the solutions you have, rather than trying to hunt down that final gap.

- Try radically different formulations. Often, there is another formulation with completely different variables, objective, and constraints that will have a much different computational experience.

3.9   CONCLUSIONS

Integer programming models represent a powerful approach to solving hard problems. The bounds generated from linear relaxations are often sufficient to greatly cut down on the search tree for these problems. Key to successful integer programming is the creation of good formulations. A good formulation is one where the linear relaxation closely resembles the underlying integer program. Improved formulations can be developed in a number of ways, including finding formulations with tight relaxations, avoiding symmetry, and creating and solving formulations that have an exponential number of variables or constraints. It is through the judicious combination of these approaches, combined with fast integer programming computer codes that the practical use of integer programming has greatly expanded in the last 20 years.

SOURCES OF ADDITIONAL INFORMATION

Integer programming has existed for more than 50 years and has developed a huge literature. This bibliography therefore makes no effort to be comprehensive, but rather provides initial pointers for further investigation.

General Integer Programming There are a number of excellent recent monographs on integer programming. The classic is Nemhauser and Wolsey (1988). A book updating much of the material is Wolsey (1998). Schrijver (1998) is an outstanding reference book, covering the theoretical underpinnings of integer programming.

Integer Programming Formulations There are relatively few books on formulating problems. An exception is Williams (1999). In addition, most operations research textbooks offer examples and exercises on formulations, though many of the examples are not of realistic size. Some choices are Winston (1997), Taha (2002), and Hillier and Lieberman (2002).

Branch and Bound Branch and bound traces back to the 1960s and the work of Land and Doig (1960). Most basic textbooks (see above) give an outline of the method (at the level given in this tutorial).

Branch and Cut The cutting plane approach dates back to the late 1950s and the work of Gomory (1958), whose cutting planes are applicable to any integer program. Juenger et al. (1995) provides a survey of the use of cutting plane algorithms for specialized problem classes.

As a computational technique, the work of Crowder et al. (1983) showed how cuts could greatly improve basic branch-and-bound.

For an example of the success of such approaches for solving extremely large optimization problems, see Applegate et al. (1998).

Branch and Price Barnhart et al. (1998) is an excellent survey of this approach.

Implementations There are a number of very good implementations that allow the optimization of realistic integer programs. Some of these are commercial, like the CPLEX implementation of ILOG, Inc. (CPLEX, 2004). Bixby et al. (1999) gives a detailed description of the advances that this software has made.

Another commercial product is Xpress-MP from Dash, with the textbook by Gueret et al. (2002) providing a very nice set of examples and applications.

COIN-OR (2004) provides an open-source initiative for optimization. Other approaches are described by Ralphs and Ladanyi (1999) and by Cordier et al. (1999).

References

Applegate, D., Bixby, R., Chvatal, V. and Cook, W., 1998, On the solution of traveling salesman problems, in: *Proc. Int. Congress of Mathematicians, Doc. Math. J. DMV*, Vol. 645.

Barnhart, C., Johnson, E. L., Nemhauser, G. L., Savelsbergh, M. W. P. and Vance, P. H., 1998, Branch-and-price: column generation for huge integer programs, *Oper. Res.* **46**:316.

Bixby, R. E., Fenelon, M., Gu, Z., Rothberg, E. and Wunderling, R., 1999, *MIP: Theory and Practice—Closing the Gap, Proc. 19th IFIP TC7 Conf. on System Modelling,* Kluwer, Dordrecht, pp. 19–50.

Common Optimization INterface for Operations Research (COIN), 2004, at http://www.coin-or.org

Cordier, C., Marchand, H., Laundy, R. and Wolsey, L. A., 1999, bc-opt: a branch-and-cut code for mixed integer programs, *Math. Program.* **86**:335.

Crowder, H., Johnson, E. L. and Padberg, M. W., 1983, Solving large scale zero-one linear programming problems, *Oper. Res.* **31**:803–834.

Gomory, R. E., 1958, Outline of an algorithm for integer solutions to linear programs, *Bulletin AMS* **64**:275–278.

Gueret, C., Prins, C. and Sevaux, M., 2002, *Applications of Optimization with Xpress-MP,* S. Heipcke, transl., Dash Optimization, Blisworth, UK.

Hillier, F. S. and Lieberman, G. J., 2002, *Introduction to Operations Research,* McGraw-Hill, New York.

ILOG CPLEX 9.0 Reference Manual, 2004, ILOG.

Juenger, M., Reinelt, G. and Thienel, S., 1995, *Practical Problem Solving with Cutting Plane Algorithms in Combinatorial Optimization,* DIMACS Series in Discrete Mathematics and Theoretical Computer Science, Vol. 111, American Mathematical Society, Providence, RI.

Land, A. H. and Doig, A. G., 1960, An Automatic Method for Solving Discrete Programming Problems, *Econometrica* **28**:83–97.

Nemhauser, G. L. and Wolsey, L. A., 1998, *Integer and Combinatorial Optimization,* Wiley, New York.

Ralphs, T. K. and Ladanyi, L., 1999, *SYMPHONY: A Parallel Framework for Branch and Cut,* White paper, Rice University.

Schrijver, A., 1998, *Theory of Linear and Integer Programming,* Wiley, New York.

Taha, H. A., 2002, *Operations Research: An Introduction*, Prentice-Hall, New York.

Williams, H. P., 1999, *Model Building in Mathematical Programming,* Wiley, New York.

Winston, W., 1997, *Operations Research: Applications and Algorithms,* Thomson, New York.

Wolsey, L. A., 1998, *Integer Programming,* Wiley, New York.

XPRESS-MP Extended Modeling and Optimisation Subroutine Library, Reference Manual, 2004, Dash Optimization, Blisworth, UK.

Chapter 4

GENETIC ALGORITHMS

Kumara Sastry, David Goldberg
University of Illinois, USA

Graham Kendall
University of Nottingham, UK

4.1   INTRODUCTION

Genetic algorithms (GAs) are search methods based on principles of natural selection and genetics (Fraser, 1957; Bremermann, 1958; Holland, 1975). We start with a brief introduction to simple genetic algorithms and associated terminology.

GAs encode the decision variables of a search problem into finite-length strings of alphabets of certain cardinality. The strings which are candidate solutions to the search problem are referred to as *chromosomes*, the alphabets are referred to as *genes* and the values of genes are called *alleles*. For example, in a problem such as the traveling salesman problem, a chromosome represents a route, and a gene may represent a city. In contrast to traditional optimization techniques, GAs work with coding of parameters, rather than the parameters themselves.

To evolve good solutions and to implement natural selection, we need a measure for distinguishing good solutions from bad solutions. The measure could be an *objective* function that is a mathematical model or a computer simulation, or it can be a *subjective* function where humans choose better solutions over worse ones. In essence, the fitness measure must determine a candidate solution's relative fitness, which will subsequently be used by the GA to guide the evolution of good solutions.

Another important concept of GAs is the notion of population. Unlike traditional search methods, genetic algorithms rely on a population of candidate solutions. The population size, which is usually a user-specified parameter, is one of the important factors affecting the scalability and performance of genetic algorithms. For example, small population sizes might lead to premature

convergence and yield substandard solutions. On the other hand, large population sizes lead to unnecessary expenditure of valuable computational time.

Once the problem is encoded in a chromosomal manner and a fitness measure for discriminating good solutions from bad ones has been chosen, we can
start to *evolve* solutions to the search problem using the following steps:

1 *Initialization.* The initial population of candidate solutions is usually
  generated randomly across the search space. However, domain-specific
  knowledge or other information can be easily incorporated.

2 *Evaluation.* Once the population is initialized or an offspring population
  is created, the fitness values of the candidate solutions are evaluated.

3 *Selection.* Selection allocates more copies of those solutions with higher
  fitness values and thus imposes the survival-of-the-fittest mechanism on
  the candidate solutions. The main idea of selection is to prefer better solutions to worse ones, and many selection procedures have been
  proposed to accomplish this idea, including roulette-wheel selection,
  stochastic universal selection, ranking selection and tournament selection, some of which are described in the next section.

4 *Recombination.* Recombination combines parts of two or more parental
  solutions to create new, possibly better solutions (i.e. offspring). There
  are many ways of accomplishing this (some of which are discussed in
  the next section), and competent performance depends on a properly
  designed recombination mechanism. The offspring under recombination
  will not be identical to any particular parent and will instead combine
  parental traits in a novel manner (Goldberg, 2002).

5 *Mutation.* While recombination operates on two or more parental chromosomes, mutation locally but randomly modifies a solution. Again, there
  are many variations of mutation, but it usually involves one or more
  changes being made to an individual's trait or traits. In other words,
  mutation performs a random walk in the vicinity of a candidate solution.

6 *Replacement.* The offspring population created by selection, recombination, and mutation replaces the original parental population. Many
  replacement techniques such as elitist replacement, generation-wise replacement and steady-state replacement methods are used in GAs.

7 Repeat steps 2–6 until a terminating condition is met.

Goldberg (1983, 1999a, 2002) has likened GAs to mechanistic versions of
certain modes of human innovation and has shown that these operators when
analyzed individually are ineffective, but when combined together they can

work well. This aspect has been explained with the concepts of the *fundamental intuition* and *innovation intuition*. The same study compares a combination of selection and mutation to *continual improvement* (a form of hill climbing), and the combination of selection and recombination to *innovation* (*cross-fertilizing*). These analogies have been used to develop a design-decomposition methodology and so-called *competent* GAs—that solve hard problems quickly, reliably, and accurately—both of which are discussed in the subsequent sections.

This chapter is organized as follows. The next section provides details of individual steps of a typical genetic algorithm and introduces several popular genetic operators. Section 4.1.2 presents a principled methodology of designing competent genetic algorithms based on decomposition principles. Section 4.1.3 gives a brief overview of designing principled efficiency-enhancement techniques to speed up genetic and evolutionary algorithms.

4.1.1   Basic Genetic Algorithm Operators

In this section we describe some of the selection, recombination, and mutation operators commonly used in genetic algorithms.

4.1.1.1   Selection Methods.   Selection procedures can be broadly classified into two classes as follows.

Fitness Proportionate Selection This includes methods such as roulette-wheel selection (Holland, 1975; Goldberg, 1989b) and stochastic universal selection (Baker, 1985; Grefenstette and Baker, 1989). In roulette-wheel selection, each individual in the population is assigned a roulette wheel slot sized in proportion to its fitness. That is, in the biased roulette wheel, good solutions have a larger slot size than the less fit solutions. The roulette wheel is spun to obtain a reproduction candidate. The roulette-wheel selection scheme can be implemented as follows:

1 Evaluate the fitness, f_i, of each individual in the population.

2 Compute the probability (slot size), p_i, of selecting each member of the population: $p_i = f_i / \sum_{j=1}^{n} f_j$, where n is the population size.

3 Calculate the cumulative probability, q_i, for each individual: $q_i = \sum_{j=1}^{i} p_j$.

4 Generate a uniform random number, $r \in (0, 1]$.

5 If $r < q_1$ then select the first chromosome, x_1, else select the individual x_i such that $q_{i-1} < r \leq q_i$.

6 Repeat steps 4–5 n times to create n candidates in the mating pool.

To illustrate, consider a population with five individuals ($n = 5$), with the fitness values as shown in the table below. The total fitness, $\sum_{j=1}^{n} f_j$ = $28 + 18 + 14 + 9 + 26 = 95$. The probability of selecting an individual and the corresponding cumulative probabilities are also shown in the table below.

Chromosome #	1	2	3	4	5
Fitness, f	28	18	14	9	26
Probability, p_i	28/95 = 0.295	0.189	0.147	0.095	0.274
Cumulative probability, q_i	0.295	0.484	0.631	0.726	1.000

Now if we generate a random number r, say 0.585, then the third chromosome is selected as $q_2 = 0.484 < 0.585 \leq q_3 = 0.631$.

Ordinal Selection This includes methods such as tournament selection (Goldberg et al., 1989b), and truncation selection (Mühlenbein and Schlierkamp-Voosen, 1993). In tournament selection, s chromosomes are chosen at random (either with or without replacement) and entered into a tournament against each other. The fittest individual in the group of k chromosomes wins the tournament and is selected as the parent. The most widely used value of s is 2. Using this selection scheme, n tournaments are required to choose n individuals. In truncation selection, the top ($1/s$)th of the individuals get s copies each in the mating pool.

4.1.1.2   Recombination (Crossover) Operators.   After selection, individuals from the mating pool are recombined (or crossed over) to create new, hopefully better, offspring. In the GA literature, many crossover methods have been designed (Goldberg, 1989b; Booker et al., 1997; Spears, 1997) and some of them are described in this section. Many of the recombination operators used in the literature are problem-specific and in this section we will introduce a few generic (problem independent) crossover operators. It should be noted that while for *hard* search problems, many of the following operators are not scalable, they are very useful as a first option. Recently, however, researchers have achieved significant success in designing scalable recombination operators that adapt linkage which will be briefly discussed in Section 4.1.2.

In most recombination operators, two individuals are randomly selected and are recombined with a probability p_c, called the crossover probability. That is, a uniform random number, r, is generated and if $r \leq p_c$, the two randomly selected individuals undergo recombination. Otherwise, that is, if $r > p_c$, the two offspring are simply copies of their parents. The value of p_c can either be set experimentally, or can be set based on schema-theorem principles (Goldberg, 1989b, 2002; Goldberg and Sastry, 2001).

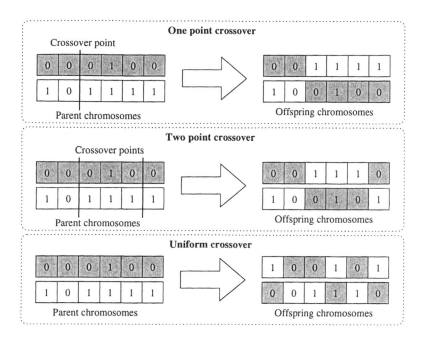

Figure 4.1. One-point, two-point, and uniform crossover methods.

k-point Crossover   One-point, and two-point crossovers are the simplest and most widely applied crossover methods. In one-point crossover, illustrated in Figure 4.1, a crossover site is selected at random over the string length, and the alleles on one side of the site are exchanged between the individuals. In two-point crossover, two crossover sites are randomly selected. The alleles between the two sites are exchanged between the two randomly paired individuals. Two-point crossover is also illustrated in Figure 4.1. The concept of one-point crossover can be extended to k-point crossover, where k crossover points are used, rather than just one or two.

Uniform Crossover   Another common recombination operator is uniform crossover (Syswerda, 1989; Spears and De Jong, 1994). In uniform crossover, illustrated in Figure 4.1, every allele is exchanged between the a pair of randomly selected chromosomes with a certain probability, p_e, known as the swapping probability. Usually the swapping probability value is taken to be 0.5.

Uniform Order-Based Crossover   The k-point and uniform crossover methods described above are not well suited for search problems with permutation codes such as the ones used in the traveling salesman problem. They often create offspring that represent invalid solutions for the search problem. Therefore,

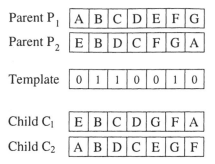

Figure 4.2. Illustration of uniform order crossover.

when solving search problems with permutation codes, a problem-specific repair mechanism is often required (and used) in conjunction with the above recombination methods to always create valid candidate solutions.

Another alternative is to use recombination methods developed specifically for permutation codes, which always generate valid candidate solutions. Several such crossover techniques are described in the following paragraphs starting with the uniform order-based crossover.

In uniform order-based crossover, two parents (say P_1 and P_2) are randomly selected and a random binary template is generated (see Figure 4.2). Some of the genes for offspring C_1 are filled by taking the genes from parent P_1 where there is a one in the template. At this point we have C_1 partially filled, but it has some "gaps". The genes of parent P_1 in the positions corresponding to zeros in the template are taken and sorted in the same order as they appear in parent P_2. The sorted list is used to fill the gaps in C_1. Offspring C_2 is created by using a similar process (see Figure 4.2).

Order-Based Crossover   The order-based crossover operator (Davis, 1985) is a variation of the uniform order-based crossover in which two parents are randomly selected and two random crossover sites are generated (see Figure 4.3). The genes between the cut points are copied to the children. Starting from the second crossover site copy the genes that are not already present in the offspring from the alternative parent (the parent other than the one whose genes are copied by the offspring in the initial phase) in the order they appear. For example, as shown in Figure 4.3, for offspring C_1, since alleles C, D, and E are copied from the parent P_1, we get alleles B, G, F, and A from the parent P_2. Starting from the second crossover site, which is the sixth gene, we copy alleles B and G as the sixth and seventh genes respectively. We then wrap around and copy alleles F and A as the first and second genes.

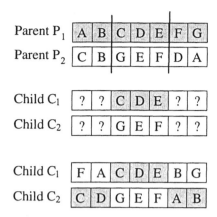

Figure 4.3. Illustration of order-based crossover.

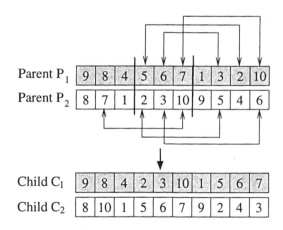

Figure 4.4. Illustration of partially matched crossover.

Partially Matched Crossover (PMX)   Apart from always generating valid offspring, the PMX operator (Goldberg and Lingle, 1985) also preserves orderings within the chromosome. In PMX, two parents are randomly selected and two random crossover sites are generated. Alleles within the two crossover sites of a parent are exchanged with the alleles corresponding to those mapped by the other parent. For example, as illustrated in Figure 4.4 (reproduced from Goldberg (1989b) with permission), looking at parent P_1, the first gene within the two crossover sites, 5, maps to 2 in P_2. Therefore, genes 5 and 2 are swapped in P_1. Similarly we swap 6 and 3, and 10 and 7 to create the offspring C_1. After all exchanges it can be seen that we have achieved a duplication of the ordering of one of the genes in between the crossover point within the opposite chromosome, and vice versa.

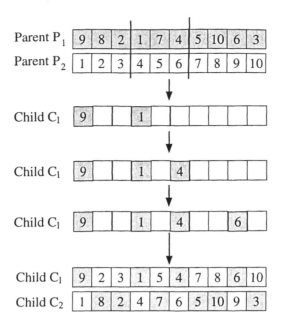

Figure 4.5. Illustration of cycle crossover.

Cycle Crossover (CX)   We describe cycle crossover (Oliver et al., 1987) with help of a simple illustration (reproduced from Goldberg (1989b) with permission). Consider two randomly selected parents P_1 and P_2 as shown in Figure 4.5 that are solutions to a traveling salesman problem. The offspring C_1 receives the first variable (representing city 9) from P_1. We then choose the variable that maps onto the same position in P_2. Since city 9 is chosen from P_1 which maps to city 1 in P_2, we choose city 1 and place it into C_1 in the same position as it appears in P_1 (fourth gene), as shown in Figure 4.5. City 1 in P_1 now maps to city 4 in P_2, so we place city 4 in C_1 in the same position it occupies in P_1 (sixth gene). We continue this process once more and copy city 6 to the ninth gene of C_1 from P_1. At this point, since city 6 in P_1 maps to city 9 in P_2, we should take city 9 and place it in C_1, but this has already been done, so we have completed a cycle; which is where this operator gets its name. The missing cities in offspring C_1 is filled from P_2. Offspring C_2 is created in the same way by starting with the first city of parent P_2 (see Figure 4.5).

4.1.1.3   Mutation Operators.   If we use a crossover operator, such as one-point crossover, we may get better and better chromosomes but the problem is, if the two parents (or worse, the entire population) has the same allele at a given gene then one-point crossover will not change that. In other words, that gene will have the same allele forever. Mutation is designed to overcome this problem in order to add diversity to the population and ensure that it is possible to explore the entire search space.

In evolutionary strategies, mutation is the primary variation/search operator. For an introduction to evolutionary strategies see, for example, Bäck et al. (1997). Unlike evolutionary strategies, mutation is often the secondary operator in GAs, performed with a low probability. One of the most common mutations is the bit-flip mutation. In bitwise mutation, each bit in a binary string is changed (a 0 is converted to 1, and vice versa) with a certain probability, p_m, known as the mutation probability. As mentioned earlier, mutation performs a random walk in the vicinity of the individual. Other mutation operators, such as problem-specific ones, can also be developed and are often used in the literature.

4.1.1.4   Replacement.   Once the new offspring solutions are created using crossover and mutation, we need to introduce them into the parental population. There are many ways we can approach this. Bear in mind that the parent chromosomes have already been selected according to their fitness, so we are hoping that the children (which includes parents which did not undergo crossover) are among the fittest in the population and so we would hope that the population will gradually, on average, increase its fitness. Some of the most common replacement techniques are outlined below.

Delete-all This technique deletes all the members of the current population and replaces them with the same number of chromosomes that have just been created. This is probably the most common technique and will be the technique of choice for most people due to its relative ease of implementation. It is also parameter-free, which is not the case for some other methods.

Steady-state This technique deletes n old members and replaces them with n new members. The number to delete and replace, n, at any one time is a parameter to this deletion technique. Another consideration for this technique is deciding which members to delete from the current population. Do you delete the worst individuals, pick them at random or delete the chromosomes that you used as parents? Again, this is a parameter to this technique.

Steady-state-no-duplicates This is the same as the steady-state technique but the algorithm checks that no duplicate chromosomes are added to the population. This adds to the computational overhead but can mean that more of the search space is explored.

4.1.2   Competent Genetic Algorithms

While using innovation for explaining the working mechanisms of GAs is very useful, as a design metaphor it poses difficulty as the processes of innovation are themselves not well understood. However, if we want GAs to successfully solve increasingly difficult problems across a wide spectrum of areas, we need a principled, but mechanistic way of designing genetic algorithms. The last few decades have witnessed great strides toward the development of so-called *competent* genetic algorithms—GAs that solve hard problems, quickly, reliably, and accurately (Goldberg, 1999a). From a computational standpoint, the existence of competent GAs suggests that many difficult problems can be solved in a scalable fashion. Furthermore, it significantly reduces the burden on a user to decide on a good coding or a good genetic operator that accompanies many GA applications. If the GA can adapt to the problem, there is less reason for the user to have to adapt the problem, coding, or operators to the GA.

In this section we briefly review some of the key lessons of competent GA design. Specifically, we restrict the discussion to selectorecombinative GAs and focus on the cross-fertilization type of innovation and briefly discuss key facets of competent GA design. Using Holland's notion of a building block (Holland, 1975), Goldberg proposed decomposing the problem of designing a competent selectorecombinative GA (Goldberg et al., 1992a). This design decomposition has been explained in detail elsewhere (Goldberg, 2002), but is briefly reviewed below.

Know that GAs Process Building Blocks   The primary idea of selectorecombinative GA theory is that genetic algorithms work through a mechanism of *decomposition* and *reassembly*. Holland (1975) called well-adapted sets of features that were components of effective solutions *building blocks* (BBs). The basic idea is that GAs (1) implicitly identify building blocks or sub-assemblies of good solutions, and (2) recombine different sub-assemblies to form very high performance solutions.

Understand BB Hard Problems   From the standpoint of cross-fertilizing innovation, problems that are hard have BBs that are hard to acquire. This may be because the BBs are complex, hard to find, or because different BBs are hard to separate, or because low-order BBs may be *misleading* or *deceptive* (Goldberg, 1987, 1989a; Goldberg et al., 1992b; Deb and Goldberg, 1994).

Understand BB Growth and Timing   Another key idea is that BBs or notions exist in a kind of competitive *market economy of ideas*, and steps must be taken to ensure that the best ones (1) grow and take over a dominant market share of the population, and (2) the growth rate can neither be too fast, nor too slow.

The growth in market share can be easily satisfied (Goldberg and Sastry, 2001) by appropriately setting the crossover probability, p_c, and the selection pressure, s, so that

$$p_c \leq \frac{1 - s^{-1}}{\epsilon} \tag{4.1}$$

where ϵ is the probability of BB disruption.

Two other approaches have been used in understanding time. It is not appropriate in a basic tutorial like this to describe them in detail, but we give a few example references for the interested reader.

Takeover time models, where the dynamics of the best individual is modeled (Goldberg and Deb, 1991; Sakamoto and Goldberg, 1997; Cantú-Paz, 1999; Rudolph, 2000).

Selection-intensity models, where approaches similar to those in quantitative genetics (Bulmer, 1985) are used and the dynamics of the average fitness of the population is modeled (Mühlenbein and Schlierkamp-Voosen, 1993; Thierens and Goldberg, 1994a, 1994b; Bäck, 1995; Miller and Goldberg, 1995, 1996a; Voigt et al., 1996).

The time models suggest that for a problem of size ℓ, with all BBs of equal importance or salience, the convergence time, t_c, of GAs is given by Miller and Goldberg (1995) to be

$$t_c = \frac{\pi}{2I}\sqrt{\ell} \tag{4.2}$$

where I is the selection intensity (Bulmer, 1985), which is a parameter dependent on the selection method and selection pressure. For tournament selection, I can be approximated in terms of s by the following relation (Blickle and Thiele, 1995):

$$I = \sqrt{2\left(\log(s) - \log\left(\sqrt{4.14\log(s)}\right)\right)} \tag{4.3}$$

On the other hand, if the BBs of a problem have different salience, then the convergence time scales-up differently. For example, when the BBs of a problem are exponentially scaled, with a particular BB being exponentially better than the others, then the convergence time, t_c, of a GA is linear with the problem size (Thierens et al., 1998) and can be represented as follows:

$$t_c = \frac{-\log 2}{\log\left(1 - 1/\sqrt{3}\right)}\ell \tag{4.4}$$

To summarize, the convergence time of GAs is $\mathcal{O}\left(\sqrt{\ell}\right)$–$\mathcal{O}\left(\ell\right)$ (see Chapter 1, Introduction, for an explanation of the $\mathcal{O}$ notation).

Understand BB Supply and Decision Making One role of the population is to ensure adequate *supply* of the raw building blocks in a population. Randomly generated populations of increasing size will, with higher probability, contain larger numbers of more complex BBs (Holland, 1975; Goldberg, 1989c; Goldberg et al., 2001). For a problem with m building blocks, each consisting of k alphabets of cardinality χ, the population size, n, required to ensure the presence of at least one copy of all the raw building blocks is given by Goldberg et al. (2001) as

$$n = \chi^k \log m + k\chi^k \log \chi \tag{4.5}$$

Just ensuring the raw supply is not enough, decision making among different, competing notions (BBs) is *statistical* in nature, and as we increase the population size, we increase the likelihood of making the best possible decisions (De Jong, 1975; Goldberg and Rudnick, 1991; Goldberg et al., 1992a; Harik et al., 1999). For an additively decomposable problem with m building blocks of size k each, the population size required to not only ensure supply, but also ensure correct decision making is approximately given by Harik et al. (1999) as

$$n = -\frac{\sqrt{\pi}}{2}\frac{\sigma_{BB}}{d}2^k\sqrt{m}\log\alpha \tag{4.6}$$

where d/σ_{BB} is the signal-to-noise ratio (Goldberg et al., 1992a), and α is the probability of incorrectly deciding among competing building blocks. In essence, the population-sizing model consists of the following components:

- *Competition complexity*, quantified by the total number of competing building blocks, 2^k.

- *Subcomponent complexity*, quantified by the number of building blocks, m.

- *Ease of decision making*, quantified by the signal-to-noise ratio, d/σ_{bb}.

- *Probabilistic safety factor*, quantified by the coefficient $-\log\alpha$.

On the other hand, if the building blocks are exponentially scaled, the population size, n, scales as (Rothlauf, 2002; Thierens et al., 1998; Goldberg, 2002)

$$n = -c_o \frac{\sigma_{BB}}{d} 2^k m \log \alpha \qquad (4.7)$$

where c_o is a constant dependent on the drift effects (Crow and Kimura, 1970; Goldberg and Segrest, 1987; Asoh and Mühlenbein, 1994).

To summarize, the complexity of the population size required by GAs is $\mathcal{O}\left(2^k \sqrt{m}\right) - \mathcal{O}\left(2^k m\right)$.

Identify BBs and Exchange Them Perhaps the most important lesson of current research in GAs is that the *identification and exchange of BBs* is the critical path to innovative success. First-generation GAs usually fail in their ability to promote this exchange reliably. The primary design challenge to achieving competence is the need to identify and promote effective BB exchange. Theoretical studies using the *facetwise* modeling approach (Thierens, 1999; Sastry and Goldberg, 2002, 2003) have shown that while fixed recombination operators such as uniform crossover, due to inadequacies of effective identification and exchange of BBs, demonstrate polynomial scalability on simple problems, they scale-up exponentially with problem size on boundedly-difficult problems. The mixing models also yield a *control map* delineating the region of good performance for a GA. Such a control map can be a useful tool in visualizing GA sweet-spots and provide insights in parameter settings (Goldberg, 1999a). This is in contrast to recombination operators that can automatically and adaptively identify and exchange BBs, which scale up polynomially (subquadratically–quadratically) with problem size.

Efforts in the principled design of effective BB identification and exchange mechanisms have led to the development of competent genetic algorithms. Competent GAs solve hard problems quickly, reliably, and accurately. Hard problems are loosely defined as those problems that have large sub-solutions that cannot be decomposed into simpler sub-solutions, or have badly scaled sub-solutions, or have numerous local optima, or are subject to a high stochastic noise. While designing a competent GA, the objective is to develop an algorithm that can solve problems with bounded difficulty and exhibit a polynomial (usually subquadratic) scale-up with the problem size.

Interestingly, the mechanics of competent GAs vary widely, but the principles of innovative success are invariant. Competent GA design began with the development of the *messy genetic algorithm* (Goldberg et al., 1989), culminating in 1993 with the *fast messy GA* (Goldberg et al., 1993). Since those early scalable results, a number of competent GAs have been constructed using different mechanism styles. We will categorize these approaches and provide some references for the interested reader, but a detailed treatment is beyond the scope of this tutorial.

Perturbation techniques, such as the messy GA (Goldberg et al., 1989), the fast messy GA (Goldberg et al., 1993), the gene expression messy GA (Kargupta, 1996), the linkage identification by nonlinearity check/linkage identification by detection GA (Munetomo and Goldberg, 1999; Heckendorn and Wright, 2004), and the dependency structure matrix driven genetic algorithm (Yu et al., 2003).

Linkage adaptation techniques, such as the linkage learning GA (Harik and Goldberg, 1997; Harik, 1997).

Probabilistic model building techniques, such as population based incremental learning (Baluja, 1994), the univariate model building algorithm (Mühlenbein and Paaß, 1996), the compact GA (Harik et al., 1998), the extended compact GA (Harik, 1999), the Bayesian optimization algorithm (Pelikan et al., 2000), the iterated distribution estimation algorithm (Bosman and Thierens, 1999), and the hierarchical Bayesian optimization algorithm (Pelikan and Goldberg, 2001). More details regarding these algorithms are given elsewhere (Pelikan et al., 2002; Larrañaga and Lozano, 2002; Pelikan, 2005).

4.1.3   Enhancement of Genetic Algorithms to Improve Efficiency and/or Effectiveness

The previous section presented a brief account of competent GAs. These GA designs have shown promising results and have successfully solved hard problems requiring only a subquadratic number of function evaluations. In other words, competent GAs usually solve an ℓ-variable search problem, requiring only $\mathcal{O}(\ell^2)$ number of function evaluations. While competent GAs take problems that were intractable with first-generation GAs and render them tractable, for large-scale problems, the task of computing even a subquadratic number of function evaluations can be daunting. If the fitness function is a complex simulation, model, or computation, then a single evaluation might take hours, even days. For such problems, even a subquadratic number of function evaluations is very high. For example, consider a 20-bit search problem and assume that a fitness evaluation takes one hour. We will require about half a month to solve the problem. This places a premium on a variety of *efficiency enhancement techniques*. Also, it is often the case that a GA needs to be integrated with problem-specific methods in order to make the approach really effective for a particular problem. The literature contains a very large number of papers which discuss enhancements of GAs. Once again, a detailed discussion is well beyond the scope of the tutorial, but we provide four broad categories of GA enhancement and examples of appropriate references for the interested reader.

Parallelization, where GAs are run on multiple processors and the computational resource is distributed among these processors (Cantú-Paz, 1997, 2000). Evolutionary algorithms are by *nature* parallel, and many different parallelization approaches can be used, such as a simple master–slave parallel GA (Grefenstette, 1981), a coarse-grained architecture (Pettey et al., 1987), a fine-grained architecture (Robertson, 1987; Gorges-Schleuter, 1989; Manderick and Spiessens, 1989), or a hierarchical architecture (Goldberg, 1989b; Gorges-Schleuter, 1997; Lin et al., 1997). Regardless of how parallelization is carried out, the key idea is to distribute the computational load on several processors thereby speeding-up the overall GA run. Moreover, there exists a principled design theory for developing an efficient parallel GA and optimizing the key facts of parallel architecture, connectivity, and deme size (Cantú-Paz, 2000).

For example, when the function evaluation time, T_f, is much greater than the communication time, T_c, which is very often the case, then a simple master–slave parallel GA—where the fitness evaluations are distributed over several processors and the rest of the GA operations are performed on a single processor—can yield linear speed-up when the number of processors is less than or equal to $\sqrt[3]{\frac{T_f}{T_c}n}$, and optimal speed-up when the number of processors equals $\sqrt{\frac{T_f}{T_c}n}$, where n is the population size.

Hybridization can be an extremely effective way of improving the performance and effectiveness of Genetic Algorithms. The most common form of hybridization is to couple GAs with local search techniques and to incorporate domain-specific knowledge into the search process. A common form of hybridization is to incorporate a local search operator into the Genetic Algorithm by applying the operator to each member of the population after each generation. This hybridization is often carried out in order to produce stronger results than the individual approaches can achieve on their own. However, this improvement in solution quality usually comes at the expense of increased computational time (e.g. Burke et al., 2001). Such approaches are often called Memetic Algorithms in the literature. This term was first used by Moscato (1989) and has since been employed very widely. For more details about memetic algorithms in general, see Krasnogor and Smith (2005), Krasnogor et al. (2004), Moscato and Cotta (2003) and Moscato (1999).

Of course, the hybridization of GAs can take other forms. Examples include:

- Initializing a GA population: e.g. Burke et al. (1998), Fleurent and Ferland (1994), Watson et al. (1999).

- Repairing infeasible solutions into legal ones: e.g. Ibaraki (1997).

- Developing specialized heuristic recombination operators: e.g. Burke et al. (1995).

- Incorporating a case-based memory (experience of past attempts) into the GA process (Louis and McDonnell, 2004).

- Heuristically decomposing large problems into smaller sub-problems before employing a memetic algorithm: e.g. Burke and Newall (1999).

Hybrid genetic algorithm and memetic approaches have demonstrated significant success in difficult real word application areas. A very small number of examples are included below (many more examples can be seen in the wider literature):

- University timetabling: examination timetabling (Burke et al., 1996, 1998; Burke and Newall, 1999) and course timetabling (Paechter et al., 1995, 1996).

- Machine scheduling (Cheng and Gen, 1997).

- Electrical power systems: unit commitment problems (Valenzuala and Smith, 2002); electricity transmission network maintenance scheduling (Burke and Smith, 1999); thermal generator mainte-nance scheduling (Burke and Smith, 2000).

- Sports scheduling (Costa, 1995).

- Nurse rostering (Burke et al., 2001).

- Warehouse scheduling (Watson et al., 1999).

While GA practitioners have often understood that real-world or com-mercial applications often require hybridization, there has been limited effort devoted to developing a theoretical underpinning of genetic algo-rithm hybridization. However, the following list contains examples of work which has aimed to answer critical issues such as

- the optimal division of labor between global and local searchers (or the right mix of exploration and exploitation) (Goldberg and Voessner, 1999);

- the effect of local search on sampling (Hart and Belew, 1996);

- hybrid GA modeling issues (Whitely, 1995).

The papers cited in this section are only a tiny proportion of the literature on hybrid genetic algorithms but they should provide a starting point for the interested reader. However, although there is a significant body of literature existing on the subject, there are many research directions still to be explored. Indeed, considering the option of hybridizing a GA with other approaches is one of the suggestions we give in the *Tricks of the Trade* section at the end of the chapter.

Time continuation, where the capabilities of both mutation and recombination are utilized to obtain a solution of as high quality as possible with a given limited computational resource (Goldberg, 1999b; Srivastava and Goldberg, 2001; Sastry and Goldberg, 2004a, 2004b). Time utilization (or continuation) exploits the tradeoff between the search for solutions with a large population and a single convergence epoch or using a small population with multiple convergence epochs.

Early theoretical investigations indicate that when the BBs are of equal (or nearly equal) salience and both recombination and mutation operators have the linkage information, then a small population with multiple convergence epochs is more efficient. However, if the fitness function is noisy or has overlapping building blocks, then a large population with a single convergence epoch is more efficient (Sastry and Goldberg, 2004a, 2004b). On the other hand, if the BBs of the problem are of non-uniform salience, which essentially means that they require serial processing, then a small population with multiple convergence epochs is more efficient (Goldberg, 1999b). Nevertheless, much work needs to be done to develop a principled design theory for efficiency enhancement via time continuation and to design competent continuation operators to reinitialize populations between epochs.

Evaluation relaxation, where an accurate, but computationally expensive fitness evaluation is replaced with a less accurate, but computationally inexpensive fitness estimate. The low-cost, less-accurate fitness estimate can either be (1) *exogenous*, as in the case of surrogate (or approximate) fitness functions (Jin, 2003), where external means can be used to develop the fitness estimate, or (2) *endogenous*, as in the case of *fitness inheritance* (Smith et al., 1995) where the fitness estimate is computed internally and is based on parental fitnesses.

Evaluation relaxation in GAs dates back to early, largely empirical work of Grefenstette and Fitzpatrick (1985) in image registration (Fitzpatrick et al., 1984) where significant speed-ups were obtained by reduced random sampling of the pixels of an image. Approximate evaluation has since been used extensively to solve complex optimization problems across many applications, such as structural engineering (Barthelemy and Haftka, 1993) and warehouse scheduling at Coors Brewery (Watson et al., 1999).

While early evaluation relaxation studies were largely empirical in nature, design theories have since been developed to understand the effect of approximate surrogate functions on population sizing and convergence time and to optimize speed-ups in approximate fitness functions with known variance (Miller and Goldberg, 1996b) in, for example, simple functions of known variance or known bias (Sastry, 2001), and in fitness inheritance (Sastry et al., 2001, 2004; Pelikan and Sastry, 2004).

4.2   TRICKS OF THE TRADE

In this section we present some suggestions for the reader who is new to the area of genetic algorithms and wants to know how best to get started. Fortunately, the ideas behind genetic algorithms are intuitive and the basic algorithm is not complex. Here are some basic *tips*.

- Start by using an "off the shelf" genetic algorithm. It is pointless developing a complex GA, if your problem can be solved using a simple and standard implementation.

- There are many excellent software packages that allow you to implement a genetic algorithm very quickly. Many of the introductory texts are supplied with a GA implementation and GA-LIB is probably seen as the software of choice for many people (see below).

- Consider carefully your representation. In the early days, the majority of implementations used a *bit representation* which was easy to implement. Crossover and mutation were simple. However, many other representations are now used, some utilizing complex data structures. You should carry out some research to determine what is the best representation for your particular problem.

- A basic GA will allow you to implement the algorithm and the only thing you have to supply is an evaluation function. If you can achieve this, then this is the fastest way to get a prototype system up and running. However, you may want to include some problem specific data in your algorithm. For example, you may want to include your own crossover operators (in order to guide the search) or you may want to produce the initial population using a constructive heuristic (to give the GA a good starting point).

- In recent times, many researchers have hybridized GAs with other search methods (see Section 4.1.3). Perhaps the most common method is to include a local searcher after the crossover and mutation operators (sometimes known as a memetic algorithm). This local searcher might be something as simple as a hill climber, which acts on each chromosome to ensure it is at a local optimum before the evolutionary process starts again.

- There are many parameters required to run a genetic algorithm (which can be seen as one of the shortcomings). At a *minimum* you have the population size, the mutation probability, and the crossover probability. The problem with having so many parameters to set is that it can take a lot of experimentation to find a set of values which solves your particular problem to the required quality. A broad *rule of thumb*, to start with, is to use a mutation probability of 0.05 (De Jong, 1975), a crossover rate of 0.6 (De Jong, 1975) and a population size of about 50. These three parameters are just an example of the many choices you are going to have to make to get your GA implementation working. To provide just a small sample: which crossover operator should you use?...which mutation operator?...Should the crossover/mutation rates be dynamic and change as the run progresses? Should you use a local search operator? If so, which one, and how long should that be allowed to run for? What selection technique should you use? What replacement strategy should you use? Fortunately, many researchers have investigated many of these issues and the additional sources section below provides many suitable references.

SOURCES OF ADDITIONAL INFORMATION
Software

- GALib, http://lancet.mit.edu/ga/. If you want GA software then GALIB should probably be your first port of call. The description (from the web page) says

> GAlib contains a set of C++ genetic algorithm objects. The library includes tools for using genetic algorithms to do optimization in any C++ program using any representation and genetic operators. The documentation includes an extensive overview of how to implement a genetic algorithm as well as examples illustrating customizations to the GAlib classes.

- GARAGe, http://garage.cps.msu.edu/. Genetic Algorithms Research and Applications Group.

- LGADOS in Coley (1999).

- NeuroDimension, http://www.nd.com/genetic/

- Simple GA (SGA) in Goldberg (1989b).

- Solver.com, http://www.solver.com/

- Ward Systems Group Inc., http://www.wardsystems.com/

- Other packages, http://www-2.cs.cmu.edu/afs/cs/project/
  ai-repository/ai/areas/genetic/ga/systems/0.html.   This URL contains
  links to a number of genetic algorithm software libraries.

Introductory Material

There are many publications which give excellent introductions to genetic algorithms: see Holland (1975), Davis (1987), Goldberg (1989b), Davis (1991), Beasley et al. (1993), Forrest (1993), Reeves (1995), Michalewicz (1996), Mitchell (1996), Falkenauer (1998), Coley (1999), and Man et al. (1999).

Memetic Algorithms

There are some excellent introductory texts for memetic algorithms: see Radcliffe and Surry (1994), Moscato (1999, 2001), Moscato and Cotta (2003), Hart et al. (2004), Krasnogor et al. (2004), Krasnogor and Smith (2005).
You might also like to refer to the Memetic Algorithms Home Page at

- http://www.densis.fee.unicamp.br/~moscato/memetic_home.html

Historical Material

An excellent work which brings together the early pioneering work in the field is Fogel (1998).

Conferences and Journals

There are a number of journals and conferences which publish papers concerned with genetic algorithms. The key conferences and journals are listed below, but remember that papers on Genetic Algorithms are published in many other outlets too.

Journals

- Evolutionary Computation, http://mitpress.mit.edu/
  catalog/item/default.asp?tid=25&ttype=4

- Genetic Programming and Evolvable Machines,
  http://www.kluweronline.com/issn/1389-2576/contents

- IEEE Transactions on Evolutionary Computation,
  http://www.ieee-nns.org/pubs/tec/

Conferences

- Congress on Evolutionary Computation (CEC)

- Genetic and Evolutionary Computation Conference (GECCO)

- Parallel Problem Solving in Nature (PPSN)

- Simulated Evolution and Learning (SEAL)

References

Asoh, H. and Mühlenbein, H., 1994, On the mean convergence time of evolutionary algorithms without selection and mutation, *Parallel Problem Solving from Nature III*, Lecture Notes in Computer Science, Vol. 866, pp. 98–107.

Bäck, T., 1995, Generalized convergence models for tournament—and (μ, λ)—selection, *Proc. 6th Int. Conf. on Genetic Algorithms,* pp. 2–8.

Bäck, T., Fogel, D. B. and Michalewicz, Z., 1997, Handbook of Evolutionary Computation, Oxford University Press, Oxford.

Baker, J. E., 1985, Adaptive selection methods for genetic algorithms, *Proc. Int. Conf. on Genetic Algorithms and Their Applications,* pp. 101–111.

Baluja, S., 1994, Population-based incremental learning: A method of integrating genetic search based function optimization and competitive learning, *Technical Report* CMU-CS-94-163, Carnegie Mellon University.

Barthelemy, J.-F. M. and Haftka, R. T., 1993, Approximation concepts for optimum structural design—a review, *Struct. Optim.* **5**:129–144.

Beasley, D., Bull, D. R. and Martin, R. R., 1993, An overview of genetic algorithms: Part 1, fundamentals, *Univ. Comput.* **15**:58–69.

Blickle, T. and Thiele, L., 1995, A mathematical analysis of tournament selection, *Proc. 6th Int. Conf. on Genetic Algorithms,* pp. 9–16.

Booker, L. B., Fogel, D. B., Whitley, D. and Angeline, P. J., 1997, Recombination, in: *The Handbook of Evolutionary Computation,* T. Bäck, D. B. Fogel, and Z. Michalewicz, eds, chapter E3.3, pp. C3.3:1–C3.3:27, IOP Publishing and Oxford University Press, Philadelphia, PA.

Bosman, P. A. N. and Thierens, D., 1999, Linkage information processing in distribution estimation algorithms, *Proc. 1999 Genetic and Evolutionary Computation Conf.,* pp. 60–67.

Bremermann, H. J., 1958, The evolution of intelligence. The nervous system as a model of its environment, *Technical Report* No. 1, Department of Mathematics, University of Washington, Seattle, WA.

Bulmer, M. G., 1985, *The Mathematical Theory of Quantitative Genetics,* Oxford University Press, Oxford.

Burke, E. K. and Newall, J. P., 1999, A multi-stage evolutionary algorithm for the timetable problem, *IEEE Trans. Evol. Comput.* **3**:63–74.

Burke, E. K. and Smith, A. J., 1999, A memetic algorithm to schedule planned maintenance, *ACM J. Exp. Algor.* **41**, www.jea.acm.org/1999/BurkeMemetic/ ISSN 1084-6654.

Burke, E. K. and Smith, A. J., 2000, Hybrid Evolutionary Techniques for the Maintenance Scheduling Problem, *IEEE Trans. Power Syst.* **15**:122–128.

Burke, E. K., Elliman, D. G. and Weare, R.F., 1995, Specialised recombinative operators for timetabling problems, in: *Evolutionary Computing: AISB Workshop 1995* T. Fogarty, ed., Lecture Notes in Computer Science, Vol. 993, pp. 75–85, Springer, Berlin.

Burke, E. K., Newall, J. P. and Weare, R. F., 1996, A memetic algorithm for university exam timetabling, in: *The Practice and Theory of Automated Timetabling I*, E. K. Burke and P. Ross, eds, Lecture Notes in Computer Science, Vol. 1153, pp. 241–250, Springer, Berlin.

Burke, E. K., Newall, J. P. and Weare, R. F., 1998, Initialisation strategies and diversity in evolutionary timetabling, *Evol. Comput. J.* (special issue on Scheduling) **6**:81–103.

Burke, E. K., Cowling, P. I., De Causmaecker, P. and Vanden Berghe, G., 2001, A memetic approach to the nurse rostering problem, *Appl. Intell.* **15**:199–214.

Cantü-Paz, E., 1997, A summary of research on parallel genetic algorithms *IlliGAL Report* No. 97003, General Engineering Department, University of Illinois at Urbana-Champaign, Urbana, IL.

Cantú-Paz, E., 1999, Migration policies and takeover times in parallel genetic algorithms, in: *Proc. Genetic and Evolutionary Computation Conf.*, p. 775, Morgan Kaufmann, San Francisco.

Cantú-Paz, E., 2000, *Efficient and Accurate Parallel Genetic Algorithms,* Kluwer, Boston, MA.

Cheng, R. W. and Gen, M., 1997, Parallel machine scheduling problems using memetic algorithms, *Comput. Indust. Eng.*, **33**:761–764.

Coley, D. A., 1999, *An Introduction to Genetic Algorithms for Scientists and Engineers,* World Scientific, New York.

Costa, D., 1995, An evolutionary tabu search algorithm and the nhl scheduling problem, *INFOR* **33**:161–178.

Crow, J. F. and Kimura, M., 1970, *An Introduction of Population Genetics Theory,* Harper and Row, New York.

Davis, L., 1985, Applying algorithms to epistatic domains, in: *Proc. Int. Joint Conf. on Artifical Intelligence*, pp. 162–164.

Davis, L. D. (ed), 1987, Genetic Algorithms and Simulated Annealing, Pitman, London.

Davis, L. (ed), 1991, *Handbook of Genetic Algorithms,* Van Nostrand Rein-hold, New York.

De Jong, K. A., 1975, An analysis of the behavior of a class of genetic adaptive systems, *Doctoral Dissertation,* University of Michigan, Ann Arbor, MI (University Microfilms No. 76-9381) (*Dissertation Abs. Int.* **36**:5140B).

Deb, K. and Goldberg, D. E., 1994, Sufficient conditions for deceptive and easy binary functions, *Ann. Math. Artif. Intell.* **10**:385–408.

Falkenauer E., 1998, Genetic Algorithms and Grouping Problems, Wiley, New York.

Fitzpatrick, J. M., Grefenstette, J. J. and Van Gucht, D., 1984, Image registration by genetic search, in: *Proc. IEEE Southeast Conf.,* IEEE, Piscataway, NJ, pp. 460–464.

Fleurent, C. and Ferland, J., 1994, Genetic hybrids for the quadratic assignment problem, in: *DIMACS Series in Mathematics and Theoretical Computer Science,* Vol. 16, pp. 190–206.

Fogel, D. B., 1998, *Evolutionary Computation: The Fossil Record,* IEEE, Piscataway, NJ.

Forrest, S., 1993, Genetic algorithms: Principles of natural selection applied to computation, *Science* **261**:872–878.

Fraser, A. S., 1957, Simulation of genetic systems by automatic digital computers. II: Effects of linkage on rates under selection, *Austral. J. Biol. Sci.* **10**:492–499.

Goldberg, D. E., 1983, Computer-aided pipeline operation using genetic algorithms and rule learning, *Doctoral Dissertation,.* University of Michigan, Ann Arbor, MI.

Goldberg, D. E., 1987, Simple genetic algorithms and the minimal deceptive problem, in: *Genetic Algorithms and Simulated Annealing,* L. Davis, ed., chapter 6, pp. 74–88, Morgan Kaufmann, Los Altos, CA.

Goldberg, D. E., 1989a, Genetic algorithms and Walsh functions: Part II, deception and its analysis, *Complex Syst.* **3**:153–171.

Goldberg, D. E., 1989b, *Genetic Algorithms in Search Optimization and Machine Learning,* Addison-Wesley, Reading, MA.

Goldberg, D. E., 1989c, Sizing populations for serial and parallel genetic algorithms, in: *Proc. 3rd Int. Conf. on Genetic Algorithms,* pp. 70–79.

Goldberg, D. E., 1999a, The race, the hurdle, and the sweet spot: Lessons from genetic algorithms for the automation of design innovation and creativity, in: *Evolutionary Design by Computers,* P. Bentley, ed., chapter 4, pp. 105–118, Morgan Kaufmann, San Mateo, CA.

Goldberg, D. E., 1999b, Using time efficiently: Genetic-evolutionary algorithms and the continuation problem, in: *Proc. Genetic and Evolutionary Computation Conf.,* pp. 212–219.

Goldberg, D. E., 2002, *Design of Innovation: Lessons From and For Competent Genetic Algorithms,* Kluwer, Boston, MA.

Goldberg, D. E. and Deb, K., 1991, A comparative analysis of selection schemes used in genetic algorithms, *Foundations of Genetic Algorithms,* G. J. E. Rawlins, ed., pp. 69–93.

Goldberg, D. E., Deb, K. and Clark, J. H., 1992a, Genetic algorithms, noise, and the sizing of populations, *Complex Syst.* **6**:333–362.

Goldberg, D. E., Deb, K. and Horn, J., 1992b, Massive multimodality, deception, and genetic algorithms, *Parallel Problem Solving from Nature II,* pp. 37–46, Elsevier, New York.

Goldberg, D. E., Deb, K., Kargupta, H. and Harik, G., 1993, Rapid, accurate optimization of difficult problems using fast messy genetic algorithms, in: *Proc. Int. Conf. on Genetic Algorithms,* pp. 56–64.

Goldberg, D. E., Korb, B. and Deb, K., 1989, Messy genetic algorithms: Motivation, analysis, and first results. *Complex Syst.* **3**:493–530.

Goldberg, D. E. and Lingle, R., 1985, Alleles, loci, and the TSP, in: *Proc. 1st Int. Conf. on Genetic Algorithms,* pp. 154–159.

Goldberg, D. E. and Rudnick, M., 1991, Genetic algorithms and the variance of fitness, *Complex Syst.* **5**:265–278.

Goldberg, D. E. and Sastry, K., 2001, A practical schema theorem for genetic algorithm design and tuning, in: *Proc. of the Genetic and Evolutionary Computation Conf.,* pp. 328–335.

Goldberg, D. E., Sastry, K. and Latoza, T., 2001, On the supply of building blocks, in: *Proc. of the Genetic and Evolutionary Computation Conf.,* pp. 336–342.

Goldberg, D. E. and Segrest, P., 1987, Finite Markov chain analysis of genetic algorithms, in: *Proc. 2nd Int. Conf. on Genetic Algorithms,* pp. 1–8.

Goldberg, D. E. and Voessner, S., 1999, Optimizing global-local search hybrids, in: *Proc. of the Genetic and Evolutionary Computation Conf.,* pp. 220–228.

Gorges-Schleuter, M., 1989, ASPARAGOS: An asynchronous parallel genetic optimization strategy, in: *Proc. 3rd Int. Conf. on Genetic Algorithms,* pp. 422–428.

Gorges-Schleuter, M., 1997, ASPARAGOS96 and the traveling salesman problem, in: *Proc. IEEE Int. Conf. on Evolutionary Computation,* pp. 171–174.

Grefenstette, J. J., 1981, Parallel adaptive algorithms for function optimization, *Technical Report* No. CS-81-19, Computer Science Department, Vanderbilt University, Nashville, TN.

Grefenstette, J. J. and Baker, J. E., 1989, How genetic algorithms work: A critical look at implicit parallelism, in: *Proc. 3rd Int. Conf. on Genetic Algorithms,* pp. 20–27.

Grefenstette, J. J. and Fitzpatrick, J. M., 1985, Genetic search with approximate function evaluations, in: *Proc. Int. Conf. on Genetic Algorithms and Their Applications,* pp. 112–120.

Harik, G. R., 1997, Learning linkage to eficiently solve problems of bounded difficulty using genetic algorithms, *Doctoral Dissertation,* University of Michigan, Ann Arbor, MI.

Harik, G., 1999, Linkage learning via probabilistic modeling in the ECGA, *IlliGAL Report* No. 99010, University of Illinois at Urbana-Champaign, Urbana, IL.

Harik, G., Cantú-Paz, E., Goldberg, D. E. and Miller, B. L., 1999, The gambler's ruin problem, genetic algorithms, and the sizing of populations, *Evol. Comput.* **7**:231–253.

Harik, G. and Goldberg, D. E., 1997, Learning linkage, *Foundations of Genetic Algorithms,* **4**:247–262.

Harik, G., Lobo, F. and Goldberg, D. E., 1998, The compact genetic algorithm, in: *Proc. IEEE Int. Conf. on Evolutionary Computation,* pp. 523–528.

Hart, W. E. and Belew, R. K., 1996, Optimization with genetic algorithm hybrids using local search, in: *Adaptive Individuals in Evolving Populations,* R. K. Belew, and M. Mitchell, eds, pp. 483–494, Addison-Wesley, Reading, MA.

Hart, W., Krasnogor, N. and Smith, J. E. (eds), 2004, Special issue on memetic algorithms, *Evol. Comput.* **12** No. 3.

Heckendorn, R. B. and Wright, A. H., 2004, Efficient linkage discovery by limited probing, *Evol. Comput.* **12**:517–545.

Holland, J. H., 1975, *Adaptation in Natural and Artificial Systems,* University of Michigan Press, Ann Arbor, MI.

Ibaraki, T., 1997, Combinations with other optimization methods, in: *Handbook of Evolutionary Computation,* T. Bäck, D. B. Fogel, and Z. Michalewicz, eds, pp. D3:1–D3:2, Institute of Physics Publishing and Oxford University Press, Bristol and New York.

Jin, Y., 2003, A comprehensive survey of fitness approximation in evolutionary computation, *Soft Comput. J.* (in press).

Kargupta, H., 1996, The gene expression messy genetic algorithm, in: *Proc. Int. Conf. on Evolutionary Computation,* pp. 814–819.

Krasnogor, N., Hart, W. and Smith, J. (eds), 2004, *Recent Advances in Memetic Algorithms,* Studies in Fuzziness and Soft Computing, Vol. 166, Springer, Berlin.

Krasnogor, N. and Smith, J. E., 2005, A tutorial for competent memetic algorithms: model, taxonomy and design issues, *IEEE Trans. Evol. Comput.,* accepted for publication.

Louis, S. J. and McDonnell, J., 2004, Learning with case injected genetic algorithms, *IEEE Trans. Evol. Comput.* **8**:316–328.

Larrañaga, P. and Lozano, J. A. (eds), 2002, *Estimation of Distribution Algorithms,* Kluwer, Boston, MA.

Lin, S.-C., Goodman, E. D. and Punch, W. F., 1997, Investigating parallel genetic algorithms on job shop scheduling problem, *6th Int. Conf. on Evolutionary Programming,* pp. 383–393.

Man, K. F., Tang, K. S. and Kwong, S., 1999, Genetic Algorithms: Concepts and Design, Springer, London.

Manderick, B. and Spiessens, P., 1989, Fine-grained parallel genetic algorithms, in: *Proc. 3rd Int. Conf. on Genetic Algorithms,* pp. 428–433.

Memetic Algorithms Home Page:
http://www.densis.fee.unicamp.br/~moscato/memetic_home.html

Michalewicz, Z., 1996, *Genetic Algorithms + Data Structures = Evolution Programs,* 3rd edn, Springer, Berlin.

Miller, B. L. and Goldberg, D. E., 1995, Genetic algorithms, tournament selection, and the effects of noise, *Complex Syst.* **9**:193–212.

Miller, B. L. and Goldberg, D. E., 1996a, Genetic algorithms, selection schemes, and the varying effects of noise, *Evol. Comput.* **4**:113–131.

Miller, B. L. and Goldberg, D. E., 1996b, Optimal sampling for genetic algorithms, *Intelligent Engineering Systems through Artificial Neural Networks (ANNIE'96),* Vol. 6, pp. 291–297, ASME Press, New York.

Mitchell, M., 1996, *Introduction to Genetic Algorithms,* MIT Press, Boston, MA.

Moscato, P., 1989, On evolution, search, optimization, genetic algorithms and martial arts: Towards memetic algorithms, *Technical Report* C3P 826, Caltech Concurrent Computation Program, California Institute of Technology, Pasadena, CA.

Moscato, P., 1999, Part 4: Memetic algorithms, in: *New Ideas in Optimization,* D. Corne, M. Dorigo and F. Glover, eds, pp. 217–294, McGraw-Hill, New York.

Moscato, P., 2001, Memetic algorithms, in: *Handbook of Applied Optimization,* Section 3.6.4, P. M. Pardalos and M. G. C. Resende, eds, Oxford University Press, Oxford.

Moscato, P. and Cotta, C., 2003, A gentle introduction to memetic algorithms, in: *Handbook of Metaheuristics,* F. Glover and G. Kochenberger, eds, Chapter 5, Kluwer, Norwell, MA.

Mühlenbein, H. and Paaß, G., 1996, From recombination of genes to the estimation of distributions I. Binary parameters, in: *Parallel Problem Solving from Nature IV,* Lecture Notes in Computer Science, Vol. 1141, Springer, Berlin.

Mühlenbein, H. and Schlierkamp-Voosen, D., 1993, Predictive models for the breeder genetic algorithm: I. continous parameter optimization, *Evol. Comput.* **1**:25–49.

Munetomo, M. and Goldberg, D. E., 1999, Linkage identification by non-monotonicity detection for overlapping functions, *Evol. Comput.* 7:377–398.

Oliver, J. M., Smith, D. J. and Holland, J. R. C., 1987, A study of permutation crossover operators on the travelling salesman problem, in: *Proc. 2nd Int. Conf. on Genetic Algorithms*, pp. 224–230.

Paechter, B., Cumming, A., Norman, M. G. and Luchian, H., 1996, Extensions to a memetic timetabling system, *The Practice and Theory of Automated Timetabling I*, E. K. Burke and P. Ross, eds, Lecture Notes in Computer Science, Vol. 1153, Springer, Berlin, pp. 251–265.

Paechter, B., Cumming, A. and Luchian, H., 1995, The use of local search suggestion lists for improving the solution of timetable problems with evolutionary algorithms, *Evolutionary Computing: AISB Workshop 1995*, T. Fogarty, ed., Lecture Notes in Computer Science, Vol. 993, Springer, Berlin, pp. 86–93.

Pelikan, M., 2005, *Hierarchical Bayesian Optimization Algorithm: Toward a New Generation of Evolutionary Algorithm,* Springer, Berlin.

Pelikan, M. and Goldberg, D. E., 2001, Escaping hierarchical traps with competent genetic algorithms, in: *Proc. Genetic and Evolutionary Computation Conf.,* pp. 511–518.

Pelikan, M., Goldberg, D. E. and Cantú-Paz, E., 2000, Linkage learning, estimation distribution, and Bayesian networks, *Evol. Comput.* **8**:314–341.

Pelikan, M., Lobo, F. and Goldberg, D. E., 2002, A survey of optimization by building and using probabilistic models, *Comput. Optim. Appl.* **21**:5–20.

Pelikan, M. and Sastry, K., 2004, Fitness inheritance in the Bayesian optimization algorithm, in: *Proc. Genetic and Evolutionary Computation Conference,* Vol. 2, pp. 48–59.

Pettey, C. C., Leuze, M. R. and Grefenstette, J. J., 1987, A parallel genetic algorithm, in: *Proc. 2nd Int. Conf. on Genetic Algorithms,* pp. 155–161.

Radcliffe, N. J. and Surry, P. D., 1994, Formal memetic algorithms, *Evolutionary Computing: AISB Workshop 1994*, T. Fogarty, ed., Lecture Notes in Computer Science, Vol. 865, pp. 1–16, Springer, Berlin.

Reeves, C. R., 1995, Genetic algorithms, in: *Modern Heuristic Techniques for Combinatorial Problems,* C. R. Reeves, ed., McGraw-Hill, New York.

Robertson, G. G., 1987, Parallel implementation of genetic algorithms in a classifier system, in: *Proc. 2nd Int. Conf. on Genetic Algorithms,* pp. 140–147.

Rothlauf, F., 2002, *Representations for Genetic and Evolutionary Algorithms,* Springer, Berlin.

Rudolph, G., 2000, Takeover times and probabilities of non-generational selection rules, in: *Proc. Genetic and Evolutionary Computation Conf.,* pp. 903–910.

Sakamoto, Y. and Goldberg, D. E., 1997, Takeover time in a noisy environment, in: *Proc. 7th Int. Conf. on Genetic Algorithms,* pp. 160–165.

Sastry, K., 2001, Evaluation-relaxation schemes for genetic and evolutionary algorithms, *Master's Thesis,* General Engineering Department, University of Illinois at Urbana-Champaign, Urbana, IL.

Sastry, K. and Goldberg, D. E., 2002, Analysis of mixing in genetic algorithms: A survey, *IlliGAL Report* No. 2002012, University of Illinois at Urbana-Champaign, Urbana, IL.

Sastry, K. and Goldberg, D. E., 2003, Scalability of selectorecombinative genetic algorithms for problems with tight linkage, in: *Proc. 2003 Genetic and Evolutionary Computation Conf.,* pp. 1332–1344.

Sastry, K. and Goldberg, D. E., 2004a, Designing competent mutation operators via probabilistic model building of neighborhoods, in: *Proc. 2004 Genetic and Evolutionary Computation Conference II,* Lecture Notes in Computer Science, Vol. 3103, Springer, Berlin, pp. 114–125.

Sastry, K. and Goldberg, D. E., 2004b, Let's get ready to rumble: Crossover versus mutation head to head, in: *Proc. 2004 Genetic and Evolutionary Computation Conf. II,* Lecture Notes in Computer Science, Vol. 3103, Springer, Berlin, pp. 126–137.

Sastry, K., Goldberg, D. E., & Pelikan, M., 2001, Don't evaluate, inherit, in: *Proc. Genetic and Evolutionary Computation Conf.,* pp. 551–558.

Sastry, K., Pelikan, M. and Goldberg, D. E., 2004, Efficiency enhancement of genetic algorithms building-block-wise fitness estimation, in: *Proc. IEEE Int. Congress on Evolutionary Computation,* pp. 720–727.

Smith, R., Dike, B. and Stegmann, S., 1995, Fitness inheritance in genetic algorithms, in: *Proc. ACM Symp. on Applied Computing,* pp. 345–350, ACM, New York.

Spears, W., 1997, Recombination parameters, in: *The Handbook of Evolutionary Computation,* T. Bäck, D. B. Fogel and Z. Michalewicz, eds, Chapter E1.3, IOP Publishing and Oxford University Press, Philadelphia, PA, pp. E1.3:1–E1.3:13.

Spears, W. M. and De Jong, K. A., 1994, On the virtues of parameterized uniform crossover, in: *Proc. 4th Int. Conf. on Genetic Algorithms.*

Srivastava, R. and Goldberg, D. E., 2001, Verification of the theory of genetic and evolutionary continuation, in: *Proc. Genetic and Evolutionary Computation Conf.,* pp. 551–558.

Syswerda, G., 1989, Uniform crossover in genetic algorithms, in: *Proc. 3rd Int. Conf. on Genetic Algorithms,* pp. 2–9.

Thierens, D., 1999, Scalability problems of simple genetic algorithms, *Evol. Comput.* 7:331–352.

Thierens, D. and Goldberg, D. E., 1994a, Convergence models of genetic algorithm selection schemes, in: *Parallel Problem Solving from Nature III*, pp. 116–121.

Thierens, D. and Goldberg, D. E., 1994b, Elitist recombination: An integrated selection recombination GA, in: *Proc. 1st IEEE Conf. on Evolutionary Computation*, pp. 508–512.

Thierens, D., Goldberg, D. E. and Pereira, A. G., 1998, Domino convergence, drift, and the temporal-salience structure of problems, in: *Proc. IEEE Int. Conf. on Evolutionary Computation*, pp. 535–540.

Valenzuala, J. and Smith, A. E., 2002, A seeded memetic algorithm for large unit commitment problems, *J. Heuristics*, **8**:173–196.

Voigt, H.-M., Mühlenbein, H. and Schlierkamp-Voosen, D., 1996, The response to selection equation for skew fitness distributions, in: *Proc. Int. Conf. on Evolutionary Computation*, pp. 820–825.

Watson, J. P., Rana, S., Whitely, L. D. and Howe, A. E., 1999, The impact of approximate evaluation on the performance of search algorithms for warehouse scheduling, *J. Scheduling*, **2**:79–98.

Whitley, D., 1995, Modeling hybrid genetic algorithms, in *Genetic Algorithms in Engineering and Computer Science*, G. Winter, J. Periaux, M. Galan and P. Cuesta, eds, Wiley, New York, pp. 191–201.

Yu, T.-L., Goldberg, D. E., Yassine, A. and Chen, Y.-P., 2003, A genetic algorithm design inspired by organizational theory: Pilot study of a dependency structure matrix driven genetic algorithm, *Artificial Neural Networks in Engineering (ANNIE 2003)*, pp. 327–332.

Chapter 5

GENETIC PROGRAMMING

John R. Koza
Stanford University
Stanford, CA, USA

Riccardo Poli
Department of Computer Science
University of Essex, UK

5.1   INTRODUCTION

The goal of getting computers to automatically solve problems is central to artificial intelligence, machine learning, and the broad area encompassed by what Turing called "machine intelligence" (Turing, 1948, 1950). In his talk entitled *AI: Where It Has Been and Where It Is Going*, machine learning pioneer Arthur Samuel stated the main goal of the fields of machine learning and artificial intelligence:

> [T]he aim [is] ... to get machines to exhibit behavior, which if done by humans, would be assumed to involve the use of intelligence.

> (Samuel, 1983)

Genetic programming is a systematic method for getting computers to automatically solve a problem starting from a high-level statement of what needs to be done. Genetic programming is a domain-independent method that genetically breeds a population of computer programs to solve a problem. Specifically, genetic programming iteratively transforms a population of computer programs into a new generation of programs by applying analogs of naturally occurring genetic operations. This process is illustrated in Figure 5.1.

The genetic operations include crossover (sexual recombination), mutation, reproduction, gene duplication, and gene deletion. Analogs of developmental processes are sometimes used to transform an embryo into a fully developed structure. Genetic programming is an extension of the genetic algorithm (Holland, 1975), see Chapter 4, in which the *structures* in the population are not

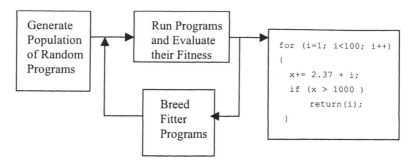

Figure 5.1. Main loop of genetic programming.

fixed-length character strings that encode candidate solutions to a problem, but *programs* that, when executed, *are* the candidate solutions to the problem.

Programs are expressed in genetic programming as *syntax trees* rather than as lines of code. For example, the simple expression

$$max(x * x, x + 3 * y)$$

is represented as shown in Figure 5.2. The tree includes *nodes* (which we will also call *points*) and *links*. The nodes indicate the instructions to execute. The links indicate the arguments for each instruction. In the following the internal nodes in a tree will be called *functions*, while the tree's leaves will be called *terminals*.

In more advanced forms of genetic programming, programs can be composed of multiple components (e.g. subroutines). In this case the representation used in genetic programming is a set of trees (one for each component) grouped together under a special node called *root*, as illustrated in Figure 5.3. We will call these (sub)trees *branches*. The number and type of the branches in a program, together with certain other features of the structure of the branches, form the *architecture* of the program.

Genetic programming trees and their corresponding expressions can equivalently be represented in *prefix notation* (e.g. as Lisp S-expressions). In prefix notation, functions always precede their arguments. For example, $max(x * x, x + 3 * y)$ becomes

$$(max(* x x)(+ x (* 3 y)))$$

In this notation, it is easy to see the correspondence between expressions and their syntax trees. Simple recursive procedures can convert prefix-notation expressions into infix-notation expressions and vice versa. Therefore, in the following, we will use trees and their corresponding prefix-notation expressions interchangeably.

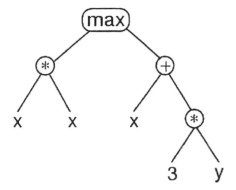

Figure 5.2. Basic tree-like program representation used in genetic programming.

5.2 PREPARATORY STEPS OF GENETIC PROGRAMMING

Genetic programming starts from a high-level statement of the requirements of a problem and attempts to produce a computer program that solves the problem.

The human user communicates the high-level statement of the problem to the genetic programming algorithm by performing certain well-defined preparatory steps.

The five major preparatory steps for the basic version of genetic programming require the human user to specify

1. the set of terminals (e.g., the independent variables of the problem, zero-argument functions, and random constants) for each branch of the to-be-evolved program,

2. the set of primitive functions for each branch of the to-be-evolved program,

3. the fitness measure (for explicitly or implicitly measuring the fitness of individuals in the population),

4. certain parameters for controlling the run, and

5. the termination criterion and method for designating the result of the run.

The first two preparatory steps specify the ingredients that are available to create the computer programs. A run of genetic programming is a competitive search among a diverse population of programs composed of the available functions and terminals.

The identification of the function set and terminal set for a particular problem (or category of problems) is usually a straightforward process. For some

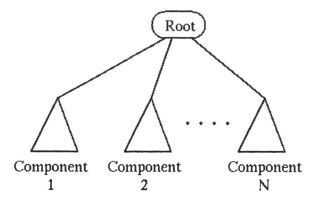

Figure 5.3. Multi-tree program representation.

problems, the function set may consist of merely the arithmetic functions of addition, subtraction, multiplication, and division as well as a conditional branching operator. The terminal set may consist of the program's external inputs (independent variables) and numerical constants.

For many other problems, the ingredients include specialized functions and terminals. For example, if the goal is to get genetic programming to automatically program a robot to mop the entire floor of an obstacle-laden room, the human user must tell genetic programming what the robot is capable of doing. For example, the robot may be capable of executing functions such as moving, turning, and swishing the mop.

If the goal is the automatic creation of a controller, the function set may consist of integrators, differentiators, leads, lags, gains, adders, subtractors, and the like and the terminal set may consist of signals such as the reference signal and plant output.

If the goal is the automatic synthesis of an analog electrical circuit, the function set may enable genetic programming to construct circuits from components such as transistors, capacitors, and resistors. Once the human user has identified the primitive ingredients for a problem of circuit synthesis, the same function set can be used to automatically synthesize an amplifier, computational circuit, active filter, voltage reference circuit, or any other circuit composed of these ingredients.

The third preparatory step concerns the fitness measure for the problem. The fitness measure specifies what needs to be done. The fitness measure is the primary mechanism for communicating the high-level statement of the problem's requirements to the genetic programming system. For example, if the goal is to get genetic programming to automatically synthesize an amplifier, the fitness function is the mechanism for telling genetic programming to synthesize a circuit that amplifies an incoming signal (as opposed to, say, a circuit that suppresses the low frequencies of an incoming signal or that computes the square root of the incoming signal). The first two preparatory steps define the search space whereas the fitness measure implicitly specifies the search's desired goal.

The fourth and fifth preparatory steps are administrative. The fourth preparatory step entails specifying the control parameters for the run. The most important control parameter is the population size. Other control parameters include the probabilities of performing the genetic operations, the maximum size for programs, and other details of the run.

The fifth preparatory step consists of specifying the termination criterion and the method of designating the result of the run. The termination criterion may include a maximum number of generations to be run as well as a problem-specific success predicate. The single best-so-far individual is then harvested and designated as the result of the run.

5.3 EXECUTIONAL STEPS OF GENETIC PROGRAMMING

After the user has performed the preparatory steps for a problem, the run of genetic programming can be launched. Once the run is launched, a series of well-defined, problem-independent steps is executed.

Genetic programming typically starts with a population of randomly generated computer programs composed of the available programmatic ingredients (as provided by the human user in the first and second preparatory steps).

Genetic programming iteratively transforms a population of computer programs into a new generation of the population by applying analogs of naturally occurring genetic operations. These operations are applied to individual(s) selected from the population. The individuals are probabilistically selected to participate in the genetic operations based on their fitness (as measured by the fitness measure provided by the human user in the third preparatory step). The iterative transformation of the population is executed inside the main generational loop of the run of genetic programming.

The executional steps of genetic programming are as follows:

1 Randomly create an initial population (generation 0) of individual computer programs composed of the available functions and terminals.

2 Iteratively perform the following sub-steps (called a *generation*) on the population until the termination criterion is satisfied:

   (a) Execute each program in the population and ascertain its fitness (explicitly or implicitly) using the problem's fitness measure.

   (b) Select one or two individual program(s) from the population with a probability based on fitness (with reselection allowed) to participate in the genetic operations in (c).

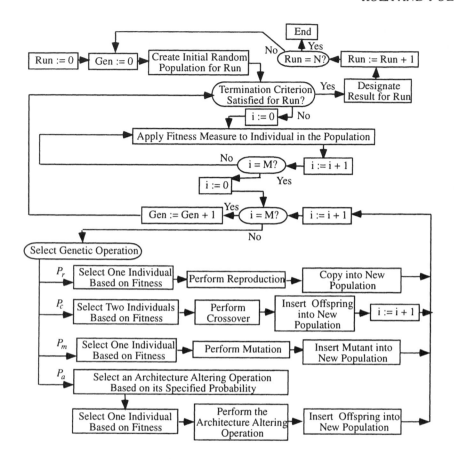

Figure 5.4. Flowchart of genetic programming.

(c) Create new individual program(s) for the population by applying the following genetic operations with specified probabilities:

  i *Reproduction:* Copy the selected individual program to the new population.

  ii *Crossover:* Create new offspring program(s) for the new population by recombining randomly chosen parts from two selected programs.

  iii *Mutation:* Create one new offspring program for the new population by randomly mutating a randomly chosen part of one selected program.

  iv *Architecture-altering operations:* Choose an architecture-altering operation from the available repertoire of such operations and create one new offspring program for the new pop-

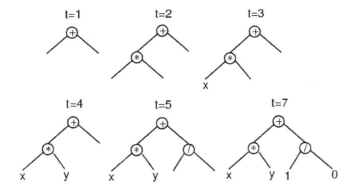

Figure 5.5. Creation of a seven-point tree using the "Full" initialization method (*t* = time).

ulation by applying the chosen architecture-altering operation to one selected program.

3 After the termination criterion is satisfied, the single best program in the population produced during the run (the best-so-far individual) is harvested and designated as the result of the run. If the run is successful, the result may be a solution (or approximate solution) to the problem.

Figure 5.4 is a flowchart of genetic programming showing the genetic operations of crossover, reproduction, and mutation as well as the architecture-altering operations. This flowchart shows a two-offspring version of the crossover operation.

The preparatory steps specify what the user must provide in advance to the genetic programming system. Once the run is launched, the executional steps as shown in the flowchart (Figure 5.4) are executed. Genetic programming is problem-independent in the sense that the flowchart specifying the basic sequence of executional steps is not modified for each new run or each new problem.

There is usually no discretionary human intervention or interaction during a run of genetic programming (although a human user may exercise judgment as to whether to terminate a run).

Genetic programming starts with an initial population of computer programs composed of functions and terminals appropriate to the problem. The individual programs in the initial population are typically generated by recursively generating a rooted point-labeled program tree composed of random choices of the primitive functions and terminals (provided by the user as part of the first and second preparatory steps). The initial individuals are usually generated subject to a pre-established maximum size (specified by the user as a minor parameter as part of the fourth preparatory step). For example, in the *"Full" initialization method* nodes are taken from the function set until a maximum

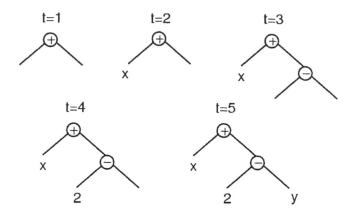

Figure 5.6. Creation of a five-point tree using the "Grow" initialization method ($t =$ time).

tree depth is reached. Beyond that depth only terminals can be chosen. Figure 5.5 shows several snapshots of this process. A variant of this, the *"Grow" initialization method,* allows the selection of nodes from the whole primitive set until the depth limit is reached. Thereafter, it behaves like the "Full" method. Figure 5.6 illustrates this process. Pseudo-code for a recursive implementation of both the "Full" and the "Grow" methods is given in Figure 5.7. The code assumes that programs are represented as prefix-notation expressions.

In general, after the initialization phase, the programs in the population are of different size (number of functions and terminals) and of different shape (the particular graphical arrangement of functions and terminals in the program tree).

Each individual program in the population is either measured or compared in terms of how well it performs the task at hand (using the fitness measure provided in the third preparatory step). For many problems, this measurement yields a single explicit numerical value, called *fitness*. Normally, fitness evaluation requires executing the programs in the population, often multiple times, *within* the genetic programming system. A variety of execution strategies exist, including the (relatively uncommon) off-line or on-line compilation and linking and the (relatively common) virtual-machine-code compilation and interpretation.

Interpreting a program tree means executing the nodes in the tree in an order that guarantees that nodes are not executed before the value of their arguments (if any) is known. This is usually done by traversing the tree in a recursive way starting from the root node, and postponing the evaluation of each node until the value of its children (arguments) is known. This process is illustrated in Figure 5.8, where the numbers to the right of internal nodes represent the results of evaluating the subtrees rooted at such nodes. In this example, the

```
procedure: gen_rnd_expr
   arguments:
      func_set        /* A function set */
      term_set        /* A terminal set */
      max_d           /* Maximum depth for expressions */
      method          /* Either "Full" or "Grow" */
   results:
      expr            /* An expression in prefix notation */
   begin
      if max_d = 0 or method = "Grow" and random digit = 1 then
         expr = choose_random_element( term_set )
      else
         func = choose_random_element( func_set )
         for i = 1 to arity(func):
            arg_i = gen_rnd_expr(func_set, term_set, max_d - 1, method );
            expr = (func, arg_1, arg_2, ...);
      endif
   end
```

Figure 5.7. Pseudo-code for recursive program generation with the "Full" and "Grow" methods.

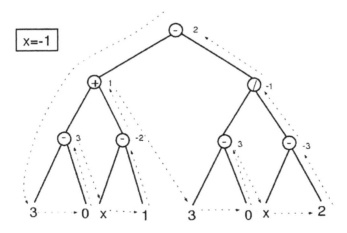

Figure 5.8. Example interpretation of a syntax tree (terminal x is a variable with value -1).

independent variable X evaluates to -1. Figure 5.9 gives a pseudo-code implementation of the interpretation procedure. The code assumes that programs are represented as prefix-notation expressions and that such expressions can be treated as lists of components (where a construct like *expr(i)* can be used to read or set component i of expression *expr*).

Irrespective of the execution strategy adopted, the fitness of a program may be measured in many different ways, including, for example, in terms of the amount of error between its output and the desired output, the amount of time (fuel, money, etc) required to bring a system to a desired target state, the accu-

```
procedure: eval
   arguments:
      expr      /* An expression in prefix notation */
   results:
      value     /* A number */
   begin
      if expr is a list then /* Non-terminal */
         proc = expr(1)
         value = proc(eval(expr(2)),eval(expr(3)),...)
      else /* Terminal */
         if expr is a variable or a constant then
            value = expr
         else /* 0-arity function */
            value = expr()
         endif
      endif
   end
```

Figure 5.9. Typical interpreter for genetic programming.

racy of the program in recognizing patterns or classifying objects into classes, the payoff that a game-playing program produces, or the compliance of a complex structure (such as an antenna, circuit, or controller) with user-specified design criteria. The execution of the program sometimes returns one or more explicit values. Alternatively, the execution of a program may consist only of side effects on the state of a world (e.g., a robot's actions). Alternatively, the execution of a program may yield both return values and side effects.

The fitness measure is, for many practical problems, multi-objective in the sense that it combines two or more different elements. In practice, the different elements of the fitness measure are in competition with one another to some degree.

For many problems, each program in the population is executed over a representative sample of different *fitness cases*. These fitness cases may represent different values of the program's input(s), different initial conditions of a system, or different environments. Sometimes the fitness cases are constructed probabilistically.

The creation of the initial random population is, in effect, a blind random search of the search space of the problem. It provides a baseline for judging future search efforts. Typically, the individual programs in generation 0 all have exceedingly poor fitness. Nonetheless, some individuals in the population are (usually) more fit than others. The differences in fitness are then exploited by genetic programming. Genetic programming applies Darwinian selection and the genetic operations to create a new population of offspring programs from the current population.

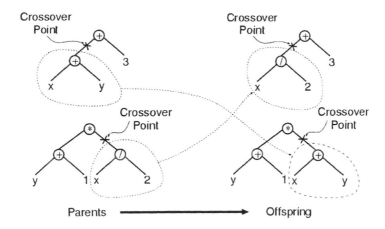

Figure 5.10. Example of two-child crossover between syntax trees.

The genetic operations include crossover (sexual recombination), mutation, reproduction, and the architecture-altering operations. Given copies of two parent trees, typically, *crossover* involves randomly selecting a crossover point (which can equivalently be thought of as either a node or a link between nodes) in each parent tree and swapping the sub-trees rooted at the crossover points, as exemplified in Figure 5.10. Often crossover points are not selected with uniform probability. A frequent strategy is, for example, to select internal nodes (functions) 90% of the times, and any node for the remaining 10% of the times. Traditional *mutation* consists of randomly selecting a mutation point in a tree and substituting the sub-tree rooted there with a randomly generated sub-tree, as illustrated in Figure 5.11. Mutation is sometimes implemented as crossover between a program and a newly generated random program (this is also known as *"headless chicken" crossover*). *Reproduction* involves simply copying certain individuals into the new population. Architecture altering operations will be discussed later in this chapter.

The genetic operations described above are applied to individual(s) that are probabilistically selected from the population based on fitness. In this probabilistic selection process, better individuals are favored over inferior individuals. However, the best individual in the population is not necessarily selected and the worst individual in the population is not necessarily passed over.

After the genetic operations are performed on the current population, the population of offspring (i.e. the new generation) replaces the current population (i.e. the now-old generation). This iterative process of measuring fitness and performing the genetic operations is repeated over many generations.

The run of genetic programming terminates when the termination criterion (as provided by the fifth preparatory step) is satisfied. The outcome of the run

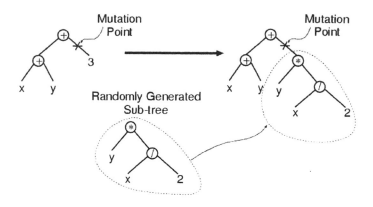

Figure 5.11. Example of sub-tree mutation.

is specified by the method of result designation. The best individual ever encountered during the run (i.e. the best-so-far individual) is typically designated as the result of the run.

All programs in the initial random population (generation 0) of a run of genetic programming are syntactically valid, executable programs. The genetic operations that are performed during the run (i.e. crossover, mutation, reproduction, and the architecture-altering operations) are designed to produce offspring that are syntactically valid, executable programs. Thus, every individual created during a run of genetic programming (including, in particular, the best-of-run individual) is a syntactically valid, executable program.

There are numerous alternative implementations of genetic programming that vary from the preceding brief description.

5.4   EXAMPLE OF A RUN OF GENETIC PROGRAMMING

To provide concreteness, this section contains an illustrative run of genetic programming in which the goal is to automatically create a computer program whose output is equal to the values of the quadratic polynomial x^2+x+1 in the range from -1 to $+1$. That is, the goal is to automatically create a computer program that matches certain numerical data. This process is sometimes called *system identification* or *symbolic regression*.

We begin with the five preparatory steps. The purpose of the first two preparatory steps is to specify the ingredients of the to-be-evolved program. Because the problem is to find a mathematical function of one independent variable, the terminal set (inputs to the to-be-evolved program) includes the independent variable, x. The terminal set also includes numerical constants.

That is, the terminal set, T, is

$$T = \{X, \Re\}$$

Here $\Re$ denotes constant numerical terminals in some reasonable range (say from -5.0 to $+5.0$).

The preceding statement of the problem is somewhat flexible in that it does not specify what functions may be employed in the to-be-evolved program. One possible choice for the function set consists of the four ordinary arithmetic functions of addition, subtraction, multiplication, and division. This choice is reasonable because mathematical expressions typically include these functions. Thus, the function set, F, for this problem is

$$F = \{+, -, *, \%\}$$

The two-argument $+$, $-$, $*$, and $\%$ functions add, subtract, multiply, and divide, respectively. To avoid run-time errors, the division function $\%$ is protected: it returns a value of 1 when division by 0 is attempted (including 0 divided by 0), but otherwise returns the quotient of its two arguments.

Each individual in the population is a composition of functions from the specified function set and terminals from the specified terminal set.

The third preparatory step involves constructing the fitness measure. The purpose of the fitness measure is to specify what the human wants. The high-level goal of this problem is to find a program whose output is equal to the values of the quadratic polynomial $x^2 + x + 1$. Therefore, the fitness assigned to a particular individual in the population for this problem must reflect how closely the output of an individual program comes to the target polynomial $x^2 + x + 1$. The fitness measure could be defined as the value of the integral (taken over values of the independent variable x between -1.0 and $+1.0$) of the absolute value of the differences (errors) between the value of the individual mathematical expression and the target quadratic polynomial $x^2 + x + 1$. A smaller value of fitness (error) is better. A fitness (error) of zero would indicate a perfect fit.

For most problems of symbolic regression or system identification, it is not practical or possible to analytically compute the value of the integral of the absolute error. Thus, in practice, the integral is numerically approximated using dozens or hundreds of different values of the independent variable x in the range between -1.0 and $+1.0$.

The population size in this small illustrative example will be just four. In actual practice, the population size for a run of genetic programming consists of thousands or millions of individuals. In actual practice, the crossover operation is commonly performed on about 90% of the individuals in the population; the reproduction operation is performed on about 8% of the population; the mutation operation is performed on about 1% of the population; and the

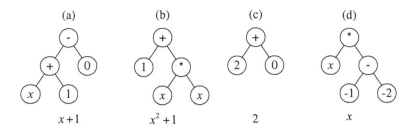

Figure 5.12. Initial population of four randomly created individuals of generation 0.

architecture-altering operations are performed on perhaps 1% of the population. Because this illustrative example involves an abnormally small population of only four individuals, the crossover operation will be performed on two individuals and the mutation and reproduction operations will each be performed on one individual. For simplicity, the architecture-altering operations are not used for this problem.

A reasonable termination criterion for this problem is that the run will continue from generation to generation until the fitness of some individual gets below 0.01. In this contrived example, the run will (atypically) yield an algebraically perfect solution (for which the fitness measure attains the ideal value of zero) after merely one generation.

Now that we have performed the five preparatory steps, the run of genetic programming can be launched. That is, the executional steps shown in the flowchart of Figure 5.4 are now performed.

Genetic programming starts by randomly creating a population of four individual computer programs. The four programs are shown in Figure 5.12 in the form of trees.

The first randomly constructed program tree (Figure 5.12(a)) is equivalent to the mathematical expression $x + 1$. A program tree is executed in a depth-first way, from left to right, in the style of the LISP programming language. Specifically, the addition function (+) is executed with the variable x and the constant value 1 as its two arguments. Then, the two-argument subtraction function (−) is executed. Its first argument is the value returned by the just-executed addition function. Its second argument is the constant value 0. The overall result of executing the entire program tree is thus $x + 1$.

The first program (Figure 5.12(a)) was constructed using the "Grow" method, by first choosing the subtraction function for the root (top point) of the program tree. The random construction process continued in a depth-first fashion (from left to right) and chose the addition function to be the first argument of the subtraction function. The random construction process then chose the terminal x to be the first argument of the addition function (thereby termi-

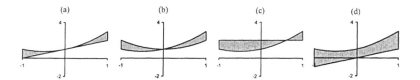

Figure 5.13. The fitness of each of the four randomly created individuals of generation 0 is equal to the area between two curves.

nating the growth of this path in the program tree). The random construction process then chose the constant terminal 1 as the second argument of the addition function (thereby terminating the growth along this path). Finally, the random construction process chose the constant terminal 0 as the second argument of the subtraction function (thereby terminating the entire construction process).

The second program (Figure 5.12(b)) adds the constant terminal 1 to the result of multiplying x by x and is equivalent to x^2+1. The third program (Figure 5.12(c)) adds the constant terminal 2 to the constant terminal 0 and is equivalent to the constant value 2. The fourth program (Figure 5.12(d)) is equivalent to x.

Randomly created computer programs will, of course, typically be very poor at solving the problem at hand. However, even in a population of randomly created programs, some programs are better than others. The four random individuals from generation 0 in Figure 5.12 produce outputs that deviate from the output produced by the target quadratic function $x^2 + x + 1$ by different amounts. In this particular problem, fitness can be graphically illustrated as the area between two curves. That is, fitness is equal to the area between the parabola $x^2 + x + 1$ and the curve representing the candidate individual. Figure 5.13 shows (as shaded areas) the integral of the absolute value of the errors between each of the four individuals in Figure 5.12 and the target quadratic function $x^2 + x + 1$. The integral of absolute error for the straight line $x + 1$ (the first individual) is 0.67 (Figure 5.13(a)). The integral of absolute error for the parabola $x^2 + 1$ (the second individual) is 1.0 (Figure 5.13(b)). The integrals of the absolute errors for the remaining two individuals are 1.67 (Figure 5.13(c)) and 2.67 (Figure 5.13(d)), respectively.

As can be seen in Figure 5.13, the straight line $x + 1$ (Figure 5.13(a)) is closer to the parabola $x^2 + x + 1$ in the range from -1 to $+1$ than any of its three cohorts in the population. This straight line is, of course, not equivalent to the parabola x^2+x+1. This best-of-generation individual from generation 0 is not even a quadratic function. It is merely the best candidate that happened to emerge from the blind random search of generation 0. In the valley of the blind, the one-eyed man is king.

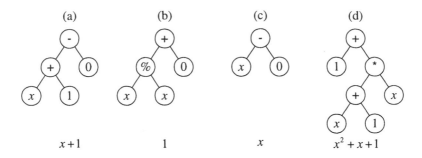

Figure 5.14. Population of generation 1 (after one reproduction, one mutation, and one two-offspring crossover operation).

After the fitness of each individual in the population is ascertained, genetic programming then probabilistically selects relatively more fit programs from the population. The genetic operations are applied to the selected individuals to create offspring programs. The most commonly employed methods for selecting individuals to participate in the genetic operations are tournament selection and fitness-proportionate selection. In both methods, the emphasis is on selecting relatively fit individuals. An important feature common to both methods is that the selection is not greedy. Individuals that are known to be inferior will be selected to a certain degree. The best individual in the population is not guaranteed to be selected. Moreover, the worst individual in the population will not necessarily be excluded. Anything can happen and nothing is guaranteed.

We first perform the reproduction operation. Because the first individual (Figure 5.12(a)) is the most fit individual in the population, it is very likely to be selected to participate in a genetic operation. Let us suppose that this particular individual is, in fact, selected for reproduction. If so, it is copied, without alteration, into the next generation (generation 1). This is shown in Figure 5.14(a) as part of the population of the new generation.

We next perform the mutation operation. Because selection is probabilistic, it is possible that the third best individual in the population (Figure 5.12(c)) is selected. One of the three nodes of this individual is then randomly picked as the site for the mutation. In this example, the constant terminal 2 is picked as the mutation site. This program is then randomly mutated by deleting the entire subtree rooted at the picked point (in this case, just the constant terminal 2) and inserting a subtree that is randomly grown in the same way that the individuals of the initial random population were originally created. In this particular instance, the randomly grown subtree computes the quotient of x and x using the protected division operation %. The resulting individual is shown in Figure 5.14(b). This particular mutation changes the original individual from one having a constant value of 2 into one having a constant value of 1. This particular mutation improves fitness from 1.67 to 1.00.

Finally, we perform the crossover operation. Because the first and second individuals in generation 0 are both relatively fit, they are likely to be selected to participate in crossover. The selection (and reselection) of relatively more fit individuals and the exclusion and extinction of unfit individuals is a characteristic feature of Darwinian selection. The first and second programs are mated sexually to produce two offspring (using the two-offspring version of the crossover operation). One point of the first parent (Figure 5.12(a)), namely the + function, is randomly picked as the crossover point for the first parent. One point of the second parent (Figure 5.12(b)), namely its leftmost terminal x, is randomly picked as the crossover point for the second parent. The crossover operation is then performed on the two parents. The two offspring are shown in Figures 5.14(c) and 5.14(d). One of the offspring (Figure 5.14(c)) is equivalent to x and is not noteworthy. However, the other offspring (Figure 5.14(d)) is equivalent to $x^2 + x + 1$ and has a fitness (integral of absolute errors) of zero. Because the fitness of this individual is below 0.01, the termination criterion for the run is satisfied and the run is automatically terminated. This best-so-far individual (Figure 5.14(d)) is designated as the result of the run. This individual is an algebraically correct solution to the problem.

Note that the best-of-run individual (Figure 5.14(d)) incorporates a good trait (the quadratic term x^2) from the second parent (Figure 5.12(b)) with two other good traits (the linear term x and constant term of 1) from the first parent (Figure 5.12(a)). The crossover operation produced a solution to this problem by recombining good traits from these two relatively fit parents into a superior (indeed, perfect) offspring.

In summary, genetic programming has, in this example, automatically created a computer program whose output is equal to the values of the quadratic polynomial $x^2 + x + 1$ in the range from -1 to $+1$.

5.5 FURTHER FEATURES OF GENETIC PROGRAMMING

Various advanced features of genetic programming are not covered by the foregoing illustrative problem and the foregoing discussion of the preparatory and executional steps of genetic programming.

5.5.1 Constrained Syntactic Structures

For certain simple problems (such as the illustrative problem above), the search space for a run of genetic programming consists of the unrestricted set of possible compositions of the problem's functions and terminals. However, for many problems, a constrained syntactic structure imposes restrictions on how the functions and terminals may be combined. Consider, for example, a function that instructs a robot to turn by a certain angle. In a typical implementation of this hypothetical function, the function's first argument may be required to return a numerical value (representing the desired turning angle) and its second argument may be required to be a follow-up command (e.g., move, turn, stop). In other words, the functions and terminals permitted in the two argument subtrees for this particular function are restricted. These restrictions are implemented by means of syntactic rules of construction.

A *constrained syntactic structure* (sometimes called *strong typing*) is a grammar that specifies the functions or terminals that are permitted to appear as a specified argument of a specified function in the program tree.

When a constrained syntactic structure is used, there are typically multiple function sets and multiple terminal sets. The rules of construction specify where the different function sets or terminal sets may be used.

When a constrained syntactic structure is used, all the individuals in the initial random population (generation 0) are created so as to comply with the constrained syntactic structure. All genetic operations (i.e. crossover, mutation, reproduction, and the architecture-altering operations) that are performed during the run are designed to produce offspring that comply with the requirements of the constrained syntactic structure. Thus, all individuals (including, in particular, the best-of-run individual) that are produced during the run of genetic programming will necessarily comply with the requirements of the constrained syntactic structure.

5.5.2   Automatically Defined Functions

Human computer programmers organize sequences of reusable steps into subroutines. They then repeatedly invoke the subroutines—typically with different instantiations of the subroutine's dummy variables (formal parameters). Reuse eliminates the need to "reinvent the wheel" on each occasion when a particular sequence of steps may be useful. Reuse makes it possible to exploit a problem's modularities, symmetries, and regularities (and thereby potentially accelerate the problem-solving process).

Programmers commonly organize their subroutines into hierarchies.

The automatically defined function (ADF) is one of the mechanisms by which genetic programming implements the parametrized reuse and hierarchical invocation of evolved code. Each ADF resides in a separate function-defining branch within the overall multi-part computer program (see Figure 5.3). When ADFs are being used, a program consists of one (or more) function-defining branches (i.e. ADFs) as well as one or more main result-producing branches. An ADF may possess zero, one, or more dummy variables (formal parameters). The body of an ADF contains its work-performing steps. Each ADF belongs to a particular program in the population. An ADF may be called by the program's main result-producing branch, another ADF, or another type of branch (such as those described below). Recursion is sometimes allowed. Typically, the ADFs are invoked with different instantiations of their dummy variables.

The work-performing steps of the program's main result-producing branch and the work-performing steps of each ADF are automatically and simultaneously created during the run of genetic programming.

The program's main result-producing branch and its ADFs typically have different function and terminal sets. A constrained syntactic structure is used to implement ADFs.

Automatically defined functions are the focus of *Genetic Programming II: Automatic Discovery of Reusable Programs* (Koza, 1994a) and the videotape *Genetic Programming II Videotape: The Next Generation* (Koza, 1994b).

5.5.3 Automatically Defined Iterations, Loops, Recursions and Stores

Automatically defined iterations, automatically defined loops, and automatically defined recursions provide means (in addition to ADFs) to reuse code.

Automatically defined stores provide means to reuse the result of executing code.

Automatically defined iterations, automatically defined loops, automatically defined recursions, and automatically defined stores are described in *Genetic Programming III: Darwinian Invention and Problem Solving* (Koza et al., 1999a).

5.5.4 Program Architecture and Architecture-Altering Operations

The architecture of a program consists of

1 the total number of branches,

2 the type of each branch (e.g., result-producing branch, automatically defined function, automatically defined iteration, automatically defined loop, automatically defined recursion, or automatically defined store),

3 the number of arguments (if any) possessed by each branch, and

4 if there is more than one branch, the nature of the hierarchical references (if any) allowed among the branches.

There are three ways by which genetic programming can arrive at the architecture of the to-be-evolved computer program:

1 The human user may prespecify the architecture of the overall program (i.e. perform an additional architecture-defining preparatory step). That is, the number of preparatory steps is increased from the five previously itemized to six.

2 The run may employ evolutionary selection of the architecture (as described in Koza, 1994a), thereby enabling the architecture of the overall program to emerge from a competitive process during the run of genetic programming. When this approach is used, the number of preparatory steps remains at the five previously itemized.

3 The run may employ the architecture-altering operations (Koza, 1994c, 1995; Koza et al., 1999a), thereby enabling genetic programming to automatically create the architecture of the overall program dynamically during the run. When this approach is used, the number of preparatory steps remains at the five previously itemized.

5.5.5   Genetic Programming Problem Solver

The Genetic Programming Problem Solver (GPPS) is described in Koza et al. (1999a, Part 4).

If GPPS is being used, the user is relieved of performing the first and second preparatory steps (concerning the choice of the terminal set and the function set). The function set for GPPS consists of the four basic arithmetic functions (addition, subtraction, multiplication, and division) and a conditional operator (i.e. functions found in virtually every general-purpose digital computer that has ever been built). The terminal set for GPPS consists of numerical constants and a set of input terminals that are presented in the form of a vector.

By employing this generic function set and terminal set, GPPS reduces the number of preparatory steps from five to three.

GPPS relies on the architecture-altering operations to dynamically create, duplicate, and delete subroutines and loops during the run of genetic programming. Additionally, in version 2.0 of GPPS, the architecture-altering operations are used to dynamically create, duplicate, and delete recursions and internal storage. Because the architecture of the evolving program is automatically determined during the run, GPPS eliminates the need for the user to specify in advance whether to employ subroutines, loops, recursions, and internal storage in solving a given problem. It similarly eliminates the need for the user to specify the number of arguments possessed by each subroutine. And, GPPS eliminates the need for the user to specify the hierarchical arrangement of the invocations of the subroutines, loops, and recursions. That is, the use of GPPS relieves the user of performing the preparatory step of specifying the program's architecture.

Table 5.1. Eight criteria for saying that an automatically created result is human-competitive.

	Criterion
A	The result was patented as an invention in the past, is an improvement over a patented invention, or would qualify today as a patentable new invention.
B	The result is equal to or better than a result that was accepted as a new scientific result at the time when it was published in a peer-reviewed scientific journal.
C	The result is equal to or better than a result that was placed into a database or archive of results maintained by an internationally recognized panel of scientific experts.
D	The result is publishable in its own right as a new scientific result—independent of the fact that the result was mechanically created.
E	The result is equal to or better than the most recent human-created solution to a long-standing problem for which there has been a succession of increasingly better human-created solutions.
F	The result is equal to or better than a result that was considered an achievement in its field at the time it was first discovered.
G	The result solves a problem of indisputable difficulty in its field.
H	The result holds its own or wins a regulated competition involving human contestants (in the form of either live human players or human-written computer programs).

5.5.6   Developmental Genetic Programming

Developmental genetic programming is used for problems of synthesizing analog electrical circuits, as described in Part 5 of Koza et al. (1999a). When developmental genetic programming is used, a complex structure (such as an electrical circuit) is created from a simple initial structure (the embryo).

5.6   HUMAN-COMPETITIVE RESULTS PRODUCED BY GENETIC PROGRAMMING

Samuel's statement (quoted above) reflects the goal articulated by the pioneers of the 1950s in the fields of artificial intelligence and machine learning, namely to use computers to automatically produce human-like results. Indeed, getting machines to produce human-like results is *the* reason for the existence of the fields of artificial intelligence and machine learning.

Table 5.2. Thirty-six instances of human-competitive results produced by genetic programming.

	Claimed instance	Basis for claim of human-competit-iveness	Reference
1	Creation of a better-than-classical quantum algorithm for the Deutsch–Jozsa "early promise" problem	B, F	Spector et al., 1998
2	Creation of a better-than-classical quantum algorithm for Grover's database search problem	B, F	Spector et al., 1999a
3	Creation of a quantum algorithm for the depth-two AND/OR query problem that is better than any previously published result	D	Spector et al., 1999b; Barnum et al., 2000
4	Creation of a quantum algorithm for the depth-one OR query problem that is better than any previously published result	D	Barnum et al., 2000
5	Creation of a protocol for communicating information through a quantum gate that was previously thought not to permit such communication	D	Spector and Bernstein, 2002
6	Creation of a novel variant of quantum dense coding	D	Spector and Bernstein, 2002
7	Creation of a soccer-playing program that won its first two games in the Robo Cup 1997 competition	H	Luke, 1998
8	Creation of a soccer-playing program that ranked in the middle of the field of 34 human-written programs in the Robo Cup 1998 competition	H	Andre and Teller, 1999
9	Creation of four different algorithms for the transmembrane segment identification problem for proteins	B, E	Koza et al., 1994a, 1999
10	Creation of a sorting network for seven items using only 16 steps	A, D	Koza et al., 1999
11	Rediscovery of the Campbell ladder topology for low-pass and highpass filters	A, F	Koza et al., 1999, 2003
12	Rediscovery of the Zobel "M-derived half section" and "constant K" filter sections	A, F	Koza et al., 1999
13	Rediscovery of the Cauer (elliptic) topology for filters	A, F	Koza et al., 1999
14	Automatic decomposition of the problem of synthesizing a crossover filter	A, F	Koza et al., 1999
15	Rediscovery of a recognizable voltage gain stage and a Darlington emitter-follower section of an amplifier and other circuits	A, F	Koza et al., 1999
16	Synthesis of 60 and 96 decibel amplifiers	A, F	Koza et al., 1999
17	Synthesis of analog computational circuits for squaring, cubing, square root, cube root, logarithm, and Gaussian functions	A, D, G	Koza et al., 1999
18	Synthesis of a real-time analog circuit for time-optimal control of a robot	G	Koza et al., 1999
19	Synthesis of an electronic thermometer	A, G	Koza et al., 1999

Table 5.2. (Continued)

Claimed instance	Basis for claim of human-competit-iveness	Reference
20 Synthesis of a voltage reference circuit	A, G	Koza et al., 1999
21 Creation of a cellular automata rule for the majority classification problem that is better than the Gacs–Kurdyumov–Levin rule and all other known rules written by humans	D, E	Koza et al., 1999
22 Creation of motifs that detect the D–E–A–D box family of proteins and the manganese superoxide dismutase family	C	Koza et al., 1999
23 Synthesis of topology for a PID-D2 (proportional, integrative, derivative, and second derivative) controller	A, F	Koza et al., 2003
24 Synthesis of an analog circuit equivalent to Philbrick circuit	A, F	Koza et al., 2003
25 Synthesis of NAND circuit	A, F	Koza et al., 2003
26 Simultaneous synthesis of topology, sizing, placement, and routing of analog electrical circuits	G	Koza et al., 2003
27 Synthesis of topology for a PID (proportional, integrative, and derivative) controller	A, F	Koza et al., 2003
28 Rediscovery of negative feedback	A, E, F, G	Koza et al., 2003
29 Synthesis of a low-voltage balun circuit	A	Koza et al., 2003
30 Synthesis of a mixed analog–digital variable capacitor circuit	A	Koza et al., 2003
31 Synthesis of a high-current load circuit	A	Koza et al., 2003
32 Synthesis of a voltage–current conversion circuit	A	Koza ct al., 2003
33 Synthesis of a cubic signal generator	A	Koza et al., 2003
34 Synthesis of a tunable integrated active filter	A	Koza et al., 2003
35 Creation of PID tuning rules that outperform the Ziegler–Nichols and Astrom–Hagglund tuning rules	A, B, D, E, F, G	Koza et al., 2003
36 Creation of three non-PID controllers that outperform a PID controller that uses the Ziegler–Nichols or Astrom–Hagglund tuning rules	A, B, D, E, F, G	Koza et al., 2003

To make the notion of human-competitiveness more concrete, we say that a result is "human-competitive" if it satisfies one or more of the eight criteria in Table 5.1.

As can be seen from Table 5.1, the eight criteria have the desirable attribute of being at arms-length from the fields of artificial intelligence, machine learning, and genetic programming. That is, a result cannot acquire the rating of "human competitive" merely because it is endorsed by researchers *inside* the specialized fields that are attempting to create machine intelligence. Instead, a result produced by an automated method must earn the rating of "human competitive" independent of the fact that it was generated by an automated method.

Table 5.2 lists the 36 human-competitive instances (of which we are aware) where genetic programming has produced human-competitive results. Each entry in the table is accompanied by the criteria (from Table 5.1) that establish the basis for the claim of human-competitiveness.

There are now 23 instances where genetic programming has duplicated the functionality of a previously patented invention, infringed a previously patented invention, or created a patentable new invention (see criterion A in Table 5.1). Specifically, there are 15 instances where genetic programming has created an entity that either infringes or duplicates the functionality of a previously patented twentieth-century invention, six instances where genetic programming has done the same with respect to an invention patented after January 1st, 2000, and two instances where genetic programming has created a patentable new invention. The two new inventions are general-purpose controllers that outperform controllers employing tuning rules that have been in widespread use in industry for most of the twentieth century.

5.7   SOME PROMISING AREAS FOR FUTURE APPLICATION

Since its early beginnings, the field of genetic and evolutionary computation has produced a cornucopia of results.

Genetic programming and other methods of genetic and evolutionary computation may be especially productive in areas having some or all of the following characteristics:

- where the inter-relationships among the relevant variables are unknown or poorly understood (or where it is suspected that the current understanding may possibly be wrong),

- where finding the size and shape of the ultimate solution to the problem is a major part of the problem,

- where large amounts of primary data requiring examination, classification, and integration are accumulating in computer readable form,

- where there are good simulators to test the performance of tentative solutions to a problem, but poor methods to directly obtain good solutions,

- where conventional mathematical analysis does not, or cannot, provide analytic solutions,

- where an approximate solution is acceptable (or is the only result that is ever likely to be obtained), or

- where small improvements in performance are routinely measured (or easily measurable) and highly prized.

5.8 GENETIC PROGRAMMING THEORY

Genetic programming is a search technique that explores the space of computer programs. As discussed above, the search for solutions to a problem starts from a group of points (random programs) in this search space. Those points that are of above average quality are then used to generate a new generation of points through crossover, mutation, reproduction and possibly other genetic operations. This process is repeated over and over again until a termination criterion is satisfied.

If we could visualize this search, we would often find that initially the population looks a bit like a cloud of randomly scattered points, but that, generation after generation, this cloud changes shape and moves in the search space following a well defined trajectory. Because genetic programming is a stochastic search technique, in different runs we would observe different trajectories. These, however, would very likely show very clear regularities to our eye that could provide us with a deep understanding of how the algorithm is searching the program space for the solutions to a given problem. We could probably readily see, for example, why genetic programming is successful in finding solutions in certain runs and with certain parameter settings, and unsuccessful in/with others.

Unfortunately, it is normally impossible to exactly visualize the program search space due to its high dimensionality and complexity, and so we cannot just use our senses to understand and predict the behavior of genetic programming.

In this situation, one approach to gain an understanding of the behavior of a genetic programming system is to perform many real runs and record the variations of certain numerical descriptors (like the average fitness or the average size of the programs in the population at each generation, the average difference between parent and offspring fitness, etc). Then, one can try to hypothesize explanations about the behavior of the system that are compatible with (and could explain) the empirical observations.

This exercise is very error prone, though, because a genetic programming system is a complex adaptive system with zillions of degrees of freedom. So, any small number of statistical descriptors is likely to be able to capture only a tiny fraction of the complexities of such a system. This is why in order to understand and predict the behavior of genetic programming (and indeed of most other evolutionary algorithms) in precise terms we need to define and then study mathematical models of evolutionary search.

Schema theories are among the oldest, and probably the best-known classes of models of evolutionary algorithms. A *schema* (plural, schemata) is a set of points in the search space sharing some syntactic feature. Schema theories provide information about the properties of individuals of the population belonging to any schema at a given generation in terms of quantities measured at the previous generation, without having to actually run the algorithm.

For example, in the context of genetic algorithms operating on binary strings, a schema is, syntactically, a string of symbols from the alphabet {0, 1, * }, like *10*1. The character * is interpreted as a "don't care" symbol, so that, semantically, a schema represents a set of bit strings. For example the schema *10*1 represents a set of four strings: {01001, 01011, 11001, 11011}.

Typically, schema theorems are descriptions of how the number (or the proportion) of members of the population belonging to (or matching) a schema varies over time.

For a given schema H the selection/crossover/mutation process can be seen as a Bernoulli trial, because a newly created individual either samples or does not sample H. Therefore, the number of individuals sampling H at the next generation, $m(H, t+1)$ is a binomial stochastic variable. So, if we denote with $\alpha(H, t)$ the success probability of each trial (i.e. the probability that a newly created individual samples H), an *exact schema theorem* is simply

$$E[m(H, t + 1)] = M\alpha(H, t)$$

where M is the population size and $E[\,.\,]$ is the expectation operator. Holland's and other approximate schema theories (Holland, 1975; Goldberg, 1989; Whitley, 1994) normally provide a lower bound for $\alpha(H, t)$ or, equivalently, for $E[m(H, t+1)]$. For example, several schema theorems for one-point crossover and point mutation have the following form:

$$\alpha(H, t) \geq p(H, t)(1 - p_m)^{O(H)} \left[1 - p_c \frac{L(H)}{N - 1} \sigma \right]$$

where $m(H, t)$ is number of individuals in the schema H at generation t, M is the population size, $p(H, t)$ is the selection probability for strings in H at generation t, p_m is the mutation probability, $O(H)$ is the schema order, i.e. number of defining bits, p_c is the crossover probability, $L(H)$ is the defining length, i.e. distance between the furthest defining bits in H, and N is the bitstring length. The factor σ differs in the different formulation of the schema theorem: $\sigma = 1 - m(H, t)/M$ in Holland (1975), where one of the parents was chosen randomly, irrespective of fitness; $\sigma = 1$ in Goldberg (1989); and $\sigma = 1 - p(H, t)$ in Whitley (1994).

More recently, Stephens and Waelbroeck (1997, 1999) have produced exact formulations for $\alpha(H, t)$, which are now known as *"exact" schema theorems* for genetic algorithms. These, however, are beyond the scope of this chapter.

The theory of schemata in genetic programming has had a slow start, one of the difficulties being that the variable size tree structure in genetic programming makes it more difficult to develop a definition of genetic programming schema having the necessary power and flexibility. Several alternatives have been proposed in the literature, which define schemata as composed of one or multiple trees or fragments of trees. Here, however, we will focus only on a particular one, which was proposed by Poli and Langdon (1997, 1998) since this has later been used to develop an exact and general schema theory for genetic programming (Poli and McPhee, 2001; Langdon and Poli, 2002).

In this definition, syntactically, a *genetic programming schema* is a tree with some "don't care" nodes which represents exactly one primitive function or terminal. Semantically, a schema represents all programs that match its size, shape and defining (non-"don't care") nodes. For example, the schema H = (DON'T CARE x(+ y DON'T CARE)) represents the programs (+ x (+ y x)), (+ x (+ y y)), (* x (+ y x)), etc.

The exact schema theorem in Poli and McPhee (2001) gives the expected proportion of individuals matching a schema in the next generation as a function of information about schemata in the current generation. The calculation is non-trivial, but it is easier than one might think.

Let us assume, for simplicity, that only reproduction and (one-offspring) crossover are performed. Because these two operators are mutually exclusive, for a generic schema H we then have

$$\alpha(H, t) \quad = \quad \Pr\left[\text{an individual in } H \text{ is obtained via reproduction}\right]$$
$$+ \Pr\left[\text{an offspring matching } H \text{ is produced by crossover}\right]$$

Then, assuming that reproduction is performed with probability p_r and crossover with probability p_c (with $p_r + p_c = 1$), we obtain

$$\alpha(H, t) \quad = \quad p_r \times \Pr\left[\text{an individual in } H \text{ is selected for cloning}\right]$$
$$+ p_c \Pr\left[\begin{array}{l}\text{the parents and the crossover points} \\ \text{are such that the offspring matches } H\end{array}\right]$$

Clearly, the first probability in this expression is simply the selection probability for members of the schema H as dictated by, say, fitness-proportionate selection or tournament selection. So,

$$\Pr\left[\text{selecting an individual in } H \text{for cloning}\right] = p(H, t)$$

We now need to calculate the second term in $\alpha(H, t)$: that is, the probability that the parents have shapes and contents compatible with the creation of an offspring matching H, and that the crossover points in the two parents are such that exactly the necessary material to create such an offspring is swapped. This is the harder part of the calculation.

An observation that helps simplify the problem is that, although the probability of choosing a particular crossover point in a parent depends on the actual size and shape of such a parent, the process of crossover point selection is independent from the actual primitives present in the parent tree. So, for example, the probability of choosing any crossover point in the program $(+ \ x \ (+ \ y \ x))$ is identical to the probability of choosing any crossover point in the program $(\text{AND} \ \text{D1} \ (\text{OR} \ \text{D1} \ \text{D2}))$. This is because the two programs have exactly the same shape. Thanks to this observation we can write

$$
\Pr \left[\begin{array}{l} \text{the parents and the crossover points} \\ \text{are such that the offspring matches } H \end{array} \right]
$$

$$
= \underset{\substack{\text{For all pairs of} \\ \text{parent shapes } k, l}}{\sum} \quad \underset{\substack{\text{For all crossover} \\ \text{points } i, j \text{ in} \\ \text{shapes } k \text{ and } l}}{\sum} \Pr \left[\begin{array}{l} \text{Choosing crossover points} \\ i \text{ and } j \text{ in shapes } k \text{ and } l \end{array} \right]
$$

$$
\times \Pr \left[\begin{array}{l} \text{Selecting parents with shapes } k \text{ and } l, \text{ such that if} \\ \text{crossed at points } i \text{ and } j \text{ produce an offspring in } H \end{array} \right]
$$

If, for simplicity, we assume that crossover points are selected with uniform probability, then

$$
\Pr \left[\begin{array}{l} \text{Choosing crossover points} \\ i \text{ and } j \text{ in shapes } k \text{ and } l \end{array} \right] = \frac{1}{\text{nodes in shape } k} \times \frac{1}{\text{nodes in shape } l}
$$

So, we are left with the problem of calculating the probability of selecting (for crossover) parents having specific shapes while at the same time having an arrangement of primitives such that, if crossed over at certain predefined points, they produce an offspring matching a particular schema of interest.

Again, here we can simplify the problem by considering how crossover produces offspring: it excises a subtree rooted at the chosen crossover point in a parent, and replaces it with a subtree excised from the chosen crossover point in the other parent. This means that the offspring will have the right shape and primitives to match the schema of interest if and only if, after the excision of the chosen subtree, the first parent has shape and primitives compatible with the schema, and the subtree to be inserted has shape and primitives compatible with the schema. That is,

$$
\Pr \left[\begin{array}{l} \text{Selecting parents with shapes } k \text{ and } l, \text{ such that if} \\ \text{crossed over at points } i \text{ and } j \text{ produce an offspring in } H \end{array} \right]
$$

$$
= \Pr \left[\begin{array}{l} \text{Selecting a root-donating parent with shape } k \\ \text{such that its upper part w.r.t. crossover} \\ \text{point } i \text{ matches the upper part of } H \text{ w.r.t. } i \end{array} \right]
$$

$$
\times \Pr \left[\begin{array}{l} \text{Selecting a subtree-donating parent with shape } l \\ \text{such that its lower part w.r.t. crossover} \\ \text{point } j \text{ matches the lower part of } H \text{ w.r.t. } i \end{array} \right]
$$

These two selection probabilities can be calculated exactly. However, the calculation requires the introduction of several other concepts and notation, which are beyond the introductory nature of this chapter. These definitions, the complete theory and a number of examples and applications can be found in Poli (2001), Langdon and Poli (2002), and Poli and McPhee (2003a, 2003b).

Although exact schema theoretic models of genetic programming have become available only very recently, they have already started shedding some light on fundamental questions regarding the how and why genetic programming works. Importantly, other important theoretical models of genetic programming have recently been developed which add even more to our theoretical understanding of genetic programming. These, however, go well beyond the scope of this chapter. The interested reader should consult Langdon and Poli (2002) and Poli and McPhee (2003a, 2003b) for more information.

5.9   TRICKS OF THE TRADE

Newcomers to the field of genetic programming often ask themselves (and/or other more experienced genetic programmers) questions such as the following:

1 What is the best way to get started with genetic programming? Which papers should I read?

2 Should I implement my own genetic programming system or should I use an existing package? If so, what package should I use?

Let us start with the first question. A variety of sources of information about genetic programming are available (many of which are listed in the following section). Consulting information available on the Web is certainly a good way to get quick answers for a newcomer who wants to know what genetic programming is. The answer, however, will often be too shallow for someone who really wants to apply genetic programming to solve practical problems. People in this position should probably invest some time going through more detailed accounts such as Koza (1992), Banzhaf et al. (1998a) and Langdon and Poli (2002), or some of the monographs listed in the following section. Technical papers may be the next stage. The literature on genetic programming is now quite extensive. So, although this is easily accessible thanks to the complete online bibliography listed in the next section, newcomers will often need to be selective in what they read, at least initially. The objective here may be different for different types of readers. Practitioners should probably identify and read only papers which deal with the problem they are interested in. Researchers and Ph.D. students interested in developing a deeper understanding of genetic programming should also make sure they identify and read as many seminal papers as possible, including papers or books on empirical and theoretical studies on the inner mechanisms and behavior of genetic programming. These are frequently cited in other papers and so can easily be identified.

The answer to the second question depends on the particular experience and background of the questioner. Implementing a simple genetic programming system from scratch is certainly an excellent way to make sure one really understands the mechanics of genetic programming. In addition to being an exceptionally useful exercise, this will always result in programmers knowing their systems so well that they will have no problems customizing them for specific purposes (e.g., by adding new, application specific genetic operators, implementing unusual, knowledge-based initialization strategies, etc). All of this, however, requires reasonable programming skills and the will to thoroughly test the resulting system until it fully behaves as expected. If the skills or the time are not available, then the best way to get a working genetic programming application is to retrieve one of the many public-domain genetic programming implementations and adapt this for the user's purposes. This process is faster, and good implementations are often quite robust, efficient, well-documented and comprehensive. The small price to pay is the need to study the available documentation and examples. These often explain also how to modify the genetic programming system to some extent. However, deeper modifications (such as the introduction of new or unusual operators) will often require studying the actual source code of the system and a substantial amount of trial and error. Good, publicly-available GP implementations include LIL-GP from Bill Punch, ECJ from Sean Luke and DGPC from David Andre.

5.10   CONCLUSIONS

In his seminal 1948 paper entitled *Intelligent Machinery*, Turing identified three ways by which human-competitive machine intelligence might be achieved. In connection with one of those ways, Turing (1948) said:

> There is the genetical or evolutionary search by which a combination of genes is looked for, the criterion being the survival value.

Turing did not specify how to conduct the "genetical or evolutionary search" for machine intelligence. In particular, he did not mention the idea of a population-based parallel search in conjunction with sexual recombination (crossover) as described in John Holland's 1975 book *Adaptation in Natural and Artificial Systems*. However, in his 1950 paper *Computing Machinery and Intelligence*, Turing did point out that

> We cannot expect to find a good child-machine at the first attempt. One must experiment with teaching one such machine and see how well it learns. One can then try another and see if it is better or worse. There is an obvious connection between this process and evolution, by the identifications...
>
> Structure of the child machine = Hereditary material
>
> Changes of the child machine = Mutations
>
> Natural selection = Judgment of the experimenter.

That is, Turing perceived in 1948 and 1950 that one possibly productive approach to machine intelligence would involve an evolutionary process in which a description of a computer program (the hereditary material) undergoes progressive modification (mutation) under the guidance of natural selection (i.e. selective pressure in the form of what we now call "fitness").

Today, many decades later, we can see that indeed Turing was right. Genetic programming has started fulfilling Turing's dream by providing us with a systematic method, based on Darwinian evolution, for getting computers to automatically solve hard real-life problems. To do so, it simply requires a high-level statement of what needs to be done (and enough computing power).

Turing also understood the need to evaluate objectively the behavior exhibited by machines, to avoid human biases when assessing their intelligence. This led him to propose an imitation game, now know as the *Turing test for machine intelligence*, whose goals are wonderfully summarized by Arthur Samuel's position statement quoted in the introduction to this chapter.

At present, genetic programming is certainly not in a position to produce computer programs that would pass the full Turing test for machine intelligence, and it might not be ready for this immense task for centuries. Nonetheless, thanks to the constant technological improvements in genetic programming technology, in its theoretical foundations and in computing power, genetic programming has been able to solve tens of difficult problems with human-competitive results (see Table 5.2) in the recent past. These are a small step towards fulfilling Turing and Samuel's dreams, but they are also early signs of things to come. It is, indeed, arguable that in a few years' time genetic programming will be able to *routinely* and *competently* solve important problems for us in a variety of specific domains of application, even when running on a single personal computer, thereby becoming an essential collaborator for many human activities. This, we believe, will be a remarkable step forward towards achieving true, human-competitive machine intelligence.

SOURCES OF ADDITIONAL INFORMATION

Sources of information about genetic programming include the following.

- *Genetic Programming: On the Programming of Computers by Means of Natural Selection* (Koza, 1992) and the accompanying videotape *Genetic Programming: The Movie* (Koza and Rice, 1992).

- *Genetic Programming II: Automatic Discovery of Reusable Programs* (Koza, 1994a) and the accompanying videotape *Genetic Programming II Videotape: The Next Generation* (Koza, 1994b).

- *Genetic Programming III: Darwinian Invention and Problem Solving* (Koza et al., 1999a) and the accompanying videotape *Genetic Programming III Videotape: Human-Competitive Machine Intelligence* (Koza et al., 1999b).

- *Genetic Programming IV. Routine Human-Competitive Machine Intelligence* (Koza et al., 2003);

- *Genetic Programming: An Introduction* (Banzhaf et al., 1998a).

- *Genetic Programming and Data Structures: Genetic Programming + Data Structures = Automatic Programming!* (Langdon, 1998) in the series on genetic programming from Kluwer.

- *Automatic Re-engineering of Software Using Genetic Programming* (Ryan, 1999) in the series on genetic programming from Kluwer.

- *Data Mining Using Grammar Based Genetic Programming and Applications* (Wong and Leung, 2000) in the series on genetic programming from Kluwer.

- *Principia Evolvica: Simulierte Evolution mit Mathematica* (Jacob, 1997, in German) and *Illustrating Evolutionary Computation with Mathematica* (Jacob, 2001).

- *Genetic Programming* (Iba, 1996, in Japanese).

- *Evolutionary Program Induction of Binary Machine Code and Its Application* (Nordin, 1997).

- *Foundations of Genetic Programming* (Langdon and Poli, 2002).

- *Emergence, Evolution, Intelligence: Hydroinformatics* (Babovic, 1996).

- *Theory of Evolutionary Algorithms and Application to System Synthesis* (Blickle, 1997).

- edited collections of papers such as the three *Advances in Genetic Programming* books from the MIT Press (Kinnear, 1994; Angeline and Kinnear, 1996; Spector et al., 1999a).

- Proceedings of the Genetic Programming Conference (Koza et al., 1996, 1997, 1998).

- Proceedings of the Annual Genetic and Evolutionary Computation Conference (GECCO) (combining the formerly annual Genetic Programming Conference and the formerly biannual International Conference on Genetic Algorithms) operated by the International Society for Genetic and Evolutionary Computation (ISGEC) and held starting in 1999 (Banzhaf et al., 1999; Whitley et al., 2000; Spector et al., 2001; Langdon et al., 2002).

- Proceedings of the Annual Euro-GP Conferences held starting in 1998 (Banzhaf et al., 1998b; Poli et al., 1999, 2000; Miller et al., 2001; Foster et al., 2002).

- Proceedings of the Workshop of Genetic Programming Theory and Practice organized by the Centre for Study of Complex Systems of the University of Michigan (to be published by Kluwer).

- The *Genetic Programming and Evolvable Machines* journal (from Kluwer) started in April 2000.

- Web sites such as www.genetic-programming.org and www.genetic-programming.com.

- LISP code for implementing genetic programming, available in Koza (1992), and genetic programming implementations in other languages such as C, C++, or Java (web sites such as www.genetic-programming.org contain links to computer code in various programming languages).

- Early papers on genetic programming, such as the Stanford University Computer Science Department Technical Report *Genetic Programming: A Paradigm for Genetically Breeding Populations of Computer Programs to Solve Problems* (Koza, 1990) and the paper *Hierarchical Genetic Algorithms Operating on Populations of Computer Programs*, presented at the 11th International Joint Conference on Artificial Intelligence in Detroit (Koza, 1989).

- An annotated bibliography of the first 100 papers on genetic programming (other than those of which John Koza was the author or co-author) in Appendix F of *Genetic Programming II: Automatic Discovery of Reusable Programs* (Koza, 1994a).

- Langdon's bibliography at http://www.cs.bham.ac.uk/wbl/biblio/ or http://liinwww.ira.uka.de/bibliography/Ai/genetic.programming.html. This bibliography is the most extensive in the field of genetic programming and contains over 3034 papers (as of January 2003) and over 880 authors. It provides on-line access to many of the papers.

References

Andre, D. and Teller, A., 1999, Evolving team Darwin United, in: *RoboCup-98: Robot Soccer World Cup II,* M. Asada, and H. Kitano, ed., Lecture Notes in Computer Science, Vol. 1604, Springer, Berlin, pp. 346–352.

Angeline, P. J. and Kinnear Jr, K. E., eds, 1996, *Advances in Genetic Programming 2,* MIT Press, Cambridge, MA.

Babovic, V., 1996, *Emergence, Evolution, Intelligence: Hydroinformatics,* Balkema, Rotterdam.

Banzhaf, W., Daida, J., Eiben, A. E., Garzon, M. H., Honavar, V., Jakiela, M. and Smith, R. E., eds, 1999, *GECCO-99: Proc. Genetic and Evolutionary Computation Conf.* (Orlando, FL), Morgan Kaufmann, San Mateo, CA.

Banzhaf, W., Nordin, P., Keller, R. E. and Francone, F. D., 1998a, *Genetic Programming: An Introduction,* Morgan Kaufmann, San Mateo, CA.

Banzhaf, W., Poli, R., Schoenauer, M. and Fogarty, T. C., 1998b, *Genetic Programming: Proc. 1st Eur. Workshop* (Paris), Lecture Notes in Computer Science. Vol. 1391, Springer, Berlin.

Barnum, H., Bernstein, H. J., and Spector, L., 2000, Quantum circuits for OR and AND of ORs, *J. Phys. A: Math. Gen.* **33**:8047–8057.

Blickle, T., 1997, *Theory of Evolutionary Algorithms and Application to System Synthesis,* TIK-Schriftenreihe Nr. 17. Zurich, Switzerland: vdf Hochschul, AG an der ETH, Zurich.

Foster, J. A., Lutton, E., Miller, J., Ryan, C. and Tettamanzi, A. G. B., eds, 2002, *Genetic Programming: Proc. 5th Eur. Conf., EuroGP 2002* (Kinsale, Ireland).

Goldberg, D. E., 1989, *Genetic Algorithms in Search, Optimization, and Machine Learning,* Addison-Wesley, Reading, MA.

Holland, J. H., 1975, *Adaptation in Natural and Artificial Systems: An Introductory Analysis with Applications to Biology, Control, and Artificial Intelligence,* University of Michigan Press, Ann Arbor, MI (reprinted 1992, MIT Press, Cambridge, MA).

Iba, H., 1996, *Genetic Programming*, Tokyo Denki University Press, Tokyo, in Japanese.

Jacob, C., 1997, *Principia Evolvica: Simulierte Evolution mit Mathematica,* dpunkt.verlag, Heidelberg.

Jacob, C., 2001, *Illustrating Evolutionary Computation with Mathematica,* Morgan Kaufmann, San Mateo, CA.

Kinnear, K. E. Jr, ed., 1994, *Advances in Genetic Programming,* MIT Press, Cambridge, MA.

Koza, J. R., 1989, Hierarchical genetic algorithms operating on populations of computer programs, in: *Proc. 11th Int. Joint Conf. on Artificial Intelligence,* Vol. 1, Morgan Kaufmann, San Mateo, CA, pp. 768–774.

Koza, J. R., 1990, Genetic programming: a paradigm for genetically breeding populations of computer programs to solve problems, *Stanford University Computer Science Department Technical Report* STAN-CS-90-1314.

Koza, J. R., 1992, *Genetic Programming: On the Programming of Computers by Means of Natural Selection,* MIT Press, Cambridge, MA.

Koza, J. R., 1994a, *Genetic Programming II: Automatic Discovery of Reusable Programs,* MIT Press, Cambridge, MA.

Koza, J. R., 1994b, *Genetic Programming II Videotape: The Next Generation,* MIT Press, Cambridge, MA.

Koza, J. R., 1994c, Architecture-altering operations for evolving the architecture of a multi-part program in genetic programming, *Stanford University Computer Science Department Technical Report* STAN-CS-TR-94-1528.

Koza, J. R., 1995, Gene duplication to enable genetic programming to concurrently evolve both the architecture and work-performing steps of a computer program, in: *Proc. 14th Int. Joint Conf. on Artificial Intelligence,* Morgan Kaufmann, San Mateo, CA, pp. 734–740.

Koza, J. R., Banzhaf, W., Chellapilla, K., Deb, K., Dorigo, M., Fogel, D. B., Garzon, M. H., Goldberg, D. E., Iba, H. and Riolo, R., eds, 1998, *Genetic Programming 1998: Proc. 3rd Annual Conf.* (Madison, WI), Morgan Kaufmann, San Mateo, CA.

Koza, J. R., Bennett III, F. H, Andre, D. and Keane, M. A., 1999a, *Genetic Programming III: Darwinian Invention and Problem Solving,* Morgan Kaufmann, San Mateo, CA.

Koza, J. R., Bennett III, F. H, Andre, D., Keane, M. A. and Brave, S., 1999b, *Genetic Programming III Videotape: Human-Competitive Machine Intelligence,* Morgan Kaufmann, San Mateo, CA.

Koza, J. R., Deb, K., Dorigo, M., Fogel, D. B., Garzon, M., Iba, H. and Riolo, R. L., eds, *Genetic Programming 1997: Proc. 2nd Annual Conf.* (Stanford University), Morgan Kaufmann, San Mateo, CA.

Koza, J. R., Goldberg, D. E., Fogel, D. B. and Riolo, R. L., eds, 1996, *Genetic Programming 1996: Proc. 1st Annual Conf.* (Stanford University), MIT Press, Cambridge, MA.

Koza, J. R., Keane, M. A., Streeter, M. J., Mydlowec, W., Yu, J. and Lanza, G., 2003, *Genetic Programming IV: Routine Human-Competitive Machine Intelligence*, Kluwer, Dordrecht.

Koza, J. R. and Rice, J. P., 1992, *Genetic Programming: The Movie*, MIT Press, Cambridge, MA.

Langdon, W. B., 1998, *Genetic Programming and Data Structures: Genetic Programming + Data Structures = Automatic Programming!* Kluwer, Amsterdam.

Langdon, W. B., Cantu-Paz, E., Mathias, K., Roy, R., Davis, D., Poli, R., Balakrishnan, K., Honavar, V., Rudolph, G., Wegener, J., Bull, L., Potter, M. A., Schultz, A. C., Miller, J. F., Burke, E. and Jonoska, N., eds, 2002, *Proc. 2002 Genetic and Evolutionary Computation Conf.*, Morgan Kaufmann, San Mateo, CA.

Langdon, W. B. and Poli, R., 2002, *Foundations of Genetic Programming*, Springer, Berlin.

Luke, S., 1998, Genetic programming produced competitive soccer softbot teams for RoboCup97, in: *Genetic Programming 1998: Proc. 3rd Annual Conf.* (Madison, WI), J. R. Koza, W. Banzhaf, K. Chellapilla, D. Kumar, K. Deb, M. Dorigo, D. B. Fogel, M. H. Garzon, D. E. Goldberg, H. Iba and R. Riolo, eds, Morgan Kaufmann, San Mateo, CA, pp. 214–222.

Miller, J., Tomassini, M., Lanzi, P. L., Ryan, C., Tettamanzi, A. G. B. and Langdon, W. B., eds, 2001, *Genetic Programming: Proc. 4th Eur. Conf., EuroGP 2001* (Lake Como, Italy), Springer, Berlin.

Nordin, P., 1997, *Evolutionary Program Induction of Binary Machine Code and Its Application,* Krehl, Munster.

Poli, R. and Langdon, W. B., 1997, A new schema theory for genetic programming with one-point crossover and point mutation, in: *Genetic Programming 1997: Proc. 2nd Annual Conf.* (Stanford University), J. R. Koza, K. Deb, M. Dorigo, D. B. Fogel, M. Garzon, H. Iba and R. L. Riolo, R. L., eds, Morgan Kaufmann, San Mateo, CA, pp. 278–285.

Poli, R, and Langdon, W. B., 1998, Schema theory for genetic programming with one-point crossover and point mutation, *Evol. Comput.* **6**:231–252.

Poli, R, and McPhee, N. F., 2001, Exact schema theorems for GP with one-point and standard crossover operating on linear structures and their application to the study of the evolution of size, in: *Genetic Programming, Proc. EuroGP 2001, Lake Como, Italy,* J. F. Miller, M. Tomassini, P. L. Lanzi, C. Ryan, A. G. B. Tettamanzi and W. B. Langdon, eds, Lecture Notes in Computer Science, Vol. 2038, Springer, Berlin, pp. 126–142.

Poli, R. and N. F., McPhee, 2003a, General schema theory for genetic programming with subtree-swapping crossover: Part I, *Evol. Comput.* **11**:53–66.

Poli, R. and N. F., McPhee, 2003b, General schema theory for genetic programming with subtree-swapping crossover: Part II, *Evol. Comput.* **11**:169–206.

Poli, R., Nordin, P., Langdon, W. B. and Fogarty, T. C., 1999, *Genetic Programming: Proc. 2nd Eur. Workshop, EuroGP'99,* Lecture Notes in Computer Science. Vol. 1598, Springer, Berlin.

Poli, R., Banzhaf, W., Langdon, W. B., Miller, J., Nordin, P. and Fogarty, T. C., 2000, *Genetic Programming: Proc. Eur. Conf., EuroGP 2000* (Edinburgh), Lecture Notes in Computer Science. Vol. 1802, Springer, Berlin.

Ryan, C., 1999, *Automatic Re-engineering of Software Using Genetic Programming,* Kluwer, Amsterdam.

Samuel, A. L., 1983, AI: Where it has been and where it is going, in: *Proc. 8th Int. Joint Conf. on Artificial Intelligence, Los Altos, CA,* Morgan Kaufmann, San Mateo, CA, pp. 1152–1157.

Spector, L., Barnum, H. and Bernstein, H. J., 1998, Genetic programming for quantum computers, in: *Genetic Programming 1998: Proc. 3rd Annual Conf.* (Madison, WI), J. R. Koza, W. Banzhaf, K. Chellapilla, D. Kumar, K. Deb, M. Dorigo, D. B. Fogel, M. H. Garzon, D. E. Goldberg, H. Iba and R. Riolo, eds, Morgan Kaufmann, San Mateo, CA, pp. 365–373.

Spector, L., Barnum, H. and Bernstein, H. J., 1999a, Quantum computing applications of genetic programming, in: *Advances in Genetic Programming 3,* L. Spector, W. B. Langdon, U.-M. O'Reilly and P. Angeline, eds, MIT Press, Cambridge, MA, pp. 135–160.

Spector, L., Barnum, H., Bernstein, H. J. and Swamy, N., 1999b, Finding a better-than-classical quantum AND/OR algorithm using genetic programming, in: *IEEE Proc. 1999 Congress on Evolutionary Computation,* IEEE, Piscataway, NJ, pp. 2239–2246.

Spector, L. and Bernstein, H. J., 2002, Communication capacities of some quantum gates, discovered in part through genetic programming, in: *Proc. 6th Int. Conf. on Quantum Communication, Measurement, and Computing* (Rinton, Paramus, NJ).

Spector, L., Goodman, E., Wu, A., Langdon, W. B., Voigt, H.-M., Gen, M., Sen, S., Dorigo, M., Pezeshk, S., Garzon, M. and Burke, E., eds, 2001, *Proc. Genetic and Evolutionary Computation Conf., GECCO-2001,* Morgan Kaufmann, San Mateo, CA.

Stephens, C. R. and Waelbroeck, H., 1997, Effective degrees of freedom in genetic algorithms and the block hypothesis, in: *Genetic Algorithms: Proc. 7th Int. Conf.,* Thomas Back, ed., Morgan Kaufmann, San Mateo, CA, pp. 34–40.

Stephens, C. R. and Waelbroeck, H., 1999, Schemata evolution and building blocks, *Evol. Comput.* 7:109–124.

Turing, A. M., 1948, Intelligent machinery. Reprinted in: 1992, *Mechanical Intelligence: Collected Works of A. M. Turing,* D. C. Ince, ed., North-Holland, Amsterdam, pp. 107–127. Also reprinted in: 1969, *Machine Intelligence 5,* B. Meltzer, and D. Michie, ed., Edinburgh University Press, Edinburgh.

Turing, A. M., 1950, Computing machinery and intelligence, *Mind* **59**:433–460. Reprinted in: 1992, *Mechanical Intelligence: Collected Works of A. M. Turing,* D. C. Ince, ed., North-Holland, Amsterdam, pp. 133–160.

Whitley, L. D., 1994, A genetic algorithm tutorial, *Statist. Comput.* **4**:65–85.

Whitley, D., Goldberg, D., Cantu-Paz, E., Spector, L., Parmee, I. and Beyer, H.-G., eds, 2000, *GECCO-2000: Proc. Genetic and Evolutionary Computation Conf.* (Las Vegas, NV), Morgan Kaufmann, San Mateo, CA.

Wong, M. L. and Leung, K. S., 2000, *Data Mining Using Grammar Based Genetic Programming and Applications*, Kluwer, Amsterdam.

Chapter 6

TABU SEARCH

Michel Gendreau, Jean-Yves Potvin

Département d'informatique et de recherche opérationnelle
and Centre de recherche sur les transports, Université de Montréal, Canada

6.1 INTRODUCTION

Over the last 15 years, hundreds of papers presenting applications of tabu search, a heuristic method originally proposed by Glover (1986), to various combinatorial problems have appeared in the operations research literature: see, for example, Glover and Laguna (1997), Glover et al. (1993b), Osman and Kelly (1996), Pardalos and Resende (2002), Ribeiro and Hansen (2002) and Voss et al. (1999). In several cases, the methods described provide solutions very close to optimality and are among the most effective, if not the best, to tackle the difficult problems at hand. These successes have made tabu search extremely popular among those interested in finding good solutions to the large combinatorial problems encountered in many practical settings. Several papers, book chapters, special issues and books have surveyed the rich tabu search literature (a list of some of the most important references is provided at the end of this chapter). In spite of this abundant literature, there still seem to be many researchers who, while they are eager to apply tabu search to new problem settings, find it difficult to properly grasp the fundamental concepts of the method, its strengths and its limitations, and to come up with effective implementations. The purpose of this chapter is thus to focus on the fundamental concepts of tabu search. Throughout the chapter, two relatively straightforward, yet challenging and relevant, problems will be used to illustrate these concepts: the job shop scheduling problem and the capacitated plant location problem. These will be introduced in the following section.

6.2 ILLUSTRATIVE PROBLEMS

6.2.1 The Job-Shop Scheduling Problem

The job-shop scheduling problem is one of the most studied problems in combinatorial optimization and a large number of papers and books deal with the numerous procedures that have been proposed to solve it, including several tabu search implementations. Although a large number of variants are found in the literature (and even more in the real world), the "classical" problem can be stated as follows. We first assume that n jobs must be scheduled on m machines. Each job corresponds to a fixed sequence of m operations, one per machine, where each operation must be processed on a specific machine for a specified duration. Note that the processing order on the machines does not need to be the same from one job to another. Each machine can process at most one operation at a time and, once started, an operation must be completed without interruption. The goal is to assign operations to time slots on the machines in order to minimize the maximum completion time of the jobs, which is also known as the makespan. A solution to this problem can be seen as a set of m permutations of the n jobs, one for each machine, with the associated machine schedules (Anderson et al., 1997).

6.2.2   The Capacitated Plant Location Problem

The capacitated plant location problem is one of the basic problems in location theory. It is encountered in many application settings that involve locating facilities with limited capacity to provide services. The problem can be formally described as follows. A set of customers I with demands $d_i, i \in I$, for some product are to be served from plants located in a subset of sites from a given set J of "potential sites". For each site $j \in J$, the fixed cost of "opening" the plant at j is f_j and its capacity is K_j. The cost of transporting one unit of the product from site j to customer i is c_{ij}. The objective is to minimize the total cost, i.e. the sum of the fixed costs for open plants and the transportation costs.

Letting x_{ij} ($i \in I, j \in J$) denote the quantity shipped from site j to customer i (the x_{ij} are the so-called *flow variables*) and y_j ($j \in J$) be a 0–1 variable indicating whether or not the plant at site j is open (the y_j are the location variables), the problem can be formulated as the following mathematical program:

$$\text{Minimize} \quad z = \sum_{j \in J} f_j y_j + \sum_{i \in I} \sum_{j \in J} c_{ij} x_{ij}$$

$$\text{subject to} \quad \sum_{j \in J} x_{ij} = d_i, i \in I$$

$$\sum_{i \in I} x_{ij} \leq K_j y_j, j \in J$$

$$x_{ij} \geq 0, i \in I, j \in J$$

$$y_j \in \{0, 1\}, j \in J$$

REMARK 6.1 *For any vector $\tilde{y}$ of location variables, optimal (w.r.t. to this plant configuration) values for the flow variables $x(\tilde{y})$ can be retrieved by solving the associated transportation problem:*

$$\text{Minimize} \quad z(\tilde{y}) = \sum_{i \in I} \sum_{j \in J} c_{ij} x_{ij}$$

$$\text{subject to} \quad \sum_{j \in J} x_{ij} = d_i, i \in I$$

$$\sum_{i \in I} x_{ij} \leq K_j \tilde{y}_j, j \in J$$

$$x_{ij} \geq 0, i \in I, j \in J$$

If $\tilde{y} = y*$, the optimal location variable vector, the optimal solution to the original problem is simply given by $(y, x(y*))$.*

REMARK 6.2 *An optimal solution of the original problem can always be found at an extreme point of the polyhedron of feasible flow vectors defined by the constraints*

$$\sum_{j \in J} x_{ij} = d_i, i \in I$$

$$\sum_{i \in I} x_{ij} \leq K_j, j \in J$$

$$x_{ij} \geq 0, i \in I, j \in J$$

This property follows from the fact that the capacitated plant location problem can be interpreted as a fixed-charge problem defined in the space of the flow variables. This fixed-charge problem has a concave objective function that always admits an extreme point minimum. The optimal values for the location variables can easily be obtained from the optimal flow vector by setting y_j equal to 1 when $\sum_{i \in I} x_{ij} > 0$, and to 0 otherwise.

6.3 BASIC CONCEPTS

6.3.1 Historical Background

Before introducing the basic concepts of tabu search, we believe it is useful to go back in time to try to better understand the genesis of the method and how it relates to previous work.

Heuristics, i.e. approximate solution techniques, have been used since the beginnings of operations research to tackle difficult combinatorial problems. With the development of complexity theory in the early 1970s, it became clear that, since most of these problems were indeed NP-hard, there was little hope of ever finding efficient exact solution procedures for them. This realization emphasized the role of heuristics for solving the combinatorial problems that were encountered in real-life applications and that needed to be tackled, whether or not they were NP-hard. While many different approaches were proposed and experimented with, the most popular ones were based on hill climbing. The latter can roughly be summarized as an iterative search procedure that, starting from an initial feasible solution, progressively improves it by applying a series of local modifications or moves (for this reason, hill climbing is in the family of local search methods). At each iteration, the search moves to an improving feasible solution that differs only slightly from the current one. In fact, the difference between the previous and the new solution amounts to one of the local modifications mentioned above. The search terminates when no more improvement is possible. At this point, we have a local optimum with regard to the local modifications considered by the hill climbing method. Clearly, this is an important limitation of the method: unless one is extremely lucky, this local optimum will often be a fairly mediocre solution. The quality of the solution obtained and computing times are usually highly dependent upon the "richness" of the set of transformations (moves) considered at each iteration.

In 1983, a new heuristic approach called simulated annealing (see Chapter 7) (Kirkpatrick et al., 1983) was shown to converge to an optimal solution of a combinatorial problem, albeit in infinite computing time. Based on analogy with statistical mechanics, simulated annealing could be interpreted as a form of controlled random walk in the space of feasible solutions. The emergence of simulated annealing indicated that one could look for other ways to tackle combinatorial optimization problems and spurred the interest of the research community. In the following years, many other new approaches, mostly based on analogies with natural phenomena, were proposed such as tabu search (the subject of this chapter), ant systems (see Chapter 14) (Dorigo, 1992) and threshold methods (Dueck and Scheuer, 1990). Together with some older ones, in particular genetic algorithms (see Chapter 4) (Holland, 1975), they gained an increasing popularity. Now collectively known under the name of metaheuristics, a term originally coined by Glover (1986), these methods have become, over the last 15 years, the leading edge of heuristic approaches for solving combinatorial optimization problems.

6.3.2   Tabu Search

Fred Glover proposed in 1986 a new approach, which he called tabu search, to allow hill climbing to overcome local optima. In fact, many elements of this first tabu search proposal, and some elements of later elaborations, had already been introduced in Glover (1977), including short-term memory to prevent the reversal of recent moves, and longer-term frequency memory to reinforce attractive components. The basic principle of tabu search is to pursue the search whenever a local optimum is encountered by allowing non-improving moves; cycling back to previously visited solutions is prevented by the use of memories, called tabu lists, that record the recent history of the search. The key idea to exploit information to guide the search can be linked to the informed search methods proposed in the late 1970s in the field of artificial intelligence (Nilsson, 1980). It is interesting to note that, in 1986 also, Hansen proposed an approach similar to tabu search, which he named steepest ascent/mildest descent (Hansen, 1986). It is also important to remark that Glover did not see tabu search as a proper heuristic, but rather as a metaheuristic, i.e. a general strategy for guiding and controlling "inner" heuristics specifically tailored to the problems at hand.

6.3.3 Search Space and Neighborhood Structure

As just mentioned, tabu search extends hill climbing methods. In fact, the basic tabu search can be seen as simply the combination of hill climbing with short-term memories. It follows that the two first basic elements of any tabu search heuristic are the definition of its search space and its neighborhood structure.

The search space is simply the space of all possible solutions that can be considered (visited) during the search. For instance, in the job shop scheduling problem of Section 6.2.1, the search space could simply be the set of feasible solutions to the problem, where each point in the search space corresponds to a set of m machine schedules that satisfies all the specified constraints. While in that case the definition of the search space seems quite natural, it is not always so. Consider now the capacitated plant location problem of Section 6.2.2: the feasible space involves both integer location and continuous flow variables that are linked by strict conditions; moreover, as has been indicated before, for any feasible set of values for the location variables, one can fairly easily retrieve optimal values for the flow variables by solving the associated transportation problem. In this context, one could obviously use as a search space the full feasible space; this would involve manipulating both location and flow variables, which is not an easy task. A more attractive search space is the set of feasible vectors of location variables, i.e. feasible vectors in $\{0, 1\}^{|J|}$ (where $|J|$ is the cardinality of set J), any solution in that space being "completed" to yield a feasible solution to the original problem by computing the associated optimal flow variables. It is interesting to note that these two possible definitions are not the only ones. Indeed, on the basis of Remark 6.2, one could also decide to search instead the set of extreme points of the set of feasible flow vectors, retrieving the associated location variables by simply noting that a plant must be open whenever some flow is allocated to it. In fact, this type of approach was used successfully by Crainic et al. (2000) to solve the fixed charge multi-commodity network design problem, which is a more general problem that includes the capacitated plant location problem as a special case. It is also important to note that it is not always a good idea to restrict the search space to feasible solutions. In many cases, allowing the search to move to infeasible solutions is desirable, and sometimes necessary (see Section 6.4.3 for further details).

Closely linked to the definition of the search space is that of the neighborhood structure. At each iteration of tabu search, the local transformations that can be applied to the current solution, denoted S, define a set of neighboring solutions in the search space, denoted $N(S)$ (the neighborhood of S). Formally, $N(S)$ is a subset of the search space defined by

$$N(S) = \text{solutions obtained by applying a single local transformation to } S$$

In general, for any specific problem at hand, there are many more possible (and even, attractive) neighborhood structures than search space definitions. This follows from the fact that there may be several plausible neighborhood structures for a given definition of the search space. This is easily illustrated on our job shop scheduling problem. In order to simplify the discussion, we assume in the following that the search space is the feasible space.

Simple neighborhood structures for the job shop scheduling problem are obtained by considering the sequence of jobs associated with a machine schedule, where the position of a job in the sequence corresponds to its processing order on the machine. For example, one can move a job at another position in the sequence or interchange the position of two jobs. While these neighborhood structures involve only one or two jobs, the neighborhoods they define contain all the feasible schedules that can be obtained from the current one either by moving any single job at any other position or by interchanging any two jobs. Examining these neighborhoods can thus be fairly demanding. In practice, it is often possible to reduce the computational burden, by identifying a restricted subset of moves that are feasible and can lead to improvements. We refer the interested reader to Vaessens et al. (1996) and Anderson et al. (1997) for a more detailed discussion of these issues.

When different definitions of the search space are considered for a given problem, neighborhood structures will inevitably differ to a considerable degree. This can be illustrated on our capacitated plant location problem. If the search space is defined with respect to the location variables, neighborhood structures will usually involve the so-called "Add/Drop" and "Swap" moves that respectively change the status of one site (i.e. either opening a closed facility or closing an open one) and move an open facility from one site to another (this move amounts to performing simultaneously an Add move and a Drop move). If, however, the search space is the set of extreme points associated with feasible flow vectors, these moves become meaningless. One should instead consider moves defined by the application of pivots to the linear programming formulation of the transportation problem, where each pivot operation modifies the flow structure to move the current solution to an adjacent extreme point.

The preceding discussion should have clarified a major point: choosing a search space and a neighborhood structure is by far the most critical step in the design of any tabu search heuristic. It is at this step that one must make the best use of the understanding and knowledge one has of the problem at hand.

6.3.4 Tabus

Tabus are one of the distinctive elements of tabu search when compared to hill climbing. As we already mentioned, tabus are used to prevent cycling when moving away from local optima through non-improving moves. The key realization here is that when this situation occurs, something needs to be done to prevent the search from tracing back its steps to where it came from. This is achieved by making certain actions tabu. This might mean not allowing the search to return to a recently visited point in the search space or not allowing a recent move to be reversed. For example, in the job shop scheduling problem, if a job j has been moved to a new position in a machine schedule, one could declare tabu moving that job back to its previous position for some number of iterations (this number is called the tabu tenure of the move).

Tabus are stored in a short-term memory of the search (the tabu list) and usually only a fixed and fairly limited quantity of information is recorded. In any given context, there are several possibilities regarding the recorded information. One could record complete solutions, but this requires a lot of storage and makes it expensive to check whether a potential move is tabu or not; it is therefore seldom used. The most commonly used tabus involve recording the last few transformations performed on the current solution and prohibiting reverse transformations (as in the example above); others are based on key characteristics of the solutions themselves or of the moves.

To better understand how tabus work, let us go back to our reference problems. In the job shop scheduling problem, one could define tabus in several ways. To continue our example where a job j has just been moved from position p_1 to position p_2, one could declare tabu specifically moving back j to position p_1 from position p_2 and record this in the short-term memory as the triplet (j, p_2, p_1). Note that this type of tabu will not constrain the search much, but that cycling may occur if j is then moved to another position p_3 and then from p_3 to p_1. A stronger tabu would involve prohibiting moving back j

to p_1 (without consideration for its current position) and be recorded as (j, p_1). An even stronger tabu would be to disallow moving j at all, and would simply be noted as (j).

In the capacitated plant location problem, tabus on Add/Drop moves should prohibit changing the status of the affected location variable and can be recorded by noting its index. Tabus for Swap moves are more complex. They could be declared with respect to the site where the facility was closed, to the site where the facility was opened, to both locations (i.e. changing the status of both location variables is tabu), or to the specific swapping operation.

Multiple tabu lists can be used simultaneously and are sometimes advisable. For example, in the capacitated plant location problem, if one uses a neighborhood structure that contains both Add/Drop and Swap moves, it might be a good idea to keep a separate tabu list for each type of move.

Standard tabu lists are usually implemented as circular lists of fixed length. It has been shown, however, that fixed-length tabus cannot always prevent cycling, and some authors have proposed varying the tabu list length during the search (Glover, 1989, 1990; Skorin-Kapov, 1990; Taillard, 1990, 1991). Another solution is to randomly generate the tabu tenure of each move within some specified interval. Using this approach requires a somewhat different scheme for recording tabus, which are usually stored as tags in an array. The entries in this array typically record the iteration number until which a move is tabu. More details are provided in Gendreau et al. (1994).

6.3.5   Aspiration Criteria

While central to the tabu search method, tabus are sometimes too powerful. They may prohibit attractive moves, even when there is no danger of cycling, or they may lead to an overall stagnation of the search process. It is thus necessary to use algorithmic devices that will allow one to revoke (cancel) tabus. These are called aspiration criteria. The simplest and most commonly used aspiration criterion, found in almost all tabu search implementations, allows a tabu move when it results in a solution with an objective value better than that of the current best-known solution (since the new solution has obviously not been previously visited). Much more complicated aspiration criteria have been proposed and successfully implemented (see, for example, de Werra and Hertz (1989) and Hertz and de Werra (1991)), but they are rarely used. The key rule is that if cycling cannot occur, tabus can be disregarded.

6.3.6   A Template for Simple Tabu Search

We are now in the position to give a general template for tabu search, integrating the elements we have seen so far. We suppose that we are trying to minimize a function $f(S)$ (sometimes known as an objective or evaluation function) over some domain and we apply the so-called "best improvement" version of tabu search, i.e. the version in which one chooses at each iteration the best available move (even if this results in an increase in the function $f(S)$). This is the most commonly used version of tabu search.

Notation:

- S the current solution,

- S^* the best-known solution,

- f^* value of S^*,

- $N(S)$ the neighborhood of S,

- $\tilde{N}(S)$ the "admissible" subset of $N(S)$ (i.e. non-tabu or allowed by aspiration),

- T tabu list.

Initialization:

Choose (construct) an initial solution S_0.

Set $S := S_0, \ f^* := f(S_0), \ S^* := S_0, \ T := \emptyset$.

Search:

While *termination criterion not satisfied* do

- Select S in $\mathrm{argmin}[f(S')]$;

$S' \varepsilon \ \tilde{N}(S)$

- if $f(S) < f^*$, then set $f^* := f(S), S^* := S$;

- record tabu for the current move in T (delete oldest entry if necessary);

Endwhile.

In this algorithm, argmin returns the subset of solutions in $\tilde{N}(S)$ that minimizes f.

6.3.7   Termination Criteria

One may have noticed that we have not specified in our template a termination criterion. In theory, the search could go on forever, unless the optimal value of the problem at hand is known beforehand. In practice, obviously, the search has to be stopped at some point. The most commonly used stopping criteria in tabu search are

- after a fixed number of iterations (or a fixed amount of CPU time);

- after some number of consecutive iterations without an improvement in the objective function value (the criterion used in most implementations);

- when the objective function reaches a pre-specified threshold value.

6.3.8   Probabilistic Tabu Search and Candidate Lists

Normally, one must evaluate the objective function for every element of the neighborhood $N(S)$ of the current solution. This can be extremely expensive from a computational standpoint. In probabilistic tabu search, only a random sample $N'(S)$ of $N(S)$ is considered, thus significantly reducing the computational overhead. Another attractive feature is that the added randomness can act as an anti-cycling mechanism. This allows one to use shorter tabu lists than would be necessary if a full exploration of the neighborhood was performed. One the negative side, it is possible to miss excellent solutions (see Section 6.6.3 for more detail). It is also possible to probabilistically select when to apply tabu criteria.

Another way to control the number of moves examined is by means of candidate list strategies, which provide more strategic ways of generating a useful subset $N'(S)$ of $N(S)$. In fact, the probabilistic approach can be considered to be one instance of a candidate list strategy, and may also be used to modify such a strategy. Failure to adequately address the issues involved in creating effective candidate lists is one of the more conspicuous shortcomings that differentiates a naive tabu search implementation from one that is more solidly grounded. Relevant designs for candidate list strategies are discussed in Glover and Laguna (1997). We also discuss a useful type of candidate generation approach in Section 6.4.4.

6.4   EXTENSIONS TO THE BASIC CONCEPTS

Simple tabu search as described above can sometimes successfully solve difficult problems, but in most cases, additional elements have to be included in the search strategy to make it fully effective. We now briefly review the most important of these.

6.4.1   Intensification

The idea behind the concept of search intensification is that, as an intelligent human being would probably do, one should explore more thoroughly the portions of the search space that seem "promising" in order to make sure that the best solutions in these areas are found. In general, intensification is based on some intermediate-term memory, such as a recency memory, in which one records the number of consecutive iterations that various "solution components" have been present in the current solution without interruption. For instance, in the capacitated plant location problem, one could record how long each site has had an open facility. A typical approach to intensification is to restart the search from the best currently known solution and to "freeze" (fix) in it the components that seem more attractive. To continue with our capacitated plant location problem, one could freeze a number of facilities in sites that have been often selected in previous iterations and perform a restricted search on the other sites. Another technique that is often used consists of changing the neighborhood structure to one allowing more powerful or more diverse moves. In the capacitated plant location problem, if Add/Drop moves were used, Swap moves could be added to the neighborhood structure. In probabilistic tabu search, one could increase the sample size or switch to searching without sampling. Intensification is used in many tabu search implementations, although it is not always necessary. This is because there are many situations where the normal search process is thorough enough.

6.4.2   Diversification

One of the main problems of all methods based on local search, and this includes tabu search in spite of the beneficial impact of tabus, is that they tend to be too "local" (as their name implies), i.e. they tend to spend most, if not all, of their time in a restricted portion of the search space. The negative consequence of this fact is that, although good solutions may be obtained, one may fail to explore the most interesting parts of the search space and thus end up with solutions that are still far from the optimal ones. Diversification is an algorithmic mechanism that tries to alleviate this problem by forcing the search into previously unexplored areas of the search space. It is usually based on some form of long-term memory of the search, such as a frequency memory, in which one records the total number of iterations (since the beginning of the search) that various "solution components" have been present in the current solution or have been involved in the selected moves. For instance, in the capacitated plant location problem, one could record the number of iterations during which each site has had an open facility. In the job shop scheduling problem, one could note how many times each job has been moved. In cases where it is possible to identify useful "regions" of the search space, the frequency memory can be refined to track the number of iterations spent in these different regions.

There are two major diversification techniques. The first, called restart diversification, involves introducing a few rarely used components in the current solution (or the best known solution) and restarting the search from this point. In the capacitated plant location problem, one could thus open one or more facilities at locations that have seldom been used up to that point and resume searching from that plant configuration (one could also close facilities at locations that have been used the most frequently). In the job shop scheduling problem, a job that has not occupied a particular position in a machine schedule can be forced to that position. The second diversification method, called continuous diversification, integrates diversification considerations directly into the regular searching process. This is achieved by biasing the evaluation of possible moves by adding to the objective a small term related to component frequencies. An extensive discussion on these two techniques is provided by Soriano and Gendreau (1996). A third way of achieving diversification is strategic oscillation which will be discussed in the next section.

We would like to stress that ensuring proper search diversification is possibly the most critical issue in the design of tabu search heuristics. It should be addressed with extreme care fairly early in the design phase and revisited if the results obtained are not up to expectations.

6.4.3   Allowing Infeasible Solutions

Accounting for all problem constraints in the definition of the search space often restricts the searching process too much and can lead to mediocre solutions. In such cases, constraint relaxation is an attractive strategy, since it creates a larger search space that can be explored with "simpler" neighborhood structures. Constraint relaxation is easily implemented by dropping selected constraints from the search space definition and adding to the objective, weighted penalties for constraint violations. In the capacitated plant location problem, this can be done by allowing solutions with flows that exceed the capacity of one or more plants. This, however, raises the issue of finding correct weights for constraint violations. An interesting way of circumventing this problem is to use self-adjusting penalties, i.e. weights are adjusted dynamically on the basis of the recent history of the search. Weights are increased if only infeasible solutions were encountered in the last few iterations, and decreased if all recent solutions were feasible; see Gendreau et al. (1994) for further details. Penalty weights can also be modified systematically to drive the search to cross the feasibility boundary of the search space and thus induce diversification. This technique, known as strategic oscillation, was introduced in Glover (1977) and used since in several successful tabu search procedures.

An important early variant oscillates among alternative types of moves, hence neighborhood structures, while another oscillates around a selected value for a critical function.

6.4.4 Surrogate and Auxiliary Objectives

There are many problems for which the true objective function is quite costly to evaluate, a typical example being the capacitated plant location problem when one searches the space of location variables. Recall that, in this case, computing the objective value for any potential solution entails solving the associated transportation problem. When this occurs, the evaluation of moves may become prohibitive, even if sampling is used. An effective approach to handle this issue is to evaluate neighbors using a surrogate objective, i.e. a function that is correlated to the true objective, but is less computationally demanding, in order to identify a small set of promising candidates (potential solutions achieving the best values for the surrogate). The true objective is then computed for this small set of candidate moves and the best one selected to become the new current solution. An example of this approach is found in Crainic et al. (1993).

Another frequently encountered difficulty is that the objective function may not provide enough information to effectively drive the search to more interesting areas of the search space. A typical illustration of this situation is observed when the fixed costs for open plants in the capacitated plant location problem are much larger than the transportation costs. In this case, it is indicated to open as few plants as possible. It is thus important to define an auxiliary objective function to orient the search. Such a function must measure in some way the desirable attributes of the solutions. In our example, one could use a function that would favor, for the same number of open plants, solutions with plants having just a small amount of flow, thus increasing the likelihood of closing them in subsequent iterations. It should be noted that developing an effective auxiliary objective is not always easy and may require a lengthy trial and error process. In some other cases, fortunately, the auxiliary objective is obvious for anyone familiar with the problem at hand: see the work of Gendreau et al. (1993) for an illustration.

6.5 SOME PROMISING AREAS FOR FUTURE APPLICATION

The concepts and techniques described in the previous sections are sufficient to design effective tabu search heuristics for many combinatorial problems. Most early tabu search implementations, several of which were extremely successful, relied indeed almost exclusively on these algorithmic components (Friden et al., 1989; Hertz and de Werra, 1987; Skorin-Kapov, 1990;

Taillard, 1991). Nowadays, however, most leading edge research in tabu search makes use of more advanced concepts and techniques. While it is clearly beyond the scope of an introductory tutorial, such as this one, to review this type of advanced material, we would like to give readers some insight into it by briefly describing some current trends. Readers who wish to learn more about this topic should read our survey paper (Gendreau, 2002) and some of the references provided in this section.

A large part of the recent research in tabu search deals with various techniques for making the search more effective. These include methods for better exploitation of the information that becomes available during search and creating better starting points, as well as more powerful neighborhood operators and parallel search strategies. For more details, see the taxonomy of Crainic et al. (1997), and the survey of Cung et al. (2002). The numerous techniques for utilizing the information are of particular significance since they can lead to dramatic performance improvements. Many of these rely on elite solutions (the best solutions previously encountered) or on parts of these to create new solutions, the rationale being that "fragments" of excellent solutions are often identified quite early in the search process. However, the challenge is to complete these fragments or to recombine them (Glover, 1992; Glover and Laguna, 1993, 1997; Rochat and Taillard, 1995). Other methods, such as the reactive tabu search of Battiti and Tecchiolli (1994), are aimed at finding ways to move the search away from local optima that have already been visited.

Another important trend (this is, in fact, a pervasive trend in the whole metaheuristics field) is hybridization, i.e. using tabu search in conjunction with other solution approaches such as genetic algorithms (Crainic and Gendreau, 1999; Fleurent and Ferland, 1996), Lagrangean relaxation (Grünert, 2002), constraint programming (Pesant and Gendreau, 1999), and column generation (Crainic et al., 2000). A whole chapter on this topic can be found in Glover and Laguna (1997). Problem specific information and simple heuristics can also be used in conjunction with different components of tabu search. For example, in Burke et al. (1999), problem-specific heuristics are used to realize diversification.

The literature on tabu search has also started moving away from its traditional application areas (graph theory problems, scheduling, vehicle routing) to new ones: continuous optimization (Rolland, 1996), multi-objective optimization (Gandibleux et al. 2000), stochastic programming (Lokketangen and Woodruff, 1996), mixed integer programming (Crainic et al., 2000; Lokketangen and Glover, 1996), real-time decision problems (Gendreau et al., 1999), etc. These new areas confront researchers with new challenges that, in turn, call for novel and original extensions of the method.

6.6   TRICKS OF THE TRADE

6.6.1   Getting Started

Newcomers to tabu search, trying to apply the method to a problem that they wish to solve, are often confused about what they need to do to come up with a successful implementation. Basically, they do not know where to start. We believe that the following step-by-step procedure will help and provides a useful framework for getting started.

A Step-by-Step Procedure

1 *Read* one or two good introductory papers to gain some knowledge of the concepts and of the vocabulary (see the references provided in "Sources of Additional Information" at the end of this chapter).

2 *Read* several papers describing in detail applications in various areas to see how the concepts have been actually implemented by other researchers (see the references provided in "Sources of Additional Information" at the end).

3 *Think* a lot about the problem at hand, focusing on the definition of the *search space* and the *neighborhood structure*.

4 *Implement* a *simple* version based on this search space definition and this neighborhood structure.

5 *Collect statistics* on the performance of this simple heuristic. It is usually useful at this point to introduce a variety of *memories*, such as frequency and recency memories, to really track down what the heuristic does.

6 *Analyse results* and *adjust* the procedure accordingly. It is at this point that one should eventually introduce mechanisms for search intensification and diversification or other intermediate features. Special attention should be paid to *diversification*, since this is often where simple tabu search procedures fail.

6.6.2   More Tips

In spite of carefully following the procedure outlined above, it is possible to end up with a heuristic that produces mediocre results. If this occurs, the following tips may prove useful:

1 *If there are constraints, consider penalizing the violation of them.* Letting the search move to infeasible solutions is often necessary in highly constrained problems to allow for a meaningful exploration of the search space (see Section 6.4).

2 *Reconsider the neighborhood structure* and change it if necessary. Many tabu search implementations fail because the neighborhood structure is too simple. In particular, one should make sure that the chosen neighborhood structure allows for a sensible evaluation of possible moves (i.e. the moves that seem intuitively to move the search in the "right" direction should be the ones that are likely to be selected); it might also be a good idea to introduce a *surrogate objective* (see Section 6.4) to achieve this.

3 *Collect more statistics.* For example, recording the number and quality of previously visited local optima can be useful to find a good trade-off between intensification and diversification

4 *Follow the execution of the algorithm step-by-step* on some reasonably sized instances (for example: Is the algorithm behaving as expected on particular solution configurations? Is the algorithm converging prematurely?).

5 *Reconsider diversification.* As mentioned earlier, this is a critical feature in most tabu search implementations.

6 *Experiment with parameter settings.* Many tabu search procedures are extremely sensitive to parameter settings; it is not unusual to see the performance of a procedure dramatically improve after changing the value of one or two key parameters (unfortunately, it is not always obvious to determine which parameters are the key ones in a given procedure).

6.6.3   Additional Tips for Probabilistic Tabu Search

While probabilistic tabu search is an effective way of tackling many problems, it creates difficulties of its own that need to be carefully addressed. The most important of these occurs because, more often than not, the best solutions returned by probabilistic tabu search will not be local optima with respect to the neighborhood structure being used. This is particularly annoying since, when it happens, better solutions can be easily obtained, sometimes even manually. An easy way to address this is to simply perform a local improvement phase (using the same neighborhood operator) from the best found solution at the end of the tabu search itself. One could alternately switch to tabu search without sampling (again from the best found solution) for a short duration before completing the algorithm. A possibly more effective technique is to add, throughout the search, an intensification step without sampling. This will mean that the best solutions available in the various regions of the space explored by the method will be found and recorded. This is similar to the method proposed by Glover and Laguna (1993). They employed special aspiration criteria for allowing the search to reach local optima at useful junctures.

6.6.4 Parameter Calibration and Computational Testing

Parameter calibration and computational experiments are key steps in the development of any algorithm. This is particularly true in the case of tabu search, since the number of parameters required by most implementations is fairly large and the performance of a given procedure can vary quite significantly when parameter values are modified. The first step in any serious computational experimentation is to select a good set of benchmark instances (either by obtaining them from other researchers or by constructing them), preferably with some reasonable measure of their difficulty and with a wide range of size and difficulty. This set should be split into two subsets, the first one being used at the algorithmic design and parameter calibration steps, and the second reserved for performing the final computational tests that will be reported in the paper(s) describing the heuristic under development. The reason for doing so is quite simple: when calibrating parameters, one always run the risk of overfitting, i.e. finding parameter values that are excellent for the instances at hand, but poor in general, because these values provide too good a "fit" (from the algorithmic standpoint) to these instances. Methods with several parameters should thus be calibrated on much larger sets of instances than ones with few parameters to ensure a reasonable degree of robustness. The calibration process itself should proceed in several stages:

1 Perform exploratory testing to find good ranges of parameters. This can be done by running the heuristic with a variety of parameter settings.

2 Fix the value of the parameters that appear to be "robust": that is, which do not seem to have a significant impact on the performance of the algorithm.

3 Perform systematic testing for the other parameters. It is usually more efficient to test values for only a single parameter at a time, the others being fixed at what appear to be reasonable values. One must be careful, however, for cross effects between parameters. For example, assume that value x_1 for parameter p_1 leads to good results when the other parameters are fixed at their default values, and that value x_2 for parameter p_2 leads to good results when the other parameters are fixed at their default values. Then, it might happen that value x_1 for parameter p_1 *and* value x_2 for parameter p_2 lead to poor results. Where such effects exist, it can be important to jointly test pairs or triplets of parameters, which can be an extremely time-consuming task.

The paper by Crainic et al. (1993) provides a detailed description of the calibration process for a fairly complex tabu search procedure and can used as a guideline for this purpose.

6.7   CONCLUSIONS

Tabu search is a powerful algorithmic approach that has been applied with great success to many difficult combinatorial problems. A particularly nice feature of tabu search is that it can quite easily handle the "dirty" complicating constraints that are typically found in real-life applications. It is thus a really practical approach. It is not, however, a panacea: every reviewer or editor of a scientific journal has seen more than his/her share of failed tabu search heuristics. These failures stem from two major causes: an insufficient understanding of fundamental concepts of the method (and we hope that this tutorial will help in alleviating this shortcoming), but also, more often than not, a crippling lack of understanding of the problem at hand. One cannot develop a good tabu search heuristic for a problem that one does not know well! This is because significant problem knowledge is absolutely vital to perform the most basic steps of the development of any tabu search procedure, namely the choice of a search space and the choice of an effective neighborhood structure. If the search space and/or the neighborhood structure are inadequate, no amount of tabu search expertise will be sufficient to save the day. A last word of caution: to be successful, all metaheuristics need to achieve both depth and breadth in their searching process; depth is usually not a problem for tabu search, which is quite aggressive in this respect (it generally finds pretty good solutions very early in the search), but breadth can be a critical issue. To handle this, it is extremely important to develop an effective diversification scheme.

SOURCES OF ADDITIONAL INFORMATION

- Good introductory papers on tabu search may be found in Glover and Laguna (1993), Glover et al. (1993b), Hertz and de Werra (1991), de Werra and Hertz (1989) and, in French, in Soriano and Gendreau (1997).

- The book by Glover and Laguna (1997) is the ultimate reference on tabu search. Apart from the fundamental concepts of the method, it presents a considerable amount of advanced material, as well as a variety of applications. It is interesting to note that this book contains several ideas applicable to tabu search that yet remain to be fully exploited.

- Two issues of *Annals of Operations Research* devoted respectively to *Tabu Search* (Glover et al., 1993a) and *Metaheuristics in Combinatorial Optimization* (Laporte and Osman, 1996) provide a good sample of applications of tabu search.

- The books made up from selected papers presented at the Meta-Heuristics International Conferences (MIC) are also extremely valuable. At this time, the books for the 1995 Breckenridge conference (Osman and Kelly,

1996), the 1997 Sophia-Antipolis one (Voss et al., 1999) and the 1999 Angra dos Reis one (Ribeiro and Hansen, 2002) have been published. The proceedings of MIC'2001, held in Porto, are available online at http://tew.ruca.ua.ac.be/eume/MIC2001.

- Two books of interest have recently been published. The one edited by Glover and Kochenberger (2003) addresses metaheuristics in general. The other, edited by Pardalos and Resende (2002), has a broader scope but contains a nice chapter on metaheuristics.

Acknowledgments. The authors are grateful to the Canadian Natural Sciences and Engineering Council and the Quebec Fonds FCAR for their financial support. An earlier version of this chapter was published under the title "An Introduction to Tabu Search" in the Handbook of Metaheuristics, *edited by Fred Glover and Gary A. Kochenberger, published by Kluwer Academic.*

References

Anderson, E. J., Glass, C. A. and Potts, C. N., 1997, Machine scheduling, in: *Local Search in Combinatorial Optimization*, E. H. L. Aarts and J. K. Lenstra, eds, Wiley, New York, pp. 361–414.

Battiti, R. and Tecchiolli, G., 1994, The reactive tabu search, *ORSA J. Comput.* **6**:126–140.

Burke, E., De Causmaecker, P. and Vanden Berghe, G., 1999, A hybrid tabu search algorithm for the nurse rostering problem, in: *Simulated Evolution and Learning*, Selected papers from the 2nd Asia-Pacific Conf. on Simulated Evolution and Learning, SEAL '98, B. McKay, X. Yao, C. S. Newton, J.-H. Kim and T. Furuhashi, eds, Lecture Notes in Artificial Intelligence, Vol. 1585, Springer, Berlin, pp. 187–194.

Crainic, T. G. and Gendreau, M., 1999, Towards an evolutionary method—cooperative multi-thread parallel tabu search heuristic hybrid, in: *Meta-Heuristics: Advances and Trends in Local Search Paradigms for Optimization*, S. Voss, S. Martello, I. H. Osman and C. Roucairol, eds, Kluwer, Dordrecht, pp. 331–344.

Crainic, T. G., Gendreau, M. and Farvolden, J. M., 2000, Simplex-based tabu search for the multicommodity capacitated fixed charge network design problem, *INFORMS J. Comput.* **12**:223–236.

Crainic, T. G., Gendreau, M., Soriano, P. and Toulouse, M., 1993, A tabu search procedure for multicommodity location/allocation with balancing requirements, *Ann. Oper. Res.* **41**:359–383.

Crainic, T. G., Toulouse, M. and Gendreau, M., 1997, Toward a taxonomy of parallel tabu search heuristics, *INFORMS J. Comput.* **9**:61–72.

Cung, V.-D., Martins, S. L., Ribeiro, C. C. and Roucairol, C., 2002, Strategies for the parallel implementation of metaheuristics, in: *Essays and Surveys* in *Metaheuristics,* C. C. Ribeiro and P. Hansen, eds, Kluwer, Dordrecht, pp. 263–308.

Dorigo, M., 1992, Optimization, learning and natural algorithms, *Ph.D. Dissertation,* Departimento di Elettronica, Politecnico di Milano, Italy.

Dueck, G. and Scheuer, T., 1990, Threshold accepting: a general purpose optimization algorithm appearing superior to simulated annealing, *J. Comput. Phys.* **90**:161–175.

Fleurent, C. and Ferland, J. A., 1996, Genetic and hybrid algorithms for graph colouring, *Ann. Oper. Res.* **63**:437–461.

Friden, C., Hertz, A. and de Werra, D., 1989, STABULUS: A technique for finding stable sets in large graphs with tabu search, *Computing* **42**:35–44.

Gandibleux, X., Jaszkiewicz, A., Freville, A. and Slowinski, R., eds, 2000, *J. Heuristics* **6**:291–431, Special issue: Multiple Objective Metaheuristics.

Gendreau, M., 2002, Recent advances in tabu search, in: *Essays and Surveys in Metaheuristics,* C. C. Ribeiro and P. Hansen, eds, Kluwer, Dordrecht, pp. 369–377.

Gendreau, M., Guertin, F., Potvin, J.-Y. and Taillard, É. D., 1999, Parallel tabu search for real-time vehicle routing and dispatching, *Transport. Sci.* **33**:381–390.

Gendreau, M., Hertz, A. and Laporte, G., 1994, A tabu search heuristic for the vehicle routing problem, *Manage. Sci.* **40**:1276–1290.

Gendreau, M., Soriano, P. and Salvail, L., 1993, Solving the maximum clique problem using a tabu search approach, *Ann. Oper. Res.* **41**:385–403.

Glover, F., 1977, Heuristics for integer programming using surrogate constraints, *Decision Sci.* **8**:156–166.

Glover, F., 1986, Future paths for integer programming and links to artificial intelligence, *Comput. Oper. Res.* **13**:533–549.

Glover, F., 1989, Tabu search—Part I, *ORSA J. Comput.* **1**:190–206.

Glover, F., 1990, Tabu search—Part II, *ORSA J. Comput.* **2**:4–32.

Glover, F., 1992, Ejection chains, reference structures and alternating path methods for traveling salesman problems, University of Colorado paper, shortened version published in *Discr. Appl. Math.* **65**:223–253, 1996.

Glover, F. and Kochenberger, G. A., eds, 2003, *Handbook of Metaheuristics,* Kluwer, Boston.

Glover, F. and Laguna, M., 1993, Tabu search, in: *Modern Heuristic Techniques for Combinatorial Problems,* C. R. Reeves, ed., Blackwell, Oxford, pp. 70–150.

Glover, F. and Laguna, M., 1997, *Tabu Search,* Kluwer, Boston, MA.

Glover, F., Laguna, M., Taillard, É. D. and de Werra, D., eds, 1993a, *Tabu Search,* Annals of Operations Research, Vol. 41, Baltzer, Basel.

Glover, F., Taillard, É. D. and de Werra, D., 1993b, A user's guide to tabu search, *Ann. Oper. Res.* **41**:3–28.

Grünert, T., 2002, Lagrangean tabu search, in: *Essays and Surveys in Meta-heuristics*, C. C. Ribeiro and P. Hansen, eds, Kluwer, Dordrecht, pp. 379–397.

Hansen, P., 1986, The steepest ascent mildest descent heuristic for combinatorial programming, in: *Proceedings of the Congress on Numerical Methods in Combinatorial Optimization*, Capri, Italy.

Hertz, A. and de Werra, D, 1987, Using tabu search for graph coloring, *Computing* **39**:345–351.

Hertz, A. and de Werra, D., 1991, The tabu search metaheuristic: how we used it, *Ann. Math. Artif. Intell.* **1**:111–121.

Holland, J. H., 1975, *Adaptation in Natural and Artificial Systems*, University of Michigan Press, Ann Arbor, MI.

Kirkpatrick, S., Gelatt Jr., C. D. and Vecchi, M. P., 1983, Optimization by simulated annealing, *Science* **220**:671–680.

Laporte, G. and Osman, I. H., eds, 1996, *Metaheuristics in Combinatorial Optimization*, Annals of Operations Research, Vol. 63, Baltzer, Basel.

Lokketangen, A. and Glover, F., 1996, Probabilistic move selection in tabu search for 0/1 mixed integer programming problems, in: *Meta-Heuristics: Theory and Applications*, I. H. Osman and J. P. Kelly, eds, Kluwer, Dordrecht, pp. 467–488.

Lokketangen, A. and Woodruff, D. L., 1996, Progressive hedging and tabu search applied to mixed integer (0, 1) multistage stochastic programming, *J. Heuristics* **2**:111–128.

Nilsson, N. J., 1980, *Principles of Artificial Intelligence*, Morgan Kaufmann, Los Altos, CA.

Osman, I. H. and Kelly, J. P., eds, 1996, *Meta-Heuristics: Theory and Applications*, Kluwer, Boston, MA.

Pardalos, P. M. and Resende, M. G. C., eds, 2002, *Handbook of Applied Optimization*, Oxford University Press, New York.

Pesant, G. and Gendreau, M., 1999, A constraint programming framework for local search methods, *J. Heuristics* **5**:255–280.

Ribeiro, C. C. and Hansen, P., eds, 2002, *Essays and Surveys in Metaheuristics*, Kluwer, Norwell, MA.

Rochat, Y. and Taillard, É. D., 1995, Probabilistic diversification and intensification in local search for vehicle routing, *J. Heuristics* **1**:147–167.

Rolland, E., 1996, A tabu search method for constrained real-number search: applications to portfolio selection, *Working Paper,* The Gary Anderson Graduate School of Management, University of California, Riverside.

Skorin-Kapov, J., 1990, Tabu search applied to the quadratic assignment problem, *ORSA J. Comput.* **2**:33–45.

Soriano, P. and Gendreau, M., 1996, Diversification strategies in tabu search algorithms for the maximum clique problems, *Ann. Oper. Res.* **63**:189–207.

Soriano, P. and Gendreau, M., 1997, Fondements et applications des méthodes de recherche avec tabous, *RAIRO (Recherche opérationnelle)* **31**:133–159.

Taillard, É. D., 1990, Some efficient heuristic methods for the flow shop sequencing problem, *Eur. J. Oper. Res.* **47**:65–74.

Taillard, É. D., 1991, Robust taboo search for the quadratic assignment problem, *Parallel Comput.* **17**:443–455.

Vaessens, R. J. M., Aarts E. H. L. and Lenstra, J. K., 1996, Job shop scheduling by local search, *INFORMS J. Comput.* **8**:302–317.

Voss, S., Martello, S., Osman, I. H. and Roucairol, C., eds, 1999, *Meta-Heuristics: Advances and Trends in Local Search Paradigms for Optimization*, Kluwer, Boston, MA.

de Werra, D. and Hertz, A., 1989, Tabu search techniques: a tutorial and an application to neural networks, *OR Spektrum* **11**:131–141.

Chapter 7

SIMULATED ANNEALING

Emile Aarts[1,2], Jan Korst[1], Wil Michiels[1,2]

[1]*Philips Research Laboratories, Eindhoven, the Netherlands*

[2]*Eindhoven University of Technology, Eindhoven, the Netherlands*

7.1   INTRODUCTION

Many problems in engineering, planning and manufacturing can be modeled as that of minimizing or maximizing a cost function over a finite set of discrete variables. This class of so-called combinatorial optimization problems has received much attention over the last two decades and major achievements have been made in its analysis (Papadimitriou and Steiglitz, 1982). One of these achievements is the separation of this class into two subclasses. The first one contains the problems that can be solved efficiently, i.e. problems for which algorithms are known that solve each instance to optimality in polynomial time. Examples are linear programming, matching and network problems. The second subclass contains the problems that are notoriously hard, formally referred to as NP-hard (see Chapter 11 for more details). For an NP-hard algorithm it is generally believed that no algorithm exists that solves each instance in polynomial time. Consequently, there are instances that require superpolynomial or exponential time to be solved to optimality. Many known problems belong to this class and probably the best known example is the traveling salesman problem. The above-mentioned distinction is supported by a general framework in computer science called complexity theory; for a detailed introduction and an extensive listing of provably hard problems see Garey and Johnson (1979) and Ausiello et al. (1999).

Clearly, hard problems must be handled in practice. Roughly speaking this can be done by two types of algorithms of inherently different nature: either one may use *optimization algorithms* that find optimal solutions possibly using large amounts of computation time or one may use *heuristic algorithms* that find approximate solutions in small amounts of computation time. Local search algorithms are of the latter type (Aarts and Lenstra, 2003). Simulated annealing, the subject of this chapter, is among the best known local search algorithms, since it performs quite well and is widely applicable. In this chapter we present the basics of simulated annealing. First, we introduce some elementary local search concepts. We introduce basic simulated annealing as an approach following directly from the strong analogy with the physical process of the annealing of solids. We analyze the asymptotic performance of basic simulated annealing. Next, we present some cooling schedules that allow for a finite-time implementation. Finally, we discuss some issues related to the practical use of simulated annealing and conclude with some suggestions for further reading.

7.2   LOCAL SEARCH

Local search algorithms constitute a widely used, general approach to hard combinatorial optimization problems. They are typically instantiations of various general search schemes, but all have the same feature of an underlying neighborhood function, which is used to guide the search for a good solution.

An instance of a combinatorial optimization problem consists of a set S of feasible solutions and a non-negative cost function f. The problem is to find a *globally optimal solution* $i^* \in S$, i.e. a solution with optimal cost f^*. A *neighborhood function* is a mapping $N : S \to 2^S$, which defines for each solution $i \in S$ a set $N(i) \subseteq S$ of solutions that are in some sense close to i. The set $N(i)$ is called the *neighborhood* of solution i, and each $j \in N(i)$ is called a *neighbor* of i. The simplest form of local search is called *iterative improvement*. An iterative improvement algorithm starts with an initial solution and then continuously explores neighborhoods for a solution with lower cost. If such a solution is found, then the current solution is replaced by this better solution. The procedure is continued until no better solutions can be found in the neighborhood of the current solution. By definition, iterative improvement terminates in a *local optimum*, i.e. a solution $\hat{i} \in S$ that is at least as good as all its neighbors with regard to the cost. Note that the concept of local optimality depends on the neighborhood function that is used.

For many combinatorial optimization problems one can represent solutions as sequences or collections of subsets of elements that constitute the solutions; examples are tours in the traveling salesman problem (TSP), partitions in the graph partitioning problem (GPP), and schedules in the job shop scheduling problem (JSSP). These solution representations enable the use of k-change neighborhoods, i.e. neighborhoods that are obtained by defining k exchanges of elements in a given sequence or collection of subsets. These so-called k-change neighborhoods constitute a class of basic local search algorithms that is widely applicable; see Lin (1965) and Lin and Kernighan (1973) for the TSP,

Kernighan and Lin (1970) for the GPP, and Van Laarhoven et al. (1992) for the JSSP.

As an example we discuss the TSP. In an instance of the TSP we are given n cities and an $n \times n$-matrix $[d_{pq}]$, whose elements denote the distance from city p to city q for each pair p, q of cities. A tour is defined as a closed path visiting each city exactly once. The problem is to find a tour of minimal length. For this problem, a solution is given by a permutation $\pi = (\pi(1), \ldots, \pi(n))$. Each solution then corresponds uniquely to a tour. The solution space is given by

$$S = \{\text{all permutations } \pi \text{ on } n \text{ cities}\}$$

The cost function is defined as

$$f(\pi) = \sum_{i=1}^{n-1} d_{\pi(i),\pi(i+1)} + d_{\pi(n),\pi(1)}$$

i.e. $f(\pi)$ gives the length of the tour corresponding to π. Furthermore, we have $|S| = (n-1)!$

In the TSP, a neighborhood function N_k, called the k-change neighborhood, defines, for each solution i, a neighborhood S_i consisting of the set of solutions that can be obtained from the given solution i by removing $k' \le k$ edges from the tour corresponding to solution i, replacing them with k' other edges such that again a tour is obtained, and choosing the direction of the tour arbitrarily (Lin, 1965; Lin and Kernighan, 1973). The simplest non-trivial version of this is the 2-change neighborhood. In that case we have

$$N = \{\pi' \in S \mid \pi' \text{ is obtained from } \pi \text{ by a 2-change}\}$$

and

$$|N| = 2 + n(n-3), \quad \text{for all } \pi \in S$$

Furthermore, it can be shown that each solution j can be obtained from any other solution i by at most $n - 2$ successive 2-changes, which implies that the N_2 neighborhood graph is strongly connected.

In general, local search can be viewed as a walk in a *neighborhood graph*. The node set of the neighborhood graph is given by the set of solutions and there is an arc from node i to node j if j is a neighbor of i. The sequence of nodes visited by the search process defines a walk. Roughly speaking, the two main issues of a local search algorithm are the choice of the neighborhood function and the search strategy that is used. Good neighborhoods often take advantage of the combinatorial structure of the problem at hand, and are therefore typically problem dependent. A disadvantage of using iterative improvement as search strategy is that it easily gets trapped in poor local optima. To avoid this disadvantage—while maintaining the basic principle of local search algorithms, i.e. iteration among neighboring solutions—one can consider the extension of accepting, in a limited way, neighboring solutions that yield a deterioration in the value of the cost function. This in fact is the basic idea underlying simulated annealing.

7.3   BASIC SIMULATED ANNEALING

In the early 1980s Kirkpatrick et al. (1983) and independently Černý (1985) introduced the concepts of annealing in combinatorial optimization. Originally these concepts were heavily inspired by an analogy between the physical annealing process of solids and the problem of solving large combinatorial optimization problems. Since this analogy is quite appealing we use it here as a background for introducing simulated annealing.

In condensed matter physics, annealing is known as a thermal process for obtaining low energy states of a solid in a heat bath. The process consists of the following two steps (Kirkpatrick et al., 1983):

- increase the temperature of the heat bath to a maximum value at which the solid melts;

- decrease *carefully* the temperature of the heat bath until the particles arrange themselves in the ground state of the solid.

In the liquid phase, all particles arrange themselves randomly, whereas in the ground state of the solid, the particles are arranged in a highly structured lattice, for which the corresponding energy is minimal. The ground state of the solid is obtained only if the maximum value of the temperature is sufficiently high and the cooling is performed sufficiently slowly. Otherwise, the solid will be frozen into a meta-stable state rather than into the true ground state.

Metropolis et al. (1953) introduced a simple algorithm for simulating the evolution of a solid in a heat bath to thermal equilibrium. Their algorithm is based on Monte Carlo techniques (Binder, 1978) and generates a sequence of states of the solid in the following way. Given a current state i of the solid with energy E_i, then a subsequent state j is generated by applying a perturbation mechanism which transforms the current state into a next state by a small distortion, for instance by displacement of a particle. The energy of the next state is E_j. If the energy difference, $E_j - E_i$, is less than or equal to zero, the state j is accepted as the current state. If the energy difference is greater than zero, then the state j is accepted with a probability given by

$$\exp\left(\frac{E_i - E_j}{k_B T}\right)$$

where T denotes the temperature of the heat bath and k_B is a physical constant called the Boltzmann constant. The acceptance rule described above is known as the Metropolis criterion and the algorithm that goes with it is known as the Metropolis algorithm. It is known that, if the lowering of the temperature is done sufficiently slowly, the solid can reach thermal equilibrium at each temperature. In the Metropolis algorithm this is achieved by generating a large number of transitions at a given value of the temperature. Thermal equilibrium is characterized by the Boltzmann distribution, which gives the probability of the solid of being in a state i with energy E_i at temperature T, and which is given by

$$\mathbb{P}_T\{\mathbf{X} = i\} = \frac{\exp(-E_i/k_B T)}{\sum_j \exp(-E_j/k_B T)} \tag{7.1}$$

where $\mathbf{X}$ is a random variable denoting the current state of the solid and the summation extends over all possible states. As we indicate below, the Boltzmann distribution plays an essential role in the analysis of the convergence of simulated annealing.

Returning to simulated annealing, the Metropolis algorithm can be used to generate a sequence of solutions of a combinatorial optimization problem by assuming the following equivalences between a physical many-particle system and a combinatorial optimization problem:

- solutions in the combinatorial optimization problem are equivalent to states of the physical system;

- the cost of a solution is equivalent to the energy of a state.

Furthermore, we introduce a *control parameter* which plays the role of the temperature. Simulated annealing can thus be viewed as an iteration of Metropolis algorithms, executed at decreasing values of the control parameter.

We can now let go of the physical analogy and formulate simulated annealing in terms of a local search algorithm. To simplify the presentation, we assume, in the remainder of this chapter, that we are dealing with a minimization problem. The discussion easily translates to maximization problems. For an instance (S, f) of a combinatorial optimization problem and a neighborhood function N, Figure 7.1 describes simulated annealing in pseudo-code.

The meaning of the four functions in the procedure SIMULATED_ANNEALING is obvious: INITIALIZE computes a start solution and initial values of the parameters c and L; GENERATE selects a solution from the neighborhood of the current solution; CALCULATE_LENGTH and CALCULATE_CONTROL compute new values for the parameters L and c, respectively.

As already mentioned, a typical feature of simulated annealing is that, besides accepting improvements in cost, it also accepts deteriorations to a limited extent. Initially, at large values of c, large deteriorations will be accepted; as c decreases, only smaller deteriorations will be accepted and, finally, as the value

---

procedure SIMULATED_ ANNEALING;

begin

    INITIALIZE (i_{start}, c_0, L_0);
    $k := 0$;
    $i := i_{start}$;

    repeat

        for $l := 1$ to L_k do

        begin

            GENERATE (j from S_i);
            if $f(j) \le f(i)$ then $i := j$
            else
            if $\exp\left(\frac{f(i)-f(j)}{c_k}\right) > \text{random}[0, 1)$ then $i := j$

        end;

        $k := k + 1$;
        CALCULATE_ LENGTH (L_k);
        CALCULATE_ CONTROL (c_k);

    until stopcriterion

end;

Figure 7.1.   The simulated annealing algorithm in pseudo-code.

of c approaches 0, no deteriorations will be accepted at all. Furthermore, there is no limitation on the size of a deterioration with respect to its acceptance. In simulated annealing, arbitrarily large deteriorations are accepted with positive probability; for these deteriorations the acceptance probability is small, however. This feature means that simulated annealing, in contrast to iterative improvement, can escape from local minima while it still exhibits the favorable features of iterative improvement, i.e. simplicity and general applicability.

Note that the probability of accepting deteriorations is implemented by comparing the value of $\exp((f(i) - f(j))/c)$ with a random number generated from a uniform distribution on the interval [0,1). Furthermore, it should be obvious that the speed of convergence of the algorithm is determined by the choice of the parameters L_k and c_k with $k = 0, 1, \ldots$, where L_k and c_k denote the values of L and k in iteration k of the algorithm. In the next section we will argue that under certain mild conditions on the choice of the parameters simulated annealing converges asymptotically to globally optimal solutions, and that it exhibits an equilibrium behavior from which some performance characteristics can be derived. In the subsequent section we present more practical, implementation-oriented choices of the parameter values that lead to a finite-time execution of the algorithm.

Comparing simulated annealing to iterative improvement, it is evident that simulated annealing can be viewed as a generalization. Simulated annealing becomes identical to iterative improvement in the case where the value of the control parameter is taken equal to zero. With respect to a comparison between the performance of both algorithms we mention that for most problems simulated annealing performs better than iterative improvement, repeated for a number of different initial solutions. We return to this subject in the concluding sections.

7.4 MATHEMATICAL MODELING

Simulated annealing can be mathematically modeled by means of Markov chains (Feller, 1950; Isaacson and Madsen, 1976; Seneta, 1981). In this model, we view simulated annealing as a process in which a sequence of Markov chains is generated, one for each value of the control parameter. Each chain consists of a sequence of trials, where the outcomes of the trials correspond to solutions of the problem instance.

Let (S, f) be a problem instance, N a neighborhood function, and $\mathbf{X}(k)$ a stochastic variable denoting the outcome of the kth trial. Then the *transition probability* at the kth trial for each pair $i, j \in S$ of outcomes is defined as

$$P_{ij}(k) = \mathbb{P}\{\mathbf{X}(k) = j | \mathbf{X}(k-1) = i\}$$

$$= \begin{cases} G_{ij}(c_k) A_{ij}(c_k) & \text{if } i \neq j \\ 1 - \sum_{l \in S, l \neq i} G_{il}(c_k) A_{il}(c_k) & \text{if } i = j \end{cases} \tag{7.2}$$

where $G_{ij}(c_k)$ denotes the *generation probability*, i.e. the probability of generating a solution j when being at solution i, and $A_{ij}(c_k)$ denotes the *acceptance probability*, i.e. the probability of accepting solution j, once it is generated from solution i. The most frequently used choice for these probabilities is the following (Aarts and Korst, 1989):

$$G_{ij}(c_k) = \begin{cases} |N(i)|^{-1} & \text{if } j \in S_i \\ 0 & \text{if } j \notin S_i \end{cases} \tag{7.3}$$

and

$$A_{ij}(c_k) = \begin{cases} 1 & \text{if } f(j) \leq f(i) \\ \exp((f(i) - f(j))/c) & \text{if } f(j) > f(i) \end{cases} \tag{7.4}$$

For fixed values of c, the probabilities do not depend on k, in which case the resulting Markov chain is *time-independent* or *homogeneous*. Using the theory of Markov chains it is fairly straightforward to show that, under the condition that the neighborhoods are strongly connected—in which case the Markov chain is *irreducible* and *aperiodic*—there exist a unique stationary distribution of the outcomes. This distribution is the probability distribution of the solutions after an infinite number of trials and assumes the following form (Aarts and Korst, 1989).

THEOREM 7.1 *Given an instance (S, f) of a combinatorial optimization problem and a suitable neighborhood function, then, after a sufficiently large number of transitions at a fixed value c of the control parameter, applying the transition probabilities of (7.2)–(7.4), simulated annealing will find a solution $i \in S$ with a probability given by*

$$\mathbb{P}_c\{\mathbf{X} = i\} \stackrel{\text{def}}{=} q_i(c) = \frac{1}{N_0(c)} \exp\left(-\frac{f(i)}{c}\right) \tag{7.5}$$

where $\mathbf{X}$ is a stochastic variable denoting the current solution obtained by simulated annealing and

$$N_0(c) = \sum_{j \in S} \exp\left(-\frac{f(j)}{c}\right) \tag{7.6}$$

denotes a normalization constant.

A proof of this theorem is beyond the scope of this chapter. For those interested, we refer to Aarts and Korst (1989). The probability distribution of (7.5) is called the stationary or equilibrium distribution and it is the equivalent of the Boltzmann distribution of (7.1). We can now formulate the following important result.

COROLLARY 7.2 *Given an instance (S, f) of a combinatorial optimization problem and a suitable neighborhood function, and, furthermore, let the probability $q_i(c)$ that simulated annealing finds solution i after an infinite number of trials at value c of the control parameter be given by (7.5), then*[1]

$$\lim_{c \downarrow 0} q_i(c) \stackrel{\text{def}}{=} q_i^* = \frac{1}{|S^*|} \chi_{(S^*)}(i) \tag{7.7}$$

where S^* denotes the set of globally optimal solutions.

---

[1]Let A and $A' \subset A$ be two sets. Then the characteristic function $\chi_{(A')} : A \to \{0, 1\}$ of the set A' is defined as $\chi_{(A')}(a) = 1$ if $a \in A'$, and $\chi_{(A')}(a) = 0$ otherwise.

Proof. Using the fact that for all $a \leq 0$, $\lim_{x \downarrow 0} e^{\frac{a}{x}} = 1$ if $a = 0$, and 0 otherwise, we obtain

$$
\begin{aligned}
\lim_{c \downarrow 0} q_i(c) &= \lim_{c \downarrow 0} \frac{\exp\left(-\frac{f(i)}{c}\right)}{\sum_{j \in S} \exp\left(-\frac{f(j)}{c}\right)} \\
&= \lim_{c \downarrow 0} \frac{\exp\left(\frac{f^* - f(i)}{c}\right)}{\sum_{j \in S} \exp\left(\frac{f^* - f(j)}{c}\right)} \\
&= \lim_{c \downarrow 0} \frac{1}{\sum_{j \in S} \exp\left(\frac{f^* - f(j)}{c}\right)} \chi_{(S^*)}(i) \\
&\quad + \lim_{c \downarrow 0} \frac{\exp\left(\frac{f^* - f(i)}{c}\right)}{\sum_{j \in S} \exp\left(\frac{f^* - f(j)}{c}\right)} \chi_{(S \backslash S^*)}(i) \\
&= \frac{1}{|S^*|} \chi_{(S^*)}(i) + \frac{0}{|S^*|} \chi_{(S \backslash S^*)}(i)
\end{aligned}
$$

which completes the proof. □

As already mentioned, the result of this corollary is important since it guarantees asymptotic convergence of the simulated annealing algorithm to the set of globally optimal solutions under the condition that the stationary distribution of (7.5) is attained at each value of c. More specifically, it implies that asymptotically optimal solutions are obtained which can be expressed as

$$
\lim_{c \downarrow 0} \lim_{k \to \infty} \mathbb{P}_c\{\mathbf{X}(k) \in S^*\} = 1
$$

We end this section with some remarks.

- It is possible to formulate a more general class of acceptance and generation probabilities than the ones we considered above, and prove asymptotic convergence to optimality in that case. The probabilities we used above are imposed by this more general class in a natural way and used in practically all applications reported in the literature.

- The simulated annealing algorithm can also be formulated as an *inhomogeneous algorithm*, namely as a single inhomogeneous Markov chain, where the value of the control parameter c is decreased in between subsequent trials. In this case, asymptotic convergence again can be proved.

However, an additional condition on the sequence $\{c_k\}$ of values of the control parameter is needed, namely

$$c_k \geq \frac{\Gamma}{\log(k+2)} \qquad k = 0, 1, \ldots$$

for some constant Γ that can be related to the neighborhood function that is applied.

- Asymptoticity estimates of the rate of convergence show that the stationary distribution of simulated annealing can only be approximated arbitrarily closely if the number of transitions is proportional to $|S|^2$. For hard problems, $|S|$ is necessarily exponential in the size of the problem instance, thus, implying that approximating the asymptotic behavior arbitrarily close results in an exponential-time execution of simulated annealing. Similar results have been derived for the asymptotic convergence of the inhomogeneous algorithm.

Summarizing, simulated annealing can find optimal solutions with probability 1 if it is allowed an infinite number of transitions. In Section 7.6 we show how a more efficient finite-time implementation of simulated annealing can be obtained. Evidently, this will be at the cost of the guarantee of obtaining optimal solutions. Nevertheless, practice shows that high-quality solutions can be obtained in this way.

7.5   EQUILIBRIUM STATISTICS

In this section, we discuss some characteristic features of simulated annealing under the assumption that we are at equilibrium, i.e. at the stationary distribution $q(c)$ given by (7.5). The expected cost $\mathbb{E}_c(f)$ at equilibrium is defined as

$$\mathbb{E}_c(f) \stackrel{\text{def}}{=} \langle f \rangle_c$$

$$= \sum_{i \in S} f(i) \mathbb{P}_c\{\mathbf{X} = i\}$$

$$= \sum_{i \in S} f(i) q_i(c) \tag{7.8}$$

Similarly, the expected squared cost $\mathbb{E}_c(f^2)$ is defined as

$$\mathbb{E}_c(f^2) \overset{\text{def}}{=} \langle f^2 \rangle_c$$

$$= \sum_{i \in S} f^2(i) \mathbb{P}_c\{\mathbf{X} = i\}$$

$$= \sum_{i \in S} f^2(i) q_i(c) \tag{7.9}$$

Using the above definitions, the variance $\mathrm{Var}_c(f)$ of the cost is given by

$$\mathrm{Var}_c(f) \overset{\text{def}}{=} \sigma_c^2$$

$$= \sum_{i \in S} (f(i) - \mathbb{E}_c(f))^2 \mathbb{P}_c\{\mathbf{X} = i\}$$

$$= \sum_{i \in S} (f(i) - \langle f \rangle_c)^2 q_i(c)$$

$$= \langle f^2 \rangle_c - \langle f \rangle_c^2 \tag{7.10}$$

The notation $\langle f \rangle_c$, $\langle f^2 \rangle_c$, and σ_c^2 is introduced as shorthand notation, and will be used in the remainder of this paper.

COROLLARY 7.3 *Let the stationary distribution be given by (7.5), then the following relation holds:*

$$\frac{\partial}{\partial c} \langle f \rangle_c = \frac{\sigma_c^2}{c^2} \tag{7.11}$$

Proof. The relation can be straightforwardly verified by using (7.8) and by substituting the expression for the stationary distribution given by (7.5). $\square$

COROLLARY 7.4 *Let the stationary distribution be given by (7.5). Then we have*

$$\lim_{c \to \infty} \langle f \rangle_c \overset{\text{def}}{=} \langle f \rangle_\infty = \frac{1}{|S|} \sum_{i \in S} f(i) \tag{7.12}$$

$$\lim_{c \downarrow 0} \langle f \rangle_c = f^* \tag{7.13}$$

$$\lim_{c \to \infty} \sigma_c^2 \overset{\text{def}}{=} \sigma_\infty^2 = \frac{1}{|S|} \sum_{i \in S} (f(i) - \langle f \rangle_\infty)^2 \tag{7.14}$$

and

$$\lim_{c \downarrow 0} \sigma_c^2 = 0 \qquad (7.15)$$

Proof. The relations can be easily verified by using the definitions of the expected cost (7.8) and the variance (7.10), and by substituting the stationary distribution of (7.5) and applying similar arguments as in the proof of Corollary 7.11.   □

Using (7.11), it follows that during execution of simulated annealing the expected cost decreases monotonically—provided equilibrium is reached at each value of the control parameter—to its final value, i.e. f^*. The dependence of the stationary distribution of (7.5) on the control parameter c is the subject of the following corollary.

COROLLARY 7.5 *Let (S, f) denote an instance of a combinatorial optimization problem with $S^* \neq S$, and let $q_i(c)$ denote the stationary distribution associated with simulated annealing and given by (7.5). Then we have*

(i) $\forall i \in S^*$

$$\frac{\partial}{\partial c} q_i(c) < 0$$

(ii) $\forall i \in S \backslash S^*, f(i) \geq \langle f \rangle_\infty$

$$\frac{\partial}{\partial c} q_i(c) > 0$$

(iii) $\forall i \in S \backslash S^*, f(i) < \langle f \rangle_\infty, \exists \tilde{c}_i > 0$

$$\begin{aligned} \frac{\partial}{\partial c} q_i(c) \quad &< \quad 0 \ \ if \ c > \tilde{c}_i \\ &= \quad 0 \ \ if \ c = \tilde{c}_i \\ &> \quad 0 \ \ if \ c < \tilde{c}_i \end{aligned}$$

Proof. From (7.6) we can derive the following expression:

$$\frac{\partial}{\partial c} N_0(c) = \sum_{j \in S} \frac{f(j)}{c^2} \exp\left(\frac{-f(j)}{c}\right)$$

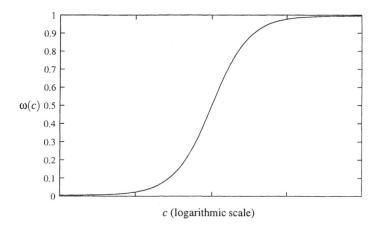

Figure 7.2. Acceptance ratio as function of the control parameter.

Hence, we obtain

$$\frac{\partial}{\partial c} q_i(c) = \frac{\partial}{\partial c} \frac{\exp\left(\frac{-f(i)}{c}\right)}{N_0(c)}$$

$$= \left\{ \frac{f(i)}{c^2} \frac{\exp\left(\frac{-f(i)}{c}\right)}{N_0(c)} - \frac{\exp\left(\frac{-f(i)}{c}\right)}{N_0^2(c)} \frac{\partial}{\partial c} N_0(c) \right\}$$

$$= \frac{q_i(c)}{c^2} f(i) - \frac{q_i(c)}{c^2} \frac{\sum_{j \in S} f(j) \exp\left(\frac{-f(j)}{c}\right)}{N_0(c)}$$

$$= \frac{q_i(c)}{c^2} (f(i) - \langle f \rangle_c) \qquad (7.16)$$

Thus, the sign of $\frac{\partial}{\partial c} q_i(c)$ is determined by the sign of $f(i) - \langle f \rangle_c$ since $\frac{q_i(c)}{c^2} > 0$, for all $i \in S$ and $c > 0$.

From (7.11)–(7.13) we have that $\langle f \rangle_c$ increases monotonically from f^* to $\langle f \rangle_\infty$ with increasing c, provided $S^* \neq S$. The remainder of the proof is now straightforward.

If $i \in S^*$ and $S \neq S^*$, then $f(i) < \langle f \rangle_c$. Hence, $\frac{\partial}{\partial c} q_i(c) < 0$ (cf. (7.16)), which completes the proof of part (i). If $i \notin S^*$, then the sign of $\frac{\partial}{\partial c} q_i(c)$ depends on the value of $\langle f \rangle_c$. Hence, using (7.16), we have that $\forall i \in S \backslash S^*$: $\frac{\partial}{\partial c} q_i(c) > 0$ if $f(i) \geq \langle f \rangle_\infty$, whereas $\forall i \in S \backslash S^*$, where $f(i) < \langle f \rangle_\infty$, there exists a $\tilde{c}_i > 0$ at which $f(i) - \langle f \rangle_c$ changes sign. Conse-

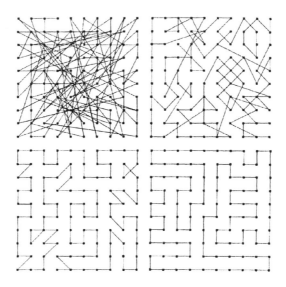

Figure 7.3. Evolution of simulated annealing for an instance with 100 cities on a regular grid.

quently, we have

$$
\frac{\partial}{\partial c} q_i(c) \quad < \quad 0 \ \text{if} \ c > \tilde{c}_i
$$
$$
= \quad 0 \ \text{if} \ c = \tilde{c}_i
$$
$$
> \quad 0 \ \text{if} \ c < \tilde{c}_i
$$

which completes the proofs of parts (ii) and (iii).   □

From Corollary 7.5 it follows that the probability of finding an optimal solution increases monotonically with decreasing c. Furthermore, for each solution, not being an optimal one, there exists a positive value of the control
parameter $\tilde{c}_i$, such that for $c < \tilde{c}_i$, the probability of finding that solution decreases monotonically with decreasing c.

We complete this section with some results that illustrate some of the elements discussed in the analysis presented above. For this we need the definition
of the *acceptance ratio* $\omega(c)$ which is defined as

$$
\omega(c) = \left. \frac{\text{number of accepted transitions}}{\text{number of proposed transitions}} \right|_c \tag{7.17}
$$

Figure 7.2 shows the behavior of the acceptance ratio as a function of the value
of the control parameter for typical implementations of simulated annealing.

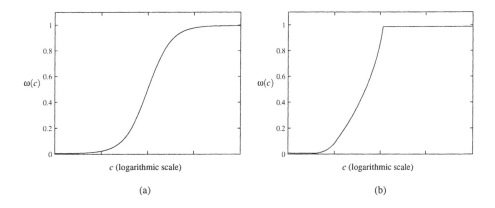

Figure 7.4. (a) Normalized average value $\frac{\langle f\rangle_c - f^*}{\langle f\rangle_\infty - f^*}$, and (b) normalized spreading $\frac{\sigma_c}{\sigma_\infty}$ of the cost function, both as a function of the control parameter.

The figure illustrates the behavior as it would be expected from the acceptance criterion given in (7.4). At large values of c, virtually all proposed transitions are accepted. As c decreases, ever fewer proposed transitions are accepted, and finally, at very small values of c, no proposed transitions are accepted at all.

Figure 7.3 shows four solutions in the evolution of simulated annealing running on a TSP instance with 100 cities on the positions of a 10×10 grid. The initial solution at the top left is given by a random sequence among 100 cities, which is evidently far from optimal. It looks very chaotic, and the corresponding value of the tour length is large. In the course of the optimization process the solutions become less and less chaotic (top right and bottom left), and the tour length decreases. Finally, the optimal solution shown at the bottom right is obtained. This solution has a highly regular pattern for which the tour length is minimal.

Figure 7.4 shows the typical behavior of (a) the normalized average cost and (b) the normalized spreading of the cost for simulated annealing as a function of the control parameter c. The typical behavior shown in this figure is observed for many different problem instances and is reported in the literature by a number of authors (Aarts et al., 1988; Hajek, 1985; Kirkpatrick et al., 1983; Van Laarhoven and Aarts, 1987; White, 1984).

From the figures we can deduce some characteristic features of the expected cost $\langle f\rangle_c$ and the variance σ_c^2 of the cost. First, it is observed that for large values of c the average and the spreading of the cost are about constant and equal to $\langle f\rangle_\infty$ and σ_∞, respectively. This behavior is directly explained from (7.12) and (7.14), from which it follows that both the average value and the spreading of the cost function are constant at large c-values.

Secondly, we observe that there exists a threshold value c_t of the control parameter for which

$$\langle f \rangle_{c_t} \approx \frac{1}{2}(\langle f \rangle_\infty + f^*) \tag{7.18}$$

and

$$\begin{aligned} \sigma_c^2 &\approx \sigma_\infty^2 && \text{if } c \geq c_t \\ &< \sigma_\infty^2 && \text{if } c < c_t \end{aligned} \tag{7.19}$$

Moreover, we mention that c_t is roughly the value of c for which $\omega(c) \approx 0.5$.

7.6   PRACTICAL APPLICATION

A finite-time implementation of simulated annealing is obtained by generating a sequence of homogeneous Markov chains of finite length at descending values of the control parameter. For this, a set of parameters must be specified that governs the convergence of the algorithm. These parameters are combined in what is called a cooling schedule. A *cooling schedule* specifies a finite sequence of values of the control parameter, and a finite number of transitions at each value of the control parameter. More precisely, it is specified by

- an *initial value* of the control parameter c_0,

- a *decrement function* for lowering the value of the control parameter,

- a *final value* of the control parameter specified by a *stop criterion*, and

- a finite *length* of each homogeneous Markov chain.

The search for adequate cooling schedules has been the subject of many studies over the past years. Reviews are given by Van Laarhoven and Aarts (1987), Collins et al. (1988), and Romeo and Sangiovanni-Vincentelli (1991). Below we discuss some results.

Most of the existing work on cooling schedules presented in the literature deals with heuristic schedules. We distinguish between two broad classes: static and dynamic schedules. In a static cooling schedule the parameters are fixed; they cannot be changed during execution of the algorithm. In a dynamic cooling schedule the parameters are adaptively changed during execution of the algorithm. Below we present some examples.

7.6.1   Static Cooling Schedules

The following simple schedule is known as the geometric schedule. It originates from the early work on cooling schedules by Kirkpatrick et al. (1983), and is still used in many practical situations.

Initial value of the control parameter. To ensure a sufficiently large value of $\omega(c_0)$, one may choose $c_0 = \Delta f_{max}$, where Δf_{max} is the maximal difference in cost between any two neighboring solutions. Exact calculation of Δf_{max} is quite time consuming in many cases. However, one often can give simple estimates of its value.

Lowering the control parameter value. A frequently used decrement function is given by

$$c_{k+1} = \alpha \cdot c_k \quad k = 0, 1, \ldots$$

where α is a positive constant smaller than but close to 1. Typical values lie between 0.8 and 0.99.

Final value of the control parameter. The final value is fixed at some small value, which may be related to the smallest possible difference in cost between two neighboring solutions.

Markov chain length. The length of Markov chains is fixed by some number that may be related to the size of the neighborhoods in the problem instance at hand.

7.6.2 Dynamic Cooling Schedules

There exist many extensions of the simple static schedule presented above that lead to a dynamic schedule. For instance, a sufficiently large value of c_0 may be obtained by requiring that the initial acceptance ratio $\omega(c_0)$ is close to 1. This can be achieved by starting off at a small positive value of c_0 and multiplying it with a constant factor, larger than 1, until the corresponding value of $\omega(c_0)$, which is calculated from a number of generated transitions, is close to 1. Typical values of $\omega(c_0)$ lie between 0.9 and 0.99. An adaptive calculation of the final value of the control parameter may be obtained by terminating the execution of the algorithm at a c_k-value for which the value of the cost function of the solution obtained in the last trial of a Markov chain remains unchanged for a number of consecutive chains. Clearly, such a value exists for each local minimum that is found. The length of Markov chains may be determined by requiring that at each value c_k, a minimum number of transitions is accepted. However, since transitions are accepted with decreasing probability, one would obtain $L_k \to \infty$ for $c_k \downarrow 0$. Therefore, L_k is usually bounded by some constant L_{max} to avoid extremely long Markov chains for small values of c_k.

In addition to this basic dynamic schedule the literature presents a number of more elaborate schedules. Most of these schedules are based on a statistical analysis of simulated annealing using the equilibrium statistics of the previous section.

7.7 TRICKS OF THE TRADE

To apply simulated annealing in practice three basic ingredients are needed: a concise problem representation, a neighborhood, and a cooling schedule. The algorithm is usually implemented as a sequence of homogeneous Markov chains of finite length, generated at descending values of the control parameter. This is specified by the cooling schedule. As for the choice of the cooling schedule, we have seen in the previous section that there exist some general guidelines. However, for the other ingredients no general rules are known that guide their choice. The way they are handled is still a matter of experience, taste, and skill left to the annealing practitioner, and we expect that this will not change in the near future.

Ever since its introduction in 1983, simulated annealing has been applied to a large number of different combinatorial optimization problems in areas as diverse as operations research, VLSI design, code design, image processing, and molecular physics. The success of simulated annealing can be characterized by the following elements:

- performance, i.e. running time and solution quality,

- ease of implementation, and

- applicability and flexibility.

With respect to the last two items we make the following remarks. It is apparent that simulated annealing is conceptually very simple and quite easy to implement. Implementation of the algorithm typically takes only a few hundred lines of computer code. Experience shows that implementations for new problems often take only a few days.

With respect to applicability and flexibility it has become obvious as a result of the overwhelming amount of practical experience that has been gathered over the past 20 years that simulated annealing can be considered as one of the most flexible and applicable algorithms that exist. However, one must bear in mind that it is not always trivial to apply the algorithm effectively to a given problem. Finding appropriate neighborhoods requires problem insight, and sometimes it is necessary to reformulate the problem or transform it into an equivalent or similar problem, before simulated annealing can be applied successfully. An example is graph coloring (Korst and Aarts, 1989).

With respect to performance, one typically trades of solution quality against running times. Performance analyses of simulated annealing algorithms have been the subject of many studies. Despite numerous studies it is still difficult to judge simulated annealing on its true merits. This is predominantly due to the fact that many of these studies lack the depth required to draw reliable conclusions. For example, results are often limited to one single run of the algorithm, instead of taking the average over a number of runs, the applied cooling schedules are often too simple and do not get the best out of the algorithm, and results are often not compared to the results obtained with other (tailored) algorithms.

Lining up the most important and consistent results from the literature allows the following general observations.

- *Mathematical problems,* predominantly based on Johnson et al. (1989, 1991), and Van Laarhoven and Aarts (1987):

  - For graph partitioning problems, simulated annealing generally performs better, both with respect to error and running time, than the classical edge-interchange algorithms introduced by Kernighan and Lin (1970).

  - For a large class of basic problems, including the graph coloring, linear arrangement, matching, quadratic assignment, and scheduling problems, simulated annealing finds solutions with an error comparable to the error of tailored approximation algorithms but at the cost of much larger running times.

  - For some basic problems such as the number partitioning problem and the traveling salesman problem, simulated annealing is outperformed by tailored heuristics, both with respect to error and running time.

- *Engineering problems* (folklore). For many engineering problems, for example problems in the field of image processing, VLSI design and code design, no tailored approximation algorithms exist. For these problems simulated annealing seems to be a panacea. For instance, for the football pool problem (Van Laarhoven et al., 1989), it was able to derive solutions better than the best solutions found so far and for the VLSI placement problem (Sechen and Sangiovanni-Vincentelli, 1985) it outperforms the time-consuming manual process. For some problems, however, the running time can be very large.

- Comparing simulated annealing to time-equivalent iterative improvement using the same neighborhood function, i.e. repeating iterative improvement with different initial solutions for an equally long time as the running time of simulated annealing and keeping the best solution, reveals that simulated annealing performs substantially better (smaller error). This difference becomes even more pronounced for larger problem instances (Van Laarhoven et al., 1992; Van Laarhoven, 1988).

- Experience shows that the performance of simulated annealing depends as much on the skill and effort that is applied to the implementation as on the algorithm itself. For instance, the choice of an appropriate neighborhood function, of an efficient cooling schedule, and of sophisticated data structures that allow fast manipulations can substantially reduce the error as well as the running time. Thus, in view of this and considering the simple nature of annealing, there lies a challenge in constructing efficient and effective implementations of simulated annealing.

7.8   CONCLUSIONS

Since its introduction in 1983, simulated annealing has been applied to a fairly large amount of different problems in many different areas. More than 20 years of experience had led to the following general observations.

- High-quality solutions can be obtained but sometimes at the cost of large amounts of computation time.

- In many practical situations, where no tailored algorithms are available, the algorithm is a real boon due to its general applicability and its ease of implementation.

So, simulated annealing is an algorithm that all practical mathematicians and computer scientists should have in their toolbox.

SOURCES OF ADDITIONAL INFORMATION

Introductory textbooks describing both theoretical and practical issues of simulated annealing are given by Aarts and Korst (1989) and Van Laarhoven and Aarts (1987). Salamon et al. (2002) present a basic text book on simulated annealing with recent improvements for practical implementations and references to software tools. Azencott (1992) presents a theoretical text book on parallelization techniques for simulated annealing for the purpose of speeding up the algorithm through effective parallel implementations.

Early proofs of the asymptotic convergence of the homogeneous Markov model for simulated annealing are presented by Aarts and Van Laarhoven (1985) and Lundy and Mees (1986). Proofs for the inhomogeneous algorithm have been published by Connors and Kumar (1987), Gidas (1985), and Mitra et al. (1986). Hajek (1988) was the first to present necessary and sufficient conditions for the asymptotic convergence of simulated annealing. Anily and Federgruen (1987) present theoretical results on the convergence of simulated annealing for a set of acceptance probabilities that are more general than the classical Metropolis acceptance probabilities. A comprehensive review of the theory of simulated annealing is given by Romeo and Sangiovanni-Vincentelli (1991).

Strenski and Kirkpatrick (1991) present an early analysis of the finite-time behavior of simulated annealing for various cooling schedules. Steinhöfel et al.

(1998) present a comparative study in which they investigate the performance of simulated annealing for different cooling schedules when applied to job shop scheduling. Nourani and Andersen (1998) present a comparative study in which they investigate the performance of simulated annealing with cooling schedules applying different types of decrement functions for lowering the value of the control parameter. Andersen (1996) elaborates on the thermodynamical analysis of finite-time implementations of simulated annealing. Orosz and Jacobson (2002) study the finite-time behavior of a special variant of simulated annealing in which the value of the control parameters is kept constant during the annealing process. Park and Kim (1998) present a systematic approach to the problem of choosing appropriate values for the parameters in a cooling schedule.

Vidal (1993) presents an edited collection of papers on practical aspects of simulated annealing, ranging from empirical studies of cooling schedules up to implementation issues of simulated annealing for problems in engineering and planning. Eglese (1990) presents a survey of the application of simulated annealing to problems in operations research. Collins et al. (1988) present an annotated bibliography with more than a thousand references to papers on simulated annealing. It is organized in two parts; one on the theory, and the other on applications. The applications range from graph-theoretic problems up to problems in engineering, biology, and chemistry. Fox (1993) discusses the integration of simulated annealing with other local search heuristics such as tabu search and genetic algorithms.

References

Aarts, E. H. L. and Korst, J. H. M., 1989, *Simulated Annealing and Boltzmann Machines,* Wiley, Chichester.

Aarts, E. H. L. and van Laarhoven, P. J. M., 1985, Statistical cooling: a general approach to combinatorial optimization problems, *Philips J. Res.* **40**:193–226.

Aarts, E. H. L., Korst, J. H. M. and van Laarhoven, P. J. M., 1988, A quantitative analysis of the simulated annealing algorithm: a case study for the traveling salesman problem, *J. Statist. Phys.* **50**:189–206.

Aarts, E. H. L. and Lenstra, J. K., eds, 2003, *Local Search in Combinatorial Optimization,* Princeton University Press, Princeton, NJ.

Andersen, B., 1996, Finite-time thermodynamics and simulated annealing, in: *Entropy and Entropy Generation,* J. S. Shiner, ed., Kluwer, Dordrecht, pp. 111–127.

Anily, S. and Federgruen, A., 1987, Simulated annealing methods with general acceptance probabilities, *J. Appl. Probab.* **24**:657–667.

Ausiello, G., Crescenzi, P., Gambosi, G., Kann, V., Marchetti-Spaccamela, A. and Protasi, M., 1999, *Complexity and Approximation: Combinatorial Optimization Problems and Their Approximability Properties,* Springer, Berlin.

Azencott, R., 1992, *Simulated Annealing: Parallelization Techniques,* Wiley, Chichester.

Binder, K., 1978, *Monte Carlo Methods in Statistical Physics,* Springer, Berlin.

Černý, V., 1985, Thermodynamical approach to the traveling salesman problem: an efficient simulation algorithm, *J. Optim. Theory Appl.* **45**:41–51.

Collins, N. E., Eglese, R. W. and Golden, B. L., 1988, Simulated annealing: An annotated bibliography, *Am. J. Math. Manage. Sci.* **8**:209–307.

Connors, D. P. and Kumar, P. R., 1987, Simulated annealing and balance of recurrence order in time-inhomogeneous Markov chains, in: *Proc. 26th IEEE Conf. on Decision and Control,* pp. 2261–2263.

Eglese, R. W., 1990, Simulated annealing: a tool for operational research, *Eur. J. Oper. Res.* **46**:271–281.

Feller, W., 1950, *An Introduction to Probability Theory and Its Applications I,* Wiley, New York.

Fox, B. L., 1993, Integrating and accelerating tabu search, simulated annealing, and genetic algorithms, in: *Tabu Search,* Annals of Operations Research, Vol. 41, F. Glover, E. Taillard, M. Laguna, and D. de Werra, eds, Baltzer, Basel, pp. 47–67.

Garey, M. R. and Johnson, D. S., 1979, *Computers and Intractability: A Guide to the Theory of NP-Completeness,* Freeman, San Francisco.

Gidas, B., 1985, Nonstationary Markov chains and convergence of the annealing algorithm, *J. Statist. Phys.* **39**:73–131.

Hajek, B., 1985, A tutorial survey of the theory and application of simulated annealing, in: *Proc. 24th IEEE Conf. on Decision and Control,* pp. 755–760.

Hajek, B., 1988, Cooling schedules for optimal annealing, *Math. Oper. Res.* **13**:311–329.

Isaacson, D. and Madsen, R., 1976, *Markov Chains,* Wiley, New York.

Johnson, D. S., Aragon, C. R., McGeoch, L. A. and Schevon, C., 1989, Optimization by simulated annealing: an experimental evaluation. I: graph partitioning, *Oper. Res.* **37**:865–892.

Johnson, D. S., Aragon, C. R., McGeoch, L. A. and Schevon, C., 1991, Optimization by simulated annealing: an experimental evaluation. II: graph coloring and number partitioning, *Oper. Res.* **39**:378–406.

Kernighan, B. W. and Lin, S., 1970, An efficient heuristic procedure for partitioning graphs, *Bell Syst. Tech. J.* **49**:291–307.

Kirkpatrick, S., Gelatt Jr, C. D. and Vecchi, M. P., 1983, Optimization by simulated annealing, *Science* **220**:671–680.

Korst, J. H. M. and Aarts, E. H. L., 1989, Combinatorial optimization on a Boltzmann machine, *J. Parallel Distrib. Comput.* **6**:331–357.

van Laarhoven, P. J. M., 1988, Theoretical and Computational Aspects of Simulated Annealing, *Ph.D. Thesis,* Erasmus University, Rotterdam.

van Laarhoven, P. J. M. and Aarts, E. H. L., 1987, *Simulated Annealing: Theory and Applications,* Reidel, Dordrecht.

van Laarhoven, P. J. M., Aarts, E. H. L. and Lenstra, J. K., 1992, Job shop scheduling by simulated annealing, *Oper. Res.* **40**:185–201.

van Laarhoven, P. J. M., Aarts, E. H. L., van Lint, J. H. and Wille, L. T., 1989, New upper bounds for the football pool problem for 6, 7 and 8 matches, *J. Combinat. Theor.* A **52**:304–312.

Lam, J. and Delosme, J.-M., 1986, Logic minimization using simulated annealing, in: *Proc. IEEE Int. Conf. on Computer-Aided Design,* pp. 348–351.

Lin, S., 1965, Computer solutions of the traveling salesman problem, *Bell Syst. Tech. J.* **44**:2245–2269.

Lin, S. and Kernighan, B. W., 1973, An effective heuristic algorithm for the traveling salesman problem, *Oper. Res.* **21**:498–516.

Lundy, M. and Mees, A., 1986, Convergence of an annealing algorithm, *Math. Programm.* **34**:111–124.

Metropolis, M., Rosenbluth, A., Rosenbluth, M., Teller, A. and Teller, E., 1953, Equation of state calculations by fast computing machines, *J. Chem. Phys.* **21**:1087–1092.

Mitra, D., Romeo, F. and Sangiovanni-Vincentelli, A. L., 1986, Convergence and finite-time behavior of simulated annealing, *Adv. Appl. Probab.* **18**:747–771.

Nourani, Y. and Andersen, B., 1998, A comparison of simulated annealing cooling strategies, *J. Phys.* A **31**:8373–8385.

Orosz, J. E. and Jacobson, S. H., 2002, Finite-time performance analysis of static simulated annealing algorithms, *Comput. Optim. Appl.* **21**:21–53.

Papadimitriou, C. H. and Steiglitz, K., 1982, *Combinatorial Optimization: Algorithms and Complexity,* Prentice-Hall, New York.

Park, M.-W. and Kim, Y.-D., 1998, A systematic procedure for setting parameters in simulated annealing algorithms, *Comput. Oper. Res.* **25**:207–217.

Romeo, F. and Sangiovanni-Vincentelli, A., 1991, A theoretical framework for simulated annealing, *Algorithmica* **6**:302–345.

Salamon, P., Sibani, P. and Frost, R., 2002, *Facts, Conjectures, and Improvements for Simulated Annealing,* SIAM Monographs.

Sechen, C. and Sangiovanni-Vincentelli, A. L., 1985, The Timber–Wolf placement and routing package, *IEEE J. Solid-State Circuits* **30**:510–522.

Seneta, E., 1981, *Non-negative Matrices and Markov Chains*, 2nd edn, Springer, New York.

Steinhöfel, K., Albrecht, A. and Wong, C. K., 1998, On various cooling schedules for simulated annealing applied to the job shop problem, in: *Randomization and Approximation Techniques in Computer Science,* Lecture Notes in Computing Science, Vol. 1518, M. Luby, J. Rolim, and M. Serna, eds, pp. 260–279.

Strenski, P. N. and Kirkpatrick, S., 1991, Analysis of finite length annealing schedules, *Algorithmica* **6**:346–366.

Vidal, R. V. V., ed., 1993, *Applied Simulated Annealing,* Lecture Notes in Economics and Mathematical Systems, Vol. 396, Springer, Berlin.

White, S. R., 1984, Concepts of scale in simulated annealing, in: *Proc. IEEE Int. Conf. on Computer Design* (Port Chester), pp. 646–651.

Chapter 8

VARIABLE NEIGHBORHOOD SEARCH

Pierre Hansen
GERAD and HEC Montreal, Canada

Nenad Mladenović
GERAD and Mathematical Institute, SANU, Belgrade, Serbia

8.1   INTRODUCTION

Variable Neighborhood Search (VNS) is a recent metaheuristic, or framework for building heuristics, which exploits systematically the idea of neighborhood change, both in the descent to local minima and in the escape from the valleys which contain them. In this tutorial we first present the ingredients of VNS, i.e. Variable Neighborhood Descent (VND) and Reduced VNS (RVNS) followed by the basic and then the general scheme of VNS itself which contain both of them. Extensions are presented, in particular Skewed VNS (SVNS) which enhances exploration of far-away valleys and Variable Neighborhood Decomposition Search (VNDS), a two-level scheme for solution of large instances of various problems. In each case, we present the scheme, some illustrative examples and questions to be addressed in order to obtain an efficient implementation.

Let us consider a combinatorial or global optimization problem

$$\min f(x) \tag{8.1}$$

subject to

$$x \in X \tag{8.2}$$

where $f(x)$ is the *objective function* to be minimized and X the set of *feasible solutions*. A solution $x^* \in X$ is *optimal* if

$$f(x^*) \le f(x), \ \forall x \in X \tag{8.3}$$

An *exact algorithm* for problem (8.1)–(8.2), if one exists, finds an optimal solution x^*, together with the proof of its optimality, or shows that there is no feasible solution, i.e. $X = \emptyset$. Moreover, in practice, the time to do so should be finite (and not too large); if one deals with a continuous function one must admit a degree of tolerance, i.e. stop when a feasible solution x^* has been found such that

$$f(x^*) < f(x) + \varepsilon, \quad \forall x \in X \qquad (8.4)$$

or

$$\frac{f(x^*) - f(x)}{f(x^*)} < \varepsilon, \quad \forall x \in X \qquad (8.5)$$

for some small positive ε.

Numerous instances of problems of the form (8.1)–(8.2), arising in Operational Research and other fields, are too large for an exact solution to be found in reasonable time. It is well known from complexity theory (Garey and Johnson, 1979; Papadimitriou, 1994) that thousands of problems are NP-hard, that no algorithm with a number of steps polynomial in the size of the instances is known and that finding one for any such problem would entail obtaining one for any and all of them. Moreover, in some cases where a problem admits a polynomial algorithm, the power of this polynomial may be so large that realistic size instances cannot be solved in reasonable time in the worst case, and sometimes also in the average case or most of the time.

So one is often forced to resort to *heuristics*, which yield quickly an approximate solution, or sometimes an optimal solution but without proof of its optimality. Some of these heuristics have a worst-case guarantee, i.e. the solution x_h obtained satisfies

$$\frac{f(x_h) - f(x)}{f(x_h)} \le \varepsilon, \quad \forall x \in X \qquad (8.6)$$

for some ε, which is however rarely small. Moreover, this ε is usually much larger than the error observed in practice and may therefore be a bad guide in selecting a heuristic. In addition to avoiding excessive computing time, heuristics address another problem: local optima. A local optimum x_L of (8.1)–(8.2) is such that

$$f(x_L) \le f(x), \quad \forall x \in N(x_L) \cap X \qquad (8.7)$$

where $N(x_L)$ denotes a neighborhood of x_L (ways to define such a neighborhood will be discussed below). If there are many local minima, the range of values they span may be large. Moreover, the globally optimum value $f(x^*)$ may differ substantially from the average value of a local minimum, or even from the best such value among many, obtained by some simple heuristic (a phenomenon called the Tchebycheff catastrophe: see Baum, 1986). There are, however, many ways to get out of local optima and, more precisely, the valleys which contain them (or set of solutions from which the descent method under consideration leads to them).

Metaheuristics are general framework to build heuristics for combinatorial and global optimization problems. For discussion of the best-known of them the reader is referred to the books of surveys edited by Reeves (1993), Glover and Kochenberger (2003) as well as to the tutorials of the present volume. Some of the many successful applications of metaheuristics are also mentioned there.

Variable Neighborhood Search (Mladenović and Hansen, 1997; Hansen and Mladenović, 1999, 2001c, 2003) is a recent metaheuristic which systematically exploits the idea of neighborhood change, both in descent to local minima and in escape from the valleys which contain them. It exploits systematically the following facts.

FACT 1 *A local minimum with respect to one neighborhood structure is not necessarily so for another;*

FACT 2 *A global minimum is a local minimum with respect to all possible neighborhood structures.*

FACT 3 *For many problems local minima with respect to one or several neighborhoods are relatively close to each other.*

This last observation, which is empirical, implies that a local optimum often provides some information about the global one. This may for instance be several variables with the same value in both. However, it is usually not known which ones are such. An organized study of the neighborhood of this local optimum is therefore in order, until a better one is found.

Unlike many other metaheuristics, the basic schemes of VNS and its extensions are simple and require few, and sometimes no, parameters. Therefore, in addition to providing very good solutions, often in simpler ways than other methods, VNS gives insight into the reasons for such a performance, which in turn can lead to more efficient and sophisticated implementations.

The tutorial is organized as follows. In the next section, we examine the preliminary problem of gathering information about the problem under study, and evaluating it. In Section 8.3 the first ingredient of VNS, Variable Neighborhood Descent (VND), which is mostly or entirely deterministic, is studied. Section 8.4 is devoted to the second ingredient, Reduced Variable Neighborhood Search (RVNS), which is stochastic. Both ingredients are merged in the basic and the general VNS schemes, described in Section 8.5. Extensions are then considered. Skewed Variable Neighborhood Search (SVNS), which addresses the problem of getting out of very large valleys is discussed in Section 8.6. Very large instances of many problems cannot be solved globally in reasonable time; Variable Neighborhood Decomposition Search (VNDS) studied in Section 8.7 is a two-level scheme which merges VNS with successive approximation (including a two-level VNS). Various tools for analyzing in detail the performance of a VNS heuristic, and then streamlining it, are presented in Section 8.8. They include *distance-to-target diagrams* and *valley profiles*. In each of these sections basic schemes, or tools, are illustrated by examples from papers by a variety of authors. Questions to be considered in order to get an efficient implementation of VNS are also systematically listed. Promising areas of research are outlined in Section 8.9. Brief conclusions complete the tutorial in Section 8.10. Finally, sources of further information are listed.

8.2   PRELIMINARIES: DOCUMENTATION

Once a problem of the form (8.1)–(8.2) has been selected for study and approximate solution by VNS, a preliminary step is to gather in a thorough way the papers written about it or closely related problems. This may be a difficult task as papers are often numerous, dispersed among many journals and volumes of proceedings and the problem may appear (usually under different names) in several fields. Tools such as the *ISI Web of Knowledge*, *NEC Research's Citeseer* or even general web browsers such as *Google* may prove to be very useful.

There are several reasons for studying the literature on the selected problem:

(i) *Evaluating its difficulty.* Is it NP-hard? Is it *strongly NP-hard*? (Does it hence admit no fully polynomial approximation scheme?) If it is in P, what is the complexity of the best-known exact algorithm, and is it sufficiently low for realistic instances to be solvable in reasonable time?

(ii) *Evaluating the performance of previous algorithms.* Are there some instances of (preferably real-world) data for the problem available (e.g. at http://www.informs.org/Resources/Resources/Problem_Instances/)? And what are the largest instances solved exactly?

(iii) *Evaluating the performance of previous heuristics.* Which metaheuristics have been applied to this problem? What are the performances of the resulting heuristics, in terms of size of problems solved, error and computing time (assuming comparison among computing environments, if needed, can be done in a fairly realistic way)?

(iv) *What steps are used in the heuristics already proposed?* What are the corresponding neighborhoods of the current solution? Are codes for these heuristics available? Are codes for simple descent methods available?

The role of question (i) is to help to assess the need for a VNS (or other) heuristic for the problem considered. Questions (ii) and (iii) aim at obtaining a benchmark to evaluate the performance of the VNS heuristic when it is

Initialization.
Choose f, X, neighborhood structure $N(x)$, initial solution x;
Current step (Repeat).
(1) Find $x' = \text{argmin}_{x \in N(x)} f(x)$;
(2) If $f(x') < f(x)$ set $x' \leftarrow x''$ and iterate; otherwise, stop.

Figure 8.1. Steepest descent heuristic.

designed and implemented: a good heuristic should obtain optimal solutions for most and preferably all instances solved by an exact algorithm (which suffers from the additional burden of having to prove optimality). Moreover, the new heuristic should do as well as previous ones on most or all instances and substantially better than them on quite a few instances to be viewed as a real progress (doing slightly better on a few instances is not sufficient).

Question (iv) aims at providing ingredients for the VNS heuristic, notably in its VND component; it also inquires indirectly about directions not yet explored. As a by-product, it raises the question of possible re-use of software, which is reasonable for standard steps: for example, a descent with Newton's method or a variant thereof.

8.3 VARIABLE NEIGHBORHOOD DESCENT

A *steepest descent* heuristic (known also as *best improvement* local search) consists of choosing an initial solution x, finding a direction of steepest descent from x, within a neighborhood $N(x)$, and moving to the minimum of $f(x)$ within $N(x)$ along that direction; if there is no direction of descent, the heuristic stops, and otherwise it is iterated. This set of rules is summarized in Figure 8.1.

Observe that a neighborhood structure $N(x)$ is defined for all $x \in X$; in discrete optimization problems it usually consists of all vectors obtained from x by some simple modification, e.g. complementing one or two components of a 0–1 vector. Then, at each step, the neighborhood $N(x)$ of x is explored completely. As this may be time-consuming, an alternative is to use the *first descent* heuristic. Vectors $x' \in N(x)$ are then enumerated systematically and a move is made as soon as a descent direction is found. This is summarized in Figure 8.2.

VND is based on Fact 1 of Section 8.1, i.e. *a local optimum for a first type of move $x \leftarrow x'$ (or heuristic, or within the neighborhood $N_1(x)$) is not necessary for another type of move $x \leftarrow \tilde{x}$ (within neighborhood $N_2(x)$)*. It may thus

Initialization.
Choose f, X, neighborhood structure $N(x)$, initial solution x; *Current step (Repeat).*
(1) Find first solution $x' \in N(x)$;
(2) If $f(x') > f(x)$, find next solution $x'' \in N(x)$; set $x' \leftarrow x''$ and iterate (2); otherwise, set $x \leftarrow x'$ and iterate (1);
(3) If all solutions of $N(x)$ have been considered, stop.

Figure 8.2. First descent heuristic.

be advantageous to combine descent heuristics. This leads to the basic VND scheme presented in Figure 8.3.

Caution should be exercised when applying that scheme. In particular, one should consider the following questions:

(i) What complexity do the different moves have?

(ii) What is the best order in applying them?

(iii) Are the moves considered sufficient to ensure a thorough exploration of the region containing x?

(iv) How precise a solution is desired?

Question (i) aims at selecting and ranking moves: if they involve too many elementary changes (e.g. complementing three components or more of a 0–1 vector), the resulting heuristic may be very slow and often take more time than an exact algorithm on small- or medium-sized examples.

Question (ii) also bears upon computing times in relation to the quality of solutions obtained. A frequent implementation consists of ranking moves by order of complexity of their application (which is often synonymous with by size of their neighborhoods $|N_\ell(x)|$), and returning to the first one each time a direction of descent is found and a step made in that direction. Alternatively, all moves may be applied in sequence as long as descent is made for some neighborhood in the series.

Question (iii) is a crucial one: for some problems elementary moves are not sufficient to leave a narrow valley, and heuristics using them only can give very poor results. This is illustrated in Example 8.2 below.

Finally, the precision desired, as asked for in question (iv), will depend upon whether VND is used alone or within some larger framework, such as VNS itself. In the former case, one will strive to obtain the best solution possible within the allocated computing time; in the latter, one may prefer to get a good solution fairly quickly by the deterministic VND and to improve it later by faster stochastic search in VNS.

Initialization.
Select the set of neighborhood structures N_ℓ, for $\ell = 1, \ldots, \ell_{max}$, that will be used in the descent; find an initial solution x (or apply the rules to a given x);
Repeat the following sequence until no improvement is obtained:
(1) Set $\ell \leftarrow 1$;
(2) *Repeat* the following steps until $\ell = \ell_{max}$:
(a) *Exploration of neighborhood.*
Find the best neighbor x' of x $(x' \in N_\ell(x))$;
(b) *Move or not.*
If the solution x' thus obtained is better than x, set $x \leftarrow x'$ and $\ell \leftarrow 1$; otherwise, set $\ell \leftarrow \ell+1$;

Figure 8.3. Steps of the basic VND.

EXAMPLE 8.1 (SIMPLE PLANT LOCATION) *(For a survey, see Cornuejols et al., 1990). The simple (or uncapacitated) plant location problem consists of locating a set of facilities i among a given set I of m potential locations, with fixed costs f_i, in order to minimize total costs for satisfying the demand of a given set of users J with delivery costs c_{ij}, $i \in I$, $j \in J$. It is expressed as follows:*

$$\min_{x,y} z_P = \sum_{i=1}^{m} f_i y_i + \sum_{i=1}^{m} \sum_{j=1}^{n} c_{ij} x_{ij} \tag{8.8}$$

s.t.

$$\sum_{i=1}^{m} x_{ij} = 1, \ \forall j \in J \tag{8.9}$$

$$y_i - x_{ij} \geq 0, \ \forall i \in I, \ \forall j \in J \tag{8.10}$$

$$y_i \in \{0, 1\}, \ \forall i \in I \tag{8.11}$$

$$x_{ij} \geq 0, \ \forall i \in I, \ \forall j \in J \tag{8.12}$$

where $y_i = 1$ if a facility is located at i, and 0 otherwise; $x_{ij} = 1$ if demand of user j is satisfied from facility i and 0 otherwise. Note that for fixed y_i, the best solution is defined by

$$x_{ij} = \begin{cases} 1 & \text{if } c_{ij} = \min_{\ell \mid y_\ell=1} c_{\ell j} \text{ (with minimum index } \ell \text{ in case of ties)} \\ 0 & \text{otherwise} \end{cases}$$

Therefore, neighborhoods can be defined on the y_i: for example, by Hamming distance (or number of components with complementary values). A first *heuristic, "greedy", proceeds by opening a facility ℓ with minimum total cost*

$$f_\ell + \sum_j c_{\ell j} = \min_i \left\{ f_i + \sum_j c_{ij} \right\} \tag{8.13}$$

then letting

$$c_{rj} = \min_{i|y_i=1} c_{ij}, \ \forall j \tag{8.14}$$

computing the gains g_i obtained by opening a facility at i

$$g_i = \sum_j \max\{c_{rj} - c_{ij}, 0\} - f_i \tag{8.15}$$

and iteratively opening the facility for which the gain is larger, as long as it is positive. Each iteration takes $O(mn)$ time.

Once the greedy heuristic has been applied, an improved solution may be obtained by the *interchange heuristic* which proceeds iteratively to the relocation of one facility at a time in the most profitable way. With an efficient implementation, the idea of which was suggested by Whitaker (1983) for the closely related *p*-median problem, an iteration of interchange can also be made in $O(mn)$ time.

Applying in turn Greedy and Interchange is a simple case of VND. Further moves in which one facility would be closed and two opened, or two closed and one opened, or two opened and two closed would be too costly if all possible exchanges are examined.

EXAMPLE 8.2 (MINIMUM SUM-OF-SQUARES CLUSTERING)   *Given N points $a_\ell \in R^p$ the minimum sum-of-squares clustering problem consists of partitioning them in M classes (or clusters) C_j such as to minimize the sum of squared distances between the points and the centroids $\overline{x}_i$ of their clusters:*

$$\min \sum_{i=1}^m \sum_{\ell : a_\ell \in C_i} \|a_\ell - \overline{x}_i\|^2 \tag{8.16}$$

where

$$\overline{x}_i = \frac{1}{|C_i|} \sum_{\ell : a_\ell \in C_i} a_\ell \tag{8.17}$$

and $\|.\|$ denotes the Euclidean norm.

Traditional heuristics for minimum sum-of-squares clustering are

- H-Means, which proceeds from an initial partition by moving one entity x_ℓ from its cluster to another one, in a greedy way, until no further move decreases the objective function value, and

- K-Means, which proceeds from an initial partition by, alternatingly, finding the centroids of its clusters, and reassigning entities to the closest centroid, until stability is attained.

Computational experiments (Hansen and Mladenović, 2001b) show that both H-Means and K-Means may lead to very poor results for instances with large M and N (the relative error being sometimes greater than 100%). This is due to bad exploration of X, or in other words, to difficulties in leaving valleys. A new "jump" move, defined as the displacement of a centroid to a point a_ℓ which does not coincide with a centroid, leads to a new VND heuristic, called J-Means, which improves very substantially on both H-Means and K-Means.

8.4 REDUCED VARIABLE NEIGHBORHOOD SEARCH

Assume a local minimum x of f has been reached. One would then like to leave its valley, and find another deeper one. In the standard versions of Variable Neighborhood Search, no previous knowledge of the landscape is assumed, or exploited. (Note that interesting hybrid techniques could be built, using also values of $f(x)$ at previous iteration points x). Then, the questions to be asked are

(i) in which direction to go?

(ii) how far?

(iii) how should one modify moves if they are not successful?

Question (i) bears upon the possibility of reaching any feasible point $x \in X$, or every valley; the simplest answer is to choose a direction at random. For problems in 0–1 variables this will amount to complementing some variables; for continuous Euclidean problems, drawing angular coefficients at random (or, in other words, choosing at random a point on the unit ball around x) takes all points of X into account.

Question (ii) is crucial. Indeed one wants to exploit to the limit Fact 2 (Section 8.1): i.e., in many combinatorial and global optimization problems, local optima tend to be close one to another and situated in one (or sometimes several) small parts of X. So once a local optimum has been reached, it contains implicit information about close better, and perhaps globally optimum, ones. It is then natural to explore first its vicinity. But, if the valley surrounding the local optimum x is large, this may not be sufficient, and what to do next is asked for in question (iii). Again, a natural answer is to go further.

These aims are pursued in the *reduced VNS*, see Figure 8.4. A set of neighborhoods $N_1(x), N_2(x), \ldots, N_{k_{max}}(x)$ will be considered around the current point x (which may be or not a local optimum). Usually, these neighborhoods

Initialization.
Select the set of neighborhood structures $\mathcal{N}_k$, for $k = 1, \ldots, k_{max}$, that will be used in the
search; find an initial solution x; choose a stopping condition;
Repeat the following sequence until the stopping condition is met:
(1) Set $k \leftarrow 1$;
(2) *Repeat* the following steps until $k = k_{max}$:
(a) *Shaking.* Generate a point x' at random from the kth neighborhood of x ($x' \in \mathcal{N}_k(x)$);
(b) *Move or not.* If this point is better than the incumbent, move there ($x \leftarrow x'$), and continue
the search with $\mathcal{N}_1$ ($k \leftarrow 1$); otherwise, set $k \leftarrow k + 1$;

Figure 8.4. Steps of the reduced VNS.

will be nested, i.e. each one contains the previous. Then a point is chosen at
random in the first neighborhood. If its value is better than that of the incumbent (i.e. $f(x') < f(x)$), the search is recentered there ($x \leftarrow x'$). Otherwise,
one proceeds to the next neighborhood. After all neighborhoods have been
considered, one begins again with the first, until a stopping condition is satisfied (usually it will be maximum computing time since the last improvement,
or maximum number of iterations).

Due to the nestedness property, the size of successive neighborhoods will be
increasing. Therefore one will explore more thoroughly close neighborhoods
of x than farther ones, but nevertheless search within these when no further
improvements are observed within the first, smaller ones.

EXAMPLE 8.3 (p-MEDIAN) *(For a survey, see Labbé et al., 1995). This is
a location problem very similar to Simple Plant Location. The differences are
that there are no fixed costs, and that the number of facilities to be opened is
set at a given value p. It is expressed as follows:*

$$\min \sum_{i=1}^{m} \sum_{j=1}^{n} c_{ij} x_{ij} \qquad (8.18)$$

subject to

$$\sum_{i=1}^{m} x_{ij} = 1, \quad \forall j \qquad (8.19)$$

$$y_i - x_{ij} \geq 0, \quad \forall i, j \qquad (8.20)$$

$$\sum_{i=1}^{m} y_i = p \qquad (8.21)$$

$$x_{ij}, y_i \in \{0, 1\} \qquad (8.22)$$

Table 8.1. 5934-customer *p*-median problem.

p	Obj. value (best known)	FI	RVNS	VNDS	FI	RVNS	VNDS
			CPU times			% Error	
100	2733 817.25	6 637.48	510.20	6 087.75	0.36	0.15	0.00
200	1809 064.38	14 966.05	663.69	14 948.37	0.79	0.36	0.00
300	1394 715.12	20 127.91	541.76	17 477.51	0.65	0.51	0.00
400	1 145 669.38	23 630.95	618.62	22 283.04	0.82	0.59	0.00
500	974 275.31	29 441.97	954.10	10 979.77	0.98	0.51	0.00
700	752 068.38	36 159.45	768.84	32 249.00	0.64	0.50	0.00
800	676 846.12	38 887.40	813.38	20 371.81	0.61	0.53	0.00
900	613 367.44	41 607.78	731.71	27 060.09	0.55	0.53	0.00
1000	558 802.38	44 176.27	742.70	26 616.96	0.73	0.66	0.00
Average		28 403.90	705.00	19 786.00	0.68	0.48	0.00

The Greedy and Interchange heuristics described above for Simple Plant Location are easily adapted to the *p*-median problem and, in fact, the latter was proposed by Teitz and Bart (1968).

Fast interchange, using Whitaker's (1983) data structure, applies here also (Hansen and Mladenović, 1997). Refinements have recently been proposed by Resende and Werneck (2003). A comparison between that approach and RVNS is made in Hansen et al. (2001), and the results are summarized in Table 8.1. It appears that RVNS gives better results than Fast Interchange in 40 times less time.

8.5   BASIC AND GENERAL VARIABLE NEIGHBORHOOD SEARCH

In the previous two sections, we examined how to use variable neighborhoods in descent to a local optimum and in finding promising regions for near-optimal solutions. Merging the tools for both tasks leads to the General Variable Neighborhood Search scheme. We first discuss how to combine a local search with systematic changes of neighborhoods around the local optimum found. We then obtain the Basic VNS scheme presented in Figure 8.5.

According to this basic scheme, a series of neighborhood structures, which define neighborhoods around any point $x \in X$ of the solution space, are first selected. Then the local search is used and leads to a local optimum x. A point x' is selected at random within the first neighborhood $\mathcal{N}_1(x)$ of x and a descent from x' is done with the local search routine. This leads to a new local minimum x''. At this point, three outcomes are possible: (i) $x'' = x$, i.e. one is again at the bottom of the same valley; in this case the procedure is iterated

Initialization. Select the set of neighborhood structures $\mathcal{N}_k$, for $k = 1, \ldots, k_{\max}$, that will be used in the search; find an initial solution x; choose a stopping condition;
Repeat the following sequence until the stopping condition is met:
(1) Set $k \leftarrow 1$;
(2) *Repeat* the following steps until $k = k_{\max}$:
(a) *Shaking.* Generate a point x' at random from the kth neighborhood of x ($x' \in \mathcal{N}_k(x)$);
(b) *Local search.* Apply some local search method with x' as initial solution; denote with x'' the so obtained local optimum;
(c) *Move or not.* If the local optimum x'' is better than the incumbent x, move there ($x \leftarrow x''$), and continue the search with $\mathcal{N}_1$ ($k \leftarrow 1$); otherwise, set $k \leftarrow k + 1$;

Figure 8.5. Steps of the basic VNS.

using the next neighborhood $\mathcal{N}_k(x)$, $k \geq 2$; (ii) $x'' \neq x$ but $f(x'') \geq f(x)$, i.e. another local optimum has been found, which is not better than the previous best solution (or incumbent); in this case too the procedure is iterated using the next neighborhood; (iii) $x'' \neq x$ and $f(x'') < f(x)$: i.e., another local optimum, better than the incumbent has been found; in this case the search is recentered around x'' and begins again with the first neighborhood. Should the last neighborhood be reached without a solution better than the incumbent being found, the search begins again at the first neighborhood $\mathcal{N}_1(x)$ until a stopping condition, e.g. a maximum time or maximum number of iterations or maximum number of iterations since the last improvement, is satisfied.

If instead of simple local search, one uses VND and if one improves the initial solution found by reduced VNS, one obtains the general VNS scheme, see Figure 8.6.

Several questions about selection of neighborhood structures are in order:

- What properties of the neighborhoods are mandatory for the resulting scheme to be able to find a globally optimal or near-optimal solution?

- What properties of the neighborhoods will favor finding a near-optimal solution?

- Should neighborhoods be nested? Otherwise how should they be ordered?

- What are desirable properties of the sizes of neighborhoods?

The first two questions bear upon the ability of the VNS heuristic to find the best valleys, and to do so fairly quickly. To avoid being blocked in a valley, while there may be deeper ones, the union of the neighborhoods around any feasible solution x should contain the whole feasible set:

$$X \subseteq \mathcal{N}_1(x) \cup \mathcal{N}_2(x) \cup \ldots \cup \mathcal{N}_{k_{\max}}(x) \qquad \forall x \in X$$

Initialization. Select the set of neighborhood structures $\mathcal{N}_k$, for $k = 1, \ldots, k_{max}$, that will be used in the shaking phase, and the set of neighborhood structures N_ℓ for $\ell = 1, \ldots, \ell_{max}$ that will be used in the local search; find an initial solution x and improve it by using RVNS; choose a stopping condition;

Repeat the following sequence until the stopping condition is met:

(1) Set $k \leftarrow 1$;

(2) *Repeat* the following steps until $k = k_{max}$:

(a) *Shaking.* Generate a point x' at random from the kth neighborhood $\mathcal{N}_k(x)$ of x;

(b) *Local search by VND.*

(b1) *Set* $\ell \leftarrow 1$;

(b2) *Repeat* the following steps until $\ell = \ell_{max}$;

· *Exploration of neighborhood.* Find the best neighbor x'' of x' in $N_\ell(x')$;

· *Move or not.* If $f(x'') < f(x')$ set $x' \leftarrow x''$ and $\ell \leftarrow 1$; otherwise set $\ell \leftarrow \ell + 1$;

(c) *Move or not.* If this local optimum is better than the incumbent, move there $(x \leftarrow x'')$, and continue the search with $\mathcal{N}_1$ $(k \leftarrow 1)$; otherwise, set $k \leftarrow k + 1$;

Figure 8.6. Steps of the general VNS.

These sets may cover X without necessarily partitioning it, which is easier to implement, e.g. when using nested neighborhoods, i.e.

$$\mathcal{N}_1(x) \subset \mathcal{N}_2(x) \subset \ldots \subset \mathcal{N}_{k_{max}}(x) \qquad X \subset \mathcal{N}_{k_{max}}(x) \qquad \forall x \in X$$

If these properties do not hold, one might still be able to explore X completely, by traversing small neighborhoods around points on some trajectory, but it is no longer guaranteed. To illustrate, as mentioned before, in minimum sum-of-squares clustering, the neighborhoods defined by moving an entity (or even a few entities) from one cluster to another one are insufficient to get out of many local optima. Moving centers of clusters does not pose a similar problem.

Nested neighborhoods are easily obtained for many combinatorial problems by defining a first neighborhood $\mathcal{N}_1(x)$ by a type of move (e.g. two-opt in the traveling salesman problem) and then iterating it k times to obtain neighborhoods $\mathcal{N}_k(x)$ for $k = 2, \ldots, k_{max}$. They have the property that their sizes are increasing. Therefore if, as is often the case, one goes many times through the whole sequence of neighborhoods the first ones will be explored more thoroughly than the last ones. This is desirable in view of Fact 3: i.e., that local optima tend to be close one from another.

Restricting moves to the feasible set X may be too constraining, particularly if this set is disconnected. Introducing some or all constraints in the objective function with Lagrangian multipliers, allows moving to infeasible solutions.

A variant of this idea is to penalize infeasibilities, such as pairs of adjacent vertices to which the same color is assigned in graph coloring: see Zufferey et al. (2003).

EXAMPLE 8.4 (SCHEDULING WORKOVER RIGS) *Many oil wells in on-shore fields rely on artificial lift methods. Maintenance services such as cleaning and others, which are essential to these wells, are performed by workover rigs. They are slow mobile units and, due to their high operation costs, there are relatively few workover rigs when compared with the number of wells demanding service. The problem of scheduling workover rigs consists in finding the best schedule S_i (i = 1, . . . , m) of the m workover rigs to attend all wells demanding maintenance services, so as to minimize the oil production loss (production before maintenance being reduced).*

In Aloise et al. (2003) a basic VNS heuristic is developed for solving the scheduling of workover rigs problem (WRP). Initial schedule S_i (where S_i is an ordered set of wells serviced by workover rig i) is obtained by a *greedy* constructive heuristic. For the shaking step $k_{max} = 9$ neighborhoods are constructed:

1 *Swap routes* (SS): the wells and the associated routes assigned to two workover rigs are interchanged;

2 *Swap wells from the same workover rig* (SWSW): the order in which two wells are serviced by the same rig is swapped;

3 *Swap wells from different workover rig* (SWDW): two wells assigned to two different workover rigs are swapped;

4 *Add/drop* (AD): a well assigned to a workover rig is reassigned to any position of the schedule of another workover rig;

5 $(SWSW)^2$: apply twice the SWSW move;

6 $(SWDW)^2$: apply twice the SWDW move;

7 $(SWDW)^3$: apply three times the SWDW move;

8 $(AD)^2$: successively apply two (AD) moves;

9 $(AD)^3$: successively apply three (AD) moves.

For local search, the neighborhood consists of all possible exchanges of pairs of wells, i.e. the union of (SWSW) and (SWDW) from above is used.

A basic VNS is compared with the genetic algorithm, the greedy randomized adaptive procedure (GRASP) and with two ant colony methods (AS and MMAS) on synthetical and real-life problems from Brazilian onshore fields.

Table 8.2. Average results with eight workover rigs over 20 runs of each synthetic test problem and three possible scenarios (from Aloise et al., 2003).

Problem	GA	GRASP	AS	MMAS	VNS
P-111	16 791.87	16 602.51	15 813.53	15 815.26	15 449.50
P-211	20 016.14	19 726.06	19 048.13	19 051.61	18 580.64
P-311	20 251.93	20 094.37	19 528.93	19 546.10	19 434.97

Initialization. Select the set of neighborhood structures $\mathcal{N}_k$, for $k = 1, \ldots, k_{max}$, that will be used in the search; find an initial solution x and its value $f(x)$; set $x_{opt} \leftarrow x$, $f_{opt} \leftarrow f(x)$: choose a stopping condition and a parameter value α;

Repeat the following until the stopping condition is met:

(1) Set $k \leftarrow 1$;

(2) *Repeat* the following steps until $k = k_{max}$:

(a) *Shaking.* Generate a point x' at random from the kth neighborhood of x;

(b) *Local search.* Apply some local search method with x' as initial solution; denote with x'' the so obtained local optimum;

(c) *Improvement or not.* If $f(x'') < f_{opt}$ set $f_{opt} \leftarrow f(x)$ and $x_{opt} \leftarrow x''$;

(d) *Move or not.* If $f(x'') - \alpha\rho(x, x'') < f(x)$ set $x \leftarrow x''$ and $k \leftarrow 1$; otherwise set $k \leftarrow k+1$.

Figure 8.7. Steps of the skewed VNS.

Some results on synthetic data are given in Table 8.2. On 27 possible scenarios in generating data sets (denoted by P-111, P-112, P-113, P-121, ..., P-333), VNS was better than others in 85% of the cases and MMAS in 15%. On real-life problems, results were much better than the gains expected. For example, a daily reduction of 109 m^3 (equivalent to 685.6 bbl) in the production losses along 15 days was obtained by VNS compared with Petrobras' previous solution. That leads to a total savings estimated at US$6600 000 a year.

8.6 SKEWED VARIABLE NEIGHBORHOOD SEARCH

VNS usually gives solutions better, or as good as, multistart, and much better ones when there are many local optima. This is due to Fact 3 (of Section 8.1): many problems have clustered local optima; often, their objective function is a globally convex one plus some noise. However, it may happen that some instances have several separated and possibly far apart valleys containing near-optimal solutions. If one considers larger and larger neighborhoods, the information related to the currently best local optimum dissolves and VNS degenerates into multistart. Moreover, if the current best local optimum is not in the deepest valley this information is in part irrelevant. It is therefore of interest to modify VNS schemes in order to explore more fully valleys which are far away from the incumbent solution. This is done by allowing a recentering of the search when a solution close to the best one known, but not necessarily as good, is found, provided that it is far from this last solution. The modified VNS scheme for this variant, called *skewed VNS* (SVNS) is presented in Figure 8.7. The relaxed rule for recentering uses an evaluation function linear in the distance from the incumbent: i.e. $f(x'')$ is replaced by $f(x'') - \alpha\rho(x, x'')$ where $\rho(x, x'')$ is the distance from x to x'' and α a parameter. A metric for the distance between solutions is usually easy to find, e.g. the Hamming distance when solutions are described by Boolean vectors or the Euclidean distance in the continuous case.

Clearly, more complicated formulae could be used for recentering; possibly, one might take into account known values at points already visited in the valley being explored. Questions to be answered when applying SVNS are the following:

■ Does the problem under consideration have a roughly convex objective function, or are there several far apart deep valleys?

■ How should α be chosen?

These questions can be answered, to some extent, by first using a multistart version of VNS, i.e. starting VNS from various random points and running it for a short time. Then one can look at the position of the best local optima found and see if they are clustered or dispersed. Further, one can plot values in function of distance from the corresponding local optima to the best known solution and choose α as a fraction of the average slope.

EXAMPLE 8.5 (WEIGHTED MAXIMUM SATISFIABILITY) *The satisfiability problem, in clausal form, consists in determining if a given set of m clauses (all in disjunctive or all in conjunctive form) built upon n logical variables has a solution or not. The maximum satisfiability problem consists in finding a solution satisfying the largest possible number of clauses. In the* weighted maximum satisfiability *problem (WMAXSAT) positive weights are assigned to the clauses and a solution maximizing the sum of weights of satisfied clauses is sought. Results of comparative experiments with VNS and TS heuristics on instances having 500 variables, 4500 clauses and three variables per clause, in direct or complemented form, are given in Table 8.3 from Hansen et al. (2001). It appears that using a restricted neighborhood consisting of a few directions of steepest descent or mildest ascent in the Shaking step does not improve results, but using this idea in conjunction with SVNS improves notably upon results of basic VNS and also upon those of a TS heuristic.*

Table 8.3. Results for GERAD test problems for WMAXSAT ($n = 500$).

	VNS	VNS-low	SVNS-low	TS
Number of instances where				
best solution is found	6	4	23	5
Average error in 10 trials (%)	0.2390	0.2702	0.0404	0.0630
Best error in 10 trials (%)	0.0969	0.1077	0.0001	0.0457
Total number of instances	25	25	25	25

8.7 VARIABLE NEIGHBORHOOD DECOMPOSITION SEARCH

The VNDS method (Hansen et al., 2001) extends the basic VNS into a two-level VNS scheme based upon decomposition of the problem. Its steps are presented in Figure 8.8.

Note that the only difference between the basic VNS and VNDS is in Step 2(b): instead of applying some local search method in the whole solution space S (starting from $x' \in \mathcal{N}_k(x)$), in VNDS we solve at each iteration a subproblem in some subspace $V_k \subseteq \mathcal{N}_k(x)$ with $x' \in V_k$. When the local search used in this step is also VNS, the two-level VNS-scheme arises.

VNDS can be viewed as embedding the classical successive approximation scheme in the VNS framework.

8.8 ANALYZING PERFORMANCE

When a first VNS heuristic has been obtained and tested, the effort should not stop there. Indeed, it is often at this point that the most creative part of the development process takes place. It exploits systematically Fact 2 (of the Introduction), i.e. that global minima are local minima for all possible neighborhoods simultaneously. The contrapositive is that if a solution $x \in X$ is a local minimum (for the current set of neighborhoods) and not a global one there are one or several neighborhoods (or moves) to be found, which will bring it to this global optimum.

The study then focuses on instances for which an optimal solution is known (or, if none or very few are available, on instances with a presumably optimal solution, i.e. the best one found by several heuristics) and compares it with the heuristic solution obtained. Visualization is helpful and make take the form of a distance-to-target diagram (Hansen and Mladenović, 2003). Then, the heuristic solutions, the optimal one and their symmetric difference (e.g. for the traveling salesman problem, TSP for short) are represented onscreen. An interactive feature allows one to follow how the heuristic works step by step.

Initialization. Select the set of neighborhood structures $\mathcal{N}_k$, for $k = 1, \ldots, k_{\max}$, that will be used in the search; find an initial solution x; choose a stopping condition;

Repeat the following sequence until the stopping condition is met:

(1) Set $k \leftarrow 1$;

(2) *Repeat* the following steps until $k = k_{\max}$:

(a) *Shaking.* Generate a point x' at random from the kth neighborhood of x ($x' \in \mathcal{N}_k(x)$); in other words, let y be a set of k solution attributes present in x' but not in x ($y = x' \setminus x$).

(b) *Local search.* Find a local optimum in the space of y either by inspection or by some heuristic; denote the best solution found with y' and with x'' the corresponding solution in the whole space S ($x'' = (x' \setminus y) \cup y'$);

(c) *Move or not.* If the solution thus obtained is better than the incumbent, move there ($x \leftarrow x''$), and continue the search with $\mathcal{N}_1$ ($k \leftarrow 1$); otherwise, set $k \leftarrow k + 1$;

Figure 8.8. Steps of the basic VNDS.

The information thus gathered is much more detailed than what one would get just from objective values and computer times if, as is often the case, the heuristic is viewed as a black box. For instance, this clearly shows that two-opt is not sufficient to get a good solution for the TSP, that moves involving three or four edges are needed and that those edges leading to an improvement may be far apart along the tour. For another application of VNS to the TSP see Burke et al. (1999).

Similarly, for location problems, one can focus on those facilities which are not at their optimal location and study why, in terms of distributions of nearby users.

Another point is to study how to get out of a large valley if there exists another promising one. *Valley* (or *mountain*) *profiles* are then useful (Hansen et al., 2001). They are obtained by drawing many points x' at random within nested neighborhoods $\mathcal{N}_1(x), \mathcal{N}_2(x), \ldots$ (or, which is equivalent, at increasing distance of a local minimum x) then performing one VND descent and plotting probabilities to get back to x, to get to another local minimum x'' with a value $f(x'') \geq f(x)$ or to get to an improved local minimum x' with $f(x'') < f(x)$. Alternatively, one may study the probabilities to go in the direction of x, i.e. $\rho(x, x'') \leq \rho(x, x')$ or towards another valley i.e. $\rho(x, x'') > \rho(x, x')$.

8.9 PROMISING AREAS OF RESEARCH

Research on VNS and its applications is currently very active. We review some of the promising areas in this section; these include a few which are barely explored yet.

A first set of areas concerns enhancements of the VNS basic scheme and ways to make various steps more efficient.

(a) *Initialization.* Both VND and VNS, as many other heuristics, require an initial solution. Two questions then arise: *How best to choose it?* and *Does it matter?* For instance, many initialization rules have been proposed for the k-Means heuristic for minimum sum-of-squares clustering, described above; 25 such rules are compared in Hansen et al. (2003c). It appears that while sensitivity of k-Means to the initial solution is considerable (best results being obtained with Ward's hierarchical clustering method), VNS results depend very little on the chosen rule. The simplest one is thus best. It would be interesting to extend and generalize this result by conducting similar experiments for other problems.

(b) *Inventory of neighborhoods.* As mentioned above, a VNS study begins by gathering material on neighborhoods used in previous heuristics for the problem under study. A systematic study of moves (or neighborhoods) used for heuristics for whole classes of problems (e.g. location, network design, routing, ...) together with the data-structures most adequate for their implementation should be of basic interest for VNS as well as for other metaheuristics. Several researchers, e.g. Ahuja et al. (2000), are working in that direction.

(c) *Distribution of neighborhoods.* When applying a general VNS scheme, neighborhoods can be used in the local search phase, in the shaking phase or in both. A systematic study of their best distribution between phases could enhance performance and provide further insight in the solution process. In particular, the trade-off between increased work in the descent, which provides better local optima, and in shaking which leads to better valleys should be focussed upon.

(d) *Ancillary tests.* VNS schemes use randomization in their attempts to find better solutions. This also avoids possible cycling. However, many moves may not lead to any improvement. This suggests the addition of an ancillary test (Hansen, 1974, 1975) the role of which is to decide if a move should be used or not, in its general or in a restricted form. Considering again minimum sum-of-squares clustering, one could try to select better the centroid to be removed from the current solution (a possible criterion being that its cluster contains a few entities only or is close to another centroid) as well as the position where it will be assigned (e.g., the location of an entity far from any other centroid and in a fairly dense region).

A second set of areas concerns changes to the basic scheme of VNS.

(e) *Use of memory.* VNS in its present form relies only on the best solutions currently known to center the search. Knowledge of previous good solutions is forgotten, but might be useful to indicate promising regions not much explored yet. Also, characteristics common to many or most good solutions, such as variables taking the same value in all or most such solutions, could be used to better focus the shaking phase. Use of memory has been much studied in tabu search and other metaheuristics. The challenge for VNS would be to introduce memory while keeping simplicity.

An interesting way to use memory to enhance performance is *reactive VNS*, explored by Braysy (2001) for the vehicle routing problem with time windows. If some constraints are hard to satisfy their violation may be penalized more frequently than for others in the solution process.

(f) *Parallel VNS.* Clearly, there are many natural ways to parallelize VNS schemes. A first one, within VND, is to perform local search in parallel. A second one, within VNS, is to assign the exploration of each neighborhood of the incumbent to a different processor. A third one, within VNDS, is to assign a different subproblem to each processor. Lopez et al. (2002) explore several options in designing a parallel VNS.

(g) *Hybrids.* Several researchers, e.g. Rodriguez et al. (1999), Festa et al. (2001), Ribeiro et al. (2001) and Drezner (2003a, 2003b), have combined VNS with other metaheuristics for various problems. Again, this is not always easy to do without losing VNS's simplicity but may lead to excellent results, particulary if the other metaheuristics are very different from VNS.

At a more general, level one might wish to explore combinations of VNS with *constraint programming*, instead of its development within mathematical programming as in the applications described above. This could be done in two directions: on the one hand, techniques from constraint programming could be applied to enhance VND; on the other hand, VNS could be applied to constraint programming by minimizing a sum of artificial variables measuring infeasibility and possibly weighted by some estimate of the difficulty of satisfying the corresponding constraints.

A third set of areas concerns new aims for VNS, i.e. non-standard uses.

(h) *Solutions with bounds on the error.* VNS, as other metaheuristics, most often provides near-optimal solutions to combinatorial problems, without bounds on their error. So while such solutions may be optimal or very close to optimality, this fact cannot be recognized. One approach to obtain such bounds is to find with VNS a heuristic solution of the primal problem, deduce from it a solution to the dual (or its continuous relaxation) and then improve this dual solution by another application of VNS. Moreover, complementary slackness conditions can be used to simplify the dual. For problems with a small duality gap this may lead to near-optimal solution guaranteed to be very close to optimality. To illustrate, recent work of Hansen et al. (2003a) on the simple plant location problem gave solutions to instances with up to 15 000 users and 15 000 possible facilities with an error bounded by 0.05%.

(i) *Using exact algorithms for mixed-integer programming.* Sophisticated algorithms for mixed-integer programming often contain various phases where heuristics are applied. This is illustrated by Desaulniers et al. (2001) for the airline crew scheduling problem.

Extending the results described in the previous section in the branch-and-bound framework led to the solution of exactly SPLP instances with up to 7000 users (Hansen et al., 2003a).

A different approach, called local branching, has been recently proposed by Fischetti and Lodi (2003) and Fischetti et al. (2003), both for exact and approximate resolution of large mixed-integer programs. At various branches in the branch-and-bound tree, cuts (which are not valid in general) are added; they express that among a given set of 0–1 variables, already at an integer value, only a few may change their value. They thus correspond to neighborhoods defined by the Hamming distance. Then CPLEX is used to find the optimal solution within the neighborhood and in this way feasible solutions are more easily obtained. Improved solutions were obtained for a series of large mixed-integer programming instances from various sources.

(j) *Artificial intelligence: enhancing graph theory with VNS.* VNS, as other metaheuristics, has been extensively used to solve a variety of optimization problems in graph theory. However, it may also be used to enhance graph theory per se, following an Artificial Intelligence approach. This is done by the AutoGraphiX (AGX) system developed by Caporossi and Hansen (2000, 2003). This system considers a graph invariant (i.e. a quantity defined for all graphs of the class under study and independent of vertex and edge labeling) or a formula involving several invariants (which is itself a graph invariant). Then AGX finds extremal or near-extremal graphs for that invariant parametrizing on a few variables, often the order n (or number of vertices) and the size m (of number of edges) of the graph. Analyzing automatically or interactively these graphs and the corresponding curves of invariant values leads to finding new conjectures, refuting, corroborating or strengthening existing ones, and obtaining hints about possible proof from the minimal list of moves needed to find the extremal graphs. To illustrate, the energy E of a graph is the sum of absolute values of the eigenvalues of its adjacency matrix. The following relations were obtained by Caporossi et al. (1999) with AGX: $E \geq 2\sqrt{m}$ and $E \geq \frac{4m}{n}$, and were easily proved. Over 70 new relations have now been obtained, in mathematics and in chemistry. Three ways to attain full automation based on the mathematics of principal component analysis, linear programming and recognition of extremal graphs together with formula manipulations are currently being studied.

8.10   TRICKS OF THE TRADE

8.10.1   Getting Started

The purpose of this section is to help students and newcomers in making a first very simple version of VNS, not necessarily competitive with later more sophisticated versions. Most of the steps are common for implementation of other metaheuristics.

A Step-by-Step Procedure

1 *Familiarization.* Think about the problem at hand; in order to understand it better, make a simple numerical example and spend some time in trying to solve it by hand in your own way. Try to understand why the problem is hard and why a heuristic is needed.

2 *Read.* Read about the problem and solution methods in the literature.

3 *Test instances.* Use your numerical example as a first instance for testing your future code, but if it is not large enough, take some data from the web, or make a routine for generating random instances. In the second case, read how to generate events using uniformly distributed numbers from $(0, 1)$ interval (each programming language has a statement for generating such random numbers).

4 *Data structure.* Think about how the solution of the problem will be represented in the memory. Consider two or more presentations of the same solution if they can reduce the complexity of some routines, i.e. analyze advantages and disadvantages of each possible presentation.

5 *Initial solution.* Having a routine for reading or generating the input data of the problem, the next step is to obtain an initial solution. For a simple version, any random feasible solution may be used, but the usual approach is to develop some *greedy* constructive heuristic, which is not hard to do.

6 *Objective value.* Devise a procedure that calculates objective function values for a given solution. Notice that at this stage, we already have all ingredients for the Monte Carlo method: generation of random solution and calculation of objective function value. Obtain the solution of your problem by the Monte Carlo heuristic (i.e. repeat steps 5 and 6 many times and keep the best one).

7 *Shaking.* Create a procedure for shaking. This is a key step of VNS. However, it is easy to implement and usually involves only a few lines of code. For example, in solving the multi-source Weber problem (see Example 2), the easiest perturbation of the current solution is to re-allocate randomly chosen entity ℓ from the cluster it belongs to another one, also chosen at random. In fact, in this case, the shaking step (or jump in the kth neighborhood) needs only three lines of code:

$$\text{For } i = 1 \text{ to } k$$
$$a(1 + n \cdot Rnd1) = 1 + m \cdot Rnd2$$
$$\text{EndFor}$$

The solution is saved in array $a(\ell) \in \{1, \ldots, m\}$ which denotes membership or allocation of entity ℓ ($\ell = 1, \ldots, n$); $Rnd1$ and $Rnd2$ denote random numbers uniformly distributed from the (0,1) interval. Compare the results of the obtained reduced VNS (take $k_{\max} = 2$) with the Monte Carlo method.

8 *Local search.* Choose an off-the-shelf local search heuristic (or develop a new one). In building a new local search, consider several usual moves that define the neighborhood of the solution *drop, add, swap, interchange, etc.* Also, for the efficiency (speed) of the method, it is important to pay special attention to *updating* of the incumbent solution. In other words, usually it is not necessary to use a procedure for calculating objective function values for each point in the neighborhood, i.e. it is possible to get those values by very simple calculation.

9 *Comparison.* Include a local search routine into RVNS to get the basic VNS, and compare it with other methods from the literature.

8.10.2   More Tips

Sometimes basic VNS does not provide very good results.

1 *First vs. best improvement.* Compare experimentally *first* and *best improvement* strategies within local search. Previous experience suggest the following: if your initial solution is chosen at random, use first improvement, but if some constructive heuristic is used, use best improvement rule.

2 *Reduce the neighborhood.* The cause of bad behavior of any local search may be unnecessary visiting to all solutions in the neighborhood. Try to identify a "promising" subset of the neighborhood and visit only them; ideally, find a rule that automatically selects solutions from the neighborhood whose objective values are not better than the current one.

3 *Intensified shaking.* In developing more effective VNS, one must spend some time in checking how sensitive is the objective function on small change (shake) of the solution. The trade-off between intensification and diversification of the search in VNS is balanced in the shaking procedure. For some problem instances completely random jump in the kth neighborhood is too diversified. In such cases, an *intensify shaking* procedure can be used to increase intensification of the search. For example, a k-interchange neighborhood may be reduced by repeating k times *random add* followed by *best drop* moves. (A special case of intensified shaking is the *large neighborhood search*, where k randomly chosen attributes of the solutions are destroyed (dropped), and then the solution is re-built in the best way—by some constructive heuristic.)

4 *VND.* Analyze several possible neighborhood structures, estimate their size, make order of them, i.e. develop VND and replace the local search routine with VND to get general VNS.

5 *Experiment with parameter settings.* The single parameter of VNS is $k_{\max}$, which should be estimated experimentally. However, usually the procedure is not very sensitive on $k_{\max}$ and, in order to create a parameter-free VNS, one can fix its value at the value of some input parameter, e.g., for the p-median (Example 3), $k_{\max} = p$; for the minimum sum-of-square clustering (Example 2) $k_{\max} = m$, etc.

8.11 CONCLUSIONS

The general schemes of VNS have been presented, discussed and illustrated by examples. References to many further successful applications are given below. In order to evaluate the VNS research program, one needs a list of desirable properties of metaheuristics. The following eight are presented in Hansen and Mladenović (2003):

(i) *Simplicity.* The metaheuristic should be based on a simple and clear principle, which should be largely applicable.

(ii) *Precision.* Steps of the metaheuristic should be formulated in precise mathematical terms, independent from the possible physical or biological analogy which was an initial source of inspiration.

(iii) *Coherence.* All steps of heuristics for particular problems should follow naturally from the metaheuristic's principle.

(iv) *Efficiency.* Heuristics for particular problems should provide optimal or near-optimal solutions for all or at least most realistic instances. Preferably, they should find optimal solutions for most problems of benchmarks for which such solutions are known, when available.

(v) *Effectiveness.* Heuristics for particular problems should take moderate computing time to provide optimal or near-optimal solutions.

(vi) *Robustness.* Performance of heuristics should be consistent over a variety of instances, i.e. not just fine-tuned to some training set and less good elsewhere.

(vii) *User-friendliness.* Heuristics should be clearly expressed, easy to understand and, most important, easy to use. This implies they should have as few parameters as possible and ideally none.

(viii) *Innovation.* Preferably, the metaheuristic's principle and/or the efficiency and effectiveness of the heuristics derived from it should lead to new types of applications.

As argued there, as well as in the more recent surveys listed below, VNS possesses, to a large extent, all of those properties. This has led to heuristics among the very best ones for several problems, but more importantly to insight into the solution process and some innovative applications.

SOURCES OF ADDITIONAL INFORMATION

Some web addresses with sources of information about VNS include

- http://www.mi.sanu.ac.yu/VNS/VNS.HTM (a working web presentation of VNS, developed by Tatjana Davidović, Ph.D. student at University of Belgrade).

- VNSHeuristic.ull.es (another web page for the VNS designed by Professor's Moreno research group from University of La Laguna).

- http://www.gerad.ca/en/publications/cahiers.php (choose "search for papers" and in the "Abstract" box type "Variable Neighborhood Search": 23 papers for downloading are returned).

- http://smg.ulb.ac.be (there are several papers on VNS in "Preprints" by Hansen, Labbé, Mélot, Mladenović, etc).

Survey papers. Hansen and Mladenović (1999, 2001a, 2001c, 2002a, 2002b, 2003), Hansen et al. (2003c), and Kochetov et al. (2003).

References

Ahuja, R. K., Orlin, J. B. and Sharma, D., 2000, Very large-scale neighborhood search, *Int. Trans. Oper. Res.* **7**:301–317.

Aloise, D. J., Aloise, D., Rocha, C. T. M., Ribeiro Filho, J. C., Moura, L. S. S. and Ribeiro, C. C., 2003, Scheduling workover rigs for onshore oil production, *Research Report,* Department of Computer Science, Catholic University of Rio de Janeiro, submitted.

Baum, E. B., 1986, Toward practical "neural" computation for combinatorial optimization problems, in: *Neural Networks for Computing*, J. Denker, ed., American Institute of Physics, New York.

Braysy, O., 2001, Local search and variable neighborhood search algorithms for vehicle routing with time windows, *Acta Wasaensia,* Vol. 87.

Burke, E. K., Cowling, P. and Keuthen, R., 1999, Effective local and guided variable neighborhood search methods for the asymmetric traveling salesman problem, in: *Proc. of the Evo Workshops,* Lecture Notes in Computer Science, Vol. 2037, Springer, Berlin, pp. 203–212.

Caporossi, G., Cvetković, D., Gutman, I. and Hansen, P., 1999, Variable neighborhood search for extremal graphs: 2. Finding graphs with extremal energy, *J. Chem. Inf. Comput. Sci.* **39**:984–996.

Caporossi, G. and Hansen, P., 2000, Variable neighborhood search for extremal graphs: 1. The AutoGraphiX system, *Discr. Math.,* **212**:29–44.

Caporossi, G. and Hansen, P., 2004, Variable neighborhood search for extremal graphs: 5. Three ways to automate conjecture finding, *Discr. Math.* **276**:81–94.

Cornuejols, G., Fisher, M. and Nemhauser, G., 1990, The uncapacitated facility location problem, in: *Discrete Location Theory,* P. Mirchandani and R. Francis, eds, Wiley, New York.

Desaulniers, G., Desrosiers, J. and Solomon, M. M., 2001, Accelerating strategies in column generation methods for vehicle routing and crew scheduling problems, *Essays and Surveys in Metaheuristics,* Kluwer, Dordrecht, pp. 309–324.

Drezner, Z., 2003a, Heuristic algorithms for the solution of the quadratic assignment problem, *J. Appl. Math. Decision Sci.,* to appear.

Drezner, Z., 2003b, A new genetic algorithm for the quadratic assignment problem, *INFORMS J. Comput.* **15**, to appear.

Festa, P., Pardalos, P., Resende, M. and Ribeiro, C., 2001, GRASP and VNS for Max-cut, *Proc. MIC'2001,* pp. 371–376.

Fischetti, M. and Lodi, A., 2003, Local branching, *Math. Program.* B, published online, 28 March.

Fischetti, M., Polo, C. and Scantamburlo, M., 2003, A local branching heuristic for mixed-integer programs with 2-level variables, *Research Report,* University of Padova.

Garey, M. R. and Johnson, D. S., 1979, *Computers and Intractability: A Guide to the Theory of NP-Completeness,* Freeman, New York.

Glover, F. and Kochenberger, G., eds, 2003, *Handbook of Metaheuristics,* Kluwer, Dordrecht.

Hansen, P., 1974, Programmes mathématiques en variables 0–1, *Thèse d'Agrégation de l'Enseignment Supérieur,* Université Libre de Bruxelles.

Hansen, P., 1975, Les procédures d'optimization et d'exploration par séparation et évaluation, in: *Combinatorial Programming,* B. Roy, ed., Reidel, Dordrecht, pp. 19–65.

Hansen, P., Brimberg, J., Urošević, D., and Mladenović, N., 2003a, Primal–dual variable neighborhood search for exact solution of the simple plant location problem (in preparation).

Hansen, P., and Mladenović, N., 1997, Variable neighborhood search for the *p*-median, *Location Sci.* **5**: 207–226.

Hansen, P., and Mladenović, N., 1999, An introduction to variable neighborhood search, in: *Metaheuristics, Advances and Trends in Local Search Paradigms for Optimization,* S. Voss et al., eds, Kluwer, Dordrecht, pp. 433–458.

Hansen, P., and Mladenović, N., 2001a, Variable neighborhood search: Principles and applications, *Eur. J. Oper. Res.* **130**:449–467.

Hansen, P., and Mladenović, N., 2001b, J-Means: A new local search heuristic for minimum sum-of-squares clustering, *Pattern Recognition* **34**:405–413.

Hansen, P. and Mladenović, N., 2001c, Developments of variable neighborhood search, in: *Essays and Surveys in Metaheuristics,* C. Ribeiro and P. Hansen, eds, Kluwer, Dordrecht, pp. 415–440.

Hansen, P. and Mladenović, N., 2002a, Variable neighborhood search, in: *Handbook of Applied Optimization,* P. Pardalos and M. Resende, Oxford University Press, New York, pp. 221–234.

Hansen, P. and Mladenović, N., 2002b, Recherche à voisinage variable in: *Optimisation Approche en Recherche Opérationnelle,* J. Teghem and M. Pirlot, eds, Lavoisier/Hermès, Paris, pp. 81–100.

Hansen, P. and Mladenović, N., 2003, Variable neighborhood search, in: *Handbook of Metaheuristics,* F. Glover and G. Kochenberger, eds, Kluwer, Dordrecht, pp. 145–184.

Hansen, P., Mladenović, N. and Moreno Pérez, J. A., 2003b, Búsqueda de entorno variable (in Spanish), *Intell. Artif.,* to appear.

Hansen, P., Mladenović, N. and Perez-Brito, D., 2001, Variable neighborhood decomposition search, *J. Heuristics* **7**:335–350.

Hansen, P., Ngai, E., Cheung, B. and Mladenović, N., 2003c, Survey and comparison of initialization methods for k-means clustering (in preparation).

Kochetov, Y., Mladenović, N. and Hansen, P., 2003, Lokalnii poisk s chereduyshimisy okrestnostyami (in Russian), *Diskretnoi analiza,* to appear.

Labbé, M., Peeters, D. and Thisse, J. F., 1995, Location on networks, in: *Network Routing,* M. Ball et al., eds, North-Holland, Amsterdam, pp. 551–624.

Lopez, F. G., Batista, B. M., Moreno Pérez, J. A. and Moreno Vega J. M., 2002, The parallel variable neighborhood search for the p-median problem, *J. Heuristics* **8**:375–388.

Mladenović, N. and Hansen, P., 1997, Variable neighborhood search, *Comput. Oper. Res.* **24**:1097–1100.

Papadimitriou, C., 1994, *Computational Complexity,* Addison-Wesley, Reading, MA.

Reeves, C.R., ed., 1993, *Modern Heuristic Techniques for Combinatorial Problems,* Blackwell, Oxford.

Resende, M. G. C., and Werneck, R., 2003, On the implementation of a swap-based local search procedure for the p-median problem, *Proc. 5th Workshop on Algorithm Engineering and Experiments (ALENEX'03),* R. E. Ladner, ed., SIAM, Philadelphia, PA, pp. 119–127.

Ribeiro, C., Uchoa, E. and Werneck, R., 2001, A hybrid GRASP with perturbations for the Steiner problem in graphs, *Technical Report,* Computer Science Department, Catholic University of Rio de Janeiro.

Rodriguez, I., Moreno-Vega, M. and Moreno-Perez, J., 1999, Heuristics for routing-median problems, *SMG Report,* Université Libre de Bruxelles, Belgium.

Teitz, M. B. and Bart, P., 1968, Heuristic methods for estimating the generalized vertex median of a weighted graph, *Oper. Res.* **16**:955–961.

Whitaker, R., 1983, A fast algorithm for the greedy interchange for large-scale clustering and median location problems, *INFOR* **21**:95–108.

Zufferey, N., Hertz A. and Avanthay, C., 2003, Variable neighborhood search for graph colouring, *Eur. J. Oper. Res.,* to appear.

Chapter 9

CONSTRAINT PROGRAMMING

Eugene C. Freuder
University College Cork, Ireland

Mark Wallace
Monash University, Australia

9.1 INTRODUCTION

Constraint satisfaction problems are ubiquitous. A simple example that we will use throughout the first half of this chapter is the following scheduling problem: Choose employees A or B for each of three tasks, X, Y, Z, subject to the work rules that the same employee cannot carry out both tasks X and Y, the same employee cannot carry out both tasks Y and Z, and only employee B is allowed to carry out task Z. (Many readers will recognize this as a simple *coloring problem*.)

This is an example of a class of problems known as constraint satisfaction problems (CSPs). CSPs consist of a set of *variables* (e.g. tasks), a *domain* of *values* (e.g. employees) for each variable, and *constraints* (e.g. work rules) among sets of variables. The constraints specify which combinations of value assignments are allowed (e.g. employee A for task X and employee B for task Y); these allowed combinations *satisfy* the constraints. A *solution* is an assignment of values to each variable such that all the constraints are satisfied (Dechter, 2003; Tsang, 1993).

We stress that the basic CSP paradigm can be *extended* in many directions: for example, variables can be added dynamically, domains of values can be continuous, constraints can have priorities, and solutions can be *optimal*, not merely satisfactory.

Some examples of constraints are

- The meeting must start at 6:30.

- The separation between the soldermasks and nets should be at least 0.15 mm.

- This model only comes in blue and green.

- This cable will not handle that much traffic.

- These sequences should align optimally.

- John prefers not to work on weekends.

- The demand will probably be for more than five thousand units in August.

Some examples of constraint satisfaction or optimization problems are

- Schedule these employees to cover all the shifts.

- Optimize the productivity of this manufacturing process.

- Configure this product to meet my needs.

- Find any violations of these design criteria.

- Optimize the use of this satellite camera.

- Align these amino acid sequences.

Many application domains (e.g. design) naturally lend themselves to modeling as CSPs. Many forms of reasoning (e.g. temporal reasoning) can be viewed as constraint reasoning. Many disciplines (e.g. operations research) have been brought to bear on these problems. Many computational "architectures" (e.g. neural networks) have been utilized for these problems. Constraint programming can solve problems in telecommunications, internet commerce, electronics, bioinformatics, transportation, network management, supply chain management, and many other fields.

Here are just a few examples of commercial application of constraint technology:

- Staff planning: BanqueBuxelles Lambert.

- Vehicle production optimization: Chrysler Corporation.

- Planning medical appointments: FREMAP.

- Task scheduling: Optichrome Computer Systems.

- Resource allocation: SNCF (French railways).

- From push to pull manufacturing: Whirlpool.

- Utility service optimization: Long Island Lighting Company.

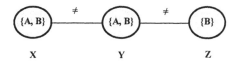

Figure 9.1. A constraint network representation of a sample constraint satisfaction problem.

- Intelligent cabling of big buildings: France Telecom.

- Financial decision support system: Caisse des Dépôts.

- Load capacity constraint regulation: Eurocontrol.

- Planning of satellites missions: Alcatel Espace.

- Optimization of configuration of telecom equipment: Alcatel CIT.

- Production scheduling of herbicides: Monsanto.

- "Just in time" transport and logistics in food industry: Sun Valley.

- Supply chain management in petroleum industry: ERG Petroli.

CSPs can be represented as *constraint networks*, where the variables correspond to nodes and the constraints to arcs. The constraint network for our sample problem is shown in Figure 9.1. Constraints involving more than two variables can be modeled with hypergraphs, but most basic CSP concepts can be introduced with binary constraints involving two variables, and that is the route we will begin with in this chapter. We will say that a value for one variable is *consistent with* a value for another if the pair of values satisfies the binary constraint between them. (This constraint could be the trivial constraint that allows all pairs of values; such constraints are not represented by arcs in the constraint network.) Note that specifying a domain of values for a variable can be viewed as providing a unary constraint on that single variable.

This chapter will focus on the methods developed in artificial intelligence and the approaches embodied in constraint programming languages. Of course, this brief chapter can only suggest some of the developments in these fields; it is not intended as a survey, only as an introduction. Rather than beginning with formal definitions, algorithms, and theorems, we will focus on introducing concepts through examples.

The constraint programming ideal is this: the programming is declarative; we simply state the problem as a CSP and powerful algorithms, provided by a constraint library or language, solve the problem. In practice, this ideal has, of course, been only partially realized, and expert constraint programmers are needed to refine modeling and solving methods for difficult problems.

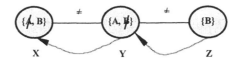

Figure 9.2. Arc consistency propagation.

9.2   INFERENCE

Inference methods make implicit constraint information explicit. Inference can reduce the effort involved in searching for solutions or even synthesize solutions without search. The most common form of inference is known as *arc consistency*. In our sample problem, we can infer that B is not a possible value for Y because there is *no* value for Z that, together with B, satisfies the constraint between Y and Z. This can be viewed as making explicit the fact that the unary constraint on the variable Y does not allow B.

This inference process can *propagate*: after deleting B from the domain of Y, there is no value remaining for Y that together with A for X will satisfy the constraint between X and Y, therefore we can delete A from the domain of X. (See Figure 9.2.) If we repeatedly eliminate inconsistent values in this fashion until any value for any variable is consistent with some value for all other variables, we have achieved arc consistency. Many algorithms have been developed to achieve arc consistency efficiently (Bessière et al., 1999; Macworth, 1977).

Eliminating inconsistent values by achieving arc consistency can greatly reduce the space we must search through for a solution. Arc consistency methods can also be interleaved with search to dynamically reduce the search space, as we shall see in the next section.

Beyond arc consistency lies a broad taxonomy of consistency methods. Many of these can be viewed as some form of (i, j)-*consistency*. A CSP is (i, j)-consistent if, given any consistent set of i values for i variables, we can find j values for any other j variables, such that the $i + j$ values together satisfy all the constraints on the $i + j$ variables. Arc consistency is $(1, 1)$-consistency. $(k - 1, 1)$-consistency, or k-*consistency*, for successive values of k constitutes an important constraint hierarchy (Freuder, 1978).

More advanced forms of consistency processing often prove impractical either because of the processing time involved or because of the space requirements. For example, 3-consistency, otherwise known as *path consistency*, is elegant because it can be shown to ensure that given values for any two variables one can find values that satisfy all the constraints forming any given path between these variables in the constraint network. However, achieving path consistency means making implicit binary constraint information explicit, and storing this information can become too costly for large problems.

For this reason, variations on *inverse consistency*, or $(1, j - 1)$-consistency, which can be achieved simply by domain reductions, have attracted some interest (Debruyne and Bessière, 2001). Various forms of *learning* achieve *partial k*-consistency during search (Dechter, 1990). For example, if we modified our sample problem to allow only A for Z, and we tried assigning B to X and A to Y during a search for a solution to this problem, we would run into a "dead end": no value would be possible for Z. From that we could learn that the constraint between X and Y should be extended to rule out the pair (B, A), achieving partial path consistency.

Interchangeability provides another form of inference, which can also eliminate values from consideration. Suppose we modify our sample problem to add employees C and D who can carry out task X. Values C and D would be interchangeable for variable X because in any solution using one we can substitute the other. Thus we can eliminate one in our search for solutions (and if we want to, just substitute it back into any solutions we find). Just as with consistency processing there is a local form of interchangeability that can be efficiently computed. In a sense, inconsistency is an extreme form of interchangeability; all inconsistent values are interchangeable in the null set of solutions that utilize them (Freuder, 1991).

9.3   MODELING

Modeling is a critical aspect of constraint satisfaction. Given a user's understanding of a problem, we must determine how to model the problem as a constraint satisfaction problem. Some models may be better suited for efficient solution than others (Régin, 2001).

Experienced constraint programmers may add constraints that are *redundant* in the sense that they do not change the set of solutions to the problem, in the hope that adding these constraints may still be cost-effective in terms of reducing problem solving effort. Added constraints that do eliminate some, but not all, of the solutions, may also be useful: for example, to break symmetries in the problem.

Specialized constraints can facilitate the process of modeling problems as CSPs, and associated specialized inference methods can again be cost-effective. For example, imagine that we have a problem with four tasks, two employees who can handle each, but three of these tasks must be undertaken simultaneously. This temporal constraint can be modeled by three separate binary inequality constraints between each pair of these tasks; arc consistency processing of these constraints will not eliminate any values from their domains. On the other hand an "all-different" constraint, that can apply to more than two variables at a time, not only simplifies the modeling of the problem, but an associated inference method can eliminate all the values from a variable domain, proving the problem unsolvable. Specialized constraints may be identified for specific problem domains: for example, scheduling problems.

It has even proven useful to maintain multiple complete models for a problem "channeling" the results of constraint processing between the two (Cheng et al., 1999). As has been noted, a variety of approaches have been brought to bear on constraint satisfaction, and it may prove useful to model part of a problem as, for example, an integer programming problem. Insight is emerging into basic modeling issues: for example, binary versus non-binary models (Bacchus et al., 2002).

In practice, modeling can be an iterative process. Users may discover that their original specification of the problem was incomplete or incorrect or simply impossible. The problems themselves may change over time.

9.4   SEARCH

In order to find solutions we generally need to conduct some form of search. One family of search algorithms attempts to build a solution by *extending* a set of consistent values for a subset of the problem variables, repeatedly adding a consistent value for one more variable, until a complete solution is reached. Another family of algorithms attempts to find a solution by *repairing* an inconsistent set of values for all the variables, repeatedly changing an inconsistent value for one variable, until a complete solution is reached. (Extension and repair techniques can also be combined.)

Often extension methods are systematic and *complete*, they will eventually try all possibilities, and thus find a solution or determine unsolvability, while often repair methods are stochastic and incomplete. The hope is that completeness can be traded off for efficiency.

9.4.1   Extension

The classic extension algorithm is *backtrack search*. Figure 9.3 shows a backtrack search tree representing a trace of a backtrack algorithm solving our sample problem.

A depth-first traversal of this tree corresponds to the order in which the algorithm tried to fit values into a solution. First the algorithm chose to try A for X, then A for Y. At this point it recognized that the choice of A for Y was inconsistent with the choice of A for X: it failed to satisfy the constraint between X and Y. Thus there was no need to try a choice for Z; instead the choice for Y was changed to B. But then B for Z was found to be inconsistent, and no other choice was available, so the algorithm "backed up" to look for

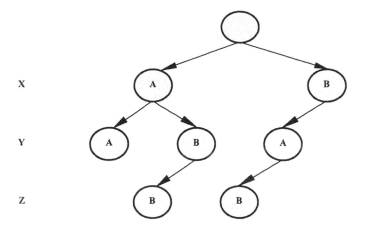

Figure 9.3. Backtrack search tree for example problem.

another choice for Y. None was available so it backed up to try B for X. This could be extended to A for Y and finally to B for Z, completing the search.

Backtrack search can prune away many potential combinations of values simply by recognizing when an assignment of values to a subset of the variables is already inconsistent and cannot be extended. However, backtrack search is still prone to "thrashing behavior". A "wrong decision" early on can require an enormous amount of "backing and filling" before it is corrected. Imagine, for example, that there were 100 other variables in our example problem, and, after initially choosing A for X and B for Y, the search algorithm tried assigning consistent values to each of those 100 variables before looking at Z. When it proved impossible to find a consistent value for Z (assuming the search was able to get that far successfully) the algorithm would begin trying different combinations of values for all those 100 variables, all in vain.

A variety of modifications to backtrack search address this problem (Kondrak and van Beek, 1997). They all come with their own overhead, but the search effort savings can make the overhead worthwhile.

Heuristics can guide the search order. For example, the "minimal domain size" heuristic suggests that as we attempt to extend a partial solution we consider the variables in order of increasing domain size; the motivation there is that we are more likely to fail with fewer values to choose from, and it is better to encounter failure higher in the search tree than lower down when it can induce more thrashing behavior. Using this heuristic in our example we would have first chosen B for Z, then proceeded to a solution without having to back up to a prior level in the search tree. While "fail first" makes sense for the order in which to consider the variables, "succeed first" makes sense for the order in which to try the values for the variables.

Various forms of inference can be used prospectively to prune the search space. For example, search choices can be interleaved with arc consistency maintenance. In our example, if we tried to restore arc consistency after choosing A for X, we would eliminate B from the domain of Z, leaving it empty. At this point we would know that A for X was doomed to failure and could immediately move on to B. Even when failure is not immediate, "look ahead" methods that infer implications of search choices can prune the remaining search space. Furthermore, "dynamic" search order heuristics can be informed by this pruning: for example, the minimal domain size heuristic can be based on the size of the domains after look-ahead pruning. Maintaining arc consistency is an extremely effective and widely used technique (Sabin and Freuder, 1997).

Memory can direct various forms of "intelligent backtracking" (Dechter and Frost, 2002). For example, suppose in our example for some reason our search heuristics directed us to start the search by choosing B for Y followed by A for X. Of course B the only choice for Z would then fail. Basic backtrack search would back up "chronologically" to then try B for X. However, if the algorithm "remembers" that failure to find a value for Z was based solely on conflict with the choice for Y, it can "jump back" to try the alternative value A at the Y level in the search tree without unnecessarily trying B for X. The benefits of maintaining arc consistency overlap with those of intelligent backtracking, and the former may make the latter unnecessary.

Search can also be reorganized to try alternatives in a top-down as opposed to bottom-up manner. This responds to the observation that heuristic choices made early in the extension process, when the remaining search space is unconstrained by the implications of many previous choices, may be most prone to failure. For example, "limited discrepancy search" iteratively restarts the search process increasing the number of "discrepancies", or deviations from heuristic advice, that are allowed, until a solution is found (Harvey and Ginsberg, 1995). (The search effort at the final discrepancy level dominates the upper bound complexity computation, so the redundant search effort is not as significant as it might seem.)

Extensional methods can be used in an incomplete manner. As a simple example, "random restart", starting the search over as soon as a dead end is reached, with a stochastic element to the search order, can be surprisingly successful (Gomes et al., 1997).

9.4.2   Repair

Repair methods start with a complete assignment of values to variables, and work by changing the value assigned to a variable in order to improve the solution. Each such change is called a *move*, and the new assignment is termed a *neighbor* of the previous assignment. Genetic algorithms (see Chapter 4), which create a new assignment by combining two previous assignments, rather than by moving to a neighbor of a single assignment, can be viewed as a form of repair.

Repair methods utilize a variety of metaphors, physical (hill climbing, simulated annealing, see Chapter 7) and biological (neural networks, genetic algorithms). For example, we might start a search on our example problem by choosing value A for each variable. Then, seeking to "hill climb" in the search space to an assignment with fewer inconsistencies, we might choose to change the value of Y to B; and we would be done. Hill climbing, is a repair-based algorithm in which each move is required to yield a neighbor with a better cost than before. It cannot, in general, guarantee to produce an optimal solution at the point where the algorithm stops because no neighbor has a better cost than the current assignment.

Repair methods can also use heuristics to guide the search. For example, the *min-conflicts* heuristic suggests finding an inconsistent value and then changing it to the alternative value that minimizes the amount of inconsistency remaining (Minton et al., 1992).

The classic repair process risks getting "stuck" at a "local maximum", where complete consistency has not been achieved, but any single change will only increase inconsistency, or "cycling" through the same set of inconsistent assignments. There are many schemes to cope. A stochastic element can be helpful. When an algorithm has to choose between equally desirable alternatives it may do so randomly. When no good alternative exists it may start over, or "jump" to a new starting point. Simulated annealing allows moves to neighbors with a worse cost with a given probability. Memory can also be utilized to guide the search and avoid cycling (tabu search, see Chapter 6).

9.5   EXAMPLE

We illustrate simple modeling, search and inference now with another example. The Queens Problem involves placing queens on a chessboard such that they do not attack one another. A simple version only uses a four-by-four corner of the chessboard to place four queens:

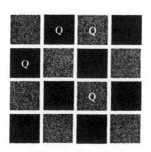

Queens in chess attack horizontally, vertically, and diagonally. So, for example, the two queens on the dark squares above attack each other diagonally, the two queens on the light squares attack vertically. One solution is

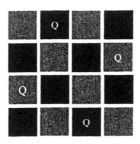

If we model this problem as a CSP where the variables are the four queens and the values for each queen are the 16 squares, we have 65 536 possible combinations to explore, looking for one where the constraints (the queens do not attack each other) are satisfied. If we observe that we can only have one queen per row, and model the problem with a variable corresponding to the queen in each row, each variable having four possible values corresponding to the squares in the row, we have only 256 possibilities to search through.

The beginning of the backtrack search tree for this example is shown in Figure 9.4. After placing the first row queen in the first column, the first successful spot for the second row queen is in the third column. However, that leaves no successful placement for the third row queen, and we need to backtrack.

In fact, there will be quite a lot of backtracking to do here before we find a solution. However, arc consistency inference can reduce the amount of search we do considerably. Consider what happens if we seek arc consistency after placing a queen in the second column of row 1. This placement directly rules out any square that can be attacked by this queen, of course, and, in fact, arc consistency propagation proceeds to rule out additional possibilities until we are left with a solution. The queens in column 3 of row 3 and column 4 of row 4 are ruled out because they are attacked by the only possibility left for row 2. After the queen in column 3 of row 3 is eliminated, the queen in column 1 of row 4 is attacked by the only remaining possibility for row 3, so it too can be eliminated.

9.6   TRACTABILITY

CSPs are in general NP-hard. Analytical and experimental progress has been made in characterizing tractable and intractable problems. The results have been used to inform algorithmic and heuristic methods.

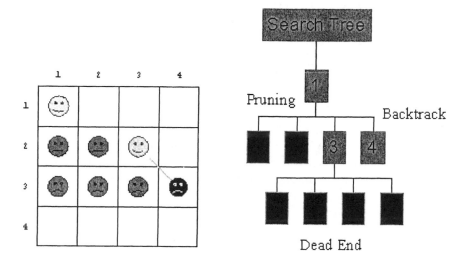

Figure 9.4. The beginning of the example backtrack search tree.

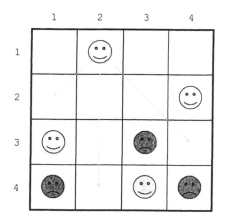

9.6.1 Theory

Tractable classes of CSPs have been identified based on the structure of the constraint network, e.g. tree structure, and on closure properties of sets of constraints (Jeavons et al., 1997), e.g. max-closed. Tractability has been associated with sets of constraints defining a specific class of problems, e.g. temporal reasoning problems defined by "simple temporal networks" (Dechter et al., 1991).

If a constraint network is tree-structured, there will be a *width-one* ordering for the variables in which each variable is directly constrained by at most one variable earlier in the ordering. In our sample problem, which has a trivial tree structure, the ordering X, Y, Z is width-one: Y is constrained by X and Z by Y; the ordering X, Z, Y is not width-one: Y is constrained by both X and Z. If we achieve arc consistency and use a width-one ordering as the order in which we consider variables when trying to extend a partial solution, backtrack search will in fact be *backtrack-free*: for each variable we will be able to find a consistent value without backing up to reconsider a previously instantiated variable (Freuder, 1982).

Max-closure means that if $(a\ b)$ and $(c\ d)$ both satisfy the constraint, then $((\max(a\ c)), (\max(b\ d)))$ will also satisfy the constraint. If all the constraints in a problem are max-closed, the problem will be tractable. The "less than" constraint is max-closed, e.g. $4 < 6$, $2 < 9$ and $4 < 9$. Thus if we replaced the inequality constraints in our sample problem by less-than constraints, we would ensure tractability, even if we gave up the tree structure by adding a constraint between X and Z. In fact simple temporal networks are max-closed.

9.6.2   Experiment

Intuitively it seems natural that many random CSPs would be relatively easy: loosely constrained problems would be easy to solve, highly constrained problems would be easy to prove unsolvable. What is more surprising is the experimental evidence that as we vary the constrainedness of the problems there is a sharp *phase transition* between solvable and unsolvable regions, which corresponds to a sharp spike of "really hard" problems (Cheeseman et al., 1991). ("Phase transition" is a metaphor for physical transitions, such as the one between water and ice.)

9.7   OPTIMIZATION

Optimization arises in a variety of contexts. If all the constraints in a problem cannot be satisfied, we can seek the "best" partial solution. If there are many solutions to a problem, we may have some criteria for distinguishing the "best" one. "Soft constraints" can allow us to express probabilities or preferences that make one solution "better" than another.

Again we face issues of modeling, inference and search. What does it mean to be the "best" solution; how do we find the "best" solution? We can assign scores to local elements of our model and combine those to obtain a global score for a proposed solution, then compare these to obtain an optimal solution for the problem.

A simple case is the *Max-CSP problem*, where we seek a solution that satisfies as many constraints as possible. Here each satisfied constraint scores 1, the score for a proposed solution is the sum of the satisfied constraints, and an optimal solution is one with a maximal score. However, there are many alternatives, such as fuzzy, probabilistic and possibilistic CSPs, many of which have been captured under the general framework of semiring-based or valued CSPs (Bistarelli et al., 1992).

Backtrack search methods can be generalized to branch and bound methods (see Chapter 2) for seeking optimal solutions, where a partial solution is abandoned when it is clear that it cannot be extended to a better solution than one already found. Inference methods can be generalized. Repair methods can often find close to optimal solutions quickly.

9.8 ALGORITHMS

9.8.1 Handling Constraints

Constraint technology is an approach that solves problems by reasoning on the constraints, and when no further inferences can be drawn, making a search step. Thus inference and search are interleaved.

For problems involving hard constraints—ones that must be satisfied in any solution to the problem—reasoning on the constraints is a very powerful technique.

The secret of much of the success of constraint technology comes from its facility to capitalize on the structure of the problem constraints. This enables "global" reasoning to be supported, which can guide a more "intelligent" search procedure than would otherwise be possible.

In this section we shall introduce some different forms of reasoning and its use in solving problems efficiently.

9.8.2 Domains, and Constraint Propagation

In general domain constraint propagation algorithms take a set of variables and the original domains as input, and either report inconsistency, or output smaller domains for the variables. Since propagation algorithms can extract more information, each time the input domains become smaller, and since the propagation behavior also makes the domains of the variables smaller, the propagation algorithms can co-operate through these domains. The output from one algorithm is input to another, whose output can in turn be input to the first algorithm. Thus many different propagation algorithms can co-operate, together yielding domain reductions which are much stronger than simply pooling the information from the separate algorithms. In the following code we constrain two variables, X and Y, to take values in the range 1–10. We also constrain X to be greater than Y, and (inconsistently!) we also constrain Y to be greater than X:

```
?- X::1..10, Y::1..10, fd:(X>Y), fd:(Y>X)
```

If the propagation algorithm for each constraint makes the bounds consistent, then the first constraint will yield new domains

```
X::2..10, Y::1..9
```

and the second constraint will yield new domains:

```
X::1..9, Y::2..10
```

Pooling the deduced information, we get the intersection of the new domains, which is

```
X::2..9, Y::2..9
```

By contrast, if the propagation algorithms communicate through the variable domains, then they will yield new domains, which are then input to the other algorithm until, at the fifth step, the inconsistency between the two constraints is detected.

The interaction of the different algorithms is predictable, even though the algorithms are completely independent, so long as they have certain natural properties. Specifically,

- the output domains must be a subset of the input domains

- if with input domains ID the algorithm produces output domains OD, then with any input domains which are a subset of ID, the output domains must be a subset of OD.

These properties guarantee that the information produced by the propagation algorithms, assuming they are executed until no new information can be deduced, is guaranteed to be the same, independent of the order in which the different algorithms (or *propagation steps*) are executed.

9.8.3   Constraints and Search

Separating Constraint Handling from Search   The constraint programming paradigm supports clarity, correctness and maintainability by separating the statement of the problem as far as possible from the details of the algorithm used to solve it. The problem is stated in terms of its decision variables, the constraints on those variables, and an expression to be optimized.

As a toy example the employee task problem introduced at the beginning of this chapter can be expressed as follows:

```
?- X::[a,b], Y::[a,b], Z::[b], % set up variables
X=\=Y, Y=\=Z % set up constraints
```

This states that the variable X can take either of the two values a or b; similarly Y can take a or b; but Z can only take the value b. Additionally, the value taken by X must be different from that taken by Y; and the values taken by Y and Z must also be different.

To map this problem statement into an algorithm, the developer must

- choose how to handle each constraint

- specify the search procedure.

For example, assuming fd is a solver which the developer chooses to handle the constraints, and labeling is a search routine, the whole program is written as follows:

```
?- X::[a,b], Y::[a,b], Z::[b],
fd:(X=$\backslash $=Y), fd:(Y=$\backslash $=Z),
labeling([X,Y,Z])
```

This code sends the constraint $X = \backslash = Y$ to the finite domain solver called fd; and the constraint $Y = \backslash = Z$ is also sent to fd. The list of variables X, Y and Z is then passed to the labeling routine.

Domain propagation algorithms take as input the domains of the variables, and yield smaller domains. For example, given that the domain of Z is [b], the fd solver for the constraint $Y = \backslash = Z$ immediately removes b from the domain of Y, yielding a new domain [a]. Every time the domain of one of those variables is reduced by another propagation algorithm, the algorithm "wakes" and reduces the domains further (possibly waking the other algorithm). In the above example, when the domain of Y is reduced to [a], the constraint $X = \backslash = Y$ wakes, and removes b from the domain of X, reducing the domain of X to [b].

The domain of a variable may be reduced either by propagation, or instead by a search decision. In this case propagation starts as before, and continues until no further information can be derived. Thus search and constraint reasoning are automatically interleaved.

Search Heuristics Exploiting Constraint Propagation   Most real applications cannot be solved to optimality because they are simply too large and complex. In such cases it is crucial that the algorithm is directed towards areas of the search space where low-cost feasible solutions are most likely to be found.

Constraint propagation can yield very useful information, which can be used to guide the search for a solution. Not only can propagation exclude impossible choices a priori, but it can also yield information about which choices would be optimal in the absence of certain (awkward) constraints.

Because constraint propagation occurs after each search step, the resulting information is used to support *dynamic* heuristics, where the next choice is contingent upon all the information about the problem gathered during the current search.

In short, incomplete extension search techniques can produce high quality solutions to large complex industrial applications in areas such as transportation and logistics. The advantage is that the solutions respect all the hard constraints and are therefore applicable in practice.

9.8.4   Global Constraints

In this section we introduce a variety of application-specific "global" constraints. These constraints achieve more, and more efficient, propagation behavior than would be possible using combinations of the standard equality and disequality constraints introduced above. We first outline two global constraints, one called `alldifferent` for handling sets of disequality constraints, and one called `schedule` for handling resource usage by a set of tasks.

Alldifferent   Consider the disequality constraint "=\=" used in the employee task example at the beginning of this chapter. Perhaps surprisingly, global reasoning on a set of such constraints brings more than local reasoning. Suppose we have a list and want to make local changes until all the elements of the list are distinct. If each element has the same domain (i.e. the same set of possible values), then it suffices to choose any element in conflict (i.e. an element whose value occurs elsewhere in the list), and change it to a new value which does not occur elsewhere. If, however, the domains of different elements of the list are different, then there is no guarantee that local improvement will converge to a list whose elements are all different. However, there is a polynomial time graph-based algorithm (Régin, 1994) which guarantees to detect if there is no solution to this problem, and otherwise it reduces the domains of all the elements until they contain only values that participate in a solution. The constraint that all the elements of a list must be distinct is usually called `alldifferent` in constraint programming. For example, consider the case

```
?- X::[a,b], Y::[a,b], Z::[a,b,c],
alldifferent([X,Y,Z])
```

In this case Régin's algorithm (Régin, 1994) for the `alldifferent` constraint will reduce the domain of Z to just [c]. As we saw above, the behavior of reducing the variables domains in this manner is called *constraint propagation*.

Global constraint propagation algorithms work co-operatively within a constraint programming framework. In the above example, the propagation algorithm for the `alldifferent` constraint takes as input the list of variables, and their original domains. As a result it outputs new, smaller, domains for those variables.

Schedule Consider a task scheduling problem comprising a set of tasks with release times (i.e. earliest start times) and due dates (i.e. latest end times), each of which requires a certain amount of resource, running on a set of machines that provide a fixed amount of resource.

The schedule constraint works, in principle, by examining each time period within the time horizon:

- First the algorithm calculates how much resource r_i each task i must necessarily take up within this period.

- If the sum $\sum r_i$ exceeds the available resource, then the constraint reports an inconsistency.

- If the sum $\sum r_i$ takes up so much resource that task j cannot be scheduled within the period, and the remaining resource within the period is r_T, then task j is constrained only to use an amount r_T of resource within this period.

For non-preemptive scheduling, this constraint may force such a task j to start a certain amount of time before the period begins, or end after it. The information propagated narrows the bounds on the start times of certain tasks.

Whilst this kind of reasoning is expensive to perform, there are many quicker, but theoretically less complete forms of reasoning, such as "edge-finding", which can be implemented in a time quadratic in the number of tasks.

Further Global Constraints The different global constraints outlined above have proven themselves in practice. Using global constraints such as schedule, constraint programming solves benchmark problems in times competitive with the best complete techniques. The main advantages of global constraints in constraint programming, in addition to their efficiency in solving standard benchmark problems, are that:

- They can be augmented by any number of application-specific "side" constraints. The constraint programming framework allows all kinds of constraints to be thrown in, without requiring any change to the algorithms handling the different constraints.

- They return high-quality information about the problem which can be used to focus the search. Not only do they work with complete search algorithms, but also they guide incomplete algorithms to return good solutions quickly.

Constraint programming systems can include a range of propagation algorithms supporting global reasoning for constraints appearing in different kinds of applications such as rostering, transportation, network optimization, and even bioinformatics.

Analysis One of the most important requirements of a programming system is support for reusability. Many complex models developed by Operations Researchers have made very little practical impact, because they are so hard to reuse. The concept of a global constraint is inherently quite simple. It is a constraint that captures a class of subproblems, with any number of variables. Global constraints have built-in algorithms, which are specialized for treating the problem class. Any new algorithm can be easily captured as a global constraint and reused. Global constraints have had a major impact, and are used widely and often as a tool in solving complex real-world problems. They are, arguably, the most important contribution that constraint programming has brought to Operations Research.

9.8.5   Different Constraint Behaviors

Constraint reasoning may derive other information than domain reductions. Indeed, any valid inference step can be made by a propagation algorithm. For example, from the constraints $X > Y$, $Y > Z$ propagation can derive that $X > Z$. To achieve co-operation between propagation algorithms, and termination of propagation sequences, constraint programming systems typically require propagation algorithms to behave in certain standard ways. Normally they are required to produce information of a certain kind: for example, domain reductions.

An alternative to propagating domain reductions is to propagate new linear constraints. Just as domain propagation ideally yields the smallest domains which include all values that could satisfy the constraint, so linear propagation ideally yields the convex hull of the set of solutions. In this ideal case, linear propagation is stronger than domain propagation, because the convex hull of the set of solutions is contained in the smallest set of variable domains (termed the *box*) that contains them.

9.8.6   Extension and Repair Search

Extension search is conservative in that, at every node of the search tree, all the problem constraints are satisfied. Repair search is optimistic, in the sense that a variable assignment at a search node may be, albeit promising, actually inconsistent with one or more problem constraints.

Constraint Reasoning and Extension Search Constraint reasoning, in the context of extension search, corresponds to logical deduction. The domain reductions, or new linear constraints yielded by propagation, are indeed a consequence of the constraint in the input state.

Constraint Reasoning and Repair Search   Constraint reasoning can also be applied in the context of repair search. In this case the constraint behavior is typically caused by the constraint being violated in the input state. Like propagation, the behavior yields new information, which is of a standard simple form that can be dealt with actively by the search procedure. We distinguish two forms of behavior: constraint generation and separation. These forms are best illustrated by example.

Constraint generation   For an example of constraint generation, consider a travelling salesman problem which is being solved by integer/linear programming. At each search node an integer/linear problem is solved, which only approximates the actual TSP constraint. Consider a search node which represents a route with a detached cycle. This violates the TSP constraint in a way that can be fixed by adding a linear constraint enforcing a unit flow out from the set of "cities" in the detached cycle. This is the *generated* constraint, at the given search node. The search is complete at the first node where the TSP constraint is no longer violated. Constraint generation can be used in case the awkward constraints can be expressed by a conjunction of easy constraints, although the number of such easy constraints in the conjunction may be too large for them all to be imposed.

Separation   Separation behavior is required to fix any violated constraint which cannot be expressed as a conjunction of easy constraints (however large). If the awkward constraint can be approximated, arbitrarily closely, by a (conjunction of) disjunction(s) of easy constraints, then separation can be used. Constraint reasoning yields one of the easy constraints—one that is violated by the current search node—and imposes it so that the algorithm which produces the next search node is guaranteed to satisfy this easy constraint. Completeness is maintained by imposing the other easy constraints in the disjunction on other branches of the search tree.

Languages and Systems   One drawback of the logical basis is that repair-based search methods have not fitted naturally into the CLP paradigm. Recently, a language has been introduced called Localizer (Michel and Van Hentenryck, 1999) which is designed specifically to support the encoding of repair-based search algorithms such as Simulated Annealing and GSAT (Selman et al., 1992). The fundamental contribution of Localizer is the concept of an "invariant", which is a constraint that retains information used during search. For GSAT, by way of example, an invariant is used to record, for each problem variable, the change in the number of satisfied propositions if the variable's value were to be changed. The invariant is specified as a constraint, but maintained by an efficient incremental algorithm. Other constraint-based languages for specifying search are SALSA (Laburthe and Caseau, 1998) and ToOLS (de Givry and Jeannin, 2003).

9.9   CONSTRAINT LANGUAGES

9.9.1   Constraint Logic Programming

The earliest constraint programming languages, such as *Ref-Arf* and *Alice*, were specialized to a particular class of algorithms. The first general purpose constraint programming languages were constraint handling systems embedded in logic programming (Jaffar and Lassez, 1987; Van Hentenryck et al., 1999), called *constraint logic programming* (CLP). Examples are CLP(fd), HAL, SICStus and ECLiPSe. Certainly logic programming is an ideal host programming paradigm for constraints, and CLP systems are widely used in industry and academia.

Logic programming is based on relations. In fact every procedure in a logic program can be read as a relation. However, the definition of a constraint is exactly the same thing—a *relation*. Consequently, the extension of logic programming to CLP is entirely natural. Logic programming also has backtrack search built in, and this is easily modified to accommodate constraint propagation. CLP has been enhanced with some high-level control primitives, allowing active constraint behaviors to be expressed with simplicity and flexibility. The direct representation of the application in terms of constraints, together with the high-level control, results in short simple programs. Since it is easy to change the model and, separately, the behavior of a program, the paradigm supports experimentation with problem solving methods. In the context of a rapid application methodology, it even supports experimentation with the problem (model) itself.

9.9.2   Modeling Languages

On the other hand, Operations Researchers have introduced a wide range of highly sophisticated specialized algorithms for different classes of problems. For many OR researchers CLP and Localizer are too powerful—they seek a modeling language rather than a computer programming language in which to encode their problems. Traditional mathematical modeling languages used by OR researchers have offered little control over the search and the constraint propagation. OPL (Van Hentenryck, 1999) is an extension of such a modeling language to give more control to the algorithm developer. It represents a step towards a full constraint programming language.

By contrast, a number of application development environments (e.g. Visual CHIP) have appeared recently that allow the developer to define and apply constraints graphically, rather than by writing a program. This represents a step in the other direction!

9.10   APPLICATIONS

9.10.1   Current Areas of Application

Constraint programming is based on logic. Consequently any formal specification of an industrial problem can be directly expressed in a constraint program. The drawbacks of earlier declarative programming paradigms have been

- that the programmer had to encode the problem in a way that was efficient to execute on a computer;

- that the end user of the application could not understand the formal specification.

The first breakthrough of constraint programming has been to separate the logical representation of the problem from the efficient encoding in the underlying constraint solvers. This separation of logic from implementation has opened up a range of applications in the area of control, verification and validation.

The second breakthrough of constraint programming has been in the area of software engineering. The constraint paradigm has proven to accommodate a wide variety of problem solving techniques, and has enabled them to be combined into hybrid techniques and algorithms, suited to whatever problem is being tackled.

As important as the algorithms to the success of constraint technology, has been the facility to link models and solutions to a graphical user interface that makes sense to the end user. Having developers display the solutions in a form intelligible to the end users, forces the developers to put themselves into the shoes of the users.

Moreover, not only are the final solutions displayed to the user: it is also possible to display intermediate solutions found during search, or even partial solutions. The ability to animate the search in a way that is intelligible to the end user means the users can put themselves into the shoes of the developers. In this way the crucial relationship and understanding between developers and end users is supported and users feel themselves involved in the development of the software that will support them in the future.

As a consequence, constraint technology has been applied very successfully in a range of combinatorial problem solving applications, extending those traditionally tackled using operations research.

The two main application areas of constraint programming are, therefore, (i) control, verification and validation, and (ii) combinatorial problem solving.

9.10.2   Applications in Control, Verification and Validation

Engineering relies increasingly on software, not only at the design stage, but also during operation. Consider the humble photocopier. Photocopiers are not as humble as they used to be—each system comprises a multitude of components, such as feeders, sorters, staplers and so on. The next generation of photocopiers will have orders of magnitude more components than now. The challenge of maintaining compatibility between the different components, and different versions of the components, has become unmanageable.

Xerox has turned to constraint technology to specify the behavior of the different components in terms of constraints. If a set of components are to be combined in a system, constraint technology is applied to determine whether the components will function correctly and coherently. The facility to specify behavior in terms of constraints has enabled engineers at Xerox not only to simulate complex systems in software but also to revise their specifications before constructing anything and achieve compatibility first time.

Control software has traditionally been expressed in terms of finite state machines. Proofs of safety and reachability are necessary to ensure that the system only moves between safe states (e.g. the lift never moves while the door is open) and that required states are reached (the lift eventually answers every request). Siemens has applied constraint technology to validate control software, using techniques such as Boolean unification to detect any errors. Similar techniques are also used by Siemens to verify integrated circuits.

Constraint technology is also used to prove properties of software. For example, abstract interpretation benefits from constraint technology in achieving the performance necessary to extract precise information about concrete program behavior.

Finally, constraints are being used not only to verify software but to monitor and restrict its behavior at runtime. *Guardian Agents* ensure that complex software, in medical applications for example, never behaves in a way that contravenes the certain safety and correctness requirements.

For applications in control, validation and verification, the role of constraints is to model properties of complex systems in terms of logic, and then to prove theorems about the systems. The main constraint reasoning used in this area is propositional theorem proving. For many applications, even temporal properties are represented in a form such that they can be proved using propositional satisfiability.

Nevertheless, the direct application of abstract interpretation to concurrent constraint programs offers another way to prove properties of complex dynamic systems.

9.10.3   Combinatorial Problem Solving

Commercially, constraint technology has made a huge impact in problem solving areas such as transportation, logistics, network optimization, scheduling and timetabling, production control, and design, and it is also showing tremendous potential in new application areas such as bio-informatics and virtual reality systems.

Starting with applications to transportation, constraint technology is used by airline, bus and railway companies, all over the world. Applications include timetabling, fleet scheduling, crew scheduling and rostering, stand, slot and platform allocation.

Constraints have been applied in the logistics area for parcel delivery, food, chilled goods, and even nuclear waste. As in other application areas, the major IT system suppliers (such as SAP and I2) are increasingly adopting constraint technology.

Constraints have been applied for Internet service planning and scheduling, for minimizing traffic in banking networks, and for optimization and control of distribution and maintenance in water and gas pipe networks. Constraints are used for network planning (bandwidth, routing, peering points), optimizing network flow and pumping energy (for gas and water), and assessing user requirements.

Constraint technology appears to have established itself as the technology of choice in the areas of short-term scheduling, timetabling and rostering. The flexibility and scalability of constraints was proven in the European market (for example at Dassault and Monsanto), but is now used worldwide.

It has been used for timetabling activities in schools and universities, for rostering staff at hospitals, call centers, banks and even radio stations. An interesting and successful application is the scheduling of satellite operations.

The chemical industry has an enormous range of complex production processes whose scheduling and control is a major challenge, currently being tackled with constraints. Oil refineries and steel plants also use constraints in controlling their production processes. Indeed, many applications of constraints to production scheduling also include production monitoring and control.

The majority of commercial applications of constraint technology have, to date, used finite domain propagation. Finite domains are a very natural way to represent the set of machines that can carry out a task, the set of vehicles that can perform a delivery, or the set of rooms/stands/platforms where an activity can be carried out. Making a choice for one task, precludes the use of the same resource for any other task which overlaps with it, and propagation captures this easily and efficiently.

Naturally, most applications involve many groups of tasks and resources with possibly complex constraints on their availability (for example personnel regulations may require that staff have two weekends off in three, that they must have a day off after each sequence of night-shifts, and that they must not work more than 40 hours a week). For complex constraints like this a number of special constraints have been introduced which not only enable these constraints to be expressed quite naturally, but also associate highly efficient specialized forms of finite domain propagation with each constraint.

9.10.4   Other Applications

Constraints and Graphics   An early use of constraints was for building graphical user interfaces. Now these interfaces are highly efficient and scalable, allowing a diagram to be specified in terms of constraints so that it still carries the same impact and meaning whatever the size or shape of the display hardware. The importance of this functionality in the context of the Internet, and mobile computing, is very clear, and constraint-based graphics is likely to make a major impact in the near future. Constraints are also used in design, involving both spatial constraints and, in the case of real-time systems design, temporal constraints.

Constraint Databases   Constraint databases have not yet made a commercial impact, but it is a good bet that future information systems will store constraints as well as data values. The first envisaged application of constraint databases is to geographical information systems. Environmental monitoring will follow, and subsequently design databases supporting both the design and maintenance of complex artifacts such as airplanes.

9.11   SOME PROMISING AREAS FOR FUTURE APPLICATIONS

There are many topics that could be addressed in additional detail. This section briefly samples a few of these.

9.11.1   Dynamic Constraints, Soft Constraints

Constraint technology was originally designed to handle "hard" constraints that had to be satisfied in every solution. Moreover, each problem had to be solved from scratch, finding values for a set of previously unassigned decision variables.

What has emerged over the years, particularly in the light of practical applications of the technology, is a need to handle "soft" constraints which should be satisfied if possible, but may be violated if necessary.

Another practical requirement is the need to handle dynamic problems, which may change while their solution is being executed (for example, due to machine breakdown, newly placed priority orders or late running).

These requirements have led to the development of new theoretical (Bistarelli et al., 1992) and practical requirements.

9.11.2   Hybrid Techniques

As constraint technology has matured, the community has recognized that it is not a standalone technology, but a weapon in an armory of mathematical tools for tackling complex problems. Indeed, an emerging role for constraint programming is as a framework for combining techniques such as constraint propagation; integer/linear and quadratic programming; interval reasoning; global optimization and metaheuristics. This role has become a focus of research in a new conference series called CPAIOR, in a current European project (Coconut) and at recent INFORMS meetings (see respective websites).

9.11.3   Knowledge Acquisition and Explanation

As constraint programming becomes increasingly commercialized, increasing attention is drawn to "human factors". Issues faced by earlier "knowledge-engineering" technologies must be faced by constraint technology.

Acquiring domain-specific knowledge is obviously a key application issue. Provision needs to be made for interactive acquisition, e.g. in electronic commerce applications. Many problems, e.g. many configuration problems, change over time. While constraint programmers tout the advantages of their declarative paradigm for maintaining programs in the face of such change, acquiring and implementing new knowledge on a large scale still presents challenges.

Users may feel more comfortable when an "explanation" can accompany a solution. Explanation is particularly important when a problem is unsolvable. The user wants to know why, and can use advice on modifying the problem to permit a solution (Amilhastre et al., 2002).

A related set of problems confronts the need constraint programmers have to better understand the solution process. Explanation and visualization of this process can assist in debugging constraint programs, computing solutions more quickly, and finding solutions closer to optimal (Deransart et al., 2000).

9.11.4   Synthesizing Models and Algorithms

Ideally, people with constraints to satisfy or optimize would simply state their problems, in a form congenial to the problem domain, and from this statement a representation suited to efficient processing and an appropriate algorithm to do the processing would be synthesized automatically. In practice, considerable human expertise is often needed to perform this synthesis. Less ambitiously, tools might be provided to assist the constraint programmer in this regard. Initial experiments with simple learning methods have proven surprisingly effective at producing efficient algorithms for specific problem classes (Minton, 1996).

9.11.5   Distributed Processing

Distributed constraint processing arises in many contexts. There are parallel algorithms for constraint satisfaction and concurrent constraint programming languages. There are applications where the problem itself is distributed in some manner. There are computing architectures that are naturally "distributed": for example, neural networks.

There is considerable interest in the synergy between constraint processing and software agents. Agents have issues that are naturally viewed in constraint-based terms, e.g. negotiation. Agents can be used to solve constraint satisfaction problems (Yokoo et al., 1998).

9.11.6   Uncertainty

Real world problems may contain elements of uncertainty. Data may be problematic. The future may not be known. For example, decisions about fuel purchases may need to be made based on uncertain demand dependent on future weather patterns. We want to model and compute with constraints in the presence of such uncertainty (Walsh, 2002).

9.12   TRICKS OF THE TRADE

The constraints community uses a variety of different tools to solve complex problems. There are a number of constraint programming systems available, which support constraint propagation, search and a variety of other techniques. For pedagogical purposes we will simply show the solution a simple problem, solved using one constraint programming system, ECLiPSe (ECLiPSe, 2005). This system is free for research use, and can be downloaded from the referenced website. If the code developed below is copied into a file, then the reader can load ECLiPSe, compile the file and run the program.

We consider a one-machine scheduling problem. The requirement is to schedule a set of tasks on a machine. Each task has a fixed duration, and each has an earliest start time (the release date) and a latest end time (the due date). How should we schedule the tasks so as to finish soonest?

In constraint programming a problem is handled in three stages:

1 Initialize the problem variables.

2 Constrain the variables.

3 Search for values for the variables that satisfy the constraints.

9.12.1 Initializing Variables

For the one-machine scheduling problem, a variable is declared to represent the start time and the end time of each task. The end time is constrained to be the start time plus the duration. The start time is constrained to be after the release date of the task, and the end time constrained to be before the due date. In the code we associate variables StartTime, EndTime and Duration with each task. Although Duration is represented by a variable, we shall assume it has a fixed value for the purposes of this example. The above constraints are expressed as follows:

```
EndTime #= StartTime + Duration,
StartTime #>= ReleaseTime,
EndTime #=<DueTime
```

The code does not fix the final end time. The "minimize" statement should be written

```
minimize( (label_starts(Tasks),
fix_to_min(FinalEndTime)), FinalEndTime )
```

Although Duration is written with a capital letter, and therefore is a variable, the program assumes that it will be supplied with a specific input value at runtime.

For the first model of the problem we impose no further constraints on the start time, until search begins.

9.12.2 Search and Propagation

When the search begins one of the tasks, is chosen to be first. This is done by the following lines of code:

```
choose(Task,Tasks,Rest),
Task = task with [start:StartTime,end:EndTime]
```

Now the following constraints are posted. The start time StartTime is constrained to take its lower bound value (in this case, its release date). This is done by the following line of code:

```
fix_to_min(StartTime)
```

The start times of each of the remaining tasks are constrained to be greater than the task end time, EndTime. This is achieved by the goal

```
demoPropagate(EndTime, Rest)
```

which is defined as follows:

```
demoPropagate(EndTime, Rest) :-
( foreach( task with start:StartTime, Tasks),
param(EndTime)
do
StartTime #>= EndTime
)
```

As a result, the lower bounds of some or all of the remaining task start times may be increased.

After having "propagated" the new constraints, by computing the new lower bounds for all the start times, search resumes. Another task is chosen to be the first of the remaining tasks, and constraints 1 and 2 are posted as before. This search procedure stops when there are no more tasks. It is defined (recursively) as follows:

```
% If no more tasks are left, then do nothing.
label_starts([]).
% Otherwise, select a task, make it start as early as
% possible and constrain the remaining tasks to start
% after it
label_starts(Tasks) :-
choose(Task,Tasks,Rest),
Task = task with [start:StartTime,end:EndTime],
fix_to_min(StartTime),
demoPropagate(EndTime,Rest),
label_starts(Rest)
```

9.12.3   Branch and Bound

When a solution is found the end time of the last task, `FinalEndTime` is recorded. The problem solving process is restarted, but now all tasks are constrained to end before `FinalEndTime`. This is captured by a constraint on each task that its `EndTime` is less than `FinalEndTime`. This behavior is implemented inside the built-in predicate minimize, therefore the code required is simply

```
minimize(label_starts(Tasks),FinalEndTime)
```

If at any time the propagation on a start time makes its lower bound larger than its upper bound, then there is no solution which extends the current task ordering. The system therefore backtracks and chooses a different task to be first at the previous choice point.

9.12.4   Code

The whole ECLiPSe program for solving the one-machine scheduling problem is as follows. To run it load ECLiPSe, compile the following code and invoke the goal:

```
Tasks = [task with [duration:4, release:1,due:10],
task with [duration:3, release:1, due:15],
{\ldots} % and all the other input tasks!
],
task_schedule(Tasks,FinalEndTime).
```

Code for One-Machine Scheduling Solution

```
% Define a data structure to hold info. about tasks
:- local struct(task(start,duration,end,release,due)).
% Load the finite domain solver
:- lib(fd).

% To solve a problem, first state the constraints and
% then encode the search procedure.
% Names starting with upper-case letters are variables.
task_schedule(Tasks,FinalEndTime) :-
constrain(Tasks,FinalEndTime),
minimize(label_starts(Tasks),FinalEndTime).
% Constrain the start and end time of each task
constrain(Tasks,FinalEndTime) :-
( foreach(Task,Tasks),
param(FinalEndTime)
do
% Each Task variable holds a data structure
% with the task details
Task = task with [release:ReleaseTime,
due:DueTime,
duration:Duration,
start:StartTime,
end:EndTime
],
% Constrain the start and end times
EndTime #= StartTime + Duration,
StartTime #>= ReleaseTime,
EndTime #=< DueTime,
% Constrain the final end time to follow the end time
% of each task
```

```
FinalEndTime #>= EndTime ).


% Stop when there are no more tasks to handle
label_starts([]).
% Select a task, make it start as early as possible
% and constrain the remaining tasks to start after it
label_starts(Tasks) :-
choose(Task,Tasks,Rest),
Task = task with [start:StartTime,end:EndTime],
fix_to_min(StartTime),
demoPropagate(EndTime,Rest),
label_starts(Rest).
% Select any task from a non-empty list.
choose(Task,[TaskTasks],Tasks).
% Alternatively choose a different task
choose(Task,[NotThisTaskTasks],[NotThisTaskRest]) :-
choose(Task,Tasks,Rest).


% Constrain the remaining tasks to start after the
% given previous end time
demoPropagate(PrevEndTime,Tasks) :-
( foreach( task with start:StartTime, Tasks),
param(PrevEndTime)
do
StartTime {\#}$>$= PrevEndTime ).


% Make the variable Time take its smallest possible
% value
fix_to_min(Time) :-
mindomain(Time,Earliest),
Time #= Earliest
```

9.12.5   Introducing Redundant Constraints

The first way to enhance this algorithm is by adding a global constraint, specialized for scheduling problems (see Section 9.4). The new constraint does not remove any solutions: it is logically redundant. However, its powerful propagation behavior enables parts of the search space, where no solutions lie, to be pruned. Consequently, the number of search steps is reduced—dramatically for larger problems! The algorithm was devised by operations researchers, but it has been encapsulated by constraint programmers as a single constraint.

9.12.6 Adding Search Heuristics

The next enhancement is to choose, at each search step, the task with the earliest due date. Whilst this does tend to yield feasible solutions, it does not necessarily produce good solutions, until the end time constraints become tight.

9.12.7 Using an Incomplete Search Technique

For very large problems, complete search may not be possible. In this case the algorithm may be controlled so as to limit the effort wasted in exploring unpromising parts of the search space. This can be done simply by limiting the number of times a non-preferred ordering of tasks is imposed during search and backtracking.

The above techniques combine very easily, and the combination is very powerful indeed. As a result constraint programming is currently the technology of choice for operational scheduling problems where task orderings are significant.

The Constraints Archive (http://www.4c.ucc.ie/archive) has pointers to constraint code libraries and constraint programming languages, both freely available software and commercial products.

SOURCES OF ADDITIONAL INFORMATION

Sources of information about constraint programming include:

- Proceedings of the International Conferences on Principles and Practice of Constraint Programming, available in the Springer LNCS series.

- The *Constraints* journal published by Kluwer.

- *Constraint Processing*, by Rina Dechter (published by Morgan Kaufmann, 2003).

- *Programming with Constraints: an Introduction*, by Kim Marriott and Peter Stuckey (MIT Press, 1998).

- On-line Guide to Constraint Programming, maintained by Roman Barták. http://kti.ms.mff.cuni.cz/~bartak/constraints/

- The comp.constraints newsgroup.

- http://carlit.toulouse.inra.fr/cgi-bin/mailman/listinfo/csp (the CSP mailing list).

Many other sources of information can be found on the web at

- Constraint Programming Online:
  http://slash.math.unipd.it/cp/index.php

- Constraints Archive: http://www.4c.ucc.ie/web/archive.

Acknowledgments. Some of this material is based upon works supported by the Science Foundation Ireland under Eugene Freuder's Grant 00/PI.1/C075; some of his contribution to this chapter was prepared while he was at the University of New Hampshire. Richard Wallace and Dan Sabin provided some assistance. The contents of this chapter overlap with a chapter by the same authors on Constraint Satisfaction in the Handbook of Metaheuristics, *edited by Fred W. Glover and Gary A. Kochenberger, and published by Kluwer.*

References

Amilhastre, J., Fargier, H. and Marquis, P., 2002, Consistency restoration and explanations in dynamic CSPs—Application to configuration, *Artif. Intell.* **135**: 199–234.

Bacchus, F., Chen, X., van Beek, P. and Walsh, T., 2002, Binary vs non-binary constraints, *Artif. Intell.* **140**:1–37.

Bessiere, C., Freuder, E. and Regin, J., 1999, Using constraint metaknowledge to reduce arc consistency computation, *Artif. Intell.* **107**:125–148.

Bistarelli, S., Fargier, H., Montanari, U., Rossi, F., Schiex, T. and Verfaille, G., 1996, Semiring-based CSPs and valued CSPs: Basic properties, in: *Over-Constrained Systems,* Lecture Notes in Computer Science, Vol. 1106, M. Jampel, E. C. Freuder, and M. Maher, eds, Springer, Berlin, pp. 111–150.

Cheng, B., Choi, K., Lee, J. and Wu, J., 1999, Increasing constraint propagation by redundant modeling: an experience report, *Constraints* **4**:167–192.

Cheeseman, P., Kanefsky, B. and Taylor, W., 1991, Where the really hard problems are, in: Proc. 12th Int. Joint Conf. in Artificial Intelligence, Morgan Kaufmann, San Mateo, CA, pp. 331–337.

www.mat.univie.ac.at/~neum/glopt/coconut/

www.informs.org/Conf/CORS-INFORMS2004/

www-sop.inria.fr/coprin/cpaior04/

Debruyne, R. and Bessière, C., 2001, Domain filtering consistencies, *J. Artif. Intell. Res.* **14**:205–230.

Dechter, R., 1990, Enhancement schemes for constraint processing: backjumping, learning, and cutset decomposition, *Artif. Intell.* **41**:273–312.

Dechter, R., 2003, *Constraint Processing,* Morgan Kaufmann, San Mateo, CA.

Dechter, R. and Frost, D., 2002, Backjump-based backtracking for constraint satisfaction problems, *Artif. Intell.* **136**:147–188.

de Givry, S. and Jeannin, L., 2004, *A Library for Partial and Hybrid Search Methods,* www.crt.umontreal.ca/cpaior/

Deransart, P., Hermenegildo, M. and Maluszynski, J., eds, 2000, *Analysis and Visualization Tools for Constraint Programming,* Lecture Notes in Computer Science, Vol. 1870, Springer, Berlin.

Dechter, R., Meiri, I. and Pearl, J., 1991, Temporal constraint networks, *Artif. Intell.* **49**:61–95.

www.icparc.ic.ac.uk/eclipse/

Freuder, E., 1978, Synthesizing constraint expressions, *Commun. ACM* **11**:958–966.

Freuder, E., 1982, A sufficient condition for backtrack-free search, *J. Assoc. Comput. Mach.* **29**:24–32.

Freuder, E., 1991, Eliminating interchangeable values in constraint satisfaction problems, in: *Proc. 9th Natl Conf. on Artificial Intelligence,* AAAI/MIT Press, Menlo Park, CA, pp. 227–233.

Gomes, C., Selman, B. and Crato, N., 1997, Heavy-tailed distributions in combinatorial search, in: *Proc. 3rd Int. Conf. on Principles and Practice of Constraint Programming (CP'97),* Lecture Notes in Computer Science, Vol. 1330, G. Smolka, ed., Springer, Berlin, pp. 121–135.

Harvey, W. and Ginsberg, M., 1995, Limited discrepancy search, in: *Proc. 14th Int. Joint Conf. on Artificial Intelligence (IJCAI-95).* Morgan Kaufmann, San Mateo, CA, pp. 607–615.

Jeavons, P., Cohen, D. and Gyssens, M., 1997, Closure properties of constraints, *J. ACM* **44**:527–548.

Jaffar, J. and Lassez, J.-L., 1987, Constraint logic programming, in: *Proc. Ann. ACM Symp. on Principles of Programming Languages (POPL),* pp. 111–119.

Kondrak, G. and van Beek, P., 1997, A theoretical evaluation of selected backtracking algorithms, *Artif. Intell.* **89**:365–387.

Laburthe, F. and Caseau, Y., 1998, SALSA: A language for search algorithms, in: *Proc. 4th Int. Conf. on the Principles and Practice of Constraint Programming (CP'98),* Pisa, Italy, Lecture Notes in Computer Science, Vol. 1520, Springer, Berlin, pp. 310–324.

Mackworth, A., 1977, Consistency in networks of relations, *Artif. Intell.* **8**:99–118.

Minton, S., 1996, Automatically configuring constraint satisfaction programs: A case study, *Constraints* **1**:7–43.

Minton, S., Johnston, M. D., Philips, A. B. and Laird, P., 1992, Minimizing conflicts: a heuristic repair method for constraint satisfaction and scheduling, *Artif. Intell.* **58**:61–205.

Michel, L. and Van Hentenryck, P., 1999, Localizer: a modeling language for local search, *INFORMS, J. Comput.* **11**:1–14.

Regin, J.-C., 1994, A filtering algorithm for constraints of difference in CSPs, in: *Proc. AAAI 12th Natl Conf. on Artificial Intelligence,* AAAI Press, Menlo Park, CA, pp. 362–367.

Régin, J.-C., 2001, Minimization of the number of breaks in sports scheduling problems using constraint programming, in: Constraint Programming and Large Scale Discrete Optimization, DIMACS Series in Discrete Mathematics and Theoretical Computer Science, Vol. 57, E. Freuder and R. Wallace, eds, American Mathematical Society, Providence, RI, pp. 115–130.

Sabin, D. and Freuder, E., 1997, Understanding and Improving the MAC Algorithm, in: *Proc. 3rd Int. Conf. on Principles and Practice of Constraint Programming (CP'97),* Lecture Notes in Computer Science, Vol. 1330, G. Smolka, ed., Springer, Berlin, pp. 167–181.

Selman, B., Levesque, H. and Mitchell, D., 1992, A new method for solving hard satisfiability problems, in: *Proc. 10th Natl Conf. on Artificial Intelligence (AAAI-92),* San Jose, CA, pp. 440–446.

Tsang, E., 1993, *Foundations of Constraint Satisfaction,* Academic, London.

Van Hentenryck, P., 1999, *The OPL Optimization Programming Language,* MIT Press, Cambridge, MA.

Van Hentenryck, P., Simonis, H. and Dincbas, M., 1992, Constraint satisfaction using constraint logic programming, *Artif. Intell.* **58**:113–159.

Veron, A., Schuerman, K., Reeve, M. and Li, L.-L., 1993, Why and How in the ElipSys OR-Parallel CLP System, in: *Proc. 5th Int. PARLE Conf.,* pp. 291–303.

Walsh, T., 2002, Stochastic constraint programming, in: *Proc. 15th ECAI,* F. van Harmelen, ed., pp. 111–115.

Yokoo, M., Durfee, E., Ishida, T. and Kuwabara, K., 1998, The distributed CSP: Formalization and algorithms, *IEEE Trans. on Knowledge and Data Engineering* **10**:673–685.

Zhang, W., 1998, Complete anytime beam search, in: *Proc. 15th Natl Conf. on Artificial Intelligence (AAAI-98),* pp. 425–430.

Chapter 10

MULTI-OBJECTIVE OPTIMIZATION

Kalyanmoy Deb
Kanpur Genetic Algorithms Laboratory (KanGAL)
Department of Mechanical Engineering
Indian Institute of Technology, Kanpur, India

10.1   INTRODUCTION

Many real-world search and optimization problems are naturally posed as non-linear programming problems having multiple objectives. Due to the lack of suitable solution techniques, such problems were artificially converted into a single-objective problem and solved. The difficulty arose because such problems give rise to a set of trade-off optimal solutions (known as *Pareto-optimal* solutions), instead of a single optimum solution. It then becomes important to find not just one Pareto-optimal solution, but as many of them as possible. This is because any two such solutions constitutes a trade-off among the objectives and users would be in a better position to make a choice when many such trade-off solutions are unveiled.

Classical methods use a very different philosophy in solving these problems, mainly because of a lack of a suitable optimization methodology to find multiple optimal solutions efficiently. They usually require repetitive applications of an algorithm to find multiple Pareto-optimal solutions and on some occasions such applications do not even guarantee finding certain Pareto-optimal solutions. In contrast, the population approach of evolutionary algorithms (EAs) allows an efficient way to find multiple Pareto-optimal solutions simultaneously in a single simulation run. This aspect has made the research and application in evolutionary multi-objective optimization (EMO) popular in the past decade. The motivated readers may explore current research issues and other important studies from various texts (Deb, 2001; Coello et al., 2002; Bagchi, 1999), conference proceedings (Fonseca et al., 2003; Zitzler et al., 2001a) and numerous research papers (archived and maintained in Coello, 2003).

In this tutorial, we discuss the fundamental differences between single- and multi-objective optimization tasks. The conditions for optimality in a multi-objective optimization problem are described and a number of state-of-the-art

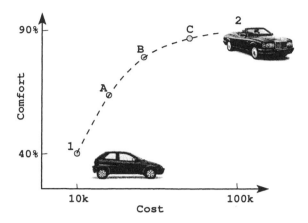

Figure 10.1. Hypothetical trade-off solutions for a car-buying decision-making problem.

multi-objective optimization techniques, including one evolutionary method, are presented. To demonstrate that the evolutionary multi-objective methods are capable and ready for solving real-world problems, we present a couple of interesting case studies. Finally, a number of important research topics in the area of evolutionary are discussed.

A multi-objective optimization problem (MOOP) deals with more than one objective function. In most practical decision-making problems, multiple objectives or multiple criteria are evident. Because of a lack of suitable solution methodologies, a MOOP has been mostly cast and solved as a single-objective optimization problem in the past. However, there exist a number of fundamental differences between the working principles of single- and multi-objective optimization algorithms because of which a MOOP must be attempted to solve using a multi-objective optimization technique. In a single-objective optimization problem, the task is to find one solution (except in some specific multi-modal optimization problems, where multiple optimal solutions are sought) which optimizes the sole objective function. Extending the idea to multi-objective optimization, it may be wrongly assumed that the task in a multi-objective optimization is to find an optimal solution corresponding to each objective function. Certainly, multi-objective optimization is much more than this simple idea. We describe the concept of multi-objective optimization by using an example problem.

Let us consider the decision-making involved in buying an automobile car. Cars are available at prices ranging from a few thousand to few hundred thousand dollars. Let us take two extreme hypothetical cars, i.e. one costing about ten thousand dollars (solution 1) and another costing about a hundred thousand dollars (solution 2), as shown in Figure 10.1. If the cost is the only objective of this decision-making process, the optimal choice is solution 1. If this were the only objective to all buyers, we would have seen only one type of car (solution 1) on the road and no car manufacturer would have produced any expensive cars. Fortunately, this decision-making process is not a single-objective one. Barring some exceptions, it is expected that an inexpensive car is likely to be less comfortable. The figure indicates that the cheapest car has a hypothetical comfort level of 40%. To rich buyers for whom comfort is the only objective of this decision-making, the choice is solution 2 (with a hypothetical maximum comfort level of 90%, as shown in the figure). This so-called two-objective optimization problem need not be considered as the two independent optimization problems, the results of which are the two extreme solutions discussed above. Between these two extreme solutions, there exist many other solutions, where a trade-off between cost and comfort exists. A number of such solutions (solutions A, B, and C) with differing costs and comfort levels are also shown in the figure. Thus, between any two such solutions, one is better in terms of one objective, but this betterment comes only from a sacrifice on the other objective. In this sense, all such trade-off solutions are optimal solutions to a multi-objective optimization problem. Often, such trade-off solutions provide a clear *front* on an objective space plotted with the objective values. This front is called the *Pareto-optimal* front and all such trade-off solutions are called Pareto-optimal solutions.

10.1.1 How Is It Different From Single-Objective Optimization?

It is clear from the above description that there exist a number of differences between a single- and a multi-objective optimization task. The latter has the following properties:

1 cardinality of the optimal set is usually more than one,

2 there are two distinct goals of optimization, instead of one, and

3 it possesses two different search spaces.

We discuss each of the above properties in the following paragraphs.

First of all, we have seen from the above car-buying example that a multi-objective optimization problem with conflicting objectives, results in a number of Pareto-optimal solutions, unlike the usual notion of only one optimal solution associated with a single-objective optimization task. However, there exist some single-objective optimization problems which also contain multiple optimal solutions (of equal or unequal importance). In a certain sense, multi-objective optimization is similar to such *multi-modal optimization* tasks. However, in principle, there is a difference, which we would like to highlight here. In most MOOPs, the Pareto-optimal solutions have certain similarities in their decision variables (Deb, 2003). On the other hand, between one local or global optimal solution and another in a multi-modal optimization problem, there may not exist any such similarity. For a number of engineering case studies (Deb, 2003), an analysis of the obtained trade-off solutions revealed the following properties:

1 Among all Pareto-optimal solutions, some decision variables take identical values. Such a property of the decision variables ensures the solution to be an optimum solution.

2 Other decision variables take different values causing the solutions to have a trade-off in their objective values.

Secondly, unlike the sole goal of finding the optimum in a single-objective optimization, here there are two distinct goals:

- convergence to the Pareto-optimal solutions and

- maintenance of a set of maximally-spread Pareto-optimal solutions.

In some sense, both the above goals are independent to each other. An optimization algorithm must have specific properties for achieving each of the goals.

One other difference between single-objective and multi-objective optimization is that in multi-objective optimization the objective functions constitute a multi-dimensional space, in addition to the usual decision variable space common to all optimization problems. This additional space is called the *objective space*, $\mathcal{Z}$. For each solution $\mathbf{x}$ in the decision variable space, there exists a point in the objective space, denoted by $\mathbf{f}(\mathbf{x}) = \mathbf{z} = (z_1, z_2, \ldots, z_M)\mathrm{T}$. The mapping takes place between an n-dimensional solution vector and an M-dimensional objective vector. Figure 10.2 illustrates these two spaces and a mapping between them. Although the search process of an algorithm takes place on the decision variables space, many interesting algorithms, particularly multi-objective EAs (MOEAs), use the objective space information in their search operators. However, the presence of two different spaces introduces a number of interesting flexibilities in designing a search algorithm for multi-objective optimization.

10.2   TWO APPROACHES TO MULTI-OBJECTIVE OPTIMIZATION

Although the fundamental difference between single and multiple objective optimization lies in the cardinality in the optimal set, from a practical standpoint a user needs only one solution, no matter whether the associated optimization problem is single-objective or multi-objective. In the case of multi-objective optimization, the user is now in a dilemma. Which of these optimal

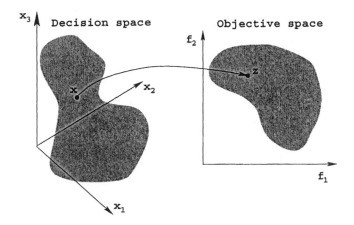

Figure 10.2. Representation of the decision variable space and the corresponding objective space.

solutions must one choose? Let us try to answer this question for the case of the car-buying problem. Knowing the number of solutions that exist in the market with different trade-offs between cost and comfort, which car does one buy? This is not an easy question to answer. It involves many other considerations, such as the total finance available to buy the car, distance to be driven each day, number of passengers riding in the car, fuel consumption and cost, depreciation value, road conditions where the car is to be mostly driven, physical health of the passengers, social status, and many other factors. Often, such higher-level information is non-technical, qualitative and experience-driven. However, if a set of trade-off solutions are already worked out or available, one can evaluate the pros and cons of each of these solutions based on all such non-technical and qualitative, yet still important, considerations and compare them to make a choice. Thus, in a multi-objective optimization, ideally the effort must be made in finding the set of trade-off optimal solutions by considering all objectives to be important. After a set of such trade-off solutions are found, a user can then use higher-level qualitative considerations to make a choice. In view of these discussions, we suggest the following principle for an *ideal multi-objective optimization procedure*:

Step 1 Find multiple trade-off optimal solutions with a wide range of values for objectives.

Step 2 Choose one of the obtained solutions using higher-level information.

Figure 10.3 shows schematically the principles in an ideal multi-objective optimization procedure. In Step 1 (vertically downwards), multiple trade-off

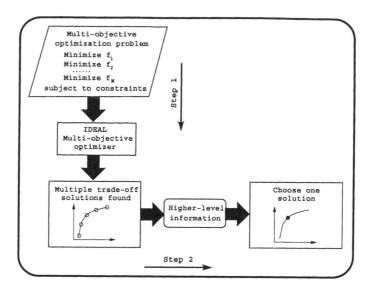

Figure 10.3. Schematic of an ideal multi-objective optimization procedure.

solutions are found. Thereafter, in Step 2 (horizontally, towards the right), higher-level information is used to choose one of the trade-off solutions. With this procedure in mind, it is easy to realize that single-objective optimization is a degenerate case of multi-objective optimization. In the case of single-objective optimization with only one global optimal solution, Step 1 will find only one solution, thereby not requiring us to proceed to Step 2. In the case of single-objective optimization with multiple global optima, both steps are necessary to first find all or many of the global optima and then to choose one from them by using the higher-level information about the problem.

If thought of carefully, each trade-off solution corresponds to a specific order of importance of the objectives. It is clear from Figure 10.1 that solution A assigns more importance to cost than to comfort. On the other hand, solution C assigns more importance to comfort than to cost. Thus, if such a relative preference factor among the objectives is known for a specific problem, there is no need to follow the above principle for solving a multi-objective optimization problem. A simple method would be to form a composite objective function as the weighted sum of the objectives, where a weight for an objective is proportional to the preference factor assigned to that particular objective. This method of scalarizing an objective vector into a single composite objective function converts the multi-objective optimization problem into a single-objective optimization problem. When such a composite objective function is optimized, in most cases it is possible to obtain one particular trade-off solution. This procedure of handling multi-objective optimization problems is much simpler,

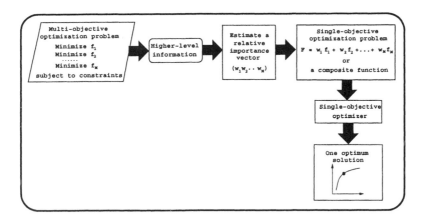

Figure 10.4. Schematic of a preference-based multi-objective optimization procedure.

though still more subjective than the above ideal procedure. We call this procedure a *preference-based* multi-objective optimization. A schematic of this procedure is shown in Figure 10.4. Based on the higher-level information, a preference vector **w** is first chosen. Thereafter, the preference vector is used to construct the composite function, which is then optimized to find a single trade-off optimal solution by a single-objective optimization algorithm. Although not often practiced, the procedure can be used to find multiple trade-off solutions by using a different preference vector and repeating the above procedure.

It is important to realize that the trade-off solution obtained by using the preference-based strategy is largely sensitive to the relative preference vector used in forming the composite function. A change in this preference vector will result in a (hopefully) different trade-off solution. Besides this difficulty, it is intuitive to realize that finding a relative preference vector itself is highly subjective and not straightforward. This requires an analysis of the non-technical, qualitative and experience-driven information to find a quantitative relative preference vector. Without any knowledge of the likely trade-off solutions, this is an even more difficult task. Classical multi-objective optimization methods which convert multiple objectives into a single objective by using a relative preference vector of objectives work according to this preference-based strategy. Unless a reliable and accurate preference vector is available, the optimal solution obtained by such methods is highly subjective to the particular user.

The ideal multi-objective optimization procedure suggested earlier is less subjective. In Step 1, a user does not need any relative preference vector information. The task there is to find as many different trade-off solutions as possible. Once a well-distributed set of trade-off solutions is found, Step 2 then requires certain problem information in order to choose one solution. It is important to mention that in Step 2, the problem information is used to evaluate and compare each of the obtained trade-off solutions. In the ideal approach, the problem information is not used to search for a *new* solution; instead, it is used to choose one solution from a set of already obtained trade-off solutions. Thus, there is a fundamental difference in using the problem information in both approaches. In the preference-based approach, a relative preference vector needs to be supplied without any knowledge of the possible consequences. However, in the proposed ideal approach, the problem information is used to choose one solution from the obtained set of trade-off solutions. We argue that the ideal approach in this matter is more methodical, more practical, and less subjective. At the same time, we highlight the fact that if a reliable relative preference vector is available to a problem, there is no reason to find other trade-off solutions. In such a case, a preference-based approach would be adequate.

In the next section, we make the above qualitative idea of multi-objective optimization more quantitative.

10.3   NON-DOMINATED SOLUTIONS AND PARETO-OPTIMAL SOLUTIONS

Most multi-objective optimization algorithms use the concept of dominance in their search. Here, we define the concept of dominance and related terms and present a number of techniques for identifying dominated solutions in a finite population of solutions.

10.3.1   Special Solutions

We first define some special solutions which are often used in multi-objective optimization algorithms.

Ideal Objective Vector   For each of the M conflicting objectives, there exists one different optimal solution. An objective vector constructed with these individual optimal objective values constitutes the ideal objective vector.

DEFINITION 1 *The mth component of the ideal objective vector* $\mathbf{z}^*$ *is the constrained minimum solution of the following problem:*

$$\left. \begin{array}{ll} \text{Minimize} & f_m(\mathbf{x}) \\ \text{subject to} & \mathbf{x} \in \mathcal{S} \end{array} \right\} \tag{10.1}$$

Thus, if the minimum solution for the mth objective function is the decision vector $\mathbf{x}^{*(m)}$ with function value f_m^*, the ideal vector is as follows:

$$\mathbf{z}^* = \mathbf{f}^* = (f_1^*, f_2^*, \dots, f_M^*)^{\mathrm{T}}$$

In general, the ideal objective vector ($\mathbf{z}^*$) corresponds to a non-existent solution (Figure 10.5). This is because the minimum solution of (10.1) for each objec-

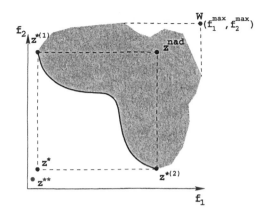

Figure 10.5. The ideal, utopian, and nadir objective vectors.

tive function need not be the same solution. The only way an ideal objective vector corresponds to a feasible solution is when the minimal solutions to all objective functions are identical. In this case, the objectives are not conflicting with each other and the minimum solution to any objective function would be the only optimal solution to the MOOP. Although the ideal objective vector is usually non-existent, it is also clear from Figure 10.5 that solutions closer to the ideal objective vector are better. Moreover, many algorithms require the knowledge of the lower bound on each objective function to normalize objective values in a common range.

Utopian Objective Vector   The ideal objective vector denotes an array of the lower bound of all objective functions. This means that for every objective function there exists at least one solution in the feasible search space sharing an identical value with the corresponding element in the ideal solution. Some algorithms may require a solution which has an objective value strictly better than (and not equal to) that of any solution in the search space. For this purpose, the utopian objective vector is defined as follows.

DEFINITION 2 *A utopian objective vector* z^{**} *has each of its components marginally smaller than that of the ideal objective vector, or* $z_i^{**} = z_i^* - \epsilon_i$ *with* $\epsilon_i > 0$ *for all* $i = 1, 2, \ldots, M$.

Figure 10.5 shows a utopian objective vector. Like the ideal objective vector, the utopian objective vector also represents a non-existent solution.

Nadir Objective Vector   Unlike the ideal objective vector which represents the lower bound of each objective in the entire feasible search space, the nadir objective vector, z^{nad}, represents the upper bound of each objective in the entire Pareto-optimal set, and not in the entire search space. A nadir objective vector must not be confused with a vector of objectives (marked as W in Figure 10.5) found by using the worst feasible function values, f_i^{max}, in the entire search space. The nadir objective vector may represent an existent or a non-existent solution, depending on the convexity and continuity of the Pareto-optimal set. In order to normalize each objective in the entire range of the Pareto-optimal region, the knowledge of nadir and ideal objective vectors can be used as follows:

$$f_i^{norm} = \frac{f_i - z_i^*}{z_i^{nad} - z_i^*} \tag{10.2}$$

10.3.2   Concept of Domination

Most multi-objective optimization algorithms use the concept of domination. In these algorithms, two solutions are compared on the basis of whether one dominates the other or not. We will describe the concept of domination in the following paragraph.

We assume that there are M objective functions. In order to cover both minimization and maximization of objective functions, we use the operator $\lhd$ between two solutions i and j as $i \lhd j$ to denote that solution i is better than solution j on a particular objective. Similarly, $i \rhd j$ for a particular objective implies that solution i is worse than solution j on this objective. For example, if an objective function is to be minimized, the operator $\lhd$ would mean the "$<$" operator, whereas if the objective function is to be maximized, the operator $\lhd$ would mean the "$>$" operator. The following definition covers mixed problems with minimization of some objective functions and maximization of the rest of them.

DEFINITION 3 *A solution* $\mathbf{x}^{(1)}$ *is said to dominate the other solution* $\mathbf{x}^{(2)}$, *if both conditions 1 and 2 are true:*

1 The solution $\mathbf{x}^{(1)}$ *is no worse than* $\mathbf{x}^{(2)}$ *in all objectives: that is,*

$$f_j(\mathbf{x}^{(1)}) \not\rhd f_j(\mathbf{x}^{(2)}) \quad \text{for all } j = 1, 2, \ldots, M$$

2 The solution $\mathbf{x}^{(1)}$ *is strictly better than* $\mathbf{x}^{(2)}$ *in at least one objective, or*

$$f_{\bar{j}}(\mathbf{x}^{(1)}) \lhd f_{\bar{j}}(\mathbf{x}^{(2)}) \text{ for at least one } \bar{j} \in \{1, 2, \ldots, M\}$$

If either of the above conditions is violated, the solution $\mathbf{x}^{(1)}$ does not dominate the solution $\mathbf{x}^{(2)}$. If $\mathbf{x}^{(1)}$ dominates the solution $\mathbf{x}^{(2)}$ (or mathematically $\mathbf{x}^{(1)} \preceq \mathbf{x}^{(2)}$), it is also customary to write any of the following:

- $\mathbf{x}^{(2)}$ is dominated by $\mathbf{x}^{(1)}$,

- $\mathbf{x}^{(1)}$ is non-dominated by $\mathbf{x}^{(2)}$, or

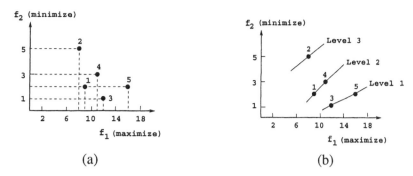

Figure 10.6. A set of five solutions and the corresponding non-dominated fronts.

- $\mathbf{x}^{(1)}$ is non-inferior to $\mathbf{x}^{(2)}$.

Let us consider a two-objective optimization problem with five different solutions shown in the objective space, as illustrated in Figure 10.6(a). Let us also assume that the objective function 1 needs to be maximized while the objective function 2 needs to be minimized. Five solutions with different objective function values are shown in this figure. Since both objective functions are of importance to us, it is usually difficult to find one solution which is best with respect to both objectives. However, we can use the above definition of domination to decide which solution is better among any two given solutions in terms of both objectives. For example, if solutions 1 and 2 are to be compared, we observe that solution 1 is better than solution 2 in objective function 1 and solution 1 is also better than solution 2 in objective function 2. Thus, both of the above conditions for domination are satisfied and we may write that solution 1 dominates solution 2. We take another instance of comparing solutions 1 and 5. Here, solution 5 is better than solution 1 in the first objective and solution 5 is no worse (in fact, they are equal) than solution 1 in the second objective. Thus, both the above conditions for domination are also satisfied and we may write that solution 5 dominates solution 1.

It is intuitive that if a solution $\mathbf{x}^{(1)}$ dominates another solution $\mathbf{x}^{(2)}$, the solution $\mathbf{x}^{(1)}$ is better than $\mathbf{x}^{(2)}$ in the parlance of multi-objective optimization. Since the concept of domination allows a way to compare solutions with multiple objectives, most multi-objective optimization methods use this domination concept to search for non-dominated solutions.

10.3.3   Properties of Dominance Relation

Definition 3 defines the dominance relation between any two solutions. There are three possibilities that can be the outcome of the dominance check between two solutions 1 and 2. i.e. (i) solution 1 dominates solution 2, (ii) solution 1 gets dominated by solution 2, or (iii) solutions 1 and 2 do not dominate each other. Let us now discuss the different binary relation properties (Cormen et al., 1990) of the dominance operator.

Reflexive The dominance relation is *not reflexive*, since any solution p does not dominate itself according to Definition 3. The second condition of dominance relation in Definition 3 does not allow this property to be satisfied.

Symmetric The dominance relation is also *not symmetric*, because $p \preceq q$ does not imply $q \preceq p$. In fact, the opposite is true. That is, if p dominates q, then q does not dominate p. Thus, the dominance relation is *asymmetric*.

Antisymmetric Since the dominance relation is not symmetric, it cannot be antisymmetric as well.

Transitive The dominance relation is *transitive*. This is because if $p \preceq q$ and $q \preceq r$, then $p \preceq r$.

There is another interesting property that the dominance relation possesses. If solution p does not dominate solution q, this does not imply that q dominates p.

In order for a binary relation to qualify as an ordering relation, it must be at least transitive (Chankong and Haimes, 1983). Thus, the dominance relation qualifies as an ordering relation. Since the dominance relation is not reflexive, it is a *strict partial order*. In general, if a relation is reflexive, antisymmetric, and transitive, it is loosely called a *partial order* and a set on which a partial order is defined is called a *partially ordered set*. However, it is important to note that the dominance relation is not reflexive and is not antisymmetric. Thus, the dominance relation is not a partial order relation in its general sense. The dominance relation is only a strict partial order relation.

10.3.4   Pareto-Optimality

Continuing with the comparisons in the previous section, let us compare solutions 3 and 5 in Figure 10.6, because this comparison reveals an interesting aspect. We observe that solution 5 is better than solution 3 in the first objective, while solution 5 is worse than solution 3 in the second objective. Thus, the first condition is not satisfied for both of these solutions. This simply suggests that we cannot conclude that solution 5 dominates solution 3, nor can we say that solution 3 dominates solution 5. When this happens, it is customary to say that solutions 3 and 5 are non-dominated with respect to each other. When both objectives are important, it cannot be said which of the two solutions 3 and 5 is better.

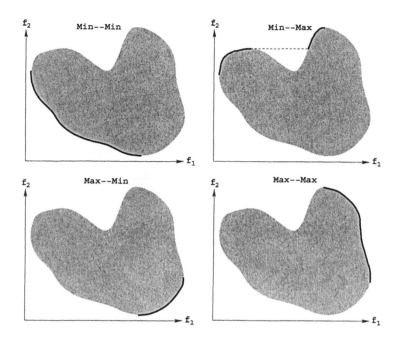

Figure 10.7. Pareto-optimal solutions are marked with continuous curves for four combinations of two types of objectives.

For a given finite set of solutions, we can perform all possible pair-wise comparisons and find which solution dominates which and which solutions are non-dominated with respect to each other. At the end, we expect to have a set of solutions, any two of which do not dominate each other. This set also has another property. For any solution outside of this set, we can always find a solution in this set which will dominate the former. Thus, this particular set has a property of dominating all other solutions which do not belong to this set. In simple terms, this means that the solutions of this set are better compared to the rest of solutions. This set is given a special name. It is called the *non-dominated set* for the given set of solutions. In the example problem, solutions 3 and 5 constitute the non-dominated set of the given set of five solutions. Thus, we define a set of non-dominated solutions as follows.

DEFINITION 4 (NON-DOMINATED SET) *Among a set of solutions P, the non-dominated set of solutions P' are those that are not dominated by any member of the set P.*

When the set P is the entire search space, or $P = S$, the resulting non-dominated set P' is called the *Pareto-optimal set*. Figure 10.7 marks the Pareto-optimal set with continuous curves for four different scenarios with two objectives. Each objective can be minimized or maximized. In the top-left figure, the task is to minimize both objectives f_1 and f_2. The solid curve

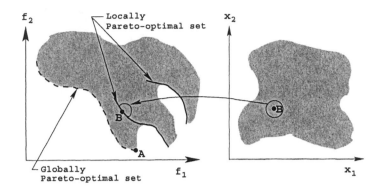

Figure 10.8. Locally and globally Pareto-optimal solutions.

marks the Pareto-optimal solution set. If f_1 is to be minimized and f_2 is to be maximized for a problem having the same search space, the resulting Pareto-optimal set is different and is shown in the top-right figure. Here, the Pareto-optimal set is a union of two disconnected Pareto-optimal regions. Similarly, the Pareto-optimal sets for two other cases, (maximizing f_1, minimizing f_2) and (maximizing f_1, maximizing f_2), are also shown in the bottom-left and bottom-right figures, respectively. In any case, the Pareto-optimal set always consists of solutions from a particular edge of the feasible search region.

It is important to note that a MOEA can be easily used to handle all of the above cases by simply using the domination definition. However, to avoid any confusion, most applications use the duality principle (Deb, 1995) to convert a maximization problem into a minimization problem and treat every problem as a combination of minimizing all objectives. Like global and local optimal solutions in the case of single-objective optimization, there could be global and local Pareto-optimal sets in multi-objective optimization.

DEFINITION 5 (GLOBALLY PARETO-OPTIMAL SET) *The non-dominated set of the entire feasible search space S is the globally Pareto-optimal set.*

DEFINITION 6 *If for every member* **x** *in a set $\underline{P}$ there exists no solution* **y** *(in the neighborhood of* **x** *such that* $\|\mathbf{y} - \mathbf{x}\|_\infty \leq \epsilon$, *where ϵ is a small positive number) dominating any member of the set $\underline{P}$, then solutions belonging to the set $\underline{P}$ constitute a locally Pareto-optimal set.*

Figure 10.8 shows two locally Pareto-optimal sets (marked with continuous curves). When any solution (say, B) in this set is perturbed locally in the decision variable space, no solution can be found dominating any member of the set. It is interesting to note that for continuous search space problems, the locally Pareto-optimal solutions need not be continuous in the decision variable space and the above definition will still hold good. Zitzler (1999) added a neighborhood constraint on the objective space in the above definition to make it more generic. By the above definition, it is also true that a globally Pareto-optimal set is also a locally Pareto-optimal set.

10.3.5   Procedure For Finding Non-dominated Solutions

Finding the non-dominated set of solutions from a given set of solutions is similar in principle to finding the minimum of a set of real numbers. In the latter case, when two numbers are compared to identify the smaller number, a "$<$" relation operation is used. In the case of finding the non-dominated set, the dominance relation $\preceq$ can be used to identify the better of two given solutions. Here, we discuss one simple procedure for finding the non-dominated set (we call it here the best non-dominated front). Many MOEAs require to find the best non-dominated solutions of a population and some MOEAs require to sort a population according to different non-domination levels. We present one algorithm for each of the tasks.

Finding the Best Non-dominated Front   In this approach, every solution from the population is checked with a partially filled population for domination. To start with, the first solution from the population is kept in an empty set P'. Thereafter, each solution i (the second solution onwards) is compared with all members of the set P', one by one. If the solution i dominates any member of P', then that solution is removed from P'. In this way non-members of the non-dominated solutions get deleted from P'. Otherwise, if solution i is dominated by any member of P', the solution i is ignored. If solution i is not dominated by any member of P', it is entered in P'. This is how the set P' grows with non-dominated solutions. When all solutions of the population are checked, the remaining members of P' constitute the non-dominated set.

Identifying the Non-dominated Set

> **Step 1** Initialize $P' = \{1\}$. Set solution counter $i = 2$.
>
> **Step 2** Set $j = 1$.
>
> **Step 3** Compare solution i with j from P' for domination.
>
> **Step 4** If i dominates j, delete the jth member from P' or update $P' = P' \backslash \{P'^{(j)}\}$. If $j < |P'|$, increment j by one and then go to Step 3. Otherwise, go to Step 5. Alternatively, if the jth member of P' dominates i, increment i by one and then go to Step 2.
>
> **Step 5** Insert i in P' or update $P' = P' \cup \{i\}$. If $i < N$, increment i by one and go to Step 2. Otherwise, stop and declare P' as the non-dominated set.

Here, we observe that the second element of the population is compared with only one solution P', the third solution with at most two solutions of P', and so on. This requires a maximum of $1+2+\cdots+(N-1)$ or $N(N-1)/2$ domination checks. Although this computation is also $O(MN^2)$ (see Chapter 11 for a description of this notation), the actual number of computations is about half of that required in Approach 1. It is interesting to note that the size of P' may not always increase (dominated solutions will get deleted from P') and not every solution in the population may be required to be checked with all solutions in the current P' set (the solution may get dominated by a solution of P'). Thus, the actual computational complexity may be smaller than the above estimate.

Another study (Kung et al., 1975) suggested a binary-search like algorithm for finding the best non-dominated front with a complexity $O\left(N(\log N)^{M-2}\right)$ for $M \geq 4$ and $O(N \log N)$ for $M = 2$ and 3.

A Non-dominated Sorting Procedure   Using the above procedure, each front can be identified with at most $O(MN^2)$ computations. In certain scenarios, this procedure may demand more than $O(MN^2)$ computational effort for the overall non-dominated sorting of a population. Here, we suggest a completely different procedure which uses a better bookkeeping strategy requiring $O(MN^2)$ overall computational complexity.

First, for each solution we calculate two entities: (i) *domination count* n_i, the number of solutions which dominate the solution i, and (ii) S_i, a set of solutions which the solution i dominates. This requires $O(MN^2)$ comparisons. At the end of this procedure, all solutions in the first non-dominated front will have their domination count as zero. Now, for each of these solutions (each solution i with $n_i = 0$), we visit each member (j) of its set S_i and reduce its domination count by one. In doing so, if for any member j the domination count becomes zero, we put it in a separate list P'. After such modifications on S_i are performed for each i with $n_i = 0$, all solutions of P' would belong to the second non-dominated front. The above procedure can be continued with each member of P' and the third non-dominated front can be identified. This process continues until all solutions are classified.

An $O(MN^2)$ Non-Dominated Sorting Algorithm

Step 1 For each $i \in P$, $n_i = 0$ and initialize $S_i = \emptyset$. For all $j \neq i$ and $j \in P$, perform Step 2 and then proceed to Step 3.

Step 2 If $i \preceq j$, update $S_p = S_p \cup \{j\}$. Otherwise, if $j \preceq i$, set $n_i = n_i + 1$.

Step 3 If $n_i = 0$, keep i in the first non-dominated front P_1 (we called this set P' in the above paragraph). Set a front counter $k = 1$.

Step 4 While $P_k \neq \emptyset$, perform the following steps.

Step 5 Initialize $Q = \emptyset$ for storing next non-dominated solutions. For each $i \in P_k$ and for each $j \in S_i$,

Step 5a Update $n_j = n_j - 1$.

Step 5b If $n_j = 0$, keep j in Q, or perform $Q = Q \cup \{j\}$.

Step 6 Set $k = k + 1$ and $P_k = Q$. Go to Step 4.

Steps 1 to 3 find the solutions in the first non-dominated front and require $O(MN^2)$ computational complexity. Steps 4 to 6 repeatedly find higher fronts and require at most $O(N^2)$ comparisons, as argued below. For each solution i in the second or higher-level of non-domination, the domination count n_i can be at most $N - 1$. Thus, each solution i will be visited at most $N - 1$ times before its domination count becomes zero. At this point, the solution is assigned a particular non-domination level and will never be visited again. Since there are at most $N - 1$ such solutions, the complexity of identifying second and more fronts is $O(N^2)$. Thus, the overall complexity of the procedure is $O(MN^2)$. It is important to note that although the time complexity has reduced to $O(MN^2)$, the storage requirement has increased to $O(N^2)$.

When the above procedure is applied to the five solutions of Figure 10.6(a), we shall obtain three non-dominated fronts as shown in Figure 10.6(b). From the dominance relations, the solutions 3 and 5 are the best, followed by solutions 1 and 4. Finally, solution 2 belongs to the worst non-dominated front. Thus, the ordering of solutions in terms of their non-domination level is as follows: ((3,5), (1,4), (2)). A recent study (Jensen, 2003b) suggested a divided-and-conquer method to reduce the complexity of sorting to $O(N \log^{M-1} N)$.

10.4   SOME APPROACHES TO MULTI-OBJECTIVE OPTIMIZATION

In this section, we briefly mention a couple of commonly-used classical multi-objective optimization methods and thereafter present a commonly-used EMO method.

10.4.1   Classical Method: Weighted-Sum Approach

The weighted sum method, as the name suggests, scalarizes a set of objectives into a single objective by pre-multiplying each objective with a user-supplied weight. This method is the simplest approach and is probably the most widely used classical approach. If we are faced with the two objectives of minimizing the cost of a product and minimizing the amount of wasted material in the process of fabricating the product, one naturally thinks of minimizing a weighted sum of these two objectives. Although the idea is simple, it introduces a not-so-simple question. What values of the weights must one use? Of course, there is no unique answer to this question. The answer depends on the importance of each objective in the context of the problem and a scaling factor. The scaling effect can be avoided somewhat by normalizing the objective functions. After the objectives are normalized, a composite objective function $F(\mathbf{x})$ can be formed by summing the weighted normalized objectives and the problem is then converted to a single-objective optimization problem as follows:

$$
\left.
\begin{array}{lll}
\text{Minimize} & F(\mathbf{x}) = \sum_{m=1}^{M} w_m f_m(\mathbf{x}) & \\
\text{subject to} & g_j(\mathbf{x}) \geq 0 & j = 1, 2, \ldots, J \\
& h_k(\mathbf{x}) = 0 & k = 1, 2, \ldots, K \\
& x_i^{(L)} \leq x_i \leq x_i^{(U)} & i = 1, 2, \ldots, n
\end{array}
\right\} \quad (10.3)
$$

Here, w_m ($\in [0, 1]$) is the weight of the mth objective function. Since the minimum of the above problem does not change if all weights are multiplied by a constant, it is the usual practice to choose weights such that their sum is one, or $\sum_{m=1}^{M} w_m = 1$.

Mathematically oriented readers may find a number of interesting theorems regarding the relationship between the optimal solution of the above problem to the true Pareto-optimal solutions in classical texts (Chankong and Haimes, 1983; Ehrgott, 2000; Miettinen, 1999).

Let us now illustrate how the weighted-sum approach can find Pareto-optimal solutions of the original problem. For simplicity, we consider the two-objective problem shown in Figure 10.9. The feasible objective space and the corresponding Pareto-optimal solution set are shown. With two objectives, there are two weights w_1 and w_2, but only one is independent. Knowing any one, the other can be calculated by simple subtraction. It is clear from the figure that a choice of a weight vector corresponds to a pre-destined optimal solution on the Pareto-optimal front, as marked by the point A. By changing the weight vector, a different Pareto-optimal point can be obtained. However, there are a couple of difficulties with this approach:

1 A uniform choice of weight vectors do not necessarily find a uniform set of Pareto-optimal solutions on the Pareto-optimal front (Deb, 2001).

2 The procedure cannot be used to find Pareto-optimal solutions which lie on the non-convex portion of the Pareto-optimal front.

The former issue makes it difficult for the weighted-sum approach to be applied reliably to any problem in order to find a good representative set of Pareto-optimal solutions. The latter issue arises due to the fact a solution lying on the non-convex Pareto-optimal front can never be the optimal solution of problem given in (10.3).

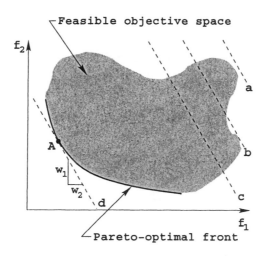

Figure 10.9. The weighted-sum approach on a convex Pareto-optimal front.

10.4.2   Classical Method: ϵ-Constraint Method

In order to alleviate the difficulties faced by the weighted-sum approach in solving problems having non-convex objective spaces, the ϵ-constraint method is used. Haimes et al. (1971) suggested reformulating the multi-objective optimization problem by just keeping one of the objectives and restricting the rest of the objectives within user-specified values. The modified problem is as follows:

$$\left.\begin{array}{lll} \text{Minimize} & f_\mu(\mathbf{x}), & \\ \text{subject to} & f_m(\mathbf{x}) \le \epsilon_m & m = 1, 2, \ldots, M \text{ and } m \neq \mu \\ & g_j(\mathbf{x}) \ge 0 & j = 1, 2, \ldots, \\ & h_k(\mathbf{x}) = 0 & k = 1, 2, \ldots, K \\ & x_i^{(L)} \le x_i \le x_i^{(U)} & i = 1, 2, \ldots, n \end{array}\right\} \quad (10.4)$$

In the above formulation, the parameter ϵ_m represents an upper bound of the value of f_m and need not necessarily mean a small value close to zero.

Let us say that we retain f_2 as an objective and treat f_1 as a constraint: $f_1(\mathbf{x}) \le \epsilon_1$. Figure 10.10 shows four scenarios with different ϵ_1 values. Let us consider the third scenario with $\epsilon_1 = \epsilon_1^c$ first. The resulting problem with this constraint divides the original feasible objective space into two portions, $f_1 \le \epsilon_1^c$ and $f_1 > \epsilon_1^c$. The left portion becomes the feasible solution of the resulting problem stated in (10.4). Now, the task of the resulting problem is to find the solution which minimizes this feasible region. From Figure 10.10, it is clear that the minimum solution is C. In this way, intermediate Pareto-optimal solutions can be obtained in the case of non-convex objective space problems by using the ϵ-constraint method.

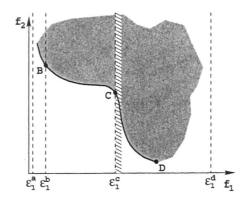

Figure 10.10. The ϵ-constraint method.

One of the difficulties of this method is that the solution to the problem stated in (10.4) largely depends on the chosen ϵ vector. Let us refer to Figure 10.10 again. Instead of choosing ϵ_1^c, if ϵ_1^a is chosen, there exists no feasible solution to the stated problem. Thus, no solution would be found. On the other hand, if ϵ_1^d is used, the entire search space is feasible. The resulting problem has the minimum at D. Moreover, as the number of objectives increases, there exist more elements in the ϵ vector, thereby requiring more information from the user.

10.4.3 Evolutionary Multi-objective Optimization Method

Over the years, researchers have suggested a number of MOEAs emphasizing non-dominated solutions in a EA population. In this section, we shall describe one state-of-the-art algorithm popularly used in EMO studies.

Elitist Non-dominated Sorting GA (NSGA-II)   The non-dominated sorting GA or NSGA-II procedure (Deb et al., 2002a) for finding multiple Pareto-optimal solutions in a multi-objective optimization problem has the following three features:

1 it uses an elitist principle,

2 it uses an explicit diversity preserving mechanism, and

3 it emphasizes the non-dominated solutions.

In NSGA-II, the offspring population Q_t is first created by using the parent population P_t and the usual genetic operators (Goldberg, 1989), see also Chapter 4. Thereafter, the two populations are combined together to form R_t of

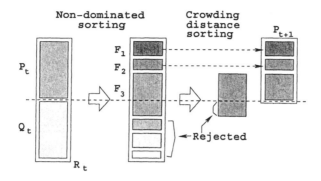

Figure 10.11. Schematic of the NSGA-II procedure.

size $2N$. Then, a non-dominated sorting is used to classify the entire population R_t. Once the non-dominated sorting is over, the new population is filled by solutions of different non-dominated fronts, one at a time. The filling starts with the best non-dominated front and continues with solutions of the second non-dominated front, followed by the third non-dominated front, and so on. Since the overall population size of R_t is $2N$, not all fronts may be accommodated in N slots available in the new population. All fronts which could not be accommodated are simply deleted. When the last allowed front is being considered, there may exist more solutions in the last front than the remaining slots in the new population. This scenario is illustrated in Figure 10.11. Instead of arbitrarily discarding some members from the last acceptable front, the solutions which will make the diversity of the selected solutions the highest are chosen. The NSGA-II procedure is outlined in the following.

NSGA-II

Step 1 Combine parent and offspring populations and create $R_t = P_t \cup Q_t$. Perform a non-dominated sorting to R_t and identify different fronts: $\mathcal{F}_i, i = 1, 2, \ldots$, etc.

Step 2 Set new population $P_{t+1} = \emptyset$. Set a counter $i = 1$.
Until $|P_{t+1}| + |\mathcal{F}_i| < N$, perform $P_{t+1} = P_{t+1} \cup \mathcal{F}_i$ and $i = i + 1$.

Step 3 Perform the Crowding-sort$(\mathcal{F}_i, <_c)$ procedure and include the most widely spread $(N - |P_{t+1}|)$ solutions by using the crowding distance values in the sorted $\mathcal{F}_i$ to P_{t+1}.

Step 4 Create offspring population Q_{t+1} from P_{t+1} by using the crowded tournament selection, crossover and mutation operators.

In Step 3, the crowding-sorting of the solutions of front i (the last front which could not be accommodated fully) is performed by using a *crowding distance metric*, which we describe a little later. The population is arranged in a descending order of magnitude of the crowding distance values. In Step 4, a crowding tournament selection operator, which also uses the crowding distance, is used.

The crowded comparison operator ($<_c$) compares two solutions and returns the winner of the tournament. It assumes that every solution i has two attributes:

1 a non-domination rank r_i in the population,

2 a local crowding distance (d_i) in the population.

The crowding distance d_i of a solution i is a measure of the search space around i which is not occupied by any other solution in the population. Based on these two attributes, we can define the crowded tournament selection operator as follows.

DEFINITION 7 (CROWDED TOURNAMENT SELECTION OPERATOR) *A solution i wins a tournament with another solution j if any of the following conditions are true:*

1 if solution i has a better rank, that is, $r_i < r_j$;

2 if they have the same rank but solution i has a better crowding distance than solution j, that is, $r_i = r_j$ and $d_i > d_j$.

The first condition ensures that the chosen solution lies on a better non-dominated front. The second condition resolves the tie of both solutions being on the same non-dominated front by deciding on their crowded distance. The one residing in a less crowded area (with a larger crowding distance d_i) wins. The crowding distance d_i can be computed in various ways. However, in NSGA-II, we use a crowding distance metric, which requires O($MN \log N$) computations.

To obtain an estimate of the density of solutions surrounding a particular solution i in the population, we take the average distance of two solutions on either side of solution i along each of the objectives. This quantity d_i serves as an estimate of the perimeter of the cuboid formed by using the nearest neighbors as the vertices (we call this the *crowding distance*). In Figure 10.12, the crowding distance of the ith solution in its front (marked with filled circles) is the average side-length of the cuboid (shown by a dashed box). The following algorithm is used to calculate the crowding distance of each point in the set $\mathcal{F}$.

Crowding Distance Assignment Procedure: `Crowding-sort`($\mathcal{F}, <_c$)

> **Step C1** Call the number of solutions in $\mathcal{F}$ as $l = |\mathcal{F}|$. For each i in the set, first assign $d_i = 0$.

> **Step C2** For each objective function $m = 1, 2, \ldots, M$, sort the set in worse order of f_m or find the sorted indices vector $I^m = \text{sort}(f_m, >)$.

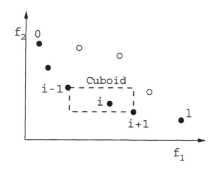

Figure 10.12. The crowding distance calculation.

Step C3 For $m = 1, 2, \ldots, M$, assign a large distance to the boundary solutions, or $d_{I_1^m} = d_{I_l^m} = \infty$, and for all other solutions $j = 2$ to $(l-1)$, assign

$$d_{I_j^m} = d_{I_j^m} + \frac{f_m^{(I_{j+1}^m)} - f_m^{(I_{j-1}^m)}}{f_m^{\max} - f_m^{\min}}$$

The index I_j denotes the solution index of the jth member in the sorted list. Thus, for any objective, I_1 and I_l denote the lowest and highest objective function values, respectively. The second term on the right-hand side of the last equation is the difference in objective function values between two neighboring solutions on either side of solution I_j. Thus, this metric denotes half of the perimeter of the enclosing cuboid with the nearest neighboring solutions placed on the vertices of the cuboid (Figure 10.12). It is interesting to note that for any solution i the same two solutions $(i + 1)$ and $(i - 1)$ need not be neighbors in all objectives, particularly for $M \geq 3$. The parameters $f_m^{\max}$ and $f_m^{\min}$ can be set as the population-maximum and population-minimum values of the mth objective function. The above metric requires M sorting calculations in Step C2, each requiring $O(N \log N)$ computations. Step C3 requires N computations. Thus, the complexity of the above distance metric computation is $O(MN \log N)$ and the overall complexity of one generation of NSGA-II is $O(MN^2)$, governed by the non-dominated sorting procedure.

10.4.4   Sample Simulation Results

In this section, we show the simulation results of NSGA-II on two test problems. The first problem (SCH1) is a simple two-objective problem with a convex Pareto-optimal front:

$$\text{SCH1}: \begin{cases} \text{Minimize} & f_1(x) = x^2 \\ \text{Minimize} & f_2(x) = (x - 2)^2 \\ & -10^3 \leq x \leq 10^3 \end{cases} \tag{10.5}$$

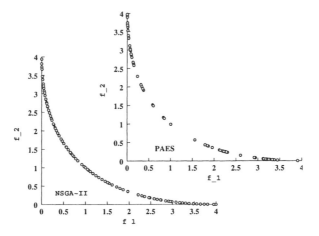

Figure 10.13. NSGA-II finds better spread of solutions than PAES on SCH.

The second problem (KUR) has a disjointed set of Pareto-optimal fronts:

$$
\text{KUR}: \begin{cases}
\text{Minimize} & f_1(\mathbf{x}) = \sum_{i=1}^{2}\left[-10\exp\left(-0.2\sqrt{x_i^2 + x_{i+1}^2}\right)\right] \\
\text{Minimize} & f_2(\mathbf{x}) = \sum_{i=1}^{3}\left[|x_i|^{0.8} + 5\sin(x_i^3)\right] \\
& -5 \le x_i \le 5, \quad i = 1, 2, 3
\end{cases}
$$

(10.6)

NSGA-II is run with a population size of 100 and for 250 generations. Figure 10.13 shows that NSGA-II converges on the Pareto-optimal front and maintains a good spread of solutions. In comparison to NSGA-II, another competing EMO method (the Pareto archived evolution strategy (PAES) (Knowles and Corne, 2000)) is run for an identical overall number of function evaluations and an inferior distribution of solutions on the Pareto-optimal front is observed.

On the KUR problem, NSGA-II is compared with another elitist EMO methodology (the strength Pareto EA or SPEA (Zitzler and Thiele, 1998)) for an identical number of function evaluations. Figures 10.14 and 10.15 clearly show the superiority of NSGA-II in achieving both tasks of convergence and maintaining diversity of optimal solutions.

10.4.5 Other State-of-the-Art MOEAs

Besides the above elitist EMO method, there exist a number of other methods which are also quite commonly used. Of them, the strength Pareto-EA or SPEA2 (Zitzler, 2001b), which uses an EA population and an archive in a synergistic manner and the Pareto envelope based selection algorithm or PESA (Corne et al., 2000), which emphasizes non-dominated solutions residing in a less-crowded hyper-box in both the selection and the offspring-acceptance operators, are common. The recently suggested ϵ-MOEA (Deb et al., 2003) is

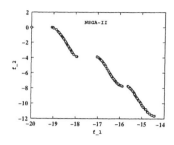

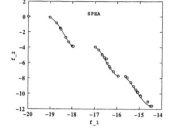

Figure 10.14. NSGA-II on KUR.   *Figure 10.15.* SPEA on KUR.

found to be a superior version of PESA, in which only one solution is allowed to occupy a hyper-box for obtaining a better distribution of solutions. In addition, the ϵ-dominance concept (Laumanns et al., 2002) makes the MOEA a practical approach for solving complex problems with a large number of objectives. The ϵ-MOEA is also demonstrated to find a well-converged and well-distributed set of solutions in a very small computational time (two to three orders of magnitude smaller) compared to a number of state-of-the-art MOEAs (Deb et al., 2003a), such as SPEA2 and PAES. There also exist other competent MOEAs, such as multi-objective messy GA (MOMGA) (Veldhuizen and Lamont, 2000), multi-objective micro-GA (Coello and Toscano, 2000), neighborhood constraint GA (Loughlin and Ranjithan, 1997), and others. Besides, there exist other EA-based methodologies, such as particle swarm EMO (Coello Coello and Salazar Lechuga, 2002; Mostaghim and Teich, 2003), ant-based EMO (McMullen, 2001; Gravel et al., 2002), and differential evolution based EMO (Babu and Jehan, 2003).

10.5 CONSTRAINT HANDLING

Constraints can be simply handled by modifying the definition of domination in an EMO method.

DEFINITION 8 *A solution* $\mathbf{x}^{(i)}$ *is said to "constrain-dominate" a solution* $\mathbf{x}^{(j)}$ *(or* $\mathbf{x}^{(i)} \preceq_c \mathbf{x}^{(j)}$), *if any of the following conditions are true:*

1 Solution $\mathbf{x}^{(i)}$ *is feasible and solution* $\mathbf{x}^{(j)}$ *is not.*

2 Solutions $\mathbf{x}^{(i)}$ *and* $\mathbf{x}^{(j)}$ *are both infeasible, but solution* $\mathbf{x}^{(i)}$ *has a smaller constraint violation.*

3 Solutions $\mathbf{x}^{(i)}$ *and* $\mathbf{x}^{(j)}$ *are feasible and solution* $\mathbf{x}^{(i)}$ *dominates solution* $\mathbf{x}^{(j)}$ *in the usual sense (see Definition 3).*

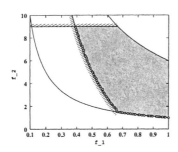

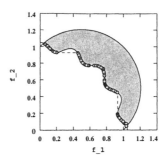

Figure 10.16. Obtained non-dominated solutions with NSGA-II on the constrained problem CONSTR.

Figure 10.17. Obtained non-dominated solutions with NSGA-II on the constrained problem TNK.

This definition allows a feasible solution to be always dominating an infeasible solution and compared two infeasible solutions based on constraint violation values and two feasible solutions in terms of their objective values.

In the following, we show simulation results of NSGA-II applied with the above constraint handling mechanism to two test problems—the CONSTR and TNK problems described below:

CONSTR

$$\text{Minimize} \quad f_1(\mathbf{x}) = x_1$$
$$\text{Minimize} \quad f_2(\mathbf{x}) = \frac{1+x_2}{x_1}$$
$$x_2 + 9x_1 \geq 6$$
$$-x_2 + 9x_1 \geq 1$$

TNK

$$\text{Minimize} \quad f_1(\mathbf{x}) = x_1$$
$$\text{Minimize} \quad f_2(\mathbf{x}) = x_2$$
$$x_1^2 + x_2^2 - 1 - \frac{1}{10}\cos\left(16\tan^{-1}\frac{x_1}{x_2}\right) \geq 0$$
$$(x_1 - 0.5)^2 + (x_2 - 0.5)^2 \leq 0.5$$

With identical parameter settings as in Section 10.4.4, NSGA-II finds a good distribution of solutions on the Pareto-optimal front in both problems (Figures 10.16 and 10.17, respectively).

10.6   SOME APPLICATIONS

Since the early development of MOEAs in 1993, they have been applied to many real-world and interesting optimization problems. Descriptions of some of these studies can be found in books (Deb, 2001; Coello et al., 2002;

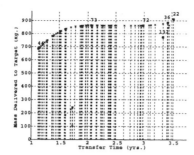

Figure 10.18. Obtained non-dominated solutions.

Osyczka, 2002), conference proceedings (Zitzler et al., 2001), and domain-specific journals and conference proceedings. In this section, we describe two case studies.

10.6.1   Spacecraft Trajectory Design

Coverstone-Carroll et al. (2000) proposed a multi-objective optimization technique using the original non-dominated sorting (NSGA) (Srinivas and Deb, 1994) to find multiple trade-off solutions in a spacecraft trajectory optimization problem. To evaluate a solution (trajectory), the SEPTOP software is called for, and the delivered payload mass and the total time of flight are calculated. In order to reduce the computational complexity, the SEPTOP program is run for a fixed number of generations. The multi-objective optimization problem had eight decision variables controlling the trajectory, three objective functions, i.e. (i) maximize the delivered payload at destination, (ii) maximize the negative of the time of flight, and (iii) maximize the total number of heliocentric revolutions in the trajectory, and three constraints, i.e. (i) limiting the SEPTOP convergence error, (ii) limiting the minimum heliocentric revolutions, and (iii) limiting the maximum heliocentric revolutions in the trajectory.

On the Earth–Mars rendezvous mission, the study found interesting trade-off solutions. Using a population of size 150, the NSGA was run for 30 generations on a Sun Ultra 10 Workstation with a 333 MHz ULTRA Sparc IIi processor. The obtained non-dominated solutions are shown in Figure 10.18 for two of the three objectives. It is clear that there exist short-time flights with smaller delivered payloads (solution marked as 44) and long-time flights with larger delivered payloads (solution marked as 36). To the surprise of the original investigators, two different types of trajectories emerged. The representative solutions of the first set of trajectories are shown in Figure 10.19. Solution 44 can deliver a mass of 685.28 kg and requires about 1.12 years. On other hand, solution 72 can deliver almost 862 kg with a travel time of

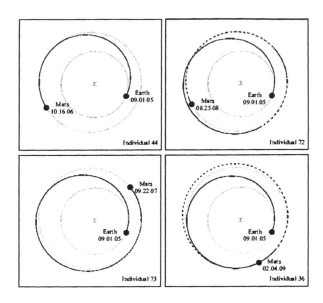

Figure 10.19.   Four trade-off trajectories.

about 3 years. In these figures, each continuous part of a trajectory represents a thrusting arc and each dashed part of a trajectory represents a coasting arc. It is interesting to note that only a small improvement in delivered mass occurs in the solutions between 73 and 72. To move to a somewhat improved delivered mass, a different strategy for the trajectory must be found. Near solution 72, an additional burn is added, causing the trajectories to have better delivered masses. Solution 36 can deliver a mass of 884.10 kg.

The scenario as in Figure 10.19 is what we envisaged in discovering in a multi-objective optimization problem while suggesting the *ideal procedure* in Figure 10.3. Once such an set of solutions with a good trade-off among objective values are obtained, one can then analyze them for choosing a particular solution. For example, in this problem context, whether the wait of an extra year to be able to carry an additional 180 kg of payload is worthwhile or not would make a decision-maker to choose between solutions 44 and 73. Without the knowledge of such a wide variety of optimal solutions, the decision-making could be difficult. Although one can set a relative weight to each objective and optimize the resulting aggregate objective function, the decision-maker will always wonder what solution would have been derived if a slightly different weight vector had been used. The ideal multi-objective optimization technique allows a flexible and a pragmatic procedure for analyzing a well-diversified set of solutions before choosing a particular solution.

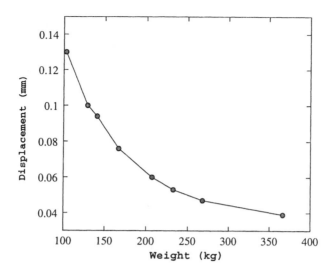

Figure 10.20. Obtained front with eight clustered solutions are shown for the cantilever plate design problem.

10.6.2 A Cantilever Plate Design

A rectangular plate (1.2×2 m^2) is fixed at one end and a 100 kN load is applied to the center element of the opposite end. The following other parameters are chosen:

Plate thickness: 20 mm	Yield strength: 150 MPa
Young's modulus: 200 GPa	Poisson's ratio: 0.25

The rectangular plate is divided into a number of grids and the presence or absence of each grid becomes a Boolean decision variable. NSGA-II is applied for 100 generations with a population size of 54 and crossover probability of 0.95. In order to increase the quality of the obtained solutions, we use an incremental grid-tuning technique. The NSGA-II and the first local search procedure are run with a coarse grid structure (6×10 or 60 elements). After the first local search procedure, each grid is divided into four equal-sized grids, thereby having a 12×20 or 240 elements. The new smaller elements inherit its parent's status of being present or absent. After the second local search is over, the elements are divided again, thereby making 24×40 or 960 elements. In all cases, an automatic mesh-generating finite element method is used to analyze the developed structure.

Figure 10.20 shows the obtained front with eight solutions. The trade-off between the weight and deflection is clear from the figure. Figure 10.21 shows the shape of these eight solutions. The solutions are arranged according to increasing weight from left to right and top to bottom. Thus, the minimum-

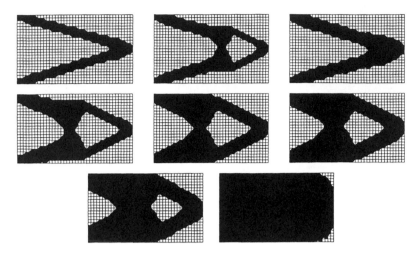

Figure 10.21. Eight trade-off solutions of the cantilever plate design problem.

weight solution is the top-left solution and the minimum-deflection solution is the right-bottom solution. An analysis of these solutions provides interesting insights about the cantilever plate design problem:

1 First, all nine solutions seem to be symmetric about the middle row of the plate. Since the loading and support are symmetrically placed around the middle row, the resulting optimum solution is also likely to be symmetric. Although this information is not explicitly coded in the hybrid techniques NSGA-II procedure, this comes out as one of the features in all optimal solutions. Although in this problem it is difficult to know the true Pareto-optimal solutions, the symmetry achieved in these solutions is an indication of their proximity to the true Pareto-optimal solutions.

2 The minimum-weight solution simply extends two arms from the extreme support nodes to reach to the element carrying the load. Since a straight line is the shortest way to join two points, this solution can be easily conceived as one close to the minimum-weight feasible solution.

3 Thereafter, to have a reduction in deflection, the weight has to be increased. This is where the hybrid procedure discovers an innovation. For a particular sacrifice in the weight, the procedure finds that the maximum reduction in deflection occurs when the two arms are connected by a *stiffener*. This is an engineering trick often used to design a stiff structure. Once again, no such information was explicitly coded in the hybrid procedure. By merely making elements on or off, the procedure has resulted in a *design innovation*.

4 Interestingly, the third solution is a thickened version of the minimum-weight solution. By making the arms thicker, the deflection can be increased maximally for a fixed change in the weight from the previous solution. Although not intuitive, this thick-arm solution is not an immediate trade-off solution to the minimum-weight solution. Although the deflection of this solution is smaller compared to the second solution, the stiffened solution is a good compromise between the thin- and thick-armed solutions.

5 Thereafter, any increase in the thickness of the two-armed solution turns out to be a suboptimal proposition and the stiffened solution is rediscovered instead. From the support until the stiffener, the arms are now thicker than before, providing a better stiffness than before.

6 In the remaining solutions, the stiffener and arms get wider and wider, finally leading to the complete plate with rounded corners. This solution is, no doubt, close to the true minimum-deflection solution.

The transition from a simple thin two-armed cantilever plate having a minimum-weight solution to a complete plate with edges rounded off having a minimum-deflection solution proceeds by discovering a vertical stiffener connecting the two arms and then by widening the arms and then by gradually thickening the stiffener. The symmetric feature of the solutions about the middle row of the plate emerges to be a *common* property to all obtained solutions. Such information about the trade-off solutions is very useful to a designer. Importantly, it is not obvious how such vital design information can be obtained by any other means and in a single simulation run.

10.7 TRICKS OF THE TRADE

Here, we discuss how to develop an ideal multi-objective optimization algorithm in a step-by-step manner. Since they can be developed by using classical or evolutionary optimization methods, we discuss each of them in turn.

10.7.1 Classical Multi-objective Optimization

We assume here that the user knows a classical optimization method to optimize a single-objective optimization problem P with constraints ($\mathbf{x} \in S$) and can find a near-optimum solution $\mathbf{x}^*$. Let us assume that the user is desired to find K different efficient solutions.

Step 1 Find individual optimum solutions $\mathbf{x}^{*i}$ for all objectives, where $i = 1, 2, \ldots, M$.

Step 2 Choose K points $\epsilon^{(k)}$ uniformly in the $(M-1)$-dimensional plane having objectives $i = 1$ to $i = M - 1$ as coordinate directions.

Step 3 Solve each subproblem ($k = 1, 2, \ldots, K$) as follows:

$$
\begin{aligned}
\text{Minimize} \quad & f_M(\mathbf{x}) \\
\text{subject to} \quad & f_i(\mathbf{x}) \leq \epsilon_i^{(k)}, \quad i = 1, 2, \ldots, (M-1) \\
& \mathbf{x} \in S
\end{aligned}
\tag{10.7}
$$

Call the optimal solution $\mathbf{x}^{*(k)}$ and corresponding objective vector $\mathbf{f}^{*(k)}$.

Step 4 Declare the non-dominated set as the set of efficient sets:

$$
F = \text{Non-dominated} \left(\mathbf{f}^{*(1)}, \ldots, \mathbf{f}^{*(K)} \right)
$$

If desired, the above ϵ-constraint method can be replaced by other conversion methods, such as weighted-sum method, Tchebyshev metric method, or others.

10.7.2 Evolutionary Multi-objective Optimization (EMO)

The bottleneck of the above method is Step 3, in which a single-objective minimizer needs to be applied K times (where K is the number of desired efficient solutions). Here, we discuss an evolutionary search principle to find a set of efficient solutions simultaneously in a synergistic manner. It must be kept in mind that the main aim in an ideal multi-objective optimization is to (i) converge close to the true Pareto-optimal front and (ii) maintain a good diversity among them. Thus, an EMO method must use specific operations to achieve each of the above goals. Usually, an emphasis to non-dominated solutions is performed to achieve the first goal and a niching operation is performed to achieve the second goal. In addition, an elite-preserving operation is used to speed up convergence.

Again, we assume that the user is familiar with a particular population-based evolutionary algorithm, in which in each generation one or more new offspring are created by means of recombination and mutation operators. We describe here a generic archive-based EMO strategy.

Step 1 A population P and an empty archive A are initialized. The non-dominated solutions of P are copied in A. Steps 2 and 3 are iterated until a termination criterion is satisfied.

Step 2 A set of λ new offspring solutions are created using P and A.

Step 3 Every new offspring solution is checked for its inclusion in A by using a archive-acceptance criterion C_A and for its inclusion in P by using a population-acceptance criterion C_P. If an offspring is not to be included to either P or A, it is deleted.

Step 4 Before termination, the non-dominated set of the archive A is declared as the efficient set.

In Step 2, random solutions from the combined set $P \cup A$ can be used to create an offspring solution or some solution from P and some other solutions from A can be used to create the offspring solution. Different archive-acceptance and population-acceptance criteria can be used. Here, we propose one criterion each. Readers can find another implementation elsewhere (Deb et al., 2003a).

Archive-Acceptance Criterion $C_A(c, A)$ The archive A has a maximum size K, but at any iteration it may not have all K members. This criterion required domination check and a niching operator which computes the niching of the archive with respect to a particular solution. For example, the crowding distance metric for a solution i in a subpopulation (suggested in Section 10.4.3) measures the objective-wise distance between two neighboring solutions of solution i in the subpopulation. The larger the crowding distance, less crowded is the solution and higher is probability of its existence in the subpopulation.

The offspring c is accepted in the archive A if any of the following conditions is true:

1 Offspring c dominates any archive member A. In this case, delete those archive members and include c in A.

2 Offspring c is non-dominated with all archive members and the archive is not *full* (that is, $|A| < K$). In this case, c is simply added to A.

3 Offspring c is non-dominated with all archive members, the archive is full, and crowding distance of c is larger than that of an archive member. In this case, that archive member is deleted and c is included in A.

Population-Acceptance Criterion $C_P(c, P)$ If the offspring is a good solution compared to the current archive, it will be included in A by the above criterion. The inclusion of the offspring in the population must be made mainly from the point of view of keeping diversity in the population. Any of the following criteria can be adopted to accept c in P:

1 Offspring c replaces the most old (in terms of its time of inclusion in the population) population member.

2 Offspring c replaces a random population member.

3 Offspring c replaces the least-used (as a parent) population member.

4 Offspring c introduces more diversity of the population compared to an existing population member. Here the crowding distance or an entropy-based metric can be used to compute the extent of diversity.

5 Offspring *c* replaces a population member *similar* (in terms of its phenotype or genotype) to itself.

6 Offspring *c* replaces a population member dominated by *c*.

It is worthwhile investigating which of the above criteria works well on standard test problems, but the maintenance of diversity in the population and search for wide-spread non-dominated solutions in the archive are two activities which should allow the combined algorithm to reach the true efficient frontier quickly and efficiently.

In the following, we suggest some important post-optimality studies which are equally important to the optimality study and are often ignored in EMO studies.

10.7.3   Post-optimality Studies

It is to be well understood that EMO methods (whether the above one or any of the existing ones) do not have guaranteed convergence properties; nor do they have any guaranteed proof for finding a diverse set of solutions. Thus, it is an onus on the part of the EMO researchers/practitioners to perform a number of post-optimality studies to ensure (or build confidence about) convergence and achievement of diversity in obtained solutions. Here, we make some suggestions.

1 Use a hybrid techniques EMO–local-search method. For each of the obtained EMO solutions, perform a single-objective search by optimizing a combined objective function (see Chapter 9.6 in Deb (2001) for more details). This will cause the EMO solutions to reach close to the true efficient frontier.

2 Obtain individual optimum solutions and compare the obtained EMO solutions with the individual optima on a plot or by means of a table. Such a visual comparison will indicate the extent of convergence as well as the extent of diversity in the obtained solutions.

3 Perform a number of ϵ-constraint studies for different values of the ϵ-vector, given in (10.7), and obtain efficient solutions. Compare these solutions with the obtained EMO solutions to get further visual confirmation of the extent of convergence and diversity of obtained EMO solutions.

4 Finally, it is advisable to also plot the initial population on the same objective space showing the efficient solutions, as this will depict the extent of optimization performed by the EMO–local-search approach.

For such a post-optimality study, refer to any of the application studies performed by the author (Deb and Jain, 2003; Deb and Tiwari, 2004; Deb et al., 2002b, 2004).

10.7.4 Evaluating a Multi-objective Optimization Algorithm

When a new algorithm is suggested to find a set of Pareto-optimal solutions in a multi-objective optimization problem, the algorithm must have to be evaluated by applying them on standard test problems and compared with existing state-of-the-art EMO algorithms applied to the identical test problems. Here, we suggest a few guidelines in this direction.

1 Test problems with 20 to 50 variables must be included in the test set.

2 Test problems with three or more objectives must be included in the test set. For scalable test problems, readers may refer to ZDT (Deb, 2001) and DTLZ (Deb et al., 2002c) test problems.

3 Test problems must include some no-linear constraints, making some portion of the unconstrained Pareto-optimal front infeasible. For a set of constrained test problems, refer to the CTP problems (Deb, 2001) or DTLZ test problems.

4 Standard EMO algorithms, such as NSGA-II, SPEA2, PESA, ϵ-MOEA, and others must have to be used for comparison purposes. See Section 10.9 for some freely downloadable codes of these algorithms.

5 A proper criterion for the comparison must be chosen. Often, the algorithms are compared based on the fixed number of evaluations. They can also be compared based on some other criterion listed elsewhere (Deb, 2001).

6 Besides static performance metrics which are applied to the final solution set, *running metrics* (Deb and Jain, 2002) may also be computed, plotted with generation number, and compared among two algorithms. The running metrics provide a dynamic (generation-wise) evaluation of the algorithm, rather than what had happened at the end of a simulation run.

10.8 FUTURE DIRECTIONS

With the growing interests in the field of multi-objective optimization particularly using evolutionary algorithms, there exist a number of research directions.

Interactive EMO   EMO methodologies have amply shown that multiple and well-spread Pareto-optimal solutions can be obtained in a single simulation run. However, this is only a part of the whole story. In practice, one needs to choose only solutions which are Pareto-optimal. The choice of a particular solution may be done as a second stage process, in which using some higher-level problem-specific information only one solution can be picked from the obtained EMO solutions. Alternatively, a completely different interactive-EMO strategy can be developed in which right from the first iteration an EMO is geared towards finding a compromised solution interactively dictated by the decision-maker. How to integrate this interactive-EMO approach in a generic manner to any problem is a matter of research and must be taken up before too long.

Handling Large Number of Objectives   So far, most studies using EMO strategies have been restricted to two or three-objective problems. In practice, there exist a considerable number of problems in which 10 or 15 objectives are commonplace. Although the objective space and corresponding dimension of the Pareto-optimal front becomes large, new and innovative principles to handle such large-scale problems must be developed. Existing EMO methodologies must have to be tested and evaluated for the abilities such large-scale problems.

Non-evolutionary Multi-objective Optimization   EMO methods now include principles of genetic algorithms, evolution strategy, genetic programming, particle swarm optimization, differential evolution and others. Besides, other non-traditional optimization techniques such as ant colony optimization, tabu search, simulated annealing can also be used for solving multi-objective optimization problems. Although there have been some research and application in this direction (Hansen, 1997; Khor et al., 2001; Balicki and Kitowski, 2001; McMullen, 2001; Gravel et al., 2002; Kumral, 2003; Parks and Suppapitnarm, 1999; Chattopadhyay and Seeley, 1994), more rigorous studies are called for and such techniques can also be suitably used to find multiple Pareto-optimal solutions.

Performance Metrics   For M objectives, the Pareto-optimal region will correspond to at most an M-dimensional surface. To compare two or more algorithms, it is then necessary to compare M-dimensional data-sets which are partially-ordered. It is not possible to have only one performance metric to compare such multi-dimensional data-sets in an unbiased manner. A recent study has shown that at least M performance metrics will be necessary to properly compare such data-sets (Zitzler et al., 2003). An alternate engineering suggestion was to compare the data-sets from a purely *functional* point of view of (i) measuring the extent of convergence to the front and (ii) measuring the extent of diversity in the obtained solutions (Deb, 2001). It then becomes a challenge to develop performance metrics for both functional goal for problems having any number of objectives.

Test Problem Design When new algorithms are designed, they need to be evaluated on test problems for which the desired Pareto-optimal solutions are known. Moreover, the test problems must be such that they are controllable to test an algorithm's ability to overcome a particular problem difficulty. Although there exist a number of such test problems (Deb, 2001; Deb et al., 2002c), more such problems providing different kinds of difficulties must be developed. Care should be taken to make sure that the test problems are scalable to any number of objectives and decision variables, so that systematic evaluation of an algorithm can be performed.

Parallel EMO Methodologies With the availability of parallel or distributed processors, it may be wise to find the complete Pareto-optimal front in a distributed manner. A recent study (Deb et al., 2003b) has suggested such a procedure based on a guided-domination concept, in which one processor focuses on finding only a portion of the Pareto-optimal front. With intermediate cross-talks among the processors, the procedure has shown that the complete Pareto-optimal front can be discovered by concatenating the solutions from a number of processors. Since each processor works on a particular region in the search space and processors communicate among themselves, a faster and parallel search is expected from such an implementation. Similar other such parallelization techniques must be attempted and evaluated.

EMO for Other Problem-Solving Over the past few years and since the development of EMO methodologies, they have been also tried to help solve a number of other optimization problems, such as (i) in reducing bloating problems commonly-found in genetic programming applications (Bleuler et al., 2001), (ii) in goal programming problems (Deb, 1999), (iii) in maintaining diversity in a single-objective EA (Jensen, 2003a), (iv) single-objective constraint-handling problems (Coello, 2000; Surry et al., 1995), and others. Because of the use of additional objectives signifying a desired effect to be achieved, the search procedure becomes more flexible. More such problems, which reportedly perform poorly due to some fixed or rigid solution procedures, must be tried using a multi-objective approach.

Theory on EMO One aspect for which the whole EMO can be criticized is the lukewarm interests among its researchers to practice much theory related to their working principles or convergence behaviors. Besides a few studies

(Rudolph, 1998; Rudolph and Agapie, 2000), this area still remains a fertile field for theoretically-oriented researchers to dive into and suggest algorithms with a good theoretical basis. Algorithms with a time complexity analysis on certain problems have just begun (Laumanns et al., 2004) and research in this area should get more popular in trying to devise problem-algorithm combinations with an estimated computational time for finding the complete Pareto-optimal set.

Real-World Applications   Although the usefulness of EMO and classical multi-objective optimization methods are increasingly being demonstrated by solving real-world problems, more complex and innovative applications would not only demonstrate the wide-spread applicability of these methods but also may open up new directions for research which were not known earlier.

10.9   CONCLUSIONS

For the past decade or so, the usual practice of treating multi-objective optimization problems by scalarizing them into a single objective and optimizing it has been seriously questioned. The presence of multiple objectives results in a number of Pareto-optimal solutions, instead of a single optimum solution. In this tutorial, we advocate the use of an ideal multi-objective optimization procedure which attempts to find a well-distributed set of Pareto-optimal solutions first. It has been argued that choosing a particular solution as a post-optimal event is a more convenient and pragmatic approach than finding an optimal solution for a particular weighted function of the objectives. Besides introducing the multi-objective optimization concepts, this tutorial also has also presented a couple of commonly-used multi-objective evolutionary algorithms.

Besides finding the multiple Pareto-optimal solutions, the suggested ideal multi-objective optimization procedure has another unique advantage. Once a set of Pareto-optimal solutions are found, they can be analyzed. The *transition* from the optimum of one objective to that of another optimum can be investigated. Since all such solutions are optimum, the transition should show interesting trade-off information of sacrificing one objective only to get a gain in other objectives.

The field of multi-objective evolutionary algorithms is fairly new. There exists a number of interesting and important research topics which must be investigated before their full potential is discovered. This tutorial has suggested some research topics in this direction to motivate the readers to pay further attention to this growing field of interest.

SOURCES OF ADDITIONAL INFORMATION

Here, we outline some dedicated literature in the area of multi-objective optimization. Further details can be found in the reference section.

Books in Print

- C. A. C. Coello, D. A. VanVeldhuizen, and G. Lamont, 2002, *Evolutionary Algorithms for Solving Multi-Objective Problems,* Kluwer, Boston, MA. (A good reference book with good citations of most EMO studies up to 2001.)

- A. Osyczka, 2002, *Evolutionary Algorithms for Single and Multicriteria Design Optimisation,* Physica, Heidelberg. (A book describing single- and multi-objective EAs with plenty of engineering applications.)

- K. Deb, 2001, *Multi-objective optimization using evolutionary algorithms,* 2nd edn (with exercise problems), Wiley, Chichester. (A comprehensive book introducing the EMO field and describing major EMO methodologies and some research directions.)

- K. Miettinen, 1999, *Nonlinear Multiobjective Optimization,* Kluwer, Boston, MA. (A good book describing classical multi-objective optimization methods and a extensive discussion on interactive methods.)

- M. Ehrgott, 2000, *Multicriteria Optimization,* Springer, Berlin. (A good book on theory of multi-objective optimization.)

Conference Proceedings

- C. Fonseca et al., eds, 2003, *Evolutionary Multi-Criterion Optimization (EMO-03) Conference Proceedings,* Springer, Berlin. (The second EMO conference proceedings, featuring EMO theory, implementation and applications papers.)

- E. Zitzler et al., eds, 2001, *Evolutionary Multi-Criterion Optimization (EMO-01) Conf. Proceedings,* Lecture Notes in Computer Science, Vol. 1993, Springer, Berlin. (The first EMO conference proceedings, featuring EMO theory, implementation and applications papers.)

- GECCO (LNCS, Springer) and CEC (IEEE Press) annual conference proceedings feature numerous research papers on EMO theory, implementation, and applications.

- MCDM conference proceedings (Springer) publish theory, implementation and application papers in the area of classical multi-objective optimization.

Mailing Lists

- emo-list@ualg.pt (EMO methodologies)

- MCRIT-L@LISTSERV.UGA.EDU (MCDM methodologies)

Public-Domain Source Codes

- NSGA-II in C: http://www.iitk.ac.in/kangal/soft.htm+

- SPEA2 in C++: http://www.tik.ee.ethz.ch/ zitzler+

- Other codes: http://www.lania.mx/ ccoello/EMOO/+

- MCDM softwares: http://www.mit.jyu.fi/MCDM/soft.html+

References

Babu, B. and Jehan, M. M. L., 2003, Differential evolution for multi-objective optimization, in: *Proc. 2003 Congress on Evolutionary Computation (CEC'2003)*, Canberra, Australia, Vol. 4, IEEE, Piscataway, NJ, pp. 2696–2703.

Bagchi, T., 1999, *Multiobjective Scheduling by Genetic Algorithms*, Kluwer, Boston, MA.

Balicki, J. and Kitowski, Z., 2001, Multicriteria evolutionary algorithm with tabu search for task assignment, in: *Proc. 1st Int. Conf. on Evolutionary Multi-Criterion Optimization (EMO-01)*, pp. 373–384.

Bleuler, S., Brack, M. and Zitzler, E., 2001, Multiobjective genetic programming: Reducing bloat using spea2, in: *Proc. 2001 Congress on Evolutionary Computation*, pp. 536–543.

Chankong, V. and Haimes, Y. Y., 1983, *Multiobjective Decision Making Theory and Methodology*, North-Holland, New York.

Chattopadhyay, A. and Seeley, C., 1994, A simulated annealing technique for multi-objective optimization of intelligent structures, *Smart Mater. Struct.* 3:98–106.

Coello, C. A. C., 2000, Treating objectives as constraints for single objective optimization, *Eng. Optim.* 32:275–308.

Coello, C. A. C., 2003, http://www.lania.mx/ ccoello/EMOO/+

Coello, C. A. C. and Toscano, G., 2000, A micro-genetic algorithm for multi-objective optimization, *Technical Report* Lania-RI-2000-06, Laboratoria Nacional de Informatica Avanzada, Xalapa, Veracruz, Mexico.

Coello, C. A. C., VanVeldhuizen, D. A. and Lamont, G., 2002, *Evolutionary Algorithms for Solving Multi-Objective Problems*, Kluwer, Boston, MA.

Coello Coello, C. A. and Salazar Lechuga, M., 2002, MOPSO: A Proposal for Multiple Objective Particle Swarm Optimization. in: *Congress on Evolu-*

tionary Computation (CEC'2002), Vol. 2, IEEE, Piscataway, NJ, pp. 1051–1056.

Cormen, T. H., Leiserson, C. E. and Rivest, R. L., 1990, *Introduction to Algorithms,* Prentice-Hall, New Delhi.

Corne, D., Knowles, J. and Oates, M., 2000, The Pareto envelope-based selection algorithm for multi-objective optimization, in: *Proc. 6th Int. Conf. on Parallel Problem Solving from Nature (PPSN-VI),* pp. 839–848.

Coverstone-Carroll, V., Hartmann, J. W. and Mason, W. J., 2000, Optimal multi-objective low-thurst spacecraft trajectories, *Comput. Methods Appl. Mech. Eng.* **186**:387–402.

Deb, K., 1995, *Optimization for Engineering Design: Algorithms and Examples,* Prentice-Hall, New Delhi.

Deb, K., 1999, Solving goal programming problems using multi-objective genetic algorithms, in: *Proc. Congress on Evolutionary Computation,* pp. 77–84.

Deb, K., 2001, *Multi-objective Optimization Using Evolutionary Algorithms,* Wiley, Chichester.

Deb, K., 2003, Unveiling innovative design principles by means of multiple conflicting objectives, *Eng. Optim.* **35**:445–470.

Deb, K., Agrawal, S., Pratap, A. and Meyarivan, T., 2002a, A fast and elitist multi-objective genetic algorithm: NSGA-II, *IEEE Trans. Evol. Comput.* **6**:182–197.

Deb, K., Jain, P., Gupta, N. and Maji, H., 2002b, Multi-objective placement of VLSI components using evolutionary algorithms, *KanGAL Technical Report* No. 2002006, Kanpur Genetic Algorithms Laboratory, Kanpur, India. Also *IEEE Trans. Components Packaging Technol.,* to appear.

Deb, K. and Jain, S., 2002, Running performance metrics for evolutionary multi-objective optimization, in: *Proc. 4th Asia-Pacific Conf. on Simulated Evolution and Learning (SEAL-02),* pp. 13–20.

Deb, K. and Jain, S., 2003, Multi-speed gearbox design using multi-objective evolutionary algorithms, *ASME Trans. Mech. Design* **125**:609–619.

Deb, K., Mitra, K., Dewri, R. and Majumdar, S., 2004, Towards a better understanding of the epoxy polymerization process using multi-objective evolutionary computation, *Chem. Eng. Sci.* **59**:4261–4277.

Deb, K., Mohan, M. and Mishra, S., 2003a, Towards a quick computation of well-spread Pareto-optimal solutions, in: *Proc. 2nd Evolutionary Multi-Criterion Optimization (EMO-03) Conf.,* Lecture Notes in Computer Science, Vol. 2632, Springer, Berlin, pp. 222–236.

Deb, K., Thiele, L., Laumanns, M. and Zitzler, E., 2002c, Scalable multi-objective optimization test problems, in: *Proc. Congress on Evolutionary Computation (CEC-2002),* pp. 825–830.

Deb, K. and Tiwari, S., 2004, Multi-objective optimization of a leg mechanism using genetic algorithms, *KanGAL Technical Report* No. 2004005, Kanpur Genetic Algorithms Laboratory, Kanpur, India.

Deb, K., Zope, P. and Jain, A., 2003b, Distributed computing of Pareto-optimal solutions using multi-objective evolutionary algorithms, in: *Proc. 2nd Evolutionary Multi-Criterion Optimization (EMO-03) Conf.,* Lecture Notes in Computer Science, Vol. 2632, Springer, Berlin, pp. 535–549.

Ehrgott, M., 2000, *Multicriteria Optimization,* Springer, Berlin.

Fonseca, C., Fleming, P., Zitzler, E., Deb, K. and Thiele, L., 2003, *Proc. 2nd Evolutionary Multi-Criterion Optimization (EMO-03) Conf.,* Lecture Notes in Computer Science, Vol. 2632, Springer, Berlin.

Goldberg, D. E., 1989, *Genetic Algorithms for Search, Optimization, and Machine Learning,* Addison-Wesley, Reading, MA.

Gravel, M., Price, W. L. and Gagné, C., 2002, Scheduling continuous casting of aluminum using a multiple objective ant colony optimization metaheuristic, *Eur. J. Oper. Res.* **143**:218–229.

Haimes, Y. Y., Lasdon, L. S. and Wismer, D. A., 1971, On a bicriterion formulation of the problems of integrated system identification and system optimization, *IEEE Trans. Syst., Man Cybernet.* **1**:296–297.

Hansen, M. P., 1997, Tabu search in multi-objective optimization: MOTS, Paper presented at The 13th Int. Conf. on Multi-Criterion Decision Making (MCDM'97), University of Cape Town.

Jensen, M. T., 2003a, Guiding single-objective optimization using multi-objective methods, in: *Applications of Evolutionary Computing. Evoworkshops 2003,* Lecture Notes in Computer Science, Vol. 2611, G. R. Raidl et al., ed., Springer, Berlin, pp. 199–210.

Jensen, M. T., 2003b, Reducing the run-time complexity of multi-objective EAs, *IEEE Trans. Evol. Comput.* **7**:503–515.

Khor, E. F., Tan, K. C. and Lee, T. H., 2001, Tabu-based exploratory evolutionary algorithm for effective multi-objective optimization, in *Proc. Evolutionary Multi-Objective Optimization (EMO-01),* pp. 344–358.

Knowles, J. D. and Corne, D. W., 2000, Approximating the non-dominated front using the Pareto archived evolution strategy, *Evol. Comput. J.* **8**:149–172.

Kumral, M., 2003, Application of chance-constrained programming based on multi-objective simulated annealing to solve a mineral blending problem, *Eng. Optim.* **35**:661–673.

Kung, H. T., Luccio, F. and Preparata, F. P., 1975, On finding the maxima of a set of vectors, *J. Assoc. Comput. Machinery* **22**:469–476.

Laumanns, M., Thiele, L., Deb, K. and Zitzler, E., 2002, Combining convergence and diversity in evolutionary multi-objective optimization, *Evol. Comput.* **10**:263–282.

Laumanns, M., Thiele, L. and Zitzler, E., 2004, Running time analysis of multi-objective evolutionary algorithms on pseudo-Boolean functions. *IEEE Trans. Evol. Comput.* **8**:170–182.

Loughlin, D. H. and Ranjithan, S., 1997, The neighborhood constraint method: A multi-objective optimization technique, in: *Proc. 7th Int. Conf. on Genetic Algorithms,* pp. 666–673.

McMullen, P. R., 2001, An ant colony optimization approach to addessing a JIT sequencing problem with multiple objectives, *Artif. Intell. Eng.* **15**:309–317.

Miettinen, K., 1999, *Nonlinear Multiobjective Optimization,* Kluwer, Boston, MA.

Mostaghim, S. and Teich, J., 2003, Strategies for finding good local guides in multi-objective particle swarm optimization (MOPSO), in: *2003 IEEE Swarm Intelligence Symp. Proc.,* Indianapolis, IN, IEEE, Piscataway, NJ, pp. 26–33.

Osyczka, A., 2002, *Evolutionary Algorithms for Single and Multicriteria Design Optimization,* Physica, Heidelberg.

Parks, G. and Suppapitnarm, A., 1999, Multiobjective optimization of PWR reload core designs using simulated annealing, in: *Proc. Int. Conf. on Mathematics and Computation, Reactor Physics and Environmental Analysis in Nuclear Applications,* Vol. 2, Madrid, Spain, pp. 1435–1444.

Rudolph, G., 1998, Evolutionary search for minimal elements in partially ordered finite sets, in: *Proc. 7th Annual Conf. on Evolutionary Programming,* Berlin, Springer, pp. 345–353.

Rudolph, G. and Agapie, A., 2000, Convergence properties of some multi-objective evolutionary algorithms, in: *Proc. 2000 Congress on Evolutionary Computation (CEC2000),* pp. 1010–1016.

Srinivas, N. and Deb, K., 1994, Multi-objective function optimization using non-dominated sorting genetic algorithms, *Evol. Comput. J.* **2**:221–248.

Surry, P. D., Radcliffe, N. J. and Boyd, I. D., 1995, A multi-objective approach to constrained optimisation of gas supply networks : The COMOGA method, in: *Evolutionary Computing, AISB Workshop,* Springer, Berlin, pp. 166–180.

Veldhuizen, D. V. and Lamont, G. B., 2000, Multiobjective evolutionary algorithms: analyzing the state-of-the-art, *Evol. Comput. J.* **8**:125–148.

Zitzler, E., 1999, Evolutionary algorithms for multi-objective optimization: methods and applications, *Ph.D. Thesis,* Swiss Federal Institute of Technology ETH, Zürich, Switzerland.

Zitzler, E., Deb, K., Thiele, L., Coello, C. A. C. and Corne, D., 2001a, *Proc. 1st Evolutionary Multi-Criterion Optimization (EMO-01) Conf.,* Lecture Notes in Computer Science, Vol. 1993, Springer, Berlin.

Zitzler, E., Laumanns, M. and Thiele, L., 2001b, SPEA2: Improving the strength Pareto evolutionary algorithm for multi-objective optimization, in: Giannakoglou, K. C., Tsahalis, D. T., Périaux, J., Papailiou, K. D. and Fogarty, T., eds, *Proc. Evolutionary Methods for Design Optimization and Control with Applications to Industrial Problems,* Athens, Greece, International Centre for Numerical Methods in Engineering (Cmine), pp. 95–100.

Zitzler, E. and Thiele, L., 1998, An evolutionary algorithm for multi-objective optimization: The strength Pareto approach, *Technical Report* No. 43, Computer Engineering and Networks Laboratory, Switzerland.

Zitzler, E., Thiele, L., Laumanns, M., Fonseca, C. M. and da Fonseca, V. G., 2003, Performance assessment of multiobjective optimizers: an analysis and review, *IEEE Trans. Evol. Comput.* **7**:117–132.

Chapter 11

COMPLEXITY THEORY AND THE NO FREE LUNCH THEOREM

Darrell Whitley
Department of Computer Science
Colorado State University, Fort Collins, CO, USA

Jean Paul Watson
Sandia National Laboratories, Albuquerque, NM, USA

11.1   INTRODUCTION

This tutorial reviews basic concepts in complexity theory, as well as various No Free Lunch results and how these results relate to computational complexity. The tutorial explains basic concepts in an informal fashion that illuminates key concepts. No Free Lunch theorems for search can be summarized by the following result:

> For all possible performance measures, no search algorithm is better than another when its performance is averaged over all possible discrete functions.

Note that No Free Lunch is often referred to simply as NFL within the heuristic search community (despite copyrights and trademarks held by the National Football League).

No Free Lunch relates to complexity theory in as much as complexity theory addresses the time and space costs of algorithms; complexity theory is also concerned with key classes of problems, such as the class of NP-complete problems that are also of interest to researchers designing search algorithms.

11.2   COMPLEXITY, P AND NP

The complexity classes denoted by P and NP are the most famous (or notorious) classes of problems in complexity theory. The problem class P is the set of problems that can be solved in polynomial time on a deterministic Turing machine. For current purposes, we can think of any computer as a surrogate for a Turing machine (except that Turing machines are assumed to have infinite memory). The P stands for polynomial. In practice, we generally think of P as representing those problems that are *tractable*, i.e. problems that can be solved in reasonable computation time (within one's lifetime, for example).

The problem class NP is the set of problems that can be solved in polynomial time on a nondeterministic Turing machine. The NP stands for nondeterministic polynomial (*not* to be confused with Not Polynomial). Nondeterminism is a little strange. In a nondeterministic machine, choices are allowed in the computation, so that some things need not be computed. In effect, the computation itself becomes a search tree. Each path in the tree represents a possible solution, but only certain paths yield an actual solution. We say that a problem is in NP if this search tree is polynomial in height, while the number of nodes in the search tree might be exponential. Thus, if we could explore all computational paths in parallel, we arrive at a solution in polynomial time. Alternatively, if we "magically" make the right choice at each decision node in the tree, then we again arrive at the desired solution in polynomial time. If we can *deterministically* find a path to a solution in polynomial time in every case, then the problem is in P. All problems in P are also in NP. Another characteristic of the class NP is that the correctness of solutions can be verified in *deterministic* polynomial time. Note that this is true, because if we have the solution in hand, we then know how to make the right choice at each decision node without needing any magical guidance.

Problems in NP that are not known to be in P are characterized by an *algorithm gap*. An algorithm gap exists when the proven difficulty of a problem (or a set of problems) has lower complexity than the best known algorithms for solving that problem. The complexity of the problem itself is algorithm independent and is a bound from below: *the problem can be proven to be at least this hard (but might be harder)*. The complexity of the algorithm is a bound from above: *the best known algorithms solves the problem this fast (but it might be done faster)*.

The complexity of sorting has been proven to be $O(N \log N)$, thus no algorithm can sort faster than $O(N \log N)$ in the worst case. Of course, there exist algorithms that sort in $O(N \log N)$ time, so sorting is said to be a *closed problem* because it does not have an algorithm gap.

If an algorithm sorts faster than $O(N \log N)$ time, then that algorithm has been designed to work on special subclasses of problems: for example, if we know that we are sorting integers from ranging from 1 to 1000, and the expected distribution of the integers is uniform, we can use a bucket sort and sort in linear, i.e. $O(N)$, time.

In contrast, an algorithm gap does exist in the well-known traveling salesman problem. Here, the only algorithm guaranteed to locate an optimal solution is, in effect, enumeration. Thus, the best known method in the worst case has complexity $O(N!)$ for an N city problem. Yet, no one has proven that the *inherent* complexity of the traveling salesman problem is such that it cannot be solved in polynomial time. And note that a solution can be verified in polynomial time. If someone has a solution that is claimed to have a particular evaluation, then that evaluation can be verified in $\mathcal{O}(N)$ time—which is polynomial, of course.

Can all the problems that are solved by a Turing machine in NP time be solved by a deterministic Turing machine using another, more clever algorithm in polynomial time? What we are really asking is whether the complexity class P = NP. The answer is unknown and is considered to be one of the most important theoretical questions in Computer Science. It is an equally important question in Operations Research. While the answer is unknown, it is widely thought that P $\neq$ NP.

Researchers have identified a very important subset of the class NP known as the class NP-complete. A problem, R, is NP-complete if (1) R is NP-hard and (2) $R \in$ NP. Informally, a problem is NP-hard if it is *at least* as hard as any other problem in NP. More formally, a problem R is NP-hard if there exists an NP-complete problem R_0 such that every instance of R_0 can be "reformulated" into an instance of R in deterministic polynomial time. R must be just as hard as R_0 since R in some sense "includes" R_0.

In a renowned theorem, Cook (1971) established that Boolean satisfiability is NP-complete by showing it is in NP and by showing that *every* problem in NP can be expressed as a Boolean satisfiability problem (also just called "SAT"). Of course SAT is a member of the set of NP problems: the nondeterministic Turing machine just selects the right assignment to the Boolean variables to make the expression true, if it is possible to do so.

Other problems in NP have been shown to be NP-complete by showing that every SAT problem can be converted into an instance of that particular problem class. Thus, every instance of SAT can be converted into an instance of the 3-CNF-SAT problem. A 3-CNF-SAT problem is a satisfiability problem, where the Boolean expression is made up of conjunctive normal form clauses. Each clause contains three Boolean literals, where a literal is a variable or its negation such as x_1 or $\tilde{x}_1$. One of the three literals must be satisfied for the clause to be satisfied (e.g. x_1 or $\tilde{x}_1$ or x_2). All of the clauses must be satisfied for the Boolean expression to be satisfied. The 3-CNF-SAT formulation allows satisfiability problems to be of regular decomposable form that can then be converted into problems such as the Hamiltonian circuit problem, which in turn can be converted to an instance of the traveling salesman problem (Corman et al., 1990). This means that all of these problems are NP-hard. Showing that they are all also in the class NP makes them NP-complete. Technically, to be NP-complete, a problem must be a decision problem. A decision problem is a problem that has a yes or no answer. Therefore, the traveling salesman problem is "NP-complete" when expressed as a decision problem (i.e., is there a tour with length $\leq X$?), but the traveling salesman problem is still said to be "NP-hard" when expressed as an optimization problem.

Given the interrelated nature of the NP-complete problems, if researchers ever discover a polynomial-time algorithm for any NP-complete problem, then it would follow that *every* problem in NP could be solved in polynomial time. In an abstract sense, this means that all problems in the NP-complete problems are all of comparable difficulty, and that the NP-complete are the most difficult problems in the set made up of all problems in NP.

11.2.1   Complexity, Search and Optimization

Since we do not know how to compute the solution to NP-hard problems in polynomial time, we have to settle for approximate solutions (which sometimes can be computed exactly in polynomial time) or use search methods to find the best solutions possible. It can be useful to think of these search methods as exploring the same decision tree that is navigated by a nondeterministic Turing machine. The solutions that are found using search methods often are not optimal, but finding sufficiently good solutions can be important for many applications.

A basic distinction can be made between search problems that are discrete versus problems that are continuous. This distinction can also be related to the difference between integers and real-valued numbers. If we ask how many integers there are in the (inclusive) interval between 1 and 10, the answer is obviously 10 different and discrete values. But if we asked how many real-valued numbers there are between 1 and 10, the answer is infinitely many.

The nondeterministic Turing machine is clearly solving a discrete problem, because there are a fixed number of decisions that must be made to reach an optimal solution. By definition, the number of decisions that must be made by the nondeterministic Turing machine must be polynomial if it is solving an NP-hard problem.

Some problems cannot be solved in polynomial time by a nondeterministic Turing machines and therefore are not in NP; we can loosely think of such problems as requiring exponential time, although in complexity theory one must worry about both space (memory) and time and balance trade-offs between space and time costs.

Consider a *parameter optimization problem* such that there is a function f that takes k parameters as inputs and returns a single value that evaluates the usefulness or goodness of those k parameters. The space of possible inputs is known as the *domain* and the space of possible outputs as the range or *co-domain* of the function. For example, we might have a parameter optimization problem that used temperature and pressure as two input control parameters for a process that produces some material (e.g. paper), where the output of the function might be the cost of the material, or some measurement of its quality.

If a parameter can be assigned any continuous real-valued number, then the input space is theoretically infinite. We will limit our attention to problems that are discrete such that the domain and therefore the co-domain are finite. Discrete parameter optimization problems are part of a larger set of discrete problems referred to as *combinatorial optimization problems.* Combinatorial optimization problems include many different types of problems, such as scheduling and resource allocation, as well as problems in graph theory and Boolean logic.

For example, we might have a scheduling problem where we want to optimize the order in which tasks are carried out. The goal might be to minimize total processing time, or to maximize work done per unit of time. For N tasks, there could be $N!$ ways to order those tasks. Or, we might want to assign truth values (0 or 1) to a Boolean expression, in which case there are 2^k assignments if there are k Boolean variables in the expression. In the first case, an input could be a permutation of tasks of length N and the evaluation might be how long it takes to process all of the N tasks. In the second case, an input might be a bit-string of length k representing the assignments made to the k Boolean variables, and the output might be a true or false (0 or 1) evaluation of the overall Boolean expression. For classic NP-hard problems, the search space is typically modeled in a general way so that the search space is exponentially large in relationship to the size of an input.

Parameter optimization problems can also be discretized. For example, a single input parameter can be restricted to a value between 0.00 and 99.99 (inclusive) where we only consider values that are increments of 0.01. In this case, there are only 10 000 possible assignments for that particular input. If all of the parameters of a parameter optimization problem are discretized in this way, then the overall search problem is discrete as well. There are a number of reasons that one might want to look at parameter optimization problems as discrete search spaces. In some cases, sensors for the inputs and/or outputs have limited precision and it does not make sense to represent and reason about extremely high precision numbers: we simply cannot measure the world that precisely. And, in general, as soon as anything is represented in a computer program it is discrete. Infinite precision is a fiction, although it is sometimes a useful fiction. But as soon as we decide to represent a parameter using a fixed-length floating point representation, the optimization problem is discrete.

This leads to the following observation. If the set of possible inputs is discrete, we can enumerate the set of inputs and label each possible input with a unique integer. We will also sort the inputs in some principled manner, so that the ith possible input is uniquely identified. This is a familiar concept in complexity, since it allows us to count all of the inputs. Thus, any particular instance of a discrete search problem using any given discrete representation can be abstractly modeled by a function

$$f(i) = j$$

where i is an integer that labels the ith input (i.e. element i of the domain) and j is a member of the set of values that make up the co-domain. This perspective also provides a general foundation for discussing the concept of No Free Lunch.

11.3   NO FREE LUNCH

In 1995, a paper by David Wolpert and William Macready caused a good deal of excitement in the search community. An updated version of the original report appeared in 1997. The paper *No Free Lunch Theorems for Search* presents proofs that can be summarized by the following No Free Lunch result:

> For all possible performance measures, no search algorithm is better than another when its performance is averaged over all possible discrete functions.

First, note that we only consider discrete functions. A performance measure includes any measurement of the quality of the solution (or set of solutions) found after sampling some fixed number of points in the search space, or how long it takes to find a solution of a particular quality. It is also implied that a performance measure is taken over the set of domain and associated co-domain values that have been sampled so far.

A key assumption behind this result is that resampling is ignored: this means that if a search algorithm samples point i and evaluates the objective function $f(i)$ then that point is never sampled again. In reality, heuristic search algorithms "focus" search toward particular regions of the search space: in other words, a focused search is one that spends more time sampling points that are near to one another in the search space. Consequently, a focused search is one that is more likely to resample previously visited points. Search algorithms that are more likely to resample points in the search space than others are in some sense "worse" than algorithms that resample less.

One of the most basic and least intelligent forms of search is random enumeration. Random enumeration means that we sample the search space randomly without replacement; this can be done using clever bookkeeping, or simply by keeping a list of visited points so that none are evaluated again. In practice, random sampling is typically unfocused, only a limited amount of the search space can be sampled, and it is reasonable to allow sampling with replacement because resampling is unlikely. When random sampling is used as a search algorithm, it provides a minimal baseline against which the performance of heuristic search algorithms can be judged. Clearly, we would expect any useful heuristic search algorithm to outperform random enumeration.

However, a startling and powerful consequence of No Free Lunch is that *no* heuristic search algorithm is better than random enumeration when compared over all possible discrete functions.

Useful search algorithms do not exhaustively enumerate the entire search space. Wolpert and Macready (1995, 1997) model a search algorithm as a procedure that searches for **m** steps. However, this does not restrict any of the No Free Lunch results.

Another issue relating to No Free Lunch involves deterministic versus stochastic search algorithms. Some algorithms make deterministic decisions, such as a steepest ascent local search algorithm: when started from the same point, steepest ascent always yields the same solution. Genetic algorithms are often implemented as largely stochastic algorithms—meaning that the search involves many random or stochastic decisions and that different runs will often produce different solutions. Wolpert and Macready present arguments showing that the No Free Lunch theorems hold for both stochastic and deterministic search algorithms. Radcliffe and Surry (1995) also point out that in practice stochastic algorithms typically employ pseudo-random number generators. Thus, if we include the random number generator and initial seed in the specification of the search algorithm, then these "stochastic" algorithms, in effect, are also deterministic.

Immediately following its introduction, researchers had two general reactions to the No Free Lunch results.

Reaction 1 Many researchers simply dismissed No Free Lunch, arguing that results concerning the set of all possible discrete functions are not applicable in the real world because this set is not representative of real-world problems. Some researchers pointed out that the set of all possible discrete functions is infinitely large and most functions are *incompressible* in that there is not a representation whose size is significantly less than the size of the function when fully enumerated. For example, if there are N values in the co-domain of a function, then writing down all of these values requires $N \log_2 N$ bits (i.e. N values, $\log_2(N)$ bits per value). In effect, this representation of the function is just a look-up table where the ith entry is the co-domain value associated with $f(i)$. If there exists no representation of a function that uses less than $\mathcal{O}(N \log_2 N)$ bits, then that function is incompressible. Even if an evaluation function only returns 0 or 1, it still requires $\mathcal{O}(N)$ bits to construct a look-up table or to enumerate the function; in this case, the look-up table is still exponentially large when N is exponentially large in relationship to the size of an input string to the evaluation function.

Of course, there are more random functions than non-random functions (English, 2000a). Furthermore, most standard textbooks on computability discuss the well-known result that the set of all possible functions is uncountably infinite (as can be shown using diagonalization arguments), while the set of all possible programs (which are just bit-strings at the lowest level) is only countably infinite (Sudcamp, 1997). So the set of all possible cost functions that can be implemented on a computer is a tiny subset of the set of all possible functions. Thus, the space of all possible discrete functions is largely composed of incompressible functions. Given these observations, "No Free Lunch is No Big Deal" seemed to be the conclusion of this point of view.

Reaction 2 The other reaction to No Free Lunch was to acknowledge that researchers trying to develop the best possible algorithm for a particular application typically need to leverage extensive problem-specific knowledge. Consequently, the No Free Lunch result seemed to be an intuitive affirmation of the idea that there are no general-purpose search methods (at least none that are very effective) and that the business of developing search algorithms is one of building special-purpose methods to solve application-specific problems. This point of view echoes a refrain from the Artificial Intelligence community: "Knowledge is Power".

Of course, there is truth in both of these views. It has taken several years for the research community to gain a deeper understanding of No Free Lunch. These investigations have led to some surprising and even fruitful results along the way. Culberson (1998) published an "algorithmic view" of No Free Lunch that added perspective to the debate; Culberson makes two important points.

First, No Free Lunch looks at search as a blind process. This means that the only information we have is the evaluation of particular points in the space. We do not have information about what a solution might look like or information about how the evaluation function is constructed that might allow us to search more intelligently. Blind search is extremely weak. Using an "adversarial argument" we can think of blind search as the process of asking an adversary to sample a point of some objective function and then return an answer. In the space of all possible discrete functions, however, the adversary is free to return any value whatsoever without regard to those values of the search space that have already been examined. In the worst case, sampled points from the search space tell us nothing about the remaining points in the search space.

Second, search is often not blind. If we construct an algorithm for the traveling salesman problem, for example, we often do exploit application-specific operators and representations. But this does not mean that we completely give up generality; our algorithms are designed to solve a particular problem, but should be general enough to solve different instances of that problem.

Radcliffe and Surry (1995) first formalized the idea that we can also include representations under No Free Lunch. That is, when we consider all possible representations of a function, No Free Lunch still holds: no search algorithm is better than another when applied to all possible representations of a function. In effect, a representation just transforms one function into another.

Not surprisingly, No Free Lunch also holds when comparing the set of possible representations under Gray codes and Binary bit encodings. However, Whitley et al. (1997) pointed out that if one selected particular subsets of problems of bounded complexity, then No Free Lunch no longer holds; Rana and Whitley (1997) and Whitley (1999) provide proofs of this related to binary representations. Droste et al. (1999) also made similar observations, indicating that one can define sets of reasonable and interesting functions where one algorithm can consistently outperform another.

If we go back in time, No Free Lunch observations were made by Greg Rawlins at the *Foundations of Genetic Algorithms* (FOGA) workshops in 1990 and 1992. In the preface to the proceedings of the 1990 FOGA workshop Rawlins (1991) makes the following observations:

> [I]t is sometimes suggested that GAs [Genetic Algorithms] are universal in that they can be used to optimize any function. These statements are true in only a very limited sense; any algorithm satisfying [these] claims can expect to do no better than random search over the space of all functions. ...
>
> ... It is now apparent that for a *fixed universal* algorithm, restricted to [bit] strings ... over the set of all possible domain functions ... it does not matter which encoding we use, since for every domain function which the encoding makes easier to solve there is another domain function that makes it more difficult to solve. Thus, changing the encoding does not affect the *expected difficulty* of solving randomly chosen domain functions.
>
> Equivalently, assume that we have a *fixed* domain function f and suppose that we choose the encoding, e, at random. ... Then, no search algorithm can expect to do better than random search, since no information is carried by e about f, except that for each string there is a value.

(Rawlins, 1991, pp. 7–8.)

Rawlins anticipated several of the consequences of No Free Lunch. Nevertheless, it was Wolpert and Macready who not only provided the first proof of No Free Lunch, but also explored many of the ramifications of the No Free Lunch Theorem.

11.3.1   No Free Lunch: Variations on a Theme

Two other common variants of NFL are as follows:

- the aggregate behavior of any two search algorithms is equivalent when compared over all possible discrete functions;

- the aggregate behavior of all possible search algorithms is equivalent when compared over any two discrete functions.

At the root of these observations is another, more concise result. Consider any algorithm A_i applied to function f_j. Let $Apply\ (A_i, f_j, m)$ represent a "meta-level" algorithm that outputs the order in which A_i visits m elements in the co-domain of f_j after m steps. For every pair of algorithms A_k and A_i and for any function f_j, there exists another function f_l such that

$$Apply\ (A_i, f_j, m) \equiv Apply\ (A_k, f_l, m)$$

The equivalence operator $\equiv$ denotes that the ordered sequence of co-domain values that is returned by "Apply" will be equivalent. We could interpret this result in another way. For every pair of functions f_j and f_l and for any algorithm A_i, there exists another algorithm A_k such that $Apply\ (A_i, f_j, m) \equiv Apply\ (A_k, f_l, m)$. In fact, if we consider the algorithms and the functions as variables that are supplied to the Apply function, then when any three of the variables are known, the fourth is immediately determined.

This also implies that we can talk about No Free Lunch in a much smaller context: for example, we can talk about any two search algorithms applied to exactly two carefully chosen paired functions.

This perspective on No Free Lunch has some rather counter-intuitive implications, which may be deeper and more profound than the general NFL result. Consider a best-first version of steepest ascent local search which restarts when a local optimum is encountered. Also consider a worst-first steepest ascent local search, also with restarts. We incorporate restarts so that these algorithms continue searching for an arbitrary number of steps. Then, for every function f_j there exists a function f_l such that

$$Apply\ (best\text{-}first, f_j, m) \equiv Apply\ (worst\text{-}first, f_l, m)$$

Virtually all researchers would accept that best-first local search is a reasonable search algorithm and that it is useful on many real-world problems. In other words, there is a subset of problems where best-first search is effective, relative to some performance measure. But there is a corresponding set of functions where worst-first local search is equally effective. What do these functions look like? They probably are "structured" in some sense, and might be compressible. Also note that if we are minimizing a function, then a worst-first local search is one that simply maximizes at each step, instead of minimizing. On the other hand, it seems reasonable that we might want to maximize one function and minimize another function. Why is best-first search generally viewed as a reasonable algorithm and worst-first as an unreasonable algorithm? This is a nagging question for which, at least formally, there are currently no good answers.

11.3.2   No Free Lunch and Permutation Closure

As has been noted, the set of all possible discrete functions is infinitely large. One easy way to see this is by considering all the functions that take K inputs: since K could be any integer from 1 to infinity, there must be infinitely many discrete functions. But even if there are exactly two inputs, the number of evaluations could be chosen from an infinite set of different possible values, resulting in infinitely many discrete functions.

Whitley et al. (1997) first explored the idea that permutations could be used to represent both algorithms and functions—and thus produce an NFL result over a finite set. This was further explored by Whitley (2000). Consider the following small example. Assume that the co-domain of our objective function consists of the set of values $\{A, B, C\}$. Let the permutation $\langle A, B, C \rangle$ represent a canonical ordering of these values. We can start by considering bijective functions, those that are one-to-one and onto: an important implication of this is that each value in the co-domain is unique. To construct a function, we need to assign values to $f(1)$, $f(2)$ and $f(3)$. Exactly 3! bijective functions can be constructed given three possible co-domain values. Additionally, only 3! *behaviors* are possible for any search algorithm, assuming that an algorithm does not resample points. Let an algorithm's behavior be represented by a permutation over the set of numbers $\{1, 2, 3\}$ which will serve as indices into the canonical permutation of co-domain values $\{A, B, C\}$. Let s_i be the ith value sampled by a search algorithm. Thus, the permutation $\langle 2, 1, 3 \rangle$ defined with respect to the canonical ordering $\langle A, B, C \rangle$ represents a search algorithm whose behavior can be described by the following sampling behavior: $s_1 = B, s_2 = A, s_3 = C$. Note that we do not need to specify a particular function to talk about behavior, we just need to define the co-domain values. In the following table, we enumerate all possible permutations over all possible functions over the co-domain $\{A, B, C\}$ as well as all possible permutations over the set of algorithm behaviors over the set of indices denoted by $\{1, 2, 3\}$.

POSSIBLE BEHAVIORS		POSSIBLE FUNCTIONS	
B1:	< 1, 2, 3 >	F1:	< A, B, C >
B2:	< 1, 3, 2 >	F2:	< A, C, B >
B3:	< 2, 1, 3 >	F3:	< B, A, C >
B4:	< 2, 3, 1 >	F4:	< B, C, A >
B5:	< 3, 1, 2 >	F5:	< C, A, B >
B6:	< 3, 2, 1 >	F6:	< C, B, A >

The implications of No Free Lunch start to become clear when one asks basic questions about the set of behaviors and the set of functions.

If we apply any two sets of behaviors to all functions, each behavior generates a set of 3! possible search behaviors which is the same as the set of all possible functions. If we apply all possible search behaviors to any two functions, for each function we again obtain a set of behaviors which, after the indices are translated into co-domain values, is the same as the set of all possible functions.

We need to be careful to distinguish between algorithms and their behaviors. There exist many algorithms (perhaps infinitely many) but once the values of the co-domain are fixed, there are only a finite number of behaviors.

Schumacher (2000) and Schumacher et al. (2001) sharpened the No Free Lunch theorem by formally relating it to the *permutation closure* of a set of functions. Let $\mathcal{X}$ and $\mathcal{Y}$ denote finite sets and let f: $\mathcal{X} \longrightarrow \mathcal{Y}$ be a function where $f(x_i) = y_i$. Let σ be a permutation such that $\sigma : \mathcal{X} \longrightarrow \mathcal{X}$. We can permute functions as follows:

$$\sigma f(x) = f(\sigma^{-1}(x))$$

Since $f(x_i) = y_i$, the permutation $\sigma f(x)$ can also be viewed as a permutation over the values that make up the co-domain (the output values) of the objective function.

We next define the permutation closure $P(F)$ of a set of functions F:

$$P(F) = \{\sigma f : f \in F \text{ and } \sigma \text{ is a permutation}\}$$

Informally, $P(F)$ is constructed by taking each function in F and reordering its co-domain values to produce a new function. This process is repeated until no new functions can be generated. This produces *closure—(* since every re-ordering of the co-domain values of any function in $P(F)$ will produce a function that is already a member of $P(F)$. Therefore, $P(F)$ is closed under permutation. This provides the foundation for the following result.

THEOREM 11.1 *The No Free Lunch theorem holds for a set of functions if and only if that set of functions is closed under permutation.*

Proofs are given by Schumacher et al. (2001). Intuitively, that NFL should hold over a set closed under permutations can be seen from Culberson's adversarial argument: any possible (remaining) value of the co-domain can occur at the next time sample. Proving that the connection between algorithm behavior and permutation closure is an *if and only if* relationship is much stronger than the observation that No Free Lunch holds over the permutation closure of a function. But if every remaining value is not equally likely at each time step, the set of functions we are sampling from is not closed under permutation and

No Free Lunch does not hold. Similar observations have also been made by Droste et al. (2002).

It is useful to view the permutation closure of a function as a table, where each row of the table is a permutation representing a function. Each row in the table also corresponds to the behavior of some optimization algorithm on some function. The *behavior* of an optimization algorithm with respect to some objective function describes the order in which the optimization algorithm samples the values that make up the co-domain of the objective function. Schumacher et al. (2001) refer to this as the *performance vector*.

This tabular representation makes it clear when NFL results hold and makes it clear why making a general declaration that one algorithm is better than another is in some sense meaningless.

Consider the following table representing the permutation closure over a function defined over a co-domain of three values.

$$< 1, 2, 3 >$$
$$< 1, 3, 2 >$$
$$< 2, 1, 3 >$$
$$< 2, 3, 1 >$$
$$< 3, 1, 2 >$$
$$< 3, 2, 1 >$$

Each column of the table represents the set of possible results at a particular time step; the rows represent all possible performance vectors. But each column is identical in its composition. The notion of robustness implies that some algorithm yields relatively good performance over a broad range of problems compared to another algorithm. This would suggest that relatively good solutions are found within some fixed (e.g. polynomial) number of time steps. Yet, if NFL holds over a set of problems, the set of co-domain values returned over all functions in the permutation closure is identical at each time step. Thus, not only are all measures of performance the same after m steps; every step of the search yields exactly the same set of co-domain samples when behavior is aggregated over all possible functions in any permutation closure.

We can now make a more precise statement about the "zero-sum" nature of No Free Lunch. If algorithm K outperforms algorithm Z on any subset of functions denoted by β, then algorithm Z will outperform algorithm K over $P(\beta) - \beta$. Differences in aggregate measures of performance such as the total number of steps taken to find a particular evaluation or the sum of the evaluations after m steps will be zero. Aggregate versus average measures of performance can be different, because the subsets are of different size. This means that No Free Lunch theorems for search apply to finite sets. These sets can in fact be quite small.

English (2000a) first pointed out that NFL can hold over sets of functions such as needle-in-a-haystack functions. A needle-in-a-haystack function is one that has the same evaluation for every point in the space except one; in effect, searching a needle-in-a-haystack function is necessarily random since there is no information about how to find the needle until after it has been found.

In the following example, NFL holds over just three functions:

$$f = \langle 0, 0, 3 \rangle$$
$$P(f) = \{\langle 0, 0, 3 \rangle, \langle 0, 3, 0 \rangle, \langle 3, 0, 0 \rangle\}$$

Clearly, NFL does not just hold over sets that are incompressible. All needle-in-a-haystack functions have a compact representation of size $\mathcal{O}(\log N)$, where $N = |\mathcal{X}|$ is the size of the search space. In effect, the evaluation function needs to indicate when the needle has been found and return a distinct evaluation.

Generally, we like to construct evaluation functions that are capable of producing a rich and discriminating set of outputs: that is, we like to have evaluation functions that tell us point i is better than point j. But it also seems reasonable to conjecture that if NFL holds over a set that is compressible, then that set has low information measure.

Schumacher et al. (2001) also note that the permutation closure has the following property:

$$P(F \cup F') = P(F) \cup P(F')$$

Given a function f and a function g, where $g \notin P(f)$, we can then construct three permutation closures: $P(f)$, $P(g)$, $P(f \cup g)$. For example, this implies that NFL holds over the following sets which are displayed in table format:

```
Set 1: {< 3,  0,  0 >,
        < 0,  3,  0 >,      Set 3: {< 3,  0,  0 >,
        < 0,  0,  3 >}              < 0,  3,  0 >,
                                    < 0,  0,  3 >,
Set 2: {< 1,  3,  2 >,              < 1,  3,  2 >,
        < 2,  1,  3 >,              < 2,  1,  3 >,
        < 2,  3,  1 >,              < 2,  3,  1 >,
        < 3,  1,  2 >,              < 3,  1,  2 >,
        < 3,  2,  1 >}              < 3,  2,  1 >}
```

We can also ask about NFL and the probability of sampling a particular function in $P(f)$. For NFL to hold, we must insist that all members of $P(f)$ for a specific function f are uniformly sampled. Otherwise, some functions are more likely to be sampled than others, and NFL breaks down. For NFL to hold over $P(g)$ the probability of sampling a function in $P(g)$ must also be uniform. But Igel and Toussaint (2004) point out that we can also have a uniform sample over $P(g)$ and a (different) uniform sample over $P(f)$ and NFL still holds. Thus, sampling need not be uniform over $P(f \cup g)$.

11.3.3   Free Lunch and Compressibility

Whitley (2000) presents the following observation (the current form is expanded to be more precise).

THEOREM 11.2 *Let $P(f)$ represent the permutation closure of the function f. If f is a bijection, or if any fixed fraction of the co-domain values of f are unique, then $|P(f)| = \mathcal{O}(N!)$ and the functions in $P(f)$ have a description length of $\mathcal{O}(N \log N)$ bits on average, where N is the number of points in the search space.*

The proof, which is sketched here, follows the well known proof demonstrating that the best sorting algorithms have complexity $\mathcal{O}(N \log N)$. We first assume that the function is a bijection and that $|P(f)| = N!$. We would like to "tag" each function in $P(f)$ with a bit string that uniquely identifies that function. We then make each of these tags a leaf in a binary tree. The tag acts as an address that tells us to go left or right at each point in the tree in order to reach a leaf node corresponding to that function. But the tag also uniquely identifies the function. The tree is constructed in a balanced fashion so that the height of the tree corresponds to the number of bits needed to tag each function. Since there are N! leaves in the tree, the height of the tree must be $\mathcal{O}(\log N!) = \mathcal{O}(N \log N)$. Thus $\mathcal{O}(N \log N)$ bits are required to uniquely label each function. (Standard binary labels can be compressed somewhat, but lexicographically ordered bit labels can be used, which cannot be compressed, so that the complexity is still $\mathcal{O}(N \log N)$.)

To construct a lookup table or a full enumeration of any permutation of N elements requires $\mathcal{O}(N \log N)$ bits, since there are N elements and $\log N$ bits are needed to distinguish each element. Thus, most of these functions have exponential description.

This is, of course, one of the major concerns about No Free Lunch theorems. Do No Free Lunch theorems really apply to sets of functions which are of practical interest? Yet this same concern is often overlooked when theoretical researchers wish to make mathematical observations about search. For example, proofs relating the number of expected optima over all possible functions (Rana and Whitley, 1998), or the expected path length to a local optimum over all possible functions (Tovey, 1985) under local search are computed with respect to the set of $N!$ functions.

Igel and Toussaint (2003) formalize the idea that if one considers all the possible ways that one can construct subsets over the set of all possible functions, then those subsets that are closed under permutation are a vanishingly small percentage. The problem with this observation is that the *a priori* probability of *any* subset of problems is vanishingly small—including any set of applications we might wish to consider. On the other hand, Droste et al. (2002) have also shown that for any function for which a given algorithm is effective, there exist related functions for which performance of the same algorithm is substantially worse. This is expressed in the *Almost No Free Lunch* (ANFL) theorem.

THEOREM 11.3 *ANFL Theorem: Let H be a randomized search strategy and* $f : \{0, 1\}^n \rightarrow \{0, 1, \ldots, N - 1\}$. *Then there exists at least* $N^{2^{n/3}-1}$ *functions* $f* : \{0, 1\} \rightarrow \{0, 1, \ldots, N\}$ *which agree with f on all but at most* $2^{n/3}$ *inputs such that H does find the optimum of f* within* $2^{n/3}$ *steps with a probability bounded above by* $2^{-n/3}$. *Exponentially many of these functions have the additional property that their evaluation time, circuit size representation, and Kolmogorov complexity is only by an additive term of* $\mathcal{O}(n)$ *larger than the corresponding complexity of f.*

For current purposes, we can think of Kolmogorov complexity as the length of the shortest program that implements a particular function. Thus, if two functions have similar Kolmogorov complexity, then there are programs or algorithms of similar size that implement those functions. The significance of the ANFL theorem is that even search algorithms designed for specific problem classes could be subject to ANFL kinds of effects.

11.3.4   No Free Lunch and NP-completeness

No Free Lunch has not been proven to hold over the set of problems in the complexity class NP. This is rather obvious if one considers the following: if No Free Lunch holds for any NP-complete problem, then it immediately follows that no algorithm is better than random enumeration on the entire class of NP-complete problems (because of the existence of a polynomial-time transformation between any two NP-complete problems). However, this would also prove that P $\neq$ NP, since it would prove that no algorithm could solve all instances of an NP-complete problem in polynomial time. This means that proofs concerning No Free Lunch do not apply to NP-complete problems unless the proofs also show (perhaps implicitly) that P $\neq$ NP.

The description length of all NP-complete problems must also be polynomial, since we need to reformulate one problem into another in polynomial time. This means that an NP-complete problem class (such as NK-landscapes: Kauffman, 1989) *cannot* be used to generate all $N!$ functions of $P(f)$ when f is a bijection, since on average the set of all possible bijective functions over a set of co-domain values do not have polynomial space descriptions.

The existence of ratio bounds also shows that NFL theorems to not hold for certain NP-complete problems. The Euclidean traveling salesman problem is a traveling salesman problem where the cities are points in real space and the distance between cities are Euclidean. For such problems, the triangle inequality holds, such that the distance from city A to city B to city C is greater than or equal to the distance from city A to city C. A greedy polynomial time approximate algorithm exists for the Euclidean traveling salesman problem that is guaranteed to yield a solution that is no worse that $2C$, when C is the cost of an optimal solution (Cormen et al., 1990). (In fact, even tighter bounds exist.) Branch and bound algorithms (Horowitz and Sahni, 1978) can use this information to compute bounds such that no solution with a cost greater than $2C$ is examined. Thus, the existence of a ratio bound means that algorithms can select which performance vectors to explore, and this excludes some search behaviors (i.e. performance vectors) that are part of the permutation closure of the objective function.

11.3.5   Evaluating Search Algorithms

From a theoretical point of view, comparative evaluation of search algorithms is a dangerous, if not dubious, enterprise. But the alternative to testing is to give up and say that all algorithms are equal—which means we have no way of recommending one algorithm over another when a search method is required to solve a problem of practical interest. The best we can do is build test functions that we believe capture some aspects of the problems we actually want to solve. But this highlights a critical question. Do benchmarks really test what we want to test? If an algorithm does well on a very simple problem— such as a linear objective function—is that good or bad? Many people have used the ONEMAX test function for testing search algorithms that use a binary representation. The objective function for ONEMAX is to maximize the number of bits set to 1 in a bit string. But should we really believe that an algorithm that does well on ONEMAX generalizes to other problems of practical interest? Theory would suggest extreme caution.

A set of benchmarks, denoted by β where $S = |\beta|$, i is really just a subset of functions. If algorithm K is better than algorithm Z on β, then algorithm Z is equally and identically better on another set of S functions drawn from $P(\beta)$.

So what does it mean to evaluate an algorithm on a set of benchmarks and compare it to another algorithm? Given the NFL theorems, comparison is meaningless unless we prove (which virtually never happens) or assume (an assumption which is rarely made explicit) that the benchmarks used in a comparison are somehow representative of a particular subclass of problems.

Benchmarks are commonly used for testing both optimization and learning algorithms. Often, the legitimacy of a new algorithm is "established" by demonstrating that it finds better solutions than existing algorithms when evaluated on a particular benchmark or collection of benchmarks. Alternatively, the new algorithm may find high-quality solutions faster than existing algorithms for one or more benchmarks.

What are some of the dangers associated with the use of benchmarks? Algorithms can be tuned so that they perform well on specific benchmarks, but fail to exhibit good performance on benchmarks with different characteristics. More importantly, there is no guarantee that algorithms developed and evaluated using synthetic benchmarks will perform well on more realistic problem instances. Furthermore, simple algorithms can often provide excellent performance on more realistic benchmarks (Watson et al., 1999).

While the dangers associated with benchmarks are well-known, most researchers continue to use benchmarks to evaluate their algorithms. This is because researchers have few alternatives. How can one algorithm be compared to another without some form of evaluation? Evaluation requires the use of either synthetic or real-world benchmarks, or at least the use of test problems drawn from problem generators so that algorithms can be compared on sets of problem instances that have similar characteristics. Researchers who develop new algorithms and do not demonstrate their merit through some form of comparative testing can expect their work to be ignored. The compulsion to develop "a new method" has resulted in the literature being full of new algorithms, most of which are never used or analyzed by anyone other than the researchers who created them.

Hooker (1995) discusses the "evils of competitive testing" and points out the difficulty of making fair comparisons of algorithm performance. Implementation details can significantly impact algorithm performance, as can the values selected for various tuning parameters. Some algorithms have been refined for years. Other algorithms have become so specialized that they only work well on specific benchmarks. Hooker argues that the evaluation of algorithms should be performed in a more scientific, hypothesis-driven manner. Barr et al. (1995) suggest guidelines for the experimental evaluation of heuristic methods. Such guidelines are for the most part useful, although rarely followed.

While evaluation is difficult, it is also important. Too many experimental papers (especially conference papers) include no comparative evaluation; researchers may present a hard problem (perhaps newly minted) and then present an algorithm to solve the problem. The question as to whether some other algorithm could have done just as well (or better!) is ignored.

11.4   TRICKS OF THE TRADE

No Free Lunch is a theoretical result about search algorithms. As such there are no specific methods or algorithms that directly follow from NFL. However, several pieces of advice do follow from No Free Lunch.

1 In most practical applications one must trade-off generality and specificity. Using simpler off-the-shelf search methods reduces time effort and cost. Simple but reasonably effective search methods, even when implemented from scratch, are often easier to work with than complex methods. Using custom-designed search methods that only work for one application will usually yield better results: but generally, one must ask how much time and money one wishes to spend and how good does the solution need to be.

2 Exploit problem-specific information when it is simple to do so. Most NP-complete problems, for example, have been studied for years and there are many problem specific methods that yield good near-optimal solutions.

3 For discrete parameter optimization problems, one has a choice of using standard binary encodings, Gray codes or real-valued representations. Gray codes are often better than binary codes when some kind of neighborhood search is used either explicitly (e.g., local search) or implicitly (e.g., via a random bit flip operator). The use of Gray codes versus real-valued is less clear, and depends on other algorithm design choices.

4 Do not assume that a search method that does well on classic benchmarks will work equally well on real-world problems. Sometimes algorithms are overly tuned to do well on benchmarks and in fact do not work well on real-world applications.

11.5 CURRENT AND FUTURE RESEARCH DIRECTIONS

Another area of research is the construction of algorithms that can provably beat random enumeration on specific subsets of problems. Christensen and Oppacher (2001) prove that No Free Lunch does not hold over sets of functions that can be described using polynomials of a single variable of bounded complexity. This also includes Fourier series of bounded complexity. (Also see the paper by English (2000a) about polynomials and No Free Lunch). They define a minimization algorithm called SubMedian-Seeker. The algorithm assumes that the target function, f, is one-dimensional and bijective and that the median value of f is known and denoted by med(f). The actual performance depends on $M(f)$, which measures the number of submedian values of f that have *successors* with supermedian values. They also define M_{crit} as the critical value of $M(f)$ such that when $M(f) < M_{crit}$ SubMedian-Seeker is better than random search. Christensen and Oppacher then prove:

If f is a uniformly sampled polynomial of degree at most k and if $M_{crit} > k/2$ then SubMedian-Seeker beats random search.

The SubMedian-Seeker is not a practical algorithm. The importance of Christensen and Oppacher's work is that it sets the stage for proving there are algorithms that are generally (if perhaps weakly) effective over a very broad class of interesting, nonrandom functions. More recently Whitley et al. (2004) have generalized these concepts to outline conditions which allow local neighborhood bit climbers to display "SubThreshold-Seeker Behavior" and then show that in practice such algorithms spend most of their time exploring the best points in the search space on common benchmarks and are obviously better than random search.

11.6   CONCLUSIONS

As in many other areas of life, extreme reactions are likely to lead to extreme errors. This is also true for No Free Lunch. It is clearly wrong to say "NFL doesn't apply to real world problems, so who cares?" It is also an error to give up on building general purpose search algorithms.

A careful consideration of the No Free Lunch theorems forces us to ask what set of problems we want to solve and how to solve them. More than this, it encourages researchers to consider more formally whether the methods they develop for particular classes of problems actually are better than other algorithms. This may involve proofs about performance behavior. In some ways, we are just starting to ask the right questions. And yet, researchers working in complexity and NP-completeness have long been concerned with algorithm performance for particular classes of problems.

Few researchers have attempted to formalize their assumptions about search problems and search algorithm behavior. But if we fail to do this, then we become trapped in a kind of empirical and experimental treadmill that leads nowhere: algorithms are developed that work on benchmarks, or on particular applications, without any evidence that such methods will work on the next problem we might wish to solve.

ADDITIONAL SOURCES OF INFORMATION

The classic textbook *Introduction to Algorithms* by Cormen et al. (1990) has a very good discussion of NP-completeness and approximate algorithms for some well-studied NP-hard problems.

Joe Culberson's 1998 paper *On the Futility of Blind Search: an Algorithmic View of No Free Lunch* helps to relate complexity theory to No Free Lunch in simple and direct terms.

Tom English has contributed several good papers to the NFL discussion (English, 2000a, 2000b). C. Igel and M. Toussaint have also contributed notable papers. Chris Schumacher's 2000 Ph.D. dissertation, *Fundamental Limitations on Search Algorithms*, deals with various issues related to No Free Lunch.

Recent work by Ingo Wegener and colleagues has focused on showing when particular methods work on particular general classes of problems, (e.g., Storch and Wegener, 2003; Fischer and Wegener, 2004) or showing the inherent complexity of particular problems for black-box optimization (Droste et al., 2003).

References

Barr, R., Golden, B., Kelly, J., Resende, M. and Stewart Jr., W., 1995, Designing and reporting on computational experiments with heuristic methods, *J. Heuristics* **1**:9–32.

Christensen, S. and Oppacher, F., 2001, What can we learn from No Free Lunch?, in: *Proc. Genetic and Evolutionary Computation Conf., GECCO-01*, Morgan Kaufmann, San Mateo, CA, pp. 1219–1226.

Cook, S., 1971, The Complexity of Theorem Proving Procedures, in: *Proc. 3rd ACM Symp. on Theory of Computing,* pp. 151–158.

Cormen, T., Leiserson, C. and Rivest, R., 1990, *Introduction to Algorithms,* McGraw-Hill, New York.

Culberson, J., 1998, On the futility of blind search, *Evol. Comput.* **6**:109–127.

Droste, S., Jansen, T. and Wegener, I., 1999, Perhaps not a free lunch, but at least a free appetizer, in: *Genetic and Evolutionary Computation Conf. (GECCO-99),* Morgan Kaufmann, San Mateo, CA, pp. 833–839.

Droste, S., Jansen, T. and Wegener, I., 2002, Optimization with randomized search heuristics; the ANFL theorem, realistic scenarios and difficult functions, *Theor. Comput. Sci.* **287**:131–144.

Droste, S., Jansen, T., Tinnefeld, K. and Wegener, I., 2003, A New framework for the valuation of algorithms for black-box optimization, in: *Foundations of Genetic Algorithms (FOGA-7),* Morgan Kaufmann, San Mateo, CA.

English, T., 2000a, Practical implications of new results in conservation of optimizer performance, in: *Parallel Problem Solving from Nature,* Vol. 6, Springer, Berlin, pp. 69–78.

English, T., 2000b, Optimization is easy and learning is hard in the typical function, in: *Proc. Congress on Evolutionary Computation (CEC-2000),* pp. 924–931.

Fischer, S. and Wegener, I., 2004, The Ising model on the ring: mutation versus recombination, in: *Genetic and Evolutionary Computation Conf., GECCO-04*, Springer, Berlin, pp. 1113–1124.

Hooker, J. N., 1995, Testing heuristics: we have it all wrong, *J. Heuristics* **1**:33–42.

Horowitz, E. and Sahni, S., 1978, *Fundamentals of Computer Algorithms,* Computer Science Press, Rockville, MD.

Igel, C. and Toussaint, M., 2003, On classes of functions for which No Free Lunch results hold, *Inform. Process. Lett.***86**:317–321.

Igel, C. and Toussaint, M., 2004, A no-free-lunch theorem for non-uniform distributions of target functions, *J. Math. Model. Algor.*, in print.

Kauffman, S. A., Adaptation on Rugged Fitness Landscapes, 1989, in: *Lectures in the Science of Complexity,* D. L. Stein, ed., Addison-Wesley, New York, pp. 527–618.

Radcliffe, N. J. and Surry, P. D., 1995, Fundamental limitations on search algorithms: Evolutionary computing in perspective, in: *Lecture Notes in Computer Science,* Vol. 1000, J. van Leeuwen, ed., Springer, Berlin.

Rana, S. and Whitley, D., 1997, Representations, search and local optima, in: *Proc. 14th National Conf. on Artificial Intelligence (AAAI-97),* MIT Press, Cambridge, MA, pp. 497–502.

Rana, S. and Whitley, D., 1998, Search, representation and counting optima, in: *Proc. IMA Workshop on Evolutionary Algorithms,* L. Davis, K. De Jong, M. Vose and D. Whitley, eds, Springer, Berlin.

Rawlins, G., ed., 1991, *Foundations of Genetic Algorithms,* Morgan Kaufmann, San Mateo, CA.

Schumacher, C., 2000, Fundamental limitations of search, *Ph.D. Thesis,* University of Tennessee, Department of Computer Sciences, Knoxville, TN.

Schumacher, C., Vose, M. and Whitley, D., 2001, The no free lunch and problem description length, in: *Genetic and Evolutionary Computation Conf. (GECCO-01),* Morgan Kaufmann, San Mateo, CA, pp. 565–570.

Storch, T. and Wegener, I., 2003, Real Royal Road Functions for Constant Population Size, in: *Genetic and Evolutionary Computation Conf. (GECCO-03),* Springer, Berlin, pp. 1406–1417.

Sudcamp, T., 1997, *Languages and Machines,* 2nd edn, Addison-Wesley, New York.

Tovey, C. A., 1985, Hill climbing and multiple local optima, *SIAM J. Algebr. Discr. Methods* **6**:384–393.

Watson, J. P., Barbulescu, L., Whitley, D. and Howe, A., 1999, Algorithm performance and problem structure for flow-shop scheduling, in: *Proc. 16th National Conf. on Artificial Intelligence.*

Whitley, D., 1999, A free lunch proof for gray versus binary encodings, in: *Genetic and Evolutionary Computation Conf. (GECCO-99),* Morgan Kaufmann, San Mateo, CA, pp. 726–733.

Whitley, D., 2000, Functions as permutations: regarding no free lunch, Walsh analysis and summary statistics, in: *Parallel Problem Solving from Nature,* Vol. 6, Schoenauer, Deb, Rudolph, Lutton, Merelo, and Schwefel, eds, Springer, Berlin, pp. 169–178.

Whitley, D., Rana, S. and Heckendorn, R., 1997, Representation issues in neighborhood search and evolutionary algorithms, in: *Genetic Algorithms and Evolution Strategies in Engineering and Computer Science,* C. Poloni, D. Quagliarella, J. Periaux, and G. Winter, eds, Wiley, New York, pp. 39–57.

Whitley, D., Rowe, J. and Bush, K., 2004, Subthreshold seeking behavior and robust local search, in: *Genetic and Evolutionary Computation Conf. (GECCO-04)*, Springer, Berlin, pp. 282–293.

Wolpert, D. H. and Macready, W. G., 1995, No free lunch theorems for search, *Technical Report* SFI-TR-95-02-010, Santa Fe Institute, NM.

Wolpert, D. H. and Macready, W. G., 1997, No free lunch theorems for optimization, *IEEE Trans. Evol. Comput.* **4**:67–82.

Chapter 12

MACHINE LEARNING

Xin Yao
School of Computer Science
The University of Birmingham, UK

Yong Liu
The University of Aizu
Tsuruga, Ikki-machi, Aizu-Wakamatsu
Fukushima, Japan

12.1   INTRODUCTION

Machine learning is a very active sub-field of artificial intelligence concerned with the development of computational models of learning. Machine learning is inspired by the work in several disciplines: cognitive sciences, computer science, statistics, computational complexity, information theory, control theory, philosophy, and biology. Simply speaking, machine learning is learning by machine. From a computational point of view, machine learning refers to the ability of a machine to improve its performance based on previous results. From a biological point of view, machine learning is the study of how to create computers that will learn from experience and modify their activity based on that learning as opposed to traditional computers whose activity will not change unless the programmer explicitly changes it.

12.1.1   Learning Models

A machine learning model has two key components: learning element and performance element, as shown in Figure 12.1. The environment supplies some information to the learning element. The learning element then uses the information to modify the performance element so that it can make better decisions. The performance element selects actions to perform its task.

A large variety of learning elements have been proposed by researchers in machine learning field. Based on the representation, there are symbolic and subsymbolic learning. Based on the algorithms, there are many different

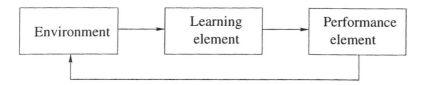

Figure 12.1. A machine learning model.

types of machine learning, such as decision tree, inductive logic programming, Bayesian learning, artificial neural networks, evolutionary learning, and reinforcement learning. Based on the feedback available, there are three different types of machine learning: supervised, unsupervised, and reinforcement learning.

The problem of supervised learning involves learning a function from a set of input–output examples. The general supervised learning model consists of two components:

1 A probability space $(\mathcal{E}, Pr)$ in which we associate each elementary event with two random variables, the input pattern x and the desired output y, where $\mathcal{E}$ is called the event set, Pr is called the probability distribution, $x \in R^p$, y is a scalar. The assumption that the output y is a scalar has been made merely to simplify exposition of ideas without loss of generality.

2 A learning machine, which is capable of implementing a set of functions $F(x, w)$, $w \in W$, where W is a set of, in general, real-valued parameters.

The purpose of supervised learning is to find the function $F(x, w)$ so that the expected squared error

$$R(w) = E\left[(F(x, w) - y)^2\right] \tag{12.1}$$

is minimized, where E represents the expectation value over the probability space $(\mathcal{E}, Pr)$.

In unsupervised learning, there is no specific output supplied. In the context of pattern classification, unsupervised learning learns to discover the statistical regularities of the patterns in the input, form internal representations for encoding features of the input, and thereby to create new classes automatically. In reinforcement learning, rather than being told what to do by a teacher, the learning of an input–output mapping is performed through continued interaction with the environment in order to minimize a scalar index of performance.

The environment can be either fully observable or partially observable. In the first case, the machine can observe the effects of its action and hence can use supervised learning methods to learn to predict them. In the second case, the immediate effects might be invisible so that reinforcement learning or unsupervised learning should be adopted.

12.1.2 Learning Tasks and Issues in Machine Learning

Machine learning can be applied to tasks in many domains. This section presents some important learning tasks and issues in machine learning.

Classification A classification task in machine learning is to take each instance and assign it to a particular class. For example, in an optical character recognition task, the machine is required to scan an image of a character and output its classification. In English language recognition, the task involves learning the classification of the digits $0 \ldots 9$ and the characters $A \ldots Z$.

Regression, Interpolation, and Density Estimation In regression, it is to learn some functional description of data in order to predict values for new input. An example of learning a regression function is predicting the future value of a share index in stock market. In interpolation, the function for certain ranges of input is known. The task is to decide the function for the intermediate ranges of input. In density estimation, the task is to estimate the density or probability that a member of a certain category will be found to have particular features.

Learning Sequence of Actions In robot learning and chess play learning, the task is to find the best strategies that can choose the optimal actions. In an example of robot navigation, a robot is assigned a task to track a colored object within a limited number of actions while avoiding obstacles and walls in an environment. There are obstacles of different shapes in the environment enclosed by the walls. To perform its task, the robot must learn the basic behavior of obstacle avoidance and moving to the target. It must also learn to co-ordinate the behavior of obstacle avoidance and the behavior of moving to the target to avoid becoming stuck due to repetition of an identical sensor–motion sequence. In chess playing, the machine must decide an action based on the state of board to move a piece in which the action will maximize its chance of winning the game.

Data Mining The problem of data mining is of searching for interesting patterns and important regularities in large databases. Many learning methods have been developed for determining general descriptions of concepts from examples in the form of relational data tables. Machine learning plays an important role in the discovery and presentation of potentially useful information from data in a form which is easily comprehensible to humans.

Issues in Machine Learning   There are many issues that need to be solved in machine learning. For example, which learning algorithm performs best for a particular learning task and representation? How many training samples are sufficient? How fast can the learning algorithms converge? When and how can prior knowledge be used in the learning process? Can a machine learn in real-time, in addition to offline learning? How do we choose from among multiple learning models that are all consistent with the data? For all of these, generalization is a key issue for any learning system. There are often two phases in the design of a learning system. The first phase is the learning. The second phase is the generalization test. The term generalization is borrowed from psychology. In neural network learning, a model is said to generalize well when it can produce correct input–output mapping for unseen test data that are not used in the learning phase.

The reminder of this chapter is organized as follows. Section 12.2 introduces a number of learning algorithms in order to give a breadth of coverage of machine learning. Section 12.3 addresses evolution and learning. Three levels of evolution can be introduced in neural network learning: the evolution of weight, the evolution of architectures, and the evolution of learning rules. Section 12.4 points out some promising areas in machine learning. Section 12.5 provides a guideline for implementing machine learning algorithms. Section 12.6 concludes with a summary of the chapter and a few remarks.

12.2   OVERVIEW OF LEARNING ALGORITHMS

This section will explore the basic ideas and the principles of a number of learning algorithms that are used for real-world applications.

12.2.1   Learning Decision Trees

The task of inductive learning is to find a function h that approximates f by a given collection of examples of f. The function h is called a hypothesis. An example is a pair $(x, f(x))$, where x is the input, and $f(x)$ is the output of the function applied to x. In decision tree learning, hypotheses are represented by decision trees.

A decision tree is a diagram representing a classification system or a predictive system. The structure of the system is a tree generated based on a sequence of simple questions. The answers to these questions trace a path down the tree. As a result, a decision tree is a collection of hierarchical rules that segment the data into groups, where a decision is made for each group. The hierarchy is called a tree, and each segment is called a node. The original segment that contains the entire data set is referred to as the root node of the tree. A node with all of its successors forms a branch of the node. The terminal nodes are called leaves that return a decision, i.e. the predicted output value for the input. The output value can be either discrete or continuous. A classification tree learns a discrete-valued function, while a regression tree learns a continuous function. Most decision learning algorithms are variations on a core algorithm that employs a top-down, greedy search through the space of possible decision trees.

A very effective decision learning algorithm, called the ID3 algorithm, was developed by Quinlan (1986). In ID3, classification trees are built by starting with the set of examples and an empty tree. An attribute test is chosen for the root of the tree, and examples are partitioned into disjoint subsets depending on the outcome of the test. The learning is then applied recursively to each of these disjoint subsets. The learning process stops when all the examples within a subset belong to the same class. At this learning stage, a leaf node is created and labeled with the class.

The method to choose the attribute test is designed to minimize the depth of the final tree. The idea is to select the attribute that can lead to an exact classification of examples as far as possible. In ID3, a statistical property, called information gain, was introduced to measure how well a given attribute separates the examples according to their target classification.

For decision tree learning, a learned classification tree has to predict what the correct classification is for a given example. Given a training set S, containing p positive examples and n negative examples, the entropy of S to this Boolean classification is

$$E(S) = -\frac{p}{p+n} \log_2 \frac{p}{p+n} - \frac{n}{p+n} \log_2 \frac{n}{p+n} \qquad (12.2)$$

In information theory, the entropy of S gives an estimate of the information contained in a correct answer before any of the attributes have been tested. Information theory measures information content in bits. After a test on a single attribute A, attribute A divides the training set S into subsets S_i, $i = 1, \ldots, v$, where A can have v distinct values. The information gain $G(S, A)$ of an attribute A, relative to a training set S, is defined as

$$G(S, A) = E(S) - \sum_{i=1}^{v} \frac{p_i + n_i}{p+n} E(S_i) \qquad (12.3)$$

where each subset S_i has p_i positive examples and n_i negative examples. The second term in (12.3) is the expected value of entropy after S is partitioned using attribute A. The expected entropy is the sum of entropies of each subset S_i, weighted by the fraction of examples in S_i. $G(S, A)$ is therefore the expected reduction in entropy caused by knowing the value of attribute A.

ID3 provides a simple and effective approach to decision tree learning. However, for the real-world applications, the algorithm needs to cope with problems such as noisy data set, missing attribute values, and attributes with continues values. How to deal with those problems was studied in ID3's successor C4.5 (Quinlan, 1993).

The first decision tree learning system, called the Elementary Perceiver and Memorizer was proposed by Feigenbaum (1961). It was studied as a cognitive-simulation model of human concept learning. The concept learning system developed by Hunt et al. (1966) used a heuristic look-ahead method to grow decision trees. ID3 (Quinlan, 1986) introduced the information content as a heuristic search. The classification and regression tree system is a widely used statistical procedure for producing classification and regression (Breiman et al., 1984). Many practical issues of decision tree induction can be found in C4.5, a decision tree learning package by Quinlan (1993).

The advantages of decision tree learning are its comprehensibility, fast classification, and mature technology. However, by using only one attribute at each internal node, decision tree learning can construct monothetic trees, which are limited to axis-parallel partitions of the instance space, rather than polythetic trees. Polythetic trees can use more than two attribute at each internal node, but are expensive to induce. The next section will introduce inductive logic programming that combines inductive learning with the power of first-order representations.

12.2.2   Inductive Logic Programming

Inductive logic programming is a combination of knowledge-based inductive learning and logic programming (Russell and Norvig, 2002). General knowledge-based inductive learning uses the kind of algorithm that satisfies the entailment constraint

$$\text{Background} \wedge \text{Hypothesis} \wedge \text{Descriptions} \models \text{Classifications} \qquad (12.4)$$

where *Classifications* denote the conjunction of all the example classifications. Given the *Background* knowledge and examples described by *Descriptions* and *Classifications*, the induction problem of knowledge-based inductive learning involves solving the entailment constraint (12.4) for the unknown *Hypothesis*.

In order to see how the background knowledge can be combined with the new hypothesis to explain examples, consider a problem of learning family relationships from examples in an extended family tree given in Figure 12.2 (Russell and Norvig, 2002).

The descriptions will be in the terms of Mother, Father, and Married relations and Male and Female properties, such as Father(Philip, Charles), Mother(Mum, Margaret), Married(Diana, Charles), Male(Philip), Female(Beatrice). Classifications depend on the target concept being learned. For learning the target concept of Grandfather, the complete set of Classifica-

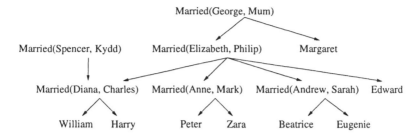

Figure 12.2. A typical family tree.

tions contains $20 \times 20 = 400$ conjuncts of the form

Grandparent(Mum, Charles)   Grandparent(Elizabeth, Beatrice)   ...

¬Grandparent(Mum, Harry)   ¬Grandparent(Spencer, Peter)   ...

Without the background knowledge, inductive learning can find the following possible *Hypothesis*:

$$
\begin{aligned}
\text{Grandparent}(x, y) \quad &\Leftrightarrow \quad [\exists z \ \text{Mother}(x, z) \wedge \text{Mother}(z, y)] \\
&\vee \quad [\exists z \ \text{Mother}(x, z) \wedge \text{Father}(z, y)] \\
&\vee \quad [\exists z \ \text{Father}(x, z) \wedge \text{Mother}(z, y)] \\
&\vee \quad [\exists z \ \text{Father}(x, z) \wedge \text{Father}(z, y)] \quad (12.5)
\end{aligned}
$$

With the help of the background knowledge represented by the sentence

$$\text{Parent}(x, y) \Leftrightarrow [\text{Mother}(x, y) \vee \text{Father}(x, y)] \quad (12.6)$$

Hypothesis could be simply defined by

$$\text{Grandparent}(x, y) \Leftrightarrow [\exists z \ \text{Parent}(x, z) \wedge \text{Parent}(z, y)] \quad (12.7)$$

By using background knowledge, we can reduce the size of hypotheses greatly.

There are two basic approaches to inductive logic programming: top-down learning of refining a very general rule and bottom-up learning of inverting the deductive process. A top-down approach will typically begin with a general clause and search the clause by adding literals so that only positive examples are entailed. First-order inductive learning (Quinlan, 1990) is such a top-down induction algorithm.

Suppose the task is to learn a definition of Grandfather(x, y) predicate in the family tree shown in Figure 12.2. Examples can be divided into positive and negative ones as in decision-tree learning. Twelve positive examples are

⟨George, Charles⟩, ⟨George, Anne⟩, ⟨George, Andrew⟩, ...

and 388 negative examples are

⟨George, Spencer⟩, ⟨George, Kydd⟩, ⟨George, Elizabeth⟩, ...

First-order inductive learning constructs a set of clauses that must classify the positive examples while ruling out the negative examples. First-order inductive learning starts with the initial clause with Grandfather(x, y) as the head, and an empty body

$$\Rightarrow \text{Grandfather}(x, y) \qquad (12.8)$$

All examples are classified as positive by this clause. To specialize it, first-order inductive learning adds literals one at a time to the clause body. Look at two clauses constructed by such addition:

$$\text{Parent}(x, z) \;\Rightarrow\; \text{Grandfather}(x, y) \qquad (12.9)$$
$$\text{Father}(x, z) \;\Rightarrow\; \text{Grandfather}(x, y) \qquad (12.10)$$

Although both clauses agree with all of the 12 positive examples, the first allows both fathers and mothers to be grandfathers and makes larger misclassification on negative examples. The second clause is chosen to be further specialized. By adding the single literal Parent(z, y), first-order inductive learning can find

$$\text{Father}(x, z) \wedge \text{Parent}(z, y) \Rightarrow \text{Grandfather}(x, y) \qquad (12.11)$$

which successfully classifies all the examples. This example gives a simple explanation how first-order inductive learning works. In real applications, first-order inductive learning generally has to search through a large number of unsuccessful clauses before finding the correct one.

Whereas first-order inductive learning (Quinlan, 1990) is a top-down approach, Cigol (logic, spelled backwards; Muggleton and Buntine, 1988), developed for inductive logic programming, worked bottom-up. Cigol incorporated a slightly incomplete version of inverse resolution and was capable of generating new predicates. A hybrid techniques (top-down and bottom-up) approach was chosen in Progol (Muggleton, 1995) with inverse entailment that has been applied to a number of practical problems. A large collection of papers on inductive logic programming can be found in Lavrač and Džeroski (1994).

Inductive logic programming provides a practical approach to the general knowledge-based inductive learning problem. The strengths of inductive logic programming lie in its firm theoretical foundations, richer hypothesis representation language, and explicit use of background knowledge. The limitations of inductive logic programming are its weak numeric representations and large search spaces.

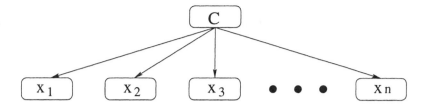

Figure 12.3. The naive Bayes model.

12.2.3   Bayesian Learning

In practice, there are cases that more than one hypothesis satisfy a given task. Because it is not certain how those hypotheses perform on unseen data, it is hard to choose the best hypothesis. Bayesian learning gives a probabilistic framework for justification. By calculating explicit probabilities for a hypothesis, Bayesian learning provides a useful perspective for understanding many learning algorithms that do not explicitly manipulate probabilities.

Let D represent all the data, and H the set of all the hypotheses h_i. The probability of each hypothesis with observed d can be calculated by Bayes' rule:

$$P(h_i \mid d) = \alpha P(d \mid h_i) P(h_i) \qquad (12.12)$$

where $P(h_i)$ is the prior probability, $P(d \mid h_i)$ denotes the probability of observed d given h_i, and $P(h_i \mid d)$ is the posterior probability of h_i.

A practical Bayesian learning used in machine learning is the naive Bayes model shown in Figure 12.3, where each instance x is described by a conjunction of attribute values $\langle x_1, x_2, \ldots, x_n \rangle$. In this model, the class variable C is the root, and the attribute values x are leaves.

According to (12.12), the probability of each class from a set of S is given by

$$P(C \mid x_1, x_2, \ldots, x_n) = \alpha P(x_1, x_2, \ldots, x_n \mid C) P(C) \qquad (12.13)$$

In the naive Bayes model, a simplified assumption is made that the attributes are conditionally independent of each other given the class. That is, the probability of the observed conjunction $x_1, x_2, \ldots, x_n$ is just the product of probabilities for the individual attributes:

$$P(x_1, x_2, \ldots, x_n \mid C) = \prod_i P(x_i \mid C) \qquad (12.14)$$

From (12.13) and (12.14), the naive Bayes model makes the prediction by choosing the most likely class:

$$C_{NB} = \operatorname{argmax}_{C \in S} P(C) \prod_i P(x_i \mid C) \qquad (12.15)$$

Table 12.1. An example for the naive Bayes model.

Diagnosis	Well	Cold	Allergy
$P(C)$	0.9	0.05	0.05
$P(\text{sneeze} \mid C)$	0.1	0.9	0.9
$P(\text{cough} \mid C)$	0.1	0.8	0.7
$P(\text{fever} \mid C)$	0.01	0.7	0.4

where C_{NB} denotes output class by the naive Bayes model.

Consider a medical diagnosis problem with three possible diagnoses (well, cold, allergy) based on three symptoms (sneeze, cough, fever). In this example, there are three attributes in which x_1 can be *sneeze* or *not sneeze*, x_2 *cough* or *not cough*, and x_3 *fever* or *not fever*, and three classes including *well*, *cold* and *allergy*. The probabilities for the three attributes and three prior class probabilities are given in Table 12.1.

Given a new $x = \langle\text{sneeze}, \text{cough}, \text{not fever}\rangle$, which class of diagnosis is most likely? First, posterior probability $P(\text{well} \mid \text{sneeze}, \text{cough}, \text{not fever})$ of *well*, *cold* and *allergy* is calculated by the product of $P(\text{sneeze} \mid \text{well})$, $P(\text{cough} \mid \text{well})$, $P(\text{not fever} \mid \text{well})$, and $P(\text{well})$:

$$P(\text{well} \mid \text{sneeze}, \text{cough}, \text{not fever}) = 0.1 \times 0.1 \times (1 - 0.01) \times 0.9$$
$$= 0.008\,91 \qquad (12.16)$$

Similarly we can obtain the posterior probability of *cold*:

$$P(\text{cold} \mid \text{sneeze}, \text{cough}, \text{not fever}) = 0.216 \qquad (12.17)$$

and the posterior probability of *allergy*:

$$P(\text{allergy} \mid \text{sneeze}, \text{cough}, \text{not fever}) = 0.378 \qquad (12.18)$$

Finally, we compare three posterior probabilities and generate output class *allergy* because the probability of allergy for the data $x = \langle\text{sneeze}, \text{cough}, \text{not fever}\rangle$ is the largest one.

The naive Bayes model has been compared with C4.5 on 28 benchmark tasks (Domingos and Pazzini, 1996). The results show that the naive Bayes model performs surprisingly well in a wide range of applications. Except for a few domains where the naive Bayes model performs poorly, it is comparable or better than C4.5.

This section just uses the naive Bayes model to introduce the idea of Bayesian learning. Heckerman (1998) gives an excellent introduction on general learning with Bayesian networks (Heckerman, 1998). Bayesian learning has been successfully applied in pattern recognition and information retrieval. Algorithms based on Bayesian learning won the 1997 and 2001 KDD Cup data mining competitions (Elkan, 1997; Cheng et al., 2002). Experimental comparisons between Bayesian learning, decision tree learning, and other algorithms have been made for a wide range of applications (Michie et al., 1994).

12.2.4   Reinforcement Learning

Reinforcement learning involves learning how to map situations to actions so as to maximize a numerical reward signal (Sutton and Barto, 1998). Unlike supervised learning, the machine is not told which actions to take but has to discover which actions yield the most reward by trying them. In the most practical cases, actions may affect both the immediate reward and the next situation throughout all subsequent rewards. Trial-and-error search and delayed reward are the two most important unique characteristics of reinforcement learning.

A central and novel idea of reinforcement learning is called temporal-difference learning (Sutton and Barto, 1998). Temporal-difference learning is a combination of Monte Carlo and dynamic programming ideas. Like Monte Carlo methods, temporal-difference learning methods can learn directly from the raw experience without a model of the environment's dynamics. Like dynamic programming methods, temporal-difference learning methods update estimates based in part on other learned estimates, without waiting for a final outcome. Temporal-difference learning works because it is possible to make local improvements. At every point in the state space, the Markov property allows actions to be chosen based only on knowledge about the current state and the states reachable by taking the actions available at that state.

Temporal-difference learning methods fall into two classes: on-policy and off-policy (Sutton and Barto, 1998). One of the most important breakthroughs in reinforcement learning was the development of an off-policy temporal-difference learning control algorithm known as Q-learning. The learned action-value function, $Q(s, a)$, directly approximates the optimal action-value function, independent of the policy being followed. The major steps of Q-learning are given as follows (Sutton and Barto, 1998):

1 Initialize $Q(s, a)$ values arbitrarily.

2 Initialize the environment.

3 Choose action a using the policy derived from $Q(s, a)$ (e.g. ϵ-greedy).

4 Take action a; Observe reward r and the next state s'.

5 Update the $Q(s, a)$ as follows:

$$Q(s, a) \leftarrow Q(s, a) + \alpha \left[r + \gamma \max_{a'} Q(s', a') - Q(s, a) \right] \quad (12.19)$$

6 Let $s \leftarrow s'$. Go to the next step if the state s is a terminal state. Otherwise, go to Step 3.

7 Repeat Steps 2–6 for a certain number of episodes.

The Sarsa learning algorithm is an on-policy temporal-difference learning method in which the action-value function Q is updated after every transition from a nonterminal state. The major steps of Sarsa learning are as follows (Sutton and Barto, 1998):

1 Initialize $Q(s, a)$ values arbitrarily.

2 Initialize the environment.

3 Choose action a using the policy derived from $Q(s, a)$ (e.g. ϵ-greedy).

4 Take action a; Observe reward r and the next state s'; Choose the next action a' using the policy derived from Q (e.g. ϵ-greedy).

5 Update the $Q(s, a)$ as follows:

$$Q(s, a) \leftarrow Q(s, a) + \alpha \left[r + \gamma Q(s', a') - Q(s, a) \right] \qquad (12.20)$$

6 Let $s \leftarrow s'$ and $a \leftarrow a'$. Go to the next step if the state s is a terminal state. Otherwise, go to Step 3.

7 Repeat Steps 2–6 for a certain number of episodes.

Sutton and Barto (1998) compared Sarsa and Q-learning. The results showed that the on-line performance of Q-learning was worse than that of Sarsa learning.

The strengths of reinforcement learning come from its firm theoretical foundation, its ability to solve broad tasks, and its easy usage of background knowledge. Work in reinforcement learning dates back to the earliest days of machine learning when Turing (1950) proposed the reinforcement learning approach, and Samuel (1959) developed his famous checkers learning program that contained most of the modern ideas in reinforcement learning, including temporal differencing and function approximation. Three threads have contributed to the modern field of reinforcement learning. The first thread is learning by trial and error, which was rooted in the psychology of animal learning, and led to the popularity of reinforcement learning in the early 1980s. The second thread arose from the problem of optimal control and its solution using value functions and dynamic programming. The third thread concerns temporal-difference methods. The survey by Kaelbling et al. (1996) provides a good introduction to the literature. The text *Reinforcement Learning: An Introduction* by Sutton and Barto (1998), two of the field's pioneers, shows architectures and algorithms of reinforcement learning in the context of learning, planning, and acting.

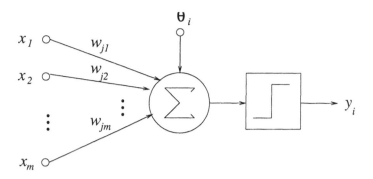

Figure 12.4. Nonlinear model of a neuron.

12.2.5 Neural Networks

Artificial neural networks, commonly referred to as neural networks, try to simulate biological brains. However, neural networks are greatly simplified in comparison with biological brains. A neural network is a parallel computational system consisting of many processing elements connected with each other in a certain way in order to perform a task. Neural networks have gained popularity because they are adaptive, robust, fault tolerant, noise tolerant, and massively parallel.

Of the many tasks that neural networks perform, the most important one is learning. A neural network can improve its performance through learning. Perceptron learning is one of the earliest learning processes developed for neural networks (Rosenblatt, 1962). Perceptrons are often used to refer to feed-forward neural networks consisting of McCulloch–Pitts (MP) neurons (McCulloch and Pitts, 1943):

$$y_i = \text{sgn}\left(\sum_j w_{ij}x_j - \theta_i\right) \tag{12.21}$$

where w_{ij} are called *weights* (synapses) and θ the *threshold*. The x_j and y_i are inputs and the output. The signum function $\text{sgn}(x)$ is defined as follows:

$$\text{sgn}(x) = \begin{cases} 1, & \text{if } x \geq 0 \\ 0, & \text{otherwise} \end{cases} \tag{12.22}$$

This is also known as the threshold function or Heaviside function, described in Figure 12.4.

Given each example to a perceptron that has one layer of neurons, perceptron learning adjusts the weights as follows until the weights converge (i.e. $\Delta w_j(t) = 0$):

$$w_j(t+1) = w_j(t) + \Delta w_j(t) \tag{12.23}$$

where

$$\Delta w_j(t) = \eta(y^p - O^p)x_j^p \tag{12.24}$$

with η the learning rate, x_j^p the jth input of the pth example, y^p the target (desired) output of the pth example, and O^p the actual output of the pth example

$$O^p = \text{sgn}\left(\sum_j w_j x_j^p - \theta\right) \tag{12.25}$$

The convergence theorem of perceptron learning states that if there exist a set of weights for a perceptron which solves a problem correctly, the perceptron learning rule will find them in a finite number of iterations (Rosenblatt, 1962). If a problem is linearly separable, then the perceptron learning rule will find a set of weights in a finite number of iterations that solves the problem correctly. A pair of linearly separable patterns means that the patterns to be classified must be sufficiently separated from each other to ensure that the decision surface consists of a hyperplane.

The perceptron learning rule, $\Delta w_j(t) = \eta(y^p - O^p)x_j^p$, is related to the Hebbian learning rule (Hebb, 1949). Hebb's postulate of learning states

> When an axon of cell A is near enough to excite a cell B and repeatedly or persistently takes part in firing it, some growth process or metabolic changes take place in one or both cells such that A's efficiency as one of the cells firing B, is increased.

In other words, if two neurons on either side of a synapse (connection) are activated simultaneously (i.e. synchronously), then the strength of that synapse is selectively increased. If two neurons on either side of a synapse are activated asynchronously, then that synapse is selectively weakened or eliminated.

It is clear from perceptron learning that the algorithm tries to minimize the difference between the actual and desired output. We can define an error function to represent such a difference:

$$E(\boldsymbol{w}) = \frac{1}{2}\sum_p (y^p - O^p)^2 \tag{12.26}$$

or

$$E(\boldsymbol{w}) = \frac{1}{2N}\sum_p (y^p - O^p)^2 \tag{12.27}$$

where N is the number of patterns. The second error function above is called the mean square error. What learning does is to minimize this error by adjusting weights $\boldsymbol{w}$.

One advantage of introducing the error function is that it can be used for any type of transfer functions, discrete or continuous. The aim of learning is to adjust w such that the error E is minimized, i.e. the network output is as close to the desired output as possible. There exist mathematical tools and algorithms which can tell us how to minimize the error function E, such as the gradient descent algorithm which is based on partial derivatives. Given a set of training examples, $\{(x_1^p, \ldots, x_m^p; y_1^p, \ldots, y_n^p)\}_p$, the gradient descent learning algorithm can be summarized as follows:

1 Construct a neural network with m inputs and n outputs.

2 Select learning rate η and the gain parameter a.

3 Generate initial weights at random in a small range, e.g. $[-0.5, 0.5]$. Note that thresholds are regarded as weights here.

4 While the neural network has not converged do: For each training example p,

   (a) Compute O_i^p. $(O_i^p = f(u_i))$

   (b) Compute $\delta_i^p = (y_i^p - O_i^p) f'(u_i)$, where $f'(u_i) = af(u_i)(1 - f(u_i))$ if the transfer function is

$$f(u_i) = \frac{1}{1 + \exp(-au_i)} \qquad (12.28)$$

   where a is a parameter determined by the user.

   (c) Compute $\Delta w_{ij} = \eta \delta_i^p x_i^p$.

   (d) Update w_{ij} for all i, j. (All weights will be updated.)

There are two modes of the gradient descent learning algorithm. One is the sequential mode of training that is also known as on-line, pattern, or stochastic mode. In this mode, weights are updated after the presentation of each example. The other is the batch mode of training in which weights are updated only after the complete presentation of all examples in the training set, i.e. only after each epoch.

The idea of the gradient descent learning algorithm for a single layer neural network can be generalized to find weights for multilayer neural networks. Multilayer feedforward neural networks can solve nonlinear problems. In fact, there are mathematical theorems which show that multilayer feedforward neural networks can approximate any input–output mapping. The backpropagation algorithm can be used to train multilayer feedforward neural networks (Rumelhart et al., 1986). Its forward pass propagates the activation values from input to output. Its backward pass propagates the errors from output to input. Backpropagation is still a gradient descent algorithm. It uses gradient information to determine how the weights should be adjusted so that the output error can be reduced. Mathematically, backpropagation uses the chain rule to figure out how to change weights in order to minimize the error.

Consider a network with M layers $m = 1, 2, \ldots, M$ and use V_i^m to represent the output of the ith unit in the mth layer. $V_i^0 = x_i$ is the ith input. Backpropagation can be described as follows:

1 Initialize the weights to small random values.

2 Choose a pattern and apply it to the input layer so that $V_i^0 = x_i^p$.

3 Propagate the signal forwards through the network using

$$V_i^m = f(u_i^m) = f\left(\sum_j w_{ij}^m V_j^{m-1}\right) \tag{12.29}$$

for each i and m until the final outputs V_i^M have all been calculated.

4 Compute the deltas for the output layer

$$\delta_i^M = f'(u_i^M)(y_i^p - V_i^M) \tag{12.30}$$

5 Compute the deltas for the preceding layers by propagating errors backwards

$$\delta_i^{m-1} = f'(u_i^{m-1})\sum_j w_{ij}^m \delta_j^m \tag{12.31}$$

for $m = M, M - 1, \ldots, 2$.

6 Update the weights according to

$$w_{ij}^{new} = w_{ij}^{old} + \Delta w_{ij} \tag{12.32}$$

where

$$\Delta w_{ij} = \eta \delta_i^m V_j^{m-1} \tag{12.33}$$

7 Goto step 2 and repeat for the next pattern. The algorithm stops when no weight changes were made for a complete epoch or the maximum number of iterations has been reached.

The study of neural networks began with the MP neuron models proposed by McCulloch and Pitts (1943). The Hebbian learning rule was introduced by Hebb (1949). Rosenblatt (1962) proposed perceptrons and proved the perceptron convergence theory. The book of Minsky and Papert (1969) showed the limitation of single-layer perceptrons, but then the field of neural networks was almost deserted in the 1970s. Hopfield published a series of papers on Hopfield networks that used the idea of an energy function to formulate a new way of understanding the computation performed by recurrent networks with symmetric synaptic connections (Hopfield, 1982; Hopfield and Tank, 1985). The two-volume "bible", *Parallel Distributed Processing: Explorations in the Microstructures of Cognition*, edited by Rumelhart and McClelland (1986) attracted a great deal of attention, after which the field of neural networks really took off.

Neural networks have been applied to solve a wide range of problems such as pattern recognition and classification, time-series prediction, function approximation, system identification, and control. Neural network applications often include two phases. The first phase is learning. The task performed by a neural network is often represented as a set of examples. The neural network is expected to learn more general concepts from these examples. The steps involved include:

1 Select a neural network architecture, where the number of input and output nodes are determined by the task. Hidden nodes and network connectivity need to be designed mostly by trial and error.

2 Train the network using a suitable training algorithm.

The second phase is the generalization test. After the neural network is trained, it is tested with new examples never encountered before to see how well it generalizes.

12.2.6 Evolutionary Learning

Evolutionary learning includes many topics, such as learning classifier systems, evolutionary neural networks, evolutionary fuzzy logic systems, co-evolutionary learning, self-adaptive systems, etc. The primary goal of evolutionary learning is the same as that of machine learning in general. Evolutionary learning can be regarded as the evolutionary computation approach to machine learning. It has been used in the framework of supervised learning, reinforcement learning and unsupervised learning, although it appears to be most promising as a reinforcement learning method.

Evolutionary computation encompasses major branches, i.e. evolution strategies, evolutionary programming, genetic algorithms (see Chapter 4) and genetic programming (see Chapter 5), due largely to historical reasons. At the philosophical level, they differ mainly in the level at which they simulate evolution. At the algorithmic level, they differ mainly in their representations of potential solutions and the operators used to modify the solutions. From a computational point of view, representation and search are two key issues.

Evolution strategies were first proposed by Rechenberg and Schwefel in the mid-1960s for numerical optimization. Real-valued vectors are used to represent individuals. Evolution strategies use both recombination and self-adaptive mutations. The original evolution strategy did not use populations. A population was introduced into evolution strategies later (Schwefel, 1981, 1995).

Evolutionary programming was first proposed by Fogel et al. (1966) for simulating intelligence. Finite-state machines were used to represent individuals, although real-valued vectors have always been used in numerical optimization. Search operators (mutations only) are applied to the phenotypic representation of individuals. There is no recombination in evolutionary programming. Tournament selection is often used in evolutionary programming.

Genetic algorithms and genetic programming are introduced in Chapters 3 and 4 of this book, respectively. Although genetic algorithms, evolutionary programming, evolution strategies, and genetic programming are different, they are all different variants of population-based generate-and-test algorithms:

Generate: Mutate and/or recombine individuals in a population.

Test: Select the next generation from the parents and offspring.

They share more similarities than differences. A better and more general term to use is evolutionary algorithms. Evolutionary algorithms have two prominent features which distinguish them from other search algorithms. First, they are all population-based. Secondly, there is communication and information exchange between individuals in a population. Such communication and information exchange is the result of selection and/or recombination in evolutionary algorithms. A general framework of evolutionary algorithms can be summarized as follows:

1 Generate the initial population $P(0)$ at random, and set $i \leftarrow 0$

2 Repeat

    (a) Evaluate the fitness of each individual in $P(i)$

    (b) Select parents from $P(i)$ based on their fitness in $P(i)$

    (c) Generate offspring from the parents using crossover and mutation to form $P(i + 1)$

    (d) $i \leftarrow i + 1$

3 Until halting criteria are satisfied

where the search operators are also called genetic operators for genetic algorithms. They are used to generate offspring (new individuals) from parents (existing individuals).

Learning classifier systems, also known as classifier systems, are probably the oldest and best known evolutionary learning systems, although they did not work very well in their classical form. Some of the recent systems have improved this situation. Due to their historical importance, a brief introduction to the classical learning classifier systems will be introduced here.

Learning classifier systems are a particular class of message-passing, rule-based systems. They can also be regarded as a type of adaptive expert system that uses a knowledge base of production rules in a low-level syntax that can be manipulated by a genetic algorithm. In a classifier system, each low-level rule is called a classifier. A general operational cycle for the classifier system is as follows:

1 Allow the detectors (input interface) to code the current environment status and place the resulting messages on the message list.

2 Determine the set of classifiers that are matched by the current messages.

3 Resolve conflicts caused by limited message list size or contradictory actions.

4 Remove those messages which match the conditions of firing classifier from the message list.

5 Add the messages suggested by the firing messages to the list.

6 Allow the effectors (output interface) that are matched by the current message list to take actions in the environment.

7 If a payoff signal is received from the environment, assign credit to the classifiers.

8 Goto Step 1.

A genetic algorithm is used in classifier systems to discover new classifiers by crossover and mutation. The strength of a classifier updated by the credit assignment scheme is used as its fitness. A classifier's strength is based on its average usefulness in the context in which it has been tried previously. Credit assignment is a very difficult task because credit must be assigned to early-acting classifiers that set the stage for a sequence of actions leading to a favorable situation. The most well known credit assignment is the bucket brigade algorithm which uses metaphors from economics.

For a classifier called middleman, its suppliers are those classifiers that have sent messages satisfying its conditions, and its consumers are those classifiers that have conditions satisfied by its message and have won their competition in turn. When a classifier wins in competition, its bid is actually apportioned to its suppliers, increasing their strengths by the amounts apportioned to them. At the same time, because the bid is treated as a payment for the right to post a message, the strength of the winning classifier is reduced by the amount of its bid. Should the classifier bid but not win, its strength remains unchanged and its suppliers receive no payment. Winning classifiers can recoup their payments from either winning consumers or the environment payoff.

The genetic algorithm is only applied to the classifiers after a certain number of operational cycles in order to approximate strengths better. There are two approaches to classifier systems; the Michigan approach and the Pitt approach. For the Michigan approach, each individual in a population is a rule. The whole population represents a complete classifier system. For the Pitt approach, each individual in a population represents a complete classifier system. The whole population includes a number of competing classifier systems.

12.3   LEARNING AND EVOLUTION

Learning and evolution are two fundamental forms of adaptation. There has been a great interest in combining learning and evolution with neural networks in recent years.

12.3.1   Evolutionary Neural Networks

Evolutionary neural networks refer to a special class of neural networks in which evolution is another fundamental form of adaptation in addition to learning (Yao, 1991, 1993a, 1994, 1995). Evolutionary algorithms are used to perform various tasks, such as connection weight training, architecture design, learning rule adaptation, input feature selection, connection weight initialization, rule extraction from neural networks, etc. One distinct feature of evolutionary neural networks is their adaptability to a dynamic environment. In other words, evolutionary neural networks can adapt to an environment as well as changes in the environment. The two forms of adaptation, i.e. evolution and learning in evolutionary neural networks, make their adaptation to a dynamic environment much more effective and efficient. In a broader sense, evolutionary neural networks can be regarded as a general framework for adaptive systems, i.e. systems that can change their architectures and learning rules appropriately without human intervention.

Evolution has been introduced into neural networks at roughly three different levels: connection weights; architectures; and learning rules.

The Evolution of Connection Weights   The evolution of connection weights introduces an adaptive and global approach to training, especially in the reinforcement learning and recurrent network learning paradigm where gradient-based training algorithms often experience great difficulties.

One way to overcome the shortcomings of gradient-descent-based training algorithms is to adopt evolutionary neural networks, i.e. to formulate the training process as the evolution of connection weights in the environment determined by the architecture and the learning task. Evolutionary algorithms can then be used effectively in the evolution to find a near-optimal set of connection weights globally without computing gradient information. The fitness of a neural network can be defined according to different needs. Two important factors which often appear in the fitness (or error) function are the error between target and actual outputs and the complexity of the neural network. Unlike the gradient-descent-based case, the fitness (or error) function does not have to be differentiable or even continuous since evolutionary algorithms do not depend on gradient information. Because evolutionary algorithms can treat large, complex, nondifferentiable, and multimodal spaces, which are the typical case in the real world, considerable research and application has been conducted on the evolution of connection weights.

The evolutionary approach to weight training in neural networks consists of two major phases. The first phase is to decide the representation of connection weights, i.e. whether in the form of binary strings or not. The second one is the evolutionary process simulated by an evolutionary algorithm, in which search operators such as crossover and mutation have to be decided in conjunction with the representation scheme. Different representations and search operators can lead to quite different training performance. A typical cycle of the evolution of connection weights is shown as follows (Yao, 1999):

1 Decode each individual (genotype) in the current generation into a set of connection weights and construct a corresponding neural network with weights.

2 Evaluate each neural network by computing its total mean square error between actual and target outputs. Other error functions can also be used. The fitness of an individual is determined by the error. The higher the error, the lower the fitness. The optimal mapping from the error to the fitness is problem dependent. A regularization term may be included in the fitness function to penalize large weights.

3 Select parents for reproduction based on their fitness.

4 Apply genetic operators, such as crossover and/or mutation, to parents to generate offspring, which form the next generation.

The evolution stops when the fitness is greater than a predefined value (i.e. the training error is smaller than a certain value) or the population has converged.

The Evolution of Architectures The evolution of architectures enables neural networks to adapt their topologies to different tasks without human intervention and thus provides an approach to automatic neural network design as both neural network connection weights and structures can be evolved.

Architecture design is crucial in the successful application of neural networks because the architecture has significant impact on a network's information processing capabilities. Given a learning task, a neural network with only a few connections and linear nodes may not be able to perform the task at all due to its limited capability, while a neural network with a large number of connections and nonlinear nodes may overfit noise in the training data and fail to have good generalization ability.

Up to now, architecture design has been very much a human expert's job. It depends heavily on the expert experience and a tedious trial-and-error process. There is no systematic way to design a near-optimal architecture for a given task automatically. Design of the optimal architecture for a neural network can be formulated as a search problem in the architecture space where each point represents an architecture. Given some performance (optimality) criteria, e.g. lowest training error, lowest network complexity, etc., about architectures, the performance level of all architectures forms a discrete surface in the space. The optimal architecture design is equivalent to finding the highest point on this surface.

Similar to the evolution of connection weights, two major phases involved in the evolution of architectures are the genotype representation scheme of architectures and the evolutionary algorithm used to evolve neural network architectures. One of the key issues in encoding neural network architectures is to decide how much information about an architecture should be encoded in the chromosome. At one extreme, all the details, i.e. every connection and node of an architecture, can be specified by the chromosome. This kind of representation scheme is called direct encoding. At the other extreme, only the most important parameters of an architecture, such as the number of hidden layers and hidden nodes in each layer, are encoded. Other details about the architecture are left to the training process to decide. This kind of representation scheme is called indirect encoding. After a representation scheme has been chosen, the evolution of architectures can progress according to the cycle as follows (Yao, 1999):

1 Decode each individual in the current generation into an architecture. If the indirect encoding scheme is used, further detail of the architecture is specified by some developmental rules or a training process.

2 Train each neural network with the decoded architecture by a pre-defined learning rule (some parameters of the learning rule could be learned during training) starting from different sets of random initial weights and, if any, learning parameters.

3 Compute the fitness of each individual (encoded architecture) according to the above training result and other performance criteria such as the complexity of the architecture.

4 Select parents from population based on their fitness.

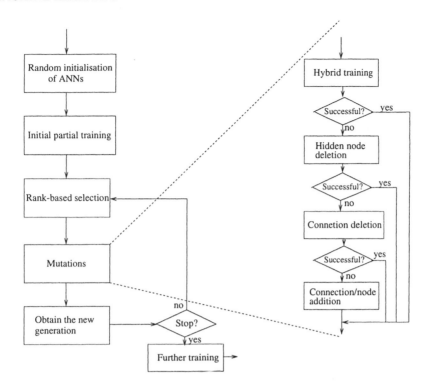

Figure 12.5. The main structure of EPNet.

5 Apply genetic operators to the parents and generate offspring which
   form the next generation.

The cycle stops when a satisfactory neural network is found.

An automatic system, EPNet (Yao and Liu, 1997, 1998), based on evolutionary programming was developed for the simultaneous evolution of neural
network architectures and connection weights. EPNet relies on a number of
mutation operators to modify architectures and weights. Behavioral (i.e. functional) evolution, rather than genetic evolution, is emphasized in EPNet. A
number of techniques were adopted to maintain the behavioral link between
a parent and its offspring (Yao and Liu, 1997). Figure 12.5 shows the main
structure of EPNet.

EPNet uses rank-based selection (Yao, 1993) and five mutations: hybrid
techniques training; node deletion; connection deletion; connection addition;
and node addition (Yao and Liu, 1997). EPNet uses a hybrid algorithm to
train the neural network for a fixed number of epochs. Such training does not
guarantee the convergence of neural network learning. Hence the training is
partial. The other four mutations are used to grow and prune hidden nodes and
connections.

The five mutations are attempted sequentially. If one mutation leads to a better offspring, it is regarded as successful. No further mutation will be applied. Otherwise the next mutation is attempted. The motivation behind ordering mutations is to encourage the evolution of compact neural networks without sacrificing generalization. A validation set is used in EPNet to measure the fitness of an individual, and another validation set to stop training in the final step. EPNet has been tested extensively on a number of benchmark problems, and very compact neural networks with good generalization ability have been evolved (Yao and Liu, 1997).

12.3.2   The Evolution of Learning Rules

The evolution of learning rules can be regarded as a process of "learning to learn" in neural networks where the adaptation of learning rules is achieved through evolution. It can also be regarded as an adaptive process of automatic discovery of novel learning rules.

The relationship between evolution and learning is extremely complex. Various models have been proposed, but most of them deal with the issue of how learning can guide evolution and the relationship between the evolution of architectures and that of connection weights connection weights (Yao, 1999). Research into the evolution of learning rules is still in its early stages. This research is important not only in providing an automatic way of optimizing learning rules and in modeling the relationship between learning and evolution, but also in modeling the creative process since newly evolved learning rules can deal with a complex and dynamic environment. This research will help us to understand better how creativity can emerge in artificial systems, like neural networks, and how to model the creative process in biological systems. A typical cycle of the evolution of learning rules can be described as follows (Yao, 1999):

1 Decode each individual in the current generation into a learning rule.

2 Construct a set of neural networks with randomly generated architectures and initial weights, and train them using the decoded learning rules.

3 Calculate the fitness of each individual (encoded learning rule) according to the average training results.

4 Select parents from the current generation according to their fitness.

5 Apply search operators to parents to generate offspring which form the next generation.

The iteration stops when the population converges or a predefined maximum number of iterations has been reached.

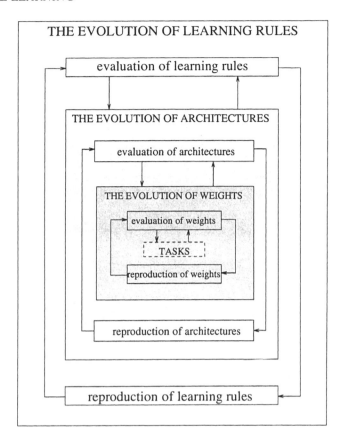

Figure 12.6. A general framework for evolutionary neural networks.

12.3.3 A General Framework for Evolutionary Neural Networks

A general framework for evolutionary neural networks can be described by Figure 12.6 (Yao, 1999). The evolution of connection weights proceeds at the lowest level on the fastest time scale in an environment determined by an architecture, a learning rule, and learning tasks. There are, however, two alternatives to decide the level of the evolution of architectures and that of learning rules: either the evolution of architectures is at the highest level and that of learning rules at the lower level or vice versa. The lower the level of evolution, the faster the time scale.

12.4   SOME PROMISING AREAS FOR FUTURE APPLICATION

Some recent trends and directions in machine learning have been summarized in Dietterich (1997) and Langley (1996). The first trend is in experimental studies of learning algorithms. Experimental studies of learning algorithms have shifted from early study of idealized, hand-crafted examples to realistic learning tasks that involve hundreds and thousands of cases. As well as robustness and generality testing of learning algorithms on a number of different data sets, comparisons between different learning algorithms on the same task domains need to be performed. It has been realized that some explicit methods for evaluating different learning algorithms should be established, and the conditions in which a learning algorithm will perform well should be identified in order to make progress in machine learning.

The second trend is in theoretical analyses of learning processes. The main goal of theoretical analysis is to find the inductive principle with the best generalization, and develop learning algorithms with such inductive principle. Early work on the study of the convergence of learning algorithms was important but gave little insight into real learning problems. A major advance was due to the introduction of the probably approximately correct model (Vapnik, 1995). This model provided, for the first time, theoretical accuracy guarantees that were based on a finite number of training samples. The resulting probably approximately correct model also served the rigorous framework that addressed the concerns arising from the real-world problems.

The third trend is in applications of machine learning. Most recent successful applications are on classification or prediction tasks. Machine learning has also been applied in the areas of configuration and layout, planning and scheduling, and execution and control. In order for machine learning to play a major role in solving problems of interest to industry and commerce, many more applications need to be pursued.

The fourth trend is in new learning algorithms. Many new learning algorithms have been studied in the past decade. For example, a support vector machine can construct a hyperplane as the decision surface in such a way that the margin of separation between positive and negative examples is maximized. A boosting algorithm trains a set of classifiers on data sets with entirely different distributions, and combines them in an elementary way to achieve near-optimal performance. The boosting algorithm was originally proposed by Schapire (1990). Schapire proved that it is theoretically possible to convert a weak learning algorithm that performs only slightly better than random guessing into one that achieves arbitrary accuracy. Boosting is a general method that can be used to improve the performance of any learning algorithm. Another ensemble learning called bagging combines models built on resamplings of data to yield a superior model (Breiman, 1996).

The fifth trend is in unified frameworks for machine learning. Machine learning has been widely studied from a variety of backgrounds. The similarities between the various approaches were often overlooked while the differences between them were emphasized. It is important to draw distinctions among different learning algorithms. However, an ultimate goal of machine learning is to study a unified framework that can explain different learning processes in terms of common underlying mechanisms. One direction towards this goal is to study hybrid techniques learning systems that incorporate aspects of different learning algorithms.

The sixth trend is in integrated cognitive architectures. This is related to the development of integrated architectures for cognition. A common implementation in early work was to design a separate system for each new task. These systems had little of the nature of intelligent behavior, and posed limitations on work in other domains. Research has now moved to the design of integrated architectures that make strong assumptions about the control structures that can support intelligence. It is clear that learning will continue to play an important role in the development of such cognitive architectures when it is necessary to acquire knowledge from the environment for long-term adaptive behavior.

These new areas will confront researchers with many more challenge problems, and novel directions will no doubt emerge when the limitations of existing learning algorithms are revealed.

12.5 TRICKS OF THE TRADE

Newcomers to the field of machine learning, applying a learning algorithm to a given problem, are often not very clear about where to start from in order to come up with a successful implementation. The following step-by-step procedure provides a guideline for implementing machine learning algorithms (Langley and Simon, 1995).

Formulating the Problem The first step is to formulate a given problem in terms of what can be dealt with by a particular learning algorithm. Often, real-world problems can be transformed into simple classification tasks. For example, the breast cancer diagnosis problem can be formulated as a classification task that classifies a tumor as either benign or malignant based on cell descriptions gathered by microscopic examination. Strategies such as divide-and-conquer can be used to decompose a complex task into a set of subproblems more amenable to the chosen learning algorithm. The strengths and limitations discussed in Section 12.2 provide a guideline for selecting an appropriate learning algorithm. Besides, based on the feedback available in the problem, supervised learning can be chosen when specific output is supplied; unsupervised learning should be adopted when there is no specific output; reinforcement learning can be applied when the environment is either fully observable or partially observable.

Choosing the Representation   The second step is to choose an appropriate representation for both the data and knowledge to be learned. The representation refers to the attributes or features describing examples rather than the representational formalism, such as decision trees or neural networks. In some real-world problems, there might be thousands of potential features describing each input. Most learning algorithms do not scale well when there are many features. Also, examples with many irrelevant and noisy input features give little information from a statistical point of view. It is essential to choose useful and important features to feed to the learning algorithms. There are three main approaches for feature selection (Dietterich, 1997). The first approach is to select a subset of the features based on some initial analysis. The second approach is to test different subsets of the features on the chosen learning algorithm and select the subsets that generate the best performance. The third approach is to automate the selection and weighting of features in the learning algorithm.

Collecting the Data   The third step is to collect data needed for the learning algorithm. In some applications, this process may be straightforward, but in others it can be very difficult. Generally speaking, the quantity of the data is decided by the chosen learning algorithm. Data preprocessing is often necessary in the learning process.

Conducting the Learning Process   Once the data are ready, the learning process can be started for finding the best learning model within a set of candidate model structures according to a certain criterion. A standard tool in statistics known as cross-validation provides a good guiding criterion. First the collected data is randomly partitioned into a training set and a test set. The training set is further divided into two disjoint subsets called estimation subset and validation subset in which the estimation subset is used for inducing the learning model, and the validation subset for validating the model. It is possible that the learned model may end up overfitting on the validation subset. Therefore, the generalization performance of the learned model is measured on the test set which is different from the validation subset. Some learning algorithms, such as inductive logic programming, rely on background knowledge available. How to obtain such helpful background knowledge is an important issue that will affect the outcome of those learning algorithms.

Analyzing and Evaluating the Learned Knowledge Empirical comparisons have often been used to evaluate the predictive performance of the various learning methods. The experiments can be conducted on simulated data sets or a real-life data set, or both. The known best predictions on some simulated data sets make it possible to compare the learned knowledge with the known knowledge. Real-life data are helpful for evaluation of the robustness and generality of different learning algorithms. Cross-validation is a method of estimating prediction error in its original form (Stone, 1974). The procedure of m-fold cross-validation is as follows:

1 Split the data into m roughly equal-sized parts.

2 For the ith part, fit the model to the other $(m - 1)$ parts of the data, and calculate the prediction error of fitted model when predicting the ith part of the data.

3 Do the above for $i = 1, \ldots, m$, and combine the m estimates of prediction error.

As well as empirical comparisons, statistical learning theory can be used to analyze the generalization ability of learning algorithms. Vapnik (1995) argued that the Vapnik–Chervonenkis dimension of the set of functions (rather than number of parameters) is responsible for the generalization ability of learning machines. This opens remarkable opportunities to overcome the "curse of dimensionality": to generalize well on the basis of a set of functions containing a huge number of parameters but possessing a small Vapnik-Chervonenkis dimension.

12.6 CONCLUSIONS

This chapter has been primarily concerned with the core learning algorithms including decision tree, inductive logic programming, Bayesian learning, neural networks, evolutionary learning, and reinforcement learning. Inevitably, some other important learning algorithms have not been covered. One important learning algorithm dealing with imprecise and uncertain knowledge and data is fuzzy logic. Imprecision is treated based on probability in statistical learning. In contrast, fuzzy logic is concerned with the use of fuzzy values that capture the meaning of words, human reasoning, and decision making. At the heart of fuzzy logic lies the concept of a linguistic variable. The values of the linguistic variable are words rather than numbers.

Generalization is one of key issues in machine learning. In neural network learning, generalization had been studied from the bias–variance trade-off point of view (Geman et al., 1992). There is usually a trade-off between bias and variance in the case of a training set with finite size: attempts to decrease bias by introducing more parameters often tend to increase variance;

attempts to reduce variance by reducing parameters often tend to increase bias. As well as the generalization issue, how to scale up learning algorithms is another important issue. Dietterich (1997) reviewed learning with a large training set and learning with many features. Even though some learning techniques could solve very large problems with millions of training examples in a reasonable amount of computer time, it is unclear whether they can successfully be applied to those problems with billions of training examples.

This chapter has also been concerned with exploring the possible benefits arising from combining learning with evolution in neural networks. Different learning algorithms have their own strengths and weaknesses. Among all the learning algorithms, there is no clear winner in terms of the best learning algorithm. The best one is always problem dependent. This is certainly true according to the no-free-lunch theorem (Wolpert and Macready, 1997). In general, hybrid techniques algorithms tend to perform better than others for a large number of problems.

SOURCES OF ADDITIONAL INFORMATION

The literature on machine learning is extensive, and has been growing rapidly.

- Mitchell's *Machine Learning* (Mitchell, 1997) and Russell and Norvig's *Artificial Intelligence: A Modern Approach* (Russell and Norvig, 2002) give good overviews of different types of learning algorithms.

- *Machine Learning*, volumes 1–3, provide the early history of machine learning development (Michalski et al., 1983, 1986; Kodratoff and Michalski, 1990).

- Some important papers in machine learning had been collected in *Readings in Machine Learning* (Shavlik and Dietterich, 1990).

- Current research in machine learning can be found in a number of journals. Major machine learning journals include *Machine Learning*, the *Journal of Machine Learning Research*, *IEEE Transactions on Neural Networks*, *IEEE Transactions on Evolutionary Computation*, and mainstream artificial intelligence journals.

- Machine learning is also covered by a number of conferences, such as the International Conference on Machine Learning, the International Joint Conference on Neural Networks, Congress on Evolutionary Computation, the IEEE International Conference on Fuzzy Systems, and the conference on Neural Information Processing Systems.

- Mlnet Online Information Service (http://www.mlnet.org/) funded by the European Commission is dedicated to the field of machine learning, knowledge discovery, case-based reasoning, knowledge acquisition, and data mining.

- Machine learning topics can also be found online at the website http://www.aaai.org/Pathfinder/html/machine.html of the American Association for Artificial Intelligence.

References

Breiman, L., Friedman, J., Olshen, R. A. and Stone, P. J., 1984, *Classification and Regression Trees,* Wadsworth, Belmont, CA.

Breiman, L., 1996, Bagging predictors, *Machine Learn.* **24**:123–140.

Cheng, J., Greiner, R., Kelly, J., Bell, D. A. and Liu, W., 2002, Learning Bayesian networks from data: an information-theory based approach, *Artif. Intell.* **137**:43–90.

Dietterich, T. G., 1997, Machine-learning research: four current directions, *AI Magazine* **18**:97–136.

Domingos, P. and Pazzani, M, 1996, Beyond independence: conditions for the optimality of the simple Bayesian classifier, in: *Proc. 13th Int. Conf. on Machine Learning,* L. Saitta, ed., Morgan Kaufmann, San Mateo, CA, pp. 105–112.

Elkan, C., 1997, Boosting and naive Bayesian learning, *Technical Report,* Department of Computer Science and Engineering, University of California.

Feigenbaum, E. A., 1961, The simulation of verbal learning behavior, in: *Proc. Western Joint Computer Conf.,* pp. 121–131.

Fogel, L. J., Owens, A. J. and Walsh, M. J., 1966, *Artificial Intelligence Through Simulated Evolution,* Wiley, New York.

Geman, S., Bienenstock, E. and Doursat, R, 1992, Neural networks and the bias/variance dilemma, *Neural Comput.* **4**:1–58.

Hebb, D. O., 1949, *The Organization of Behavior: A Neurophysiological Theory,* Wiley, New York.

Heckerman, D., 1998, A tutorial on learning with Bayesian networks, in: *Learning in Graphical Models,* M. I. Jordan, ed., Kluwer, Dordrecht.

Hopfield, J. J., 1982, Neural networks and physical systems with emergent collective computational abilities, *Proc. Natl Acad. Sci. USA* **79**:2554–2558.

Hopfield, J. J. and Tank, D. W., 1985, Neural computation of decisions in optimization problems, *Biol. Cybernet.* **52**:141–152.

Hunt, E. B., Marin, J. and Stone, P. T., 1966, *Experiments in Induction,* Academic, New York.

Kaelbling, L. P., Littman, M. L. and Moore, A. W., 1996, Reinforcement learning: a survey, *J. Artif. Intell. Res.* **4**:237–285.

Kodratoff, Y. and Michalski, R. S., eds, 1990, *Machine Learning—An Artificial Intelligence Approach,* Vol. 3, Morgan Kaufmann, San Mateo, CA.

Langley, P., 1996, *Elements of Machine Learning,* Morgan Kaufmann, San Mateo, CA.

Langley, P. and Simon, H., 1995, Applications of machine learning and rule induction, *Commun. ACM* **38**:54–64.

Lavrač, N. and Džeroski, S., 1994, *Inductive Logic Programming: Techniques and Applications,* Ellis Horwood, Chichester.

Michalski, R. S., Carbonell, J. G. and Mitchell, T. M., eds, 1983, *Machine Learning—An Artificial Intelligence Approach,* Vol. 1, Morgan Kaufmann, San Mateo, CA.

Michalski, R. S., Carbonell, J. G. and Mitchell, T. M., eds, 1986, *Machine Learning—An Artificial Intelligence Approach,* Vol. 2, Morgan Kaufmann, San Mateo, CA.

McCulloch, W. S. and Pitts, W., 1943, A logical calculus of the ideas immanent in nervous activity, *Bull. Math. Biophys.* **5**:115–137.

Michie, D., Spiegelhalter, D. J. and Taylor, C. C., 1994, *Machine Learning, Neural and Statistical Classification,* Ellis Horwood, London.

Minsky, M. L. and Papert, S., 1969, *Perceptrons: An Introduction to Computational Geometry,* MIT Press, Cambridge, MA.

Mitchell, T. M., 1997, *Machine Learning,* McGraw-Hill, New York.

Muggleton, S. H., 1995, Inverse entailment and progol, *New Generation Comput.* (Special issue on Inductive Logic Programming) **13**:245–286.

Muggleton, S. H. and Buntine, W., 1988, Machine invention of first-order predicates by inverting resolution, in: *Proc. 5th Int. Conf. on Machine Learning,* Morgan Kaufmann, San Mateo, CA, pp. 339–352.

Quinlan, J. R., 1986, Introduction to decision tree, *Machine Learn.* **1**:81–106.

Quinlan, J. R., 1990, Learning logical definitions from relations, *Machine Learn.* **5**:239–266.

Quinlan, J. R.,1993, *C4.5: Programs for Machine Learning,* Morgan Kaufmann, San Mateo, CA.

Rumelhart, D. E., Hinton, G. E. and Williams, R. J., 1986, Learning internal representations by error propagation, in: *Parallel Distributed Processing: Explorations in the Microstructures of Cognition,* Vol. 1, D. E. Rumelhart and J. L. McClelland, eds, MIT Press, Cambridge, MA, pp. 318–362.

Rosenblatt, F., 1962, *Principles of Neurodynamics: Perceptrons and the Theory of Brain Mechanisms,* Spartan, Chicago, IL.

Rumelhart, D. E. and McClelland, J. L., eds, 1986, *Parallel Distributed Processing: Explorations in the Microstructures of Cognition,* MIT Press, Cambridge, MA.

Russell, S. and Norvig, P., 2002, *Artificial Intelligence: A Modern Approach,* Prentice-Hall, Englewood Cliffs, NJ.

Samuel, A. L., 1959, Some studies in machine learning using the game of checkers, *IBM J. Res. Dev.* **3**:210–229.

Schapire, R. E., 1990, The strength of weak learnability, *Machine Learn.* **5**:197–227.

Shavlik, J. and Dietterich, T. (eds), 1990, *Readings in Machine Learning,* Morgan Kaufmann, San Mateo, CA.

Schwefel, H.-P., 1981, *Numerical Optimization of Computer Models*, Wiley, Chichester.

Schwefel, H.-P., 1995, *Evolution and Optimum Seeking,* Wiley, New York.

Stone, M., 1974, Cross-validatory choice and assessment of statistical predictions, *J. R. Statist. Soc.* **36**:111–147.

Sutton, R. S. and Barto, A. G., 1998, *Reinforcement Learning: An Introduction,* MIT Press, Cambridge, MA.

Turing, A., 1950, Computing machinery and intelligence, *Mind* **59**:433–460.

Vapnik, V. N., 1995, *The Nature of Statistical Learning Theory,* Springer, New York.

Wolpert, D. H. and Macready, W. G., 1997, No free lunch theorems for optimization, *IEEE Trans. Evol. Comput.* **1**:67–82.

Yao, X., 1991, Evolution of connectionist networks, in: *Preprints of the Int. Symp. on AI, Reasoning and Creativity* (Griffith University, Queensland, Australia), T. Dartnall, ed., pp. 49–52.

Yao, X., 1993a, A review of evolutionary artificial neural networks, *Int. J. Intell. Syst.* **8**:539–567. **28**:417–425.

Yao, X., 1993b, An empirical study of genetic operators in genetic algorithms, *Microprocess. Microprogram.* **38**:707–714.

Yao, X., 1994, The evolution of connectionist networks, in: *Artificial Intelligence and Creativity,* T. Dartnall, ed., Kluwer, Dordrecht, pp. 233–243.

Yao, X., 1995, Evolutionary artificial neural networks, in: *Encyclopedia of Computer Science and Technology,* Vol. 33, A. Kent and J. G. Williams, ed., Dekker, New York, pp. 137–170.

Yao, X., 1999, Evolving artificial neural networks, *Proc. IEEE* **87**:1423–1447.

Yao, X. and Liu, Y., 1997, A new evolutionary system for evolving artificial neural networks, *IEEE Trans. Neural Networks* **8**:694–713.

Yao, X. and Liu, Y., 1998, Making use of population information in evolutionary artificial neural networks, *IEEE Trans. Syst., Man Cybernet.* B **28**:417–425.

Chapter 13

ARTIFICIAL IMMUNE SYSTEMS

Uwe Aickelin
University of Nottingham, UK

Dipankar Dasgupta
University of Memphis, Memphis, TN 38152, USA

13.1   INTRODUCTION

The biological immune system is a robust, complex, adaptive system that defends the body from foreign pathogens. It is able to categorize all cells (or molecules) within the body as self-cells or nonself cells. It does this with the help of a distributed task force that has the intelligence to take action from a local and also a global perspective using its network of chemical messengers for communication. There are two major branches of the immune system. The innate immune system is an unchanging mechanism that detects and destroys certain invading organisms, whilst the adaptive immune system responds to previously unknown foreign cells and builds a response to them that can remain in the body over a long period of time. This remarkable information processing biological system has caught the attention of computer science in recent years.

A novel computational intelligence technique, inspired by immunology, has emerged, known as Artificial Immune Systems. Several concepts from immunology have been extracted and applied for the solution of real-world science and engineering problems. In this tutorial, we briefly describe the immune system metaphors that are relevant to existing Artificial Immune System methods. We then introduce illustrative real-world problems and give a step-by-step algorithm walkthrough for one such problem. A comparison of Artificial Immune Systems to other well-known algorithms, areas for future work, tips and tricks and a list of resources round the tutorial off. It should be noted that as Artificial Immune Systems is still a young and evolving field, there is not yet a fixed algorithm template and hence actual implementations may differ somewhat from time to time and from those examples given here.

13.2   OVERVIEW OF THE BIOLOGICAL IMMUNE SYSTEM

The biological immune system is an elaborate defense system which has evolved over millions of years. While many details of the immune mechanisms (innate and adaptive) and processes (humoral and cellular) are yet unknown (even to immunologists), it is, however, well known that the immune system uses multilevel (and overlapping) defense both in parallel and sequential fashion. Depending on the type of the pathogen, and the way it gets into the body, the immune system uses different response mechanisms (differential pathways) either to neutralize the pathogenic effect or to destroy the infected cells. A detailed overview of the immune system can be found in many textbooks, such as Kubi (2002). The immune features that are particularly relevant to our tutorial are matching, diversity and distributed control. Matching refers to the binding between antibodies and antigens. Diversity refers to the fact that, in order to achieve optimal antigen space coverage, antibody diversity must be encouraged (see Hightower et al., 1995). Distributed control means that there is no central controller; rather, the immune system is governed by local interactions between immune cells and antigens.

Two of the most important cells in this process are white blood cells, called T-cells and B-cells. Both of these originate in the bone marrow, but T-cells pass on to the thymus to mature, before circulating in the blood and lymphatic vessels.

The T-cells are of three types: helper T-cells which are essential to the activation of B-cells, killer T-cells which bind to foreign invaders and inject poisonous chemicals into them causing their destruction, and suppressor T-cells which inhibit the action of other immune cells thus preventing allergic reactions and autoimmune diseases.

B-cells are responsible for the production and secretion of antibodies, which are specific proteins that bind to the antigen. Each B-cell can only produce one particular antibody. The antigen is found on the surface of the invading organism and the binding of an antibody to the antigen is a signal to destroy the invading cell as shown in Figure 13.1.

As mentioned above, the human body is protected against foreign invaders by a multi-layered system. The immune system is composed of physical barriers such as the skin and respiratory system; physiological barriers such as destructive enzymes and stomach acids; and the immune system, which can be broadly viewed as of two types: innate (non-specific) immunity and adaptive (specific) immunity, which are inter-linked and influence each other. Adaptive immunity can again be subdivided into two types: humoral immunity and cell-mediated immunity.

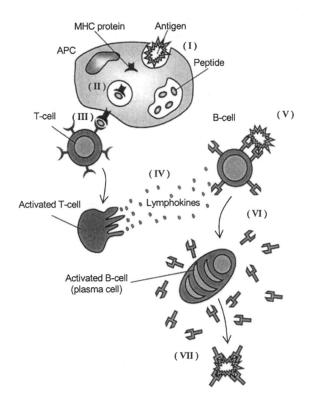

Figure 13.1. Pictorial representation of the essence of the acquired immune system mechanism (taken from de Castro and van Zuben (1999): the invader enters the body and activates T-cells, which then (in IV) activate the B-cells; V is the antigen matching, VI is the antibody production and VII is the antigen's destruction.

Innate immunity is present at birth. Physiological conditions such as pH, temperature and chemical mediators provide inappropriate living conditions for foreign organisms. Also, micro-organisms are coated with antibodies and/or complementary products (opsonization) so that they are easily recognized. Extracellular material is then ingested by macrophages by a process called phagocytosis. Also, T_{DH}-cells influence the phagocytosis of macrophages by secreting certain chemical messengers called lymphokines. The low levels of sialic acid on foreign antigenic surfaces make C_3b bind to these surfaces for a long time and thus activate alternative pathways. Thus MAC is formed, which punctures the cell surfaces and kills the foreign invader.

Adaptive immunity is the main focus of interest here as learning, adaptability, and memory are important characteristics of adaptive immunity. It is subdivided under two heads: humoral immunity and cell-mediated immunity:

1 Humoral immunity is mediated by antibodies contained in body fluids
   (known as humors). The humoral branch of the immune system involves
   interaction of B-cells with antigen and their subsequent proliferation and
   differentiation into antibody-secreting plasma cells. Antibody functions
   as the effectors of the humoral response by binding to antigen and facilitating its elimination. When an antigen is coated with antibody, it can
   be eliminated in several ways. For example, antibody can cross-link the
   antigen, forming clusters that are more readily ingested by phagocytic
   cells. Binding of antibody to antigen on a micro-organism also can activate the complement system, resulting in lysis of the foreign organism.

2 Cellular immunity is cell-mediated; effector T-cells generated in response to antigen are responsible for cell-mediated immunity. Cytotoxic
   T-lymphocytes (CTLs) participate in cell-mediated immune reactions by
   killing altered self-cells; they play an important role in the killing of
   virus-infected and tumor cells. Cytokines secreted by T_{DH} can mediate
   the cellular immunity, and activate various phagocytic cells, enabling
   them to phagocytose and kill micro-organisms more effectively. This
   type of cell-mediated immune response is especially important in host
   defense against intracellular bacteria and protozoa.

Whilst there is more than one mechanism at work (for more details see
Farmer et al., 1986; Kubi, 2002; Jerne, 1973), the essential process is the
matching of antigen and antibody, which leads to increased concentrations
(proliferation) of more closely matched antibodies. In particular, idiotypic network theory, negative selection mechanism, and the "clonal selection" and "somatic hypermutation" theories are primarily used in Artificial Immune System
models.

13.2.1   Immune Network Theory

The immune network theory was proposed by Jerne (1973). The hypothesis
was that the immune system maintains an idiotypic network of interconnected
B-cells for antigen recognition. These cells both stimulate and suppress each
other in certain ways that lead to the stabilization of the network. Two B-cells
are connected if the affinities they share exceed a certain threshold, and the
strength of the connection is directly proportional to the affinity they share.

13.2.2   Negative Selection Mechanism

The purpose of negative selection is to provide tolerance for self-cells. It
deals with the immune system's ability to detect unknown antigens while not
reacting to the self-cells. During the generation of T-cells, receptors are made
through a pseudo-random genetic rearrangement process. Then, they undergo

a censoring process in the thymus, called the negative selection. There, T-cells that react against self-proteins are destroyed; thus, only those that do not bind to self-proteins are allowed to leave the thymus. These matured T-cells then circulate throughout the body to perform immunological functions and protect the body against foreign antigens.

13.2.3   Clonal Selection Principle

The clonal selection principle describes the basic features of an immune response to an antigenic stimulus. It establishes the idea that only those cells that recognize the antigen proliferate, thus being selected against those that do not. The main features of the clonal selection theory are that

1 the new cells are copies of their parents (clone) subjected to a mutation mechanism with high rates (somatic hypermutation);

2 elimination of newly differentiated lymphocytes carrying self-reactive receptors;

3 proliferation and differentiation on contact of mature cells with antigens.

When an antibody strongly matches an antigen the corresponding B-cell is stimulated to produce clones of itself that then produce more antibodies. This (hyper) mutation, is quite rapid, often as much as "one mutation per cell division" (de Castro and Von Zuben, 1999). This allows a very quick response to the antigens. It should be noted here that in the Artificial Immune System literature, often no distinction is made between B-cells and the antibodies they produce. Both are subsumed under the word "antibody" and statements such as mutation of antibodies (rather than mutation of B-cells) are common.

There are many more features of the immune system, including adaptation, immunological memory and protection against auto-immune attacks, not discussed here. In the following sections, we will revisit some important aspects of these concepts and show how they can be modeled in "artificial" immune systems and then used to solve real-world problems. First, let us give an overview of typical problems that we believe are amenable to being solved by artificial immune systems.

13.3   ILLUSTRATIVE PROBLEMS

13.3.1   Intrusion Detection Systems

Anyone keeping up-to-date with current affairs in computing can confirm numerous cases of attacks made on computer servers of well-known companies. These attacks range from denial-of-service attacks to extracting credit-card details and sometimes we find ourselves thinking "haven't they installed a firewall?" The fact is they often have a firewall. A firewall is useful, indeed often essential, but current firewall technology is insufficient to detect and block all kinds of attacks.

On ports that need to be open to the internet, a firewall can do little to prevent attacks. Moreover, even if a port is blocked from internet access, this does not stop an attack from inside the organization. This is where intrusion detection systems come in. As the name suggests, intrusion detection systems are installed to identify (potential) attacks and to react by usually generating an alert or blocking the unscrupulous data.

The main goal of intrusion detection systems is to detect unauthorized use, misuse and abuse of computer systems by both system insiders and external intruders. Most current intrusion detection systems define suspicious signatures based on known intrusions and probes. The obvious limit of this type of intrusion detection systems is its failure in detecting previously unknown intrusions. In contrast, the human immune system adaptively generates new immune cells so that it is able to detect previously unknown and rapidly evolving harmful antigens (Forrest et al., 1994). Thus the challenge is to emulate the success of the natural systems.

13.3.2   Data Mining—Collaborative Filtering and Clustering

Collaborative filtering is the term for a broad range of algorithms that use similarity measures to obtain recommendations. The best-known example is probably the "people who bought this also bought" feature of the internet company Amazon (2004). However, any problem domain where users are required to rate items is amenable to collaborative filtering techniques. Commercial applications are usually called recommender systems (Resnick and Varian, 1997). A canonical example is movie recommendation.

In traditional collaborative filtering, the items to be recommended are treated as "black boxes". That is, your recommendations are based purely on the votes of other users, and not on the content of the item. The preferences of a user, usually a set of votes on an item, comprise a user profile, and these profiles are compared in order to build a neighborhood. The key decision is what similarity measure is used. The most common method to compare two users is a correlation-based measure like Pearson or Spearman, which gives two neighbors a matching score between -1 and 1. The canonical example is the k-nearest-neighbor algorithm, which uses a matching method to select k reviewers with high similarity measures. The votes from these reviewers, suitably weighted, are used to make predictions and recommendations.

The evaluation of a collaborative filtering algorithm usually centers on its accuracy. There is a difference between prediction (given a movie, predict a given user's rating of that movie) and recommendation (given a user, suggest movies that are likely to attract a high rating). Prediction is easier to assess quantitatively but recommendation is a more natural fit to the movie domain. A related problem to collaborative filtering is that of clustering data or users in a database. This is particularly useful in very large databases, which have become too large to handle. Clustering works by dividing the entries of the database into groups, which contain people with similar preferences or in general data of similar type.

13.4 ARTIFICIAL IMMUNE SYSTEMS BASIC CONCEPTS

13.4.1 Initialization/Encoding

To implement a basic artificial immune system, four decisions have to be made: encoding, similarity measure, selection and mutation. Once an encoding has been fixed and a suitable similarity measure is chosen, the algorithm will then perform selection and mutation, both based on the similarity measure, until stopping criteria are met. In this section, we will describe each of these components in turn.

Along with other heuristics, choosing a suitable encoding is very important for the algorithm's success. Similar to genetic algorithms, there is close interplay between the encoding and the fitness function (the latter is in artificial immune systems referred to as the "matching" or "affinity" function). Hence both ought to be thought about at the same time. For the current discussion, let us start with the encoding.

First, let us define what we mean by antigen and antibody in the context of an application domain. Typically, an antigen is the target or solution, e.g. the data item we need to check to see if it is an intrusion, or the user that we need to cluster or make a recommendation for. The antibodies are the remainder of the data, e.g. other users in the data base, a set of network traffic that has already been identified, etc. Sometimes there can be more than one antigen at a time and there are usually a large number of antibodies present simultaneously.

Antigens and antibodies are represented or encoded in the same way. For most problems the most obvious representation is a string of numbers or features, where the length is the number of variables, the position is the variable identifier and the value is the value (could be binary or real) of the variable. For instance, in a five-variable binary problem, an encoding could look like this: (10010).

We have previously mentioned data mining and intrusion detection applications. What would an encoding look like in these cases? For data mining, let us consider the problem of recommending movies. Here the encoding has to represent a user's profile with regards to the movies he has seen and how much he has (dis)liked them. A possible encoding for this could be a list of numbers, where each number represents the "vote" for an item. Votes could be binary, e.g. Did you visit this web page? (Morrison and Aickelin, 2002), but can also be integers in a range (say [0, 5]: i.e. 0, did not like the movie at all; 5, liked it very much).

Hence, for the movie recommendation, a possible encoding is

$$User = \{\{id_1, score_1\}, \{id_2, score_2\} \dots \{id_n, score_n\}\}$$

Where *id* corresponds to the unique identifier of the movie being rated and score to this user's score for that movie. This captures the essential features of the data available (Cayzer and Aickelin, 2002a).

For intrusion detection, the encoding may be to encapsulate the essence of each data packet transferred, e.g. [<protocol><source ip><source port><destination ip><destination port>]

example: [<tcp> <113.112.255.254><any><108.200.111.12><25>]

which represents an incoming data packet sent to port 25. In these scenarios, wildcards like "any port" are also often used.

13.4.2   Similarity or Affinity Measure

As mentioned above, the similarity measure or matching rule is one of the most important design choices in developing an artificial immune system algorithm, and is closely coupled to the encoding scheme.

Two of the simplest matching algorithms are best explained using binary encoding. Consider the strings (00000) and (00011). If one does a bit-by-bit comparison, the first three bits are identical and hence we could give this pair a matching score of 3. In other words, we compute the opposite of the Hamming distance (which is defined as the number of bits that have to be changed in order to make the two strings identical).

Now consider the pair (00000) and (01010). Again, simple bit matching gives us a similarity score of 3. However, the matching is quite different as the three matching bits are not connected. Depending on the problem and encoding, this might be better or worse. Thus, another simple matching algorithm is to count the number of continuous bits that match and return the length of the longest matching as the similarity measure. For the first example above this would still be 3; for the second example it would be 1.

If the encoding is non-binary, e.g. real variables, there are even more possibilities to compute the "distance" between the two strings, for instance we could compute the geometrical (Euclidian) distance.

For data mining problems, similarity often means "correlation". Take the movie recommendation problem as an example and assume that we are trying to find users in a database that are similar to the key user who's profile were are trying to match in order to make recommendations. In this case, what we are trying to measure is how similar are the two users' tastes. One of the easiest ways of doing this is to compute the Pearson correlation coefficient between the two users, i.e. if the Pearson measure is used to compare two user's u and v:

$$r = \frac{\sum_{i=1}^{n}(u_i - \bar{u})(v_i - \bar{v})}{\sqrt{\sum_{i=1}^{n}(u_i - \bar{u})^2 \sum_{i=1}^{n}(v_i - \bar{v})^2}}$$

where u and v are users, n is the number of overlapping votes (i.e. movies for which both u and v have voted), u_i is the vote of user u for movie i and $\bar{u}$ is the average vote of user u over all films (not just the overlapping votes). The measure is amended to default to a value of 0 if the two users have no films in common. During our research reported in Cayzer and Aickelin (2002a, 2002b) we also found it useful to introduce a penalty parameter (as in penalties in genetic algorithms) for users who only have very few films in common, which in essence reduces their correlation.

The outcome of this measure is a value between -1 and 1, where values close to 1 mean strong agreement, values near to -1 mean strong disagreement and values around 0 mean no correlation. From a data mining point of view, those users who score either 1 or -1 are the most useful and hence will be selected for further treatment by the algorithm.

For other applications, "matching" might not actually be beneficial and hence those items that match might be eliminated. This approach is known as "negative selection" and mirrors what is believed to happen during the maturation of B-cells who have to learn not to "match" our own tissues as otherwise we would be subject to auto-immune diseases.

Under what circumstance would a negative selection algorithm be suitable for an artificial immune system implementation? Consider the case of intrusion detection as solved by Hofmeyr and Forrest (2000). One way of solving this problem is by defining a set of "self", i.e. a trusted network, our company's computers, known partners, etc. During the initialization of the algorithm, we would then randomly create a large number of "detectors", i.e. strings that look similar to the sample intrusion detection system encoding given above. We would then subject these detectors to a matching algorithm that compares them to our "self". Any matching detector would be eliminated and hence we select those that do not match (negative selection). All non-matching detectors will then form our final detector set. This detector set is then used in the second phase of the algorithm to continuously monitor all network traffic. Should a match be found now the algorithm would report this as a possible alert or "nonself". There are a number of problems with this approach, which we discuss further in Section 13.7.

13.4.3   Negative, Clonal or Neighborhood Selection

The meaning of this step differs depending on the exact problem the Artificial Immune Systems is applied to. We have already described the concept of negative selection. For the film recommender, choosing a suitable neighborhood means choosing good correlation scores and hence we will perform "positive" selection. How would the algorithm use this?

Consider the artificial immune system to be empty at the beginning. The target user is encoded as the antigen, and all other users in the database are possible antibodies. We add the antigen to the artificial immune system and then we add one candidate antibody at a time. Antibodies will start with a certain concentration value. This value decreases over time (death rate), similar to the evaporation in ant systems. Antibodies with a sufficiently low concentration are removed from the system, whereas antibodies with a high concentration may saturate. However, an antibody can increase its concentration by matching the antigen: the better the match the higher the increase (a process called stimulation). The process of stimulation or increasing concentration can also be regarded as "cloning" if one thinks in a discrete setting. Once enough antibodies have been added to the system, it starts to iterate a loop of reducing concentration and stimulation until at least one antibody drops out. A new antibody is added and the process is repeated until the artificial immune system is stabilized, i.e. there are no more drop-outs for a certain period of time.

Mathematically, at each step (iteration) an antibody's concentration is increased by an amount dependent on its matching to each antigen. In the absence of matching, an antibody's concentration will slowly decrease over time. Hence an artificial immune system iteration is governed by the following equation, based on Farmer et al. (1986):

$$\frac{dx_i}{dt} = \left[\left(\begin{array}{c} antigens \\ recognized \end{array} \right) - \left(\begin{array}{c} death \\ rate \end{array} \right) \right]$$

$$= \left[k_2 \left(\sum_{j=1}^{N} m_{ji} x_i y_j \right) - k_3 x_i \right]$$

where N is the number of antigens, x_i is the concentration of antibody i, y_j is the concentration of antigen j, k_2 is the stimulation effect and k_3 is the death rate, and m_{ji} is the matching function between antibody i and antibody (or antigen) j.

The following pseudo-code summarizes the artificial immune system of the movie recommender:

Initialize Artificial Immune Systems
Encode user for whom to make predictions as antigen Ag
WHILE (Artificial Immune Systems not Full) & (More Antibodies) DO
    Add next user as an antibody Ab
    Calculate matching scores between Ab and Ag
    WHILE (Artificial Immune Systems at full size) & (Artificial Immune
    Systems not Stabilized) DO
        Reduce Concentration of all Abs by a fixed amount
        Match each Ab against Ag and stimulate as necessary
    OD
OD
Use final set of Antibodies to produce recommendation.

For example, the artificial immune system is considered stable after iterating for ten iterations without changing in size. Stabilization thus means that a sufficient number of "good" neighbors have been identified and therefore a prediction can be made. "Poor" neighbors would be expected to drop out of the artificial immune system after a few iterations. Once the artificial immune system has stabilized using the above algorithm, we use the antibody concentration to weigh the neighbors and then perform a weighted average type recommendation.

13.4.4 Somatic Hypermutation

The mutation most commonly used in artificial immune systems is very similar to that found in genetic algorithms, e.g. for binary strings bits are flipped, for real value strings one value is changed at random, or for others the order of elements is swapped. In addition, the mechanism is often enhanced by the somatic idea, i.e. the closer the match (or the less close the match, depending on what we are trying to achieve), the more (or less) disruptive the mutation.

However, mutating the data might not make sense for all problems considered. For instance, it would not be suitable for the movie recommender. Certainly, mutation could be used to make users more similar to the target; however, the validity of recommendations based on these artificial users is questionable and if over-done, we would end up with the target user itself. Hence for some problems, somatic hypermutation is not used, since it is not immediately obvious how to mutate the data sensibly such that these artificial entities still represent plausible data.

Nevertheless, for other problem domains, mutation might be very useful. For instance, taking the negative selection approach to intrusion detection, rather than throwing away matching detectors in the first phase of the algorithm, these could be mutated to save time and effort. Also, depending on the degree of matching, the mutation could be more or less strong. This was in fact one extension implemented by Hofmeyr and Forrest (2000).

For data mining problems, mutation might also be useful, if for instance the aim is to cluster users. Then the center of each cluster (the antibodies) could be an artificial pseudo-user that can be mutated at will until the desired degree of matching between the center and antigens in its cluster is reached. This is an approach implemented by de Castro and von Zuben (2002).

13.5 COMPARISON WITH GENETIC ALGORITHMS AND NEURAL NETWORKS

So far in this tutorial, both genetic algorithms and neural networks have been mentioned a number of times. In fact, they both have a number of ideas in common with artificial immune systems and Table 13.1 highlights their similarities and differences (Dasgupta, 1999). Evolutionary computation shares many elements, concepts like population, genotype phenotype mapping, and proliferation of the most fitted are present in different artificial immune system methods.

Artificial immune system models based on immune networks resemble the structures and interactions of connectionist models. Some works have pointed to the similarities and the differences between artificial immune systems and artificial neural networks (Dasgupta, 1999; de Castro and Von Zuben, 2002); de Castro has also used artificial immune systems to initialize the centers of radial basis function neural networks and to produce a good initial set of weights for feed-forward neural networks.

Some of the items in Table 13.1 are gross simplifications, both to benefit the design of the table and so as not to overwhelm the reader, and some of the points are debatable; however, we believe that the comparison is nevertheless valuable, to show exactly where artificial immune systems fit into the wider picture. The comparisons are based on a genetic algorithm (GA) used for optimization and a neural network (NN) used for classification.

13.6 EXTENSIONS OF ARTIFICIAL IMMUNE SYSTEMS

13.6.1 Idiotypic Networks—Network Interactions (Suppression)

The idiotypic effect builds on the premise that antibodies can match other antibodies as well as antigens. It was first proposed by Jerne (1973) and formalized into a model by Farmer et al. (1986). The theory is currently debated

Table 13.1. Comparison of artificial immune systems with genetic algorithms and neural networks.

	GA (Optimization)	NN (Classification)	Artificial immune systems
Components	Chromosome strings	Artificial neurons	Attribute strings
Location of components	Dynamic	Pre-defined	Dynamic
Structure	Discrete components	Networked components	Discrete/networked components
Knowledge storage	Chromosome strings	Connection strengths	Component concentration/network connections
Dynamics	Evolution	Learning	Evolution/learning
Meta-dynamics	Recruitment/elimination of components	Construction/pruning of connections	Recruitment/elimination of components
Interaction between components	Crossover	Network connections	Recognition/network connections
Interaction with environment	Fitness function	External stimuli	Recognition/objective function
Threshold activity	Crowding/sharing	Neuron activation	Component affinity

by immunologists, with no clear consensus yet on its effects in the humoral immune system (Kuby, 2002). The idiotypic network hypothesis builds on the recognition that antibodies can match other antibodies as well as antigens. Hence, an antibody may be matched by other antibodies, which in turn may be matched by yet other antibodies. This activation can continue to spread through the population and potentially has much explanatory power. It could, for example, help explain how the memory of past infections is maintained. Furthermore, it could result in the suppression of similar antibodies, thus encouraging diversity in the antibody pool. The idiotypic network has been formalized by a number of theoretical immunologists (Perelson and Weisbuch, 1997):

$$\frac{dx_i}{dt} = c\left[\left(\begin{array}{c}antibodies\\recognized\end{array}\right) - \left(\begin{array}{c}I\ am\\recognized\end{array}\right) + \left(\begin{array}{c}antigens\\recognized\end{array}\right)\right]$$
$$- \left(\begin{array}{c}death\\rate\end{array}\right)$$
$$= c\left[\sum_{j=1}^{N} m_{ji}x_ix_j - k_1\sum_{j-1}^{N} m_{ij}x_ix_j + \sum_{j=1}^{n} m_{ji}x_iy_j\right] - k_2x_i$$

where N is the number of antibodies and n is the number of antigens, x_i (or x_j) is the concentration of antibody i (or j), y_j is the concentration of antigen j, c is a rate constant, k_1 is a suppressive effect and k_2 is the death rate, and m_{ji} is the matching function between antibody i and antibody (or antigen) j.

As can be seen from the above equation, the nature of an idiotypic interaction can be either positive or negative. Moreover, if the matching function is symmetric, then the balance between "I am recognized" and "antibodies recognized" (parameters c and k_1 in the equation) wholly determines whether the idiotypic effect is positive or negative, and we can simplify the equation. We can further simplify (1) if we only allow one antigen in the artificial immune system. In (2), the first term is simplified as we only have one antigen, and the suppression term is normalized to allow a "like for like" comparison between the different rate constants:

$$\frac{dx_i}{dt} = k_1m_ix_iy - \frac{k_2}{n}\sum_{j=1}^{n} m_{ij}x_ix_j - k_3x_i \qquad (13.1)$$

where k_1 is stimulation, k_2 suppression, k_3 death rate, m_i is the correlation between antibody i and the (sole) antigen, x_i (or x_j) is the concentration of antibody i (or j), y is the concentration of the (sole) antigen, m_{ij} is the correlation between antibodies i and j, and n is the number of antibodies.

Why would we want to use the idotypic effect? Because it might provide us with a way of achieving "diversity", similar to "crowding" or "fitness sharing"

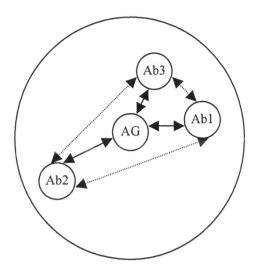

Figure 13.2. Illustration of the idiotypic effect.

in a genetic algorithm. For instance, in the movie recommender, we want to ensure that the final neighborhood population is diverse, so that we get more interesting recommendations. Hence, to use the idiotypic effect in the movie recommender system mentioned previously, the pseudo-code would be amended by adding the italicized lines as follows:

Initialize Artificial Immune Systems
Encode user for whom to make predictions as antigen Ag
WHILE (Artificial Immune Systems not Full) & (More Antibodies) DO
    Add next user as an antibody Ab
    Calculate matching scores between Ab and Ag *and Ab and other Abs*
    WHILE (Artificial Immune Systems at full size) & (Artificial Immune Systems not Stabilized) DO
        Reduce Concentration of all Abs by a fixed amount
        Match each Ab against Ag and stimulate as necessary
        Match each Ab against each other Ab and execute idiotypic effect
    OD
OD
Use final set of Antibodies to produce recommendation.

Figure 13.2 shows the idiotypic effect using dotted arrows and the standard stimulation using solid arrows. In the diagram antibodies Ab1 and Ab3 are very similar and they would have their concentrations reduced in the "iterate artificial immune systems" stage of the algorithm above.

At each iteration of the film recommendation artificial immune system the concentration of the antibodies is changed according to the formula outlined below. This will increase the concentration of antibodies that are similar to the antigen and can allow either the stimulation, suppression, or both, of antibody–antibody interactions to have an effect on the antibody concentration. More detailed discussion of these effects on recommendation problems are contained within Cayzer and Aickelin (2002a, b).

13.6.2   Danger Theory

Over the last decade, a new theory, called the Danger Theory, has become popular amongst immunologists. Its chief advocate is Matzinger (1994, 2001, 2003). A number of advantages are claimed for this theory; not least that it provides a method of "grounding" the immune response. The theory is not complete, and there are some doubts about how much it actually changes behaviour and/or structure. Nevertheless, the theory contains enough potentially interesting ideas to make it worth assessing its relevance to artificial immune systems.

To function properly, it is not simply a question of matching in the humoral immune system. It is fundamental that only the "correct" cells are matched as otherwise this could lead to a self-destructive autoimmune reaction. Classical immunology (Kuby, 2002) stipulates that an immune response is triggered when the body encounters something nonself or foreign. It is not yet fully understood how this self–nonself discrimination is achieved, but many immunologists believe that the difference between them is learnt early in life. In particular, it is thought that the maturation process plays an important role to achieve self-tolerance by eliminating those T- and B-cells that react to self. In addition, a "confirmation" signal is required: that is, for either B-cell or T- (killer) cell activation, a T- (helper) lymphocyte must also be activated. This dual activation is further protection against the chance of accidentally reacting to self.

Danger Theory debates this point of view (for a good introduction, see Matzinger, 2003). Technical overviews can be found in Matzinger (1994, 2001). She points out that there must be discrimination happening that goes beyond the self–nonself distinction described above. For instance:

1 There is no immune reaction to foreign bacteria in the gut or to the food we eat although both are foreign entities.

2 Conversely, some auto-reactive processes are useful, for example against self molecules expressed by stressed cells.

3 The definition of self is problematic—realistically, self is confined to the subset actually seen by the lymphocytes during maturation.

4 The human body changes over its lifetime and thus self changes as well. Therefore, the question arises whether defences against nonself learned early in life might be autoreactive later.

Other aspects that seem to be at odds with the traditional viewpoint are autoimmune diseases and certain types of tumors that are fought by the immune system (both attacks against self) and successful transplants (no attack against nonself).

Matzinger concludes that the immune system actually discriminates "some self from some nonself". She asserts that the Danger Theory introduces not just new labels, but a way of escaping the semantic difficulties with self and nonself, and thus provides grounding for the immune response. If we accept the Danger Theory as valid we can take care of "nonself but harmless" and of "self but harmful" invaders into our system. To see how this is possible, we will have to examine the theory in more detail.

The central idea in the Danger Theory is that the immune system does not respond to nonself but to danger. Thus, just like the self–nonself theories, it fundamentally supports the need for discrimination. However, it differs in the answer to what should be responded to. Instead of responding to foreignness, the immune system reacts to danger. This theory is borne out of the observation that there is no need to attack everything that is foreign, something that seems to be supported by the counter-examples above. In this theory, danger is measured by damage to cells indicated by distress signals that are sent out when cells die an unnatural death (cell stress or lytic cell death, as opposed to programmed cell death, or apoptosis).

Figure 13.3 depicts how we might picture an immune response according to the Danger Theory (Aickelin and Cayzer, 2002c). A cell that is in distress sends out an alarm signal, whereupon antigens in the neighborhood are captured by antigen-presenting cells such as macrophages, which then travel to the local lymph node and present the antigens to lymphocytes. Essentially, the danger signal establishes a danger zone around itself. Thus B-cells producing antibodies that match antigens within the danger zone get stimulated and undergo the clonal expansion process. Those that do not match or are too far away do not get stimulated.

Matzinger admits that the exact nature of the danger signal is unclear. It may be a "positive" signal (for example heat shock protein release) or a "negative" signal (for example lack of synaptic contact with a dendritic antigen-presenting cell). This is where the Danger Theory shares some of the problems associated with traditional self–nonself discrimination (i.e. how to discriminate danger from non-danger). However, in this case, the signal is grounded rather than being some abstract representation of danger.

How could we use the Danger Theory in artificial immune systems? The Danger Theory is not about the way artificial immune systems represent data

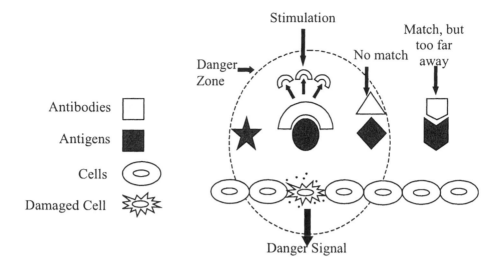

Figure 13.3. Danger theory illustration.

(Aickelin and Cayzer, 2002c). Instead, it provides ideas about which data the artificial immune systems should represent and deal with. They should focus on dangerous, i.e. interesting, data. It could be argued that the shift from non-self to danger is merely a symbolic label change that achieves nothing. We do not believe this to be the case, since danger is a grounded signal, and nonself is (typically) a set of feature vectors with no further information about whether all or some of these features are required over time. The danger signal helps us to identify which subset of feature vectors is of interest. A suitably defined danger signal thus overcomes many of the limitations of self–nonself selection. It restricts the domain of nonself to a manageable size, removes the need to screen against all self, and deals adaptively with scenarios where self (or nonself) changes over time.

The challenge is clearly to define a suitable danger signal, a choice that might prove as critical as the choice of fitness function for an evolutionary algorithm. In addition, the physical distance in the biological system should be translated into a suitable proxy measure for similarity or causality in artificial immune systems. This process is not likely to be trivial. Nevertheless, if these challenges are met, then future artificial immune system applications might derive considerable benefit, and new insights, from the Danger Theory: in particular, intrusion detection systems (Aickelin et al., 2003).

13.7   SOME PROMISING AREAS FOR FUTURE APPLICATION

It seems intuitively obvious that artificial immune systems should be most suitable for computer security problems. If the human immune system keeps our body alive and well, why can we not do the same for computers using artificial immune systems? (Aickelin et al., 2004)

We have outlined the traditional approach to do this. However, in order to provide viable intrusion detection systems, artificial immune systems must build a set of detectors that accurately match antigens. In current artificial-immune-system-based intrusion detection systems (Dasgupta and Gonzalez, 2002; Esponda et al., 2004; Hofmeyr and Forrest, 2000), both network connections and detectors are modeled as strings. Detectors are randomly created and then undergo a maturation phase where they are presented with good, i.e. self, connections. If the detectors match any of these they are eliminated, otherwise they become mature. These mature detectors start to monitor new connections during their lifetime. If these mature detectors match anything else, exceeding a certain threshold value, they become activated. This is then reported to a human operator who decides whether there is a true anomaly. If so, the detectors are promoted to memory detectors with an indefinite life span and minimum activation threshold (immunization) (Kim and Bentley, 2002).

An approach such as the above is known as negative selection as only those detectors (antibodies) that do not match live on (Forrest et al., 1994). Earlier versions of negative selection algorithm used a binary representation scheme; however, this scheme shows scaling problems when it is applied to real network traffic (Kim and Bentley, 2001). As the systems to be protected grow larger and larger so does self and nonself. Hence, it becomes more and more problematic to find a set of detectors that provides adequate coverage, whilst being computationally efficient. It is inefficient to map the entire self or nonself universe, particularly as they will be changing over time and only a minority of nonself is harmful, whilst some self might cause damage (e.g. internal attack). This situation is further aggravated by the fact that the labels self and nonself are often ambiguous and even with expert knowledge they are not always applied correctly (Kim and Bentley, 2002).

How can this problem be overcome? One approach might be to borrow ideas from the Danger Theory to provide a way of grounding the response and hence removing the necessity to map self or nonself. In our system, the correlation of low-level alerts (danger signals) will trigger a reaction (Aickelin et al, 2003). An important and recent research issue for intrusion detection systems is how to find true intrusion alerts from many thousands of false alerts generated (Hofmeyr and Forrest, 2000). Existing intrusion detection systems employ various types of sensors that monitor low-level system events. Those sensors report anomalies of network traffic patterns, unusual terminations of UNIX processes, memory usages, the attempts to access unauthorized files, etc. (Kim and Bentley, 2001). Although these reports are useful signals of real intrusions, they are often mixed with false alerts and their unmanageable volume forces a security officer to ignore most alerts (Hoagland and Staniford, 2002). Moreover, the low level of alerts makes it very hard for a security officer to identify advancing intrusions that usually consist of different stages of attack sequences. For instance, it is well known that computer hackers use a number of preparatory stages before actual hacking. Hence, the correlations between intrusion alerts from different attack stages provide more convincing attack scenarios than detecting an intrusion scenario based on low-level alerts from individual stages. Furthermore, such scenarios allow the intrusion detection system to detect intrusions early before damage becomes serious.

To correlate intrusion detection system alerts for detection of an intrusion scenario, recent studies have employed two different approaches: a probabilistic approach (Valdes and Skinner, 2001) and an expert system approach (Ning et al., 2002). The probabilistic approach represents known intrusion scenarios as Bayesian networks. The nodes of Bayesian networks are intrusion detection system alerts and the posterior likelihood between nodes is updated as new alerts are collected. The updated likelihood can lead to conclusions about a specific intrusion scenario occurring or not. The expert system approach initially builds possible intrusion scenarios by identifying low-level alerts. These alerts consist of prerequisites and consequences, and they are represented as hypergraphs. Known intrusion scenarios are detected by observing the low-level alerts at each stage, but these approaches have the following problems (Cuppens et al., 2002):

1 handling unobserved low-level alerts that comprise an intrusion scenario,

2 handling optional prerequisite actions,

3 handling intrusion scenario variations.

The common trait of these problems is that the intrusion detection system can fail to detect an intrusion when an incomplete set of alerts comprising an intrusion scenario is reported. In handling this problem, the probabilistic approach is more advantageous than the expert system approach because in theory it allows the intrusion detection system to correlate missing or mutated alerts. The current probabilistic approach builds Bayesian networks based on the similarities between selected alert features. However, these similarities alone can fail to identify a causal relationship between prerequisite actions and actual attacks if pairs of prerequisite actions and actual attacks do not appear frequently enough to be reported. Attackers often do not repeat the same actions in order to disguise their attempts. Thus, the current probabilistic approach fails to detect intrusions that do not show strong similarities between alert features but have causal relationships leading to final attacks. This limit means that such intrusion detection systems fail to detect sophisticated intrusion scenarios.

We propose artificial immune systems based on Danger Theory ideas that can handle the above intrusion detection system alert correlation problems (Aickelin et al., 2003). The Danger Theory explains the immune response of the human body by the interaction between antigen-presenting cells and various signals. The immune response of each antigen-presenting cell is determined by the generation of danger signals through cellular stress or cell death. In particular, the balance and correlation between different danger signals depending on different cell death causes would appear to be critical to the immunological outcome. In the human immune system, antigen-presenting cells activate according to the balance of apoptotic and necrotic cells and this activation leads to protective immune responses. Similarly, the sensors in intrusion detection systems report various low-level alerts and the correlation of these alerts will lead to the construction of an intrusion scenario.

13.8   TRICKS OF THE TRADE

Are artificial immune systems suitable for pure optimization? Depending on what is meant by optimization, the answer is probably no, in the same sense as "pure" genetic algorithms are not "function optimizers". One has to keep in mind that although the immune system is about matching and survival, it is really a team effort where multiple solutions are produced all the time that together provide the answer. Hence, in our opinion artificial immune systems are probably more suited as an optimizer where multiple solutions are of benefit, either directly, e.g. because the problem has multiple objectives or indirectly, e.g. when a neighborhood of solutions is produced that is then used to generate the desired outcome. However, artificial immune systems can be made into more focused optimizers by adding hill climbing or other functions that exploit local or problem-specific knowledge, similar to the idea of augmenting genetic algorithm to memetic algorithms.

What problems are artificial immune systems most suitable for? As mentioned above, we believe that although using artificial immune systems for pure optimization, e.g. the traveling salesman problem or job shop scheduling, can be made to work, this is probably missing the point. Artificial immune systems are powerful when a population of solution is essential either during the search or as an outcome. Furthermore, the problem has to have some concept of "matching". Finally, because at their heart artificial immune systems are evolutionary algorithms, they are more suitable for problems that change over time and need to be solved again and again, rather than one-off optimizations. Hence, the evidence seems to point to data mining in its wider meaning as the best area for artificial immune systems.

How do I set the parameters? Unfortunately, there is no short answer to this question. As with the majority of other heuristics that require parameters to operate, their setting is individual to the problem solved and universal values are not available. However, it is fair to say that along with other evolutionary algorithms artificial immune systems are robust with respect to parameter values as long as they are chosen from a sensible range.

Why not use a genetic algorithm instead? Because you may miss out on the benefits of the idiotypic network effects.

Why not use a neural network instead? Because you may miss out on the benefits of a population of solutions and the evolutionary selection pressure and mutation.

Are artificial immune systems Learning Classifier Systems under a different name? No, not quite. However, to our knowledge learning classifier systems are probably the most similar of the better known metaheuristics, as they also combine some features of evolutionary algorithms and neural networks. However, these features are different. Someone who is interested in implementing artificial immune systems or learning classifier systems is likely to be well advised to read about both approaches to see which one is most suited for the problem at hand.

13.9   CONCLUSIONS

The immune system is highly distributed, highly adaptive, self-organizing in nature, maintains a memory of past encounters, and has the ability to continually learn about new encounters. The artificial immune system is an example of a system developed around the current understanding of the immune system. It illustrates how an artificial immune system can capture the basic elements of the immune system and exhibit some of its chief characteristics.

Artificial immune systems can incorporate many properties of natural immune systems, including diversity, distributed computation, error tolerance, dynamic learning and adaptation and self-monitoring. The human immune system has motivated scientists and engineers for finding powerful information processing algorithms that has solved complex engineering tasks. The artificial immune system is a general framework for a distributed adaptive system and could, in principle, be applied to many domains. The artificial immune system can be applied to classification problems, optimization tasks and other domains. Like many biologically inspired systems it is adaptive, distributed and autonomous. The primary advantages of the artificial immune system are that it only requires positive examples, and the patterns it has learnt can be explicitly examined. In addition, because it is self-organizing, it does not require effort to optimize any system parameters.

To us, the attraction of the immune system is that if an adaptive pool of antibodies can produce "intelligent" behavior, can we harness the power of this computation to tackle the problem of preference matching, recommendation and intrusion detection? Our conjecture is that if the concentrations of those antibodies that provide a better match are allowed to increase over time, we should end up with a subset of good matches. However, we are not interested in optimizing, i.e. in finding the one best match. Instead, we require a set of antibodies that are a close match but which are at the same time distinct from each other for successful recommendation. This is where we propose to harness the idiotypic effects of binding antibodies to similar antibodies to encourage diversity.

SOURCES OF ADDITIONAL INFORMATION

The following websites, books and proceedings should be an excellent starting point for those readers wishing to learn more about artificial immune systems.

1 *Artificial Immune Systems and Their Applications* by D. Dasgupta (ed.), Springer, Berlin, 1999.

2 *Artificial Immune Systems: A New Computational Intelligence Approach* by L. de Castro and J. Timmis, Springer, Berlin, 2002.

3 *Immunocomputing: Principles and Applications* by A. Tarakanov et al., Springer, Berlin, 2003.

4 *Proceedings of the International Conference on Artificial Immune Systems (ICARIS),* Springer, Berlin, 2003.

5 Artificial Immune Systems Forum Webpage: http://www.artificial-immune-systems.org/artist.htm

6 Artificial Immune Systems Bibliography:
http://issrl.cs.memphis.edu/ Artificial Immune Systems/Artificial Immune Systems_bibliography.pdf

References

Aickelin, U. and Cayzer, S., 2002c, The danger theory and its application to artificial immune systems, in: *Proc. 1st Int. Conf. on Artificial Immune Systems* (Canterbury, UK), pp. 141–148.

Aickelin, U., Bentley, P., Cayzer, S., Kim, J. and McLeod, J., 2003, Danger theory: The link between artificial immune systems and intrusion detection systems, in: *Proc. 2nd Int. Conf. on Artificial Immune Systems* (Edinburgh), Springer, Berlin, pp. 147–155.

Aickelin, U., Greensmith, J. and Twycross, J., 2004, Immune system approaches to intrusion detection—a review, in: *Proc. ICARIS-04, 3rd Int. Conf. on Artificial Immune Systems* (Catania, Italy), Lecture Notes in Computer Science, Vol. 3239, Springer, Berlin, pp. 316–329.

Amazon, 2003, Recommendations, http://www.amazon.com/

Cayzer, S. and Aickelin, U., 2002a, A recommender system based on the immune network, in: *Proc. CEC2002* (Honolulu, HI), pp. 807–813.

Cayzer, S. and Aickelin, U., 2002b, On the effects of idiotypic interactions for recommendation communities in artificial immune systems, in: *Proc. 1st Int. Conf. on Artificial Immune Systems* (Canterbury, UK), pp. 154–160.

Cuppens, F. et al., 2002, Correlation in an intrusion process, *Internet Security Communication Workshop (SECI'02)*.

Dasgupta, D., ed., 1999, *Artificial Immune Systems and Their Applications,* Springer, Berlin.

Dasgupta, D., Gonzalez, F., 2002, An immunity-based technique to characterize intrusions in computer networks, *IEEE Trans. Evol. Comput.* **6**:1081–1088.

de Castro, L. N. and Von Zuben, F. J., 2002, Learning and optimization using the clonal selection principle, *IEEE Trans. Evol. Comput.*, Special issue on artificial immune systems, **6**:239–251.

Esponda, F., Forrest, S. and Helman, P., 2004, A formal framework for positive and negative detection, *IEEE Trans. Syst., Man Cybernet.*, **34**:357–373.

Farmer, J. D., Packard, N. H. and Perelson, A. S., 1986, The immune system, adaptation, and machine learning, *Physica* **22**:187–204.

Forrest, S., Perelson, A. S., Allen, L. and Cherukuri, R., 1994, Self–nonself discrimination in a computer, in: *Proc. IEEE Symp. on Research in Security and Privacy,* (Oakland, CA), pp. 202–212.

Hightower, R. R., Forrest, S. and Perelson, A. S., 1995, The evolution of emergent organization in immune system gene libraries, *Proc. 6th Conf. on Genetic Algorithms,* pp. 344–350.

Hoagland, J. and Staniford, S., 2002, *Viewing Intrusion Detection Systems Alerts: Lessons from SnortSnarf,* http://www.silicondefense.com/ software/snortsnarf

Hofmeyr, S. and Forrest, S., 2000, Architecture for an artificial immune systems, *Evol. Comput.* **7**:1289–1296.

Jerne, N. K., 1973, Towards a network theory of the immune system, *Ann. Immunol.* **125**:373–389.

Kim, J. and Bentley, P., 1999, The artificial immune model for network intrusion detection, *Proc. 7th Eur. Congress on Intelligent Techniques and Soft Computing (EUFIT'99)*.

Kim, J. and Bentley, P., 2001, Evaluating negative selection in an artificial immune systems for network intrusion detection, *Proc. Genetic and Evolutionary Computation Conf. 2001,* pp. 1330–1337.

Kim, J. and Bentley, P., 2002, Towards an artificial immune systems for network intrusion detection: an investigation of dynamic clonal selection, *Proc. Congress on Evolutionary Computation 2002,* pp. 1015–1020.

Kubi, J., 2002, *Kubi Immunology,* 5th edn, Richard A. Goldsby, Thomas J. Kindt and Barbara A. Osborne, eds, Freeman, San Francisco.

Matzinger, P., 2003, http://cmmg.biosci.wayne.edu/asg/polly.html

Matzinger, P., 2001, The danger model in its historical context, *Scand. J. Immunol.* **54**:4–9.

Matzinger, P., 1994, Tolerance, danger and the extended family, *Ann. Rev. Immunol.* **12**:991–1045.

Morrison, T. and Aickelin, U., 2002, An artificial immune system as a recommender system for web sites, in: *Proc. 1st Int. Conf. on Artificial Immune Systems (ICARIS-2002)* (Canterbury, UK), pp. 161–169.

Ning, P., Cui, Y. and Reeves, S., 2002, Constructing attack scenarios through correlation of intrusion alerts, in: *Proc. 9th ACM Conf. on Computer and Communications Security,* pp. 245–254.

Perelson, A. S. and Weisbuch, G., 1997, Immunology for physicists, *Rev. Mod. Phys.* **69**:1219–1267.

Resnick, P. and Varian, H. R., 1997, Recommender systems, *Commun. ACM* **40**:56–58.

Valdes, A. and Skinner, K., 2001, Probabilistic alert correlation, *Proc. RAID'2001,* pp. 54–68.

Chapter 14

SWARM INTELLIGENCE

Daniel Merkle, Martin Middendorf
Department of Computer Science
University of Leipzig, Germany

14.1 INTRODUCTION

The complex and often coordinated behavior of swarms fascinates not only biologists but also computer scientists. Bird flocking and fish schooling are impressive examples of coordinated behavior that emerges without central control. Social insect colonies show complex problem-solving skills arising from the actions and interactions of nonsophisticated individuals.

Swarm Intelligence is a field of computer science that designs and studies efficient computational methods for solving problems in a way that is inspired by the behavior of real swarms or insect colonies (see e.g. Bonabeau et al., 1999; Kennedy et al., 2001). Principles of self-organization and local or indirect communication are important for understanding the complex collective behavior (Camazine et al., 2001). Examples where insights into the behavior of natural swarms has influenced the design of algorithms and systems in computer science include the following (see Bonabeau et al., 1999; Middendorf, 2002 for more information):

- Collective transport of ants has inspired the design of controllers of robots for doing coordinated work (Kube and Bonabeau, 2000).

- Brood sorting behavior of ants motivated several clustering and sorting algorithms (e.g., Handl and Meyer, 2002; Lumer and Faieta, 1994).

- The path-finding and orientation skills of the desert ant *Cataglyphis* were used as an archetype for building a robot orientation unit (Lambrinos et al., 2000).

- Models for the division of labor between members of an ant colony were used to regulate the joint work of robots (e.g. Agassounoun et al., 2001; Goldberg and Mataric, 2000).

In this chapter we focus on swarm intelligence methods for solving optimization and search problems. The two main areas of swarm intelligence that are relevant for such problems are ant colony optimization (ACO) and particle swarm optimization (PSO).

ACO is a metaheuristic for solving combinatorial optimization problems. It is inspired by the way real ants find shortest paths from their nest to food sources. An essential aspect thereby is the indirect communication of the ants via pheromone, i.e., a chemical substance which is released into the environment and that influences the behavior or development of other individuals of the same species. Ants mark their paths to the food sources by laying a pheromone trail along their way. The pheromone traces can be smelled by other ants and lead them to the food source.

PSO is a metaheuristic that is mainly used for finding maximum or minimum values of a function (Kennedy et al., 2001). PSO is inspired by the behavior of swarms of fishes or flocks of birds to find a good food place. The coordination of movements of the individuals in the swarm is the central aspect that inspires PSO.

14.2   ANT COLONY OPTIMIZATION

A renowned biological experiment called the double bridge experiment was the inspiring source for the first ACO algorithm (Dorigo et al., 1991; Dorigo, 1992). The double bridge experiment (Deneubourg et al., 1990; Goss et al., 1989) was designed to investigate the pheromone trail laying and following behavior of the Argentine ant *Iridomyrmex humilis*. In the experiment a double bridge with two branches of different lengths connected the nest of this species with a food source (see Figure 14.2). The long branch of the bridge was twice as long as the shorter branch. In most runs of this experiment it was found that after a few minutes nearly all ants use the shorter branch. This is interesting because Argentine ants cannot see very well. The explanation of this behavior has to do with the fact that the ants lay pheromone along their path. It is likely that ants which randomly choose the shorter branch arrive earlier at the food source. When they go back to the nest they smell some pheromone on the shorter branch and therefore prefer this branch. The pheromone on the shorter branch will accumulate faster than on the longer branch so that after some time the concentration of pheromone on the former is much higher and nearly all ants take the shorter branch. Similar to the experiment with branches of different lengths, when both branches have the same length, after some minutes nearly all ants use the same branch. But in several repetitions it is a random process which of the two branches will be chosen. The explanation is that when one branch has got a slightly higher pheromone concentration due to random fluctuations this branch will be preferred by the ants so that the difference in

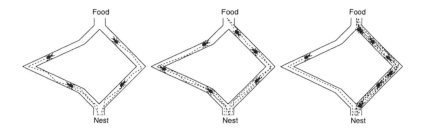

Figure 14.1. Double bridge experiment.

pheromone concentration will increase and after some time all ants take this branch.

Inspired by this experiment, Dorigo and colleagues designed an algorithm for solving the traveling salesperson problem (TSP) (see Dorigo, 1992; Dorigo et al., 1991), and initiated the field of ACO. In recent years this field of research has become quite rich and ACO algorithms have now been designed for various application problems and different types of combinatorial optimization problems including dynamic and multi-objective optimization problems (see Cordón et al., 2002; Maniezzo et al., 2001; Stützle and Dorigo, 2002b for other overviews). Some papers treat the theory of ACO and modeling of ACO algorithms (Gutjahr, 2000, 2002; Merkle and Middendorf, 2002a; Stützle and Dorigo, 2002a). An ACO metaheuristic has been formulated as a generic frame that contains most of the different ACO algorithms that have been proposed so far (see Dorigo and Di Caro, 1999).

The idea of ACO is to let artificial ants construct solutions for a given combinatorial optimization problem. A prerequisite for designing an ACO algorithm is to have a constructive method which can be used by an ant to create different solutions through a sequence of decisions. Typically an ant constructs a solution by a sequence of probabilistic decisions where every decision extends a partial solution by adding a new solution component until a complete solution is derived. The sequence of decisions for constructing a solution can be viewed as a path through a corresponding decision graph (also called construction graph). Hence, an artificial ant that constructs a solution can be viewed as walking through the decision graph. The aim is to let the artificial ants find paths through the decision graph that correspond to good solutions. This is done in an iterative process where the good solutions found by the ants of an iteration should guide the ants of following iterations. Therefore, ants that have found good solutions are allowed to mark the edges of the corresponding path in the decision graph with artificial pheromone. This pheromone guides following ants of the next iteration so that they search near the paths to good solutions. In order that pheromone from older iterations does not influence the following iterations for too long, during an update of the pheromone values some percentage of the pheromone evaporates. Thus, an ACO algorithm is an iterative process where pheromone information is transferred from one iteration to the next one. The process continues until some stopping criterion is met: e.g., a certain number of iterations has been done or a solution of a given quality has been found. A scheme of an ACO algorithm is given in the following.

ACO scheme:
  Initialize pheromone values
  repeat
    for ant $k \in \{1, \ldots, m\}$
      construct a solution
    endfor
    forall pheromone values **do**
      decrease the value by a certain percentage {evaporation}
    endfor
    forall pheromone values corresponding to good solutions
    do
      increase the value {intensification}
    endfor
  until stopping criterion is met

In what follows we illustrate how the general ACO scheme can be applied to a broad class of optimization problems by means of three examples. In the first example a more detailed ACO scheme is described and applied to the TSP. An alternative approach is contained in the second example. The third example is an application of ACO to a scheduling problem which is used in comparison to the first example to discuss some additional aspects that have to be considered for designing ACO algorithms.

14.2.1 Example 1: Basic ACO and the TSP

The objective of ACO is to find good solutions for a given combinatorial optimization problem (Dorigo, 1992; Dorigo and Di Caro, 1999; Dorigo et al., 1991). For an easier description we restrict the following description to the broad class of optimization problems which have solutions that can be expressed as permutations of a set of given items. Such problems are called permutation problems, the TSP being a well known example. After definition of the TSP we describe the elements of the ACO scheme that constitute an ACO algorithm: namely, pheromone information, solution construction, pheromone update: evaporation + intensification, and stochastic.

The TSP Problem   This problem is to find for a given set of n cities with distances d_{ij} between each pair of cities $i, j \in [1 : n]$ a shortest closed tour that contains every city exactly once. Every such tour together with a start city can be characterized by the permutation of all cities as they are visited along the tour. Vice versa, each permutation of all cities corresponds to a valid solution, i.e. a closed tour.

Pheromone Information   An important part in the design of an ACO algorithm is to find a definition of the pheromone information so that it reflects the most relevant information for the solution construction. The pheromone information for permutation problems can usually be encoded in an $n \times n$ pheromone matrix $[\tau_{ij}]$, $i, j \in [1 : n]$. For the TSP problem the pheromone value τ_{ij} expresses the desirability to assign city j after city i in the permutation. The pheromone matrix for the TSP problem is initialized so that all values τ_{ij} with $i \neq j$ are the same. Note that the values τ_{ii} are not needed because each city is selected only once.

> *TSP-ACO*:
>   Initialize pheromone values
>   **repeat**
>   **for** ant $k \in \{1, \ldots, m\}$ {solution construction}
> $S := \{1, \ldots, n\}$ {set of selectable cities}
>   choose city i with probability p_{0i}
>   **repeat**
>   choose city $j \in S$ with probability p_{ij}
> $S := S - \{j\}$
> $i := j$
>   **until** $S = \emptyset$
>   **endfor**
>   **forall** i, j **do**
> $\tau_{ij} := (1 - \rho) \cdot \tau_{ij}$ {evaporation}
>   **endfor**
>   **forall** i, j in iteration best solution **do**
> $\tau_{i,j} := \tau_{ij} + \Delta$ {intensification}
>   **endfor**
>   **until** stopping criterion is met

Solution Construction   An iterative solution construction method that can be used by the ants is to start with a random item and then always choose the next item from the set S of selectable items that have not been selected so far until no item is left. Initially, the set of selectable items S contains all items;

after each decision, the selected item is removed from S. Recall that in the case of the TSP the items are the cities. Every decision is made randomly where the probability equals the amount of pheromone relative to the sum of all pheromone values of items in the selection set S:

$$p_{ij} := \frac{\tau_{ij}}{\sum_{z \in S} \tau_{iz}} \quad \forall j \in S$$

For most optimization problems additional problem-dependent heuristic information can be used to give the ants additional hints about which item to choose next. To each pheromone value τ_{ij} there is defined a corresponding heuristic value η_{ij}. For the TSP a suitable heuristic is to prefer a next city j that is near to the current city i, for example by setting $\eta_{ij} := 1/d_{ij}$. The probability distribution when using a heuristic is

$$p_{ij} := \frac{\tau_{ij}^{\alpha} \cdot \eta_{ij}^{\beta}}{\sum_{z \in S} \tau_{iz}^{\alpha} \cdot \eta_{iz}^{\beta}} \quad \forall j \in S \tag{14.1}$$

where parameters α and β are used to determine the relative influence of pheromone values and heuristic values.

In order to better exploit the pheromone information it has been proposed that the ant follows with some probability $q_0 \in (0, 1)$ the strongest trail, i.e. the edge in the decision graph with the maximal product of pheromone value and corresponding heuristic information (Dorigo and Gambardella, 1997). For this case q_0 is a parameter of the algorithm and with probability q_0 an ant chooses next city j from the selectable cities in S which maximizes $\tau_{ij}^{\alpha} \cdot \eta_{ij}^{\beta}$. With probability $1 - q_0$ the next item is chosen according to the probability distribution determined by (14.1).

Pheromone Update   All m solutions that are constructed by the ants in one iteration are evaluated according to the respective objective function and the best solution π^* of the current iteration is determined. Then the pheromone matrix is updated in two steps:

1 Evaporation: All pheromone values are reduced by a fixed proportion $\rho \in (0, 1)$:

$$\tau_{ij} := (1 - \rho) \cdot \tau_{ij} \quad \forall i, j \in [1 : n]$$

2 Intensification: All pheromone values corresponding to the best solution π^* are increased by an absolute amount $\Delta > 0$:

$$\tau_{i\pi^*(i)} := \tau_{i\pi^*(i)} + \Delta \quad \forall i \in [1 : n]$$

Table 14.1. ACO variables and parameters.

τ_{ij}	Pheromone value
η_{ij}	Heuristic value
m	Number of ants per iteration
$\bar{m}$	Number of ants per iteration allowed to increase pheromone
α	Influence of pheromone
β	Influence of heuristic
ρ	Evaporation rate
Δ	Amount of pheromone added during pheromone intensification
q_0	Probability to follow the strongest trail
π^*	Best solution in the actual iteration
π^e	Best solution found so far (elitist solution)

Stopping Criterion The ACO algorithm executes a number of iterations until a specified stopping criterion has been met. Most commonly used stopping criteria are (possibly used in combination) that a predefined maximum number of iterations has been executed, a specific level of solution quality has been reached, or the best solution has not changed over a certain number of iterations.

A good comparison of the optimization behavior of different ACO implementations for the TSP problem can be found in Stützle and Hoos (2000). The parameters and variables of ACO algorithms introduced in this section are summarized in Table 14.1.

14.2.2 Example 2: Population-Based ACO and TSP

In standard ACO algorithms the information that is transferred from one iteration to the next is the pheromone information—in the case of permutation problems this is the pheromone matrix. An alternative approach that was proposed recently is population-based ACO (P-ACO) (see Guntsch and Middendorf, 2002b). One idea of P-ACO is to transfer less and only the most important information from one iteration to the next. This is done in the form of a small population of good solutions. In this section we describe the differences between P-ACO and standard ACO for permutation problems. It was shown that both approaches show a similar performance on the TSP (Guntsch and Middendorf, 2002b). A scheme of a P-ACO algorithm for the TSP is given in the following (compare with the scheme of ACO-TSP).

P-ACO-TSP:
  $P := \emptyset$
  Initialize pheromone values

repeat
    for ant $k \in \{1, \ldots, m\}$ {solution construction}
        $S := \{1, \ldots, n\}$ {set of selectable cities}
        choose city i with probability p_{0i}
        for $i = 1$ **to** n **do**
            choose city j with probability p_{ij}
            $S := S - \{j\}$
            $i := j$
        endfor
    endfor
    If $|P| = k$ remove the oldest solution $\bar{\pi}$ from
        the population: $P := P - \bar{\pi}$
    Determine the best solution of the iteration and add it
        to the population: $P := P + \pi^*$
    Compute the new pheromone matrix from P
until stopping criterion is met

Information Transfer and Population Matrix Instead of a complete pheromone matrix as in ACO, P-ACO transfers a small population P of the k best solutions that have been found in past iterations. Since each solution for a permutation problem is a permutation of n items, the population can be stored in an $n \times k$ matrix $P = [p_{ij}]$, where each column of P contains one solution. This matrix is called the population matrix. It contains the best solution of each of the preceding k iterations. When employing an elitism strategy, the best solution found so far in all iterations is—as in standard ACO—also always transferred to the next iteration. In that case the population matrix contains an additional column for the elitist solution.

Population Matrix Update When the ants in an iteration have constructed their solutions the population (matrix) is updated. The best solution of the current iteration is added to P. If, afterwards, P contains $k + 1$ solutions, the oldest solution is removed from P. The initial population is empty and after the first k iterations the population size remains k. Hence, for an update only one column in the population matrix has to be changed. Additionally, if elitist update is used and the best solution of the iteration is better than the elitist solution, the corresponding column is overwritten by the new solution. Note that each solution in the population has an influence on the decisions of the ants over exactly k subsequent iterations. Other schemes for deciding which solutions should enter/leave the population are discussed in Guntsch and Middendorf (2002a).

Construction of Pheromone Matrix In P-ACO a pheromone matrix (τ_{ij}) is used by the ants for solution construction in the same way as in standard ACO. But differently, in P-ACO the pheromone matrix is derived in every iteration anew from the population matrix as follows. Each pheromone value is set to an initial value $\tau_{\text{init}} > 0$ and is increased, if there are corresponding solutions in the population:

$$\tau_{ij} := \tau_{\text{init}} + \zeta_{ij} \cdot \Delta \qquad (14.2)$$

with ζ_{ij} denoting the number of solutions $\pi \in P$ with $\pi(i) = j$, i.e. $\zeta_{ij} = |\{h : p_{ih} = j\}|$. Hence, in P-ACO a pheromone value is equal to one of the following possible values $\tau_{\text{init}}, \tau_{\text{init}} + \Delta, \ldots, \tau_{\text{init}} + k \cdot \Delta$ (when using an elitist solution $\tau_{\text{init}} + (k+1) \cdot \Delta$ is also possible). An update of the pheromone values is done implicitly by a population update:

- A solution π entering the population, corresponds to a positive update:

$$\tau_{i\pi(i)} := \tau_{i\pi(i)} + \Delta$$

- A solution σ leaving the population, corresponds to a negative update:

$$\tau_{i\sigma(i)} := \tau_{i\sigma(i)} - \Delta$$

Note that a difference to the standard ACO algorithm is that no evaporation is used to reduce the pheromone values at the end of an iteration.

14.2.3 Example 3: ACO for a Scheduling Problem

In this section the ACO approach is applied to a scheduling permutation problem which is called the Single Machine Total Weighted Tardiness Problem (SMTWTP). The differences between the ACO algorithm for the SMTWTP and the TSP-ACO illuminate two important aspects for the design of ACO algorithms, namely the pheromone encoding and the pheromone evaluation. Moreover, the proper adaptation of heuristics to be used for ACO is discussed. These aspects can be arranged as follows into the list of elements that constitute an ACO algorithm:

A Pheromone information

- Pheromone encoding

B Solution construction

- Pheromone evaluation
- Adaptation of heuristics

The SMTWTP Problem   For the SMTWTP n jobs are given that have to be scheduled onto a single machine. Every job $j \in [1 : n]$ has a due date d_j, a processing time p_j, and a weight w_j. If C_j denotes the completion time of job j in a schedule, then $L_j = C_j - d_j$ defines its lateness and $T_j = \max(0, L_j)$ its tardiness. The objective is to find a schedule that minimizes the total weighted tardiness of all jobs $\sum_{j=1}^{n} w_j T_j$.

Pheromone Encoding   When designing an ACO algorithm for an optimization problem it is important to encode the pheromone information in a way that is suitable for the problem. For the TSP it is relevant which cities are next to each other in the permutation because the distance between the cities determines the quality of the solution. Therefore pheromone values τ_{ij} are used to express the desirability that city j comes after i. For the SMTWTP the relative position of a job in the schedule is much more important than its direct predecessor or its direct successor in the schedule (see also (Blum and Sampels, 2002a) for other scheduling problems). Therefore pheromone values for the SMTWTP are used differently than for the TSP. Pheromone value τ_{ij} expresses the desirability to assign item j at place i of the permutation. This pheromone matrix is of type place×item whereas the pheromone matrix used for the TSP is of type item×item. For SMTWTP an ant starts to decide which job is the first in the schedule and then always decides which job is on the next place. The pheromone matrix for the SMTWTP problem is initialized so that all values τ_{ij}, $i, j \in [1 : n]$ are the same.

Pheromone Evaluation   Another important aspect of ACO algorithms is how the pheromone information is used by the ants for their decisions. Real ants use trail pheromone only locally because they cannot smell it over long distances. The artificial ants in TSP-ACO also use the pheromone values locally which means that an ant at city i considers only the pheromone values τ_{ij} that lead to a possible next city $j \in S$. In principle a local evaluation of the pheromone values is also possible for the SMTWTP (and has been used so, see Bauer et al., 1999). An ant that has to decide which job is on the next place i in the permutation considers all values τ_{ij}, $j \in S$ which indicate how good the selectable jobs have performed on this place. But assume that for some selectable job $j \in S$ its highest pheromone value is τ_{lj} for an $l < i$. This indicates that for job j place l in the schedule is very good. But this also means that job j should not be placed much later than place l in order not to risk a due date violation. Therefore, even when the value τ_{ij} is small the ant should choose job l with high probability. Therefore, for SMTWTP a global pheromone evaluation rule has been proposed which is called summation evaluation because an ant that has to decide about place i of the permutation makes the selection probability for every selectable job dependent on the sum of all pheromone values for this job up to place i (Merkle and Middendorf, 2003b):

$$p_{ij} = \frac{\left(\sum_{l=1}^{i} \tau_{lj}\right)^{\alpha} \cdot \eta_{ij}^{\beta}}{\sum_{z \in S}\left(\sum_{l=1}^{i} \tau_{lz}\right)^{\alpha} \cdot \eta_{iz}^{\beta}} \quad \forall j \in S \qquad (14.3)$$

So far the potential of global evaluation has not been fully recognized because nearly all ACO algorithms use only local evaluation: see Bautista and Pereira (2002), Merkle et al. (2002), Merkle and Middendorf (2002b, 2003a) for other applications or other global evaluation methods.

To demonstrate the influence of the pheromone evaluation method and the change of pheromone values in ACO we show results of a very simple SMTWTP test instance (for more results see Merkle and Middendorf (2003a); another investigation of ACO on simple problems is Stützle and Dorigo (2001)). It consists of 50 jobs where job i, $i \in [1 : 50]$ has processing time $p_i = 1$, due date $d_i = i$, and weight $w_i = 1$. Clearly, to place job i, $i = 1, \ldots, n$ on place i is the only optimal solution with costs 0. Figure 14.2 shows the average change of pheromone values for ACO-SMTWTP with local and with global evaluation (no heuristic was used) for several runs with $m = 10$ ants per iteration. The figure shows clearly that for this problem summation evaluation performs much better than local evaluation. Compared to local evaluation the results of summation evaluation depicted in Figure 14.2 show a very symmetric behavior and do not have the undesired property that some of the jobs with small number are scheduled very late.

Adaptation of Heuristics For many (scheduling) problems there exist priority heuristics which can be used to decide which job is next when building a schedule. An example for the unweighted form of the SMTWTP is the modified due date (MDD) rule, i.e.

$$\eta_{ij} = \frac{1}{\max\{\mathcal{T} + p_j, d_j\}} \qquad (14.4)$$

where $\mathcal{T}$ is the total processing time of all jobs already scheduled. Observe that the heuristic prefers jobs with a small due date from all jobs that would finish before their due date when scheduled next. Furthermore, of all those jobs that will finish after their due date the jobs with short processing times are preferred. Some care has to be taken when using standard priority heuristics for scheduling problems in an ACO algorithm because the heuristic values might not properly reflect the relative influence they should have on the decisions of the ants. In the case of the MDD heuristic the problem occurs that the values of $\max\{\mathcal{T} + p_j, d_j\}$ become much larger—due to $\mathcal{T}$—when deciding about jobs to place further at the end of the schedule. As a consequence, the heuristic differences between the jobs are, in general, small at the end of the schedule. This

Local evaluation

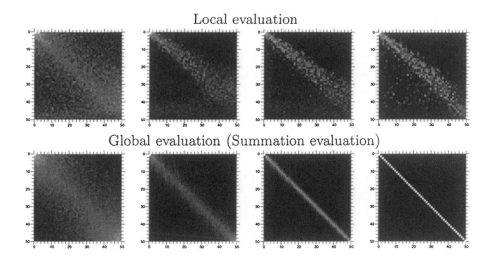

Global evaluation (Summation evaluation)

Figure 14.2. Comparison between SMTWTP-ACO with local evaluation and summation evaluation: Pheromone matrices averaged over 25 runs in iterations 800, 1500, 2200, and 2900: brighter gray colors indicate higher pheromone values.

means that the ants cannot really differentiate between the various alternatives. To avoid this effect an accordingly adapted heuristics should be used (Merkle and Middendorf, 2003b): for example,

$$\eta_{ij} = \frac{1}{\max\{\mathcal{T} + p_j, d_j\} - \mathcal{T}} \qquad (14.5)$$

To illustrate the effect of an adapted heuristic together with global pheromone evaluation, some test results for benchmark problems with $n = 100$ jobs from the OR-Library, 2004 are given. The ACO parameters used are $m = 20$ ants per generation $\alpha = 1$, $\beta = 1$, $\rho = 0.1$, $q_0 = 0.9$, and local optimization was applied to solutions found by the ants (see Merkle and Middendorf, 2003b for more details). Table 14.2 compares the behavior of the algorithm using non-adapted heuristic (14.4) and local pheromone evaluation with the algorithms that use one or both of adapted heuristic (14.5) and global pheromone evaluation. The results clearly show that using an adapted heuristic (14.5) or the global pheromone evaluation improves the results significantly and using both is best.

Table 14.2. Influence of global pheromone evaluation and adapted heuristic on solution quality for SMTWTP. Difference of total tardiness to total tardiness of best results from the literature average over 125 instances (for details see Merkle and Middendorf, 2003b). Σ with summation evaluation; H with adapted heuristic.

ACO-ΣH	ACO-Σ	ACO-H	ACO
79.5	200.0	204.5	1198.6

14.2.4 Advanced Features of ACO

In this section several variations and extension of the ACO algorithms are described that often lead to an increased search efficiency and better optimization results.

Variants of Pheromone Update Several variations of pheromone update have been proposed in the literature:

- *Quality-dependent pheromone update.* In some ACO algorithms not only the best solution, but the $\bar{m} < m$ best solutions of each iteration are allowed to increase the pheromone values. In addition the amount of pheromone that is added can be made dependent on the quality of the solution so that the more pheromone is added the better the solution is Dorigo et al. (1996). For the TSP this means that for shorter tours more pheromone is added.

- *Rank-based pheromone update.* Here the $\bar{m} \leq m$ best ants of an iteration are allowed to increase the pheromone. The amount of pheromone an ant is allowed to add depends on its rank within the $\bar{m}$ best solutions and the quality of the solution (Bullnheimer et al., 1998).

- *Elitist solution pheromone update.* It can be advantageous to enforce the influence of the best solution π^e that has been found so far over all iterations, called the elitist solution (Dorigo et al., 1996). This is done by adding pheromone during pheromone intensification also according to this solution. Several variations have been studied: e.g. to let randomly update either the iteration best or the elitist solution with increasing probability for an elitist update (Stützle and Hoos, 2000) or to apply elitist pheromone update but to forget the elitist solution after several iterations by replacing it with the iteration best solution (Merkle et al., 2002).

- *Best–worst pheromone update.* This pheromone update method in addition to the standard pheromone update reduces the pheromone values according to the worst solution of an iteration provided that a pheromone value does not also correspond to the elitist solution (Cordón et al., 2000). A problem of this method is that often a decision which can lead to bad solutions can also lead to a very good solution. In that case there is a danger that the corresponding pheromone value decreases too fast so that the very good solution is not found by the ants.

- *Online step-by-step pheromone update (Dorigo and Gambardella, 1997; Dorigo et al., 1991).* This means that an ants adds or removes pheromone from an edge in the decision graph it has chosen immediately after the decision was done (see Dorigo and Di Caro (1999) for more details). One motivation to use online step-by-step pheromone update in addition to the standard update is to remove pheromone to increase the variability in the choices of the ants during an iteration.

- *Moving average pheromone update.* A pheromone update scheme where each constructed solution is allowed to update the pheromone (Maniezzo, 1999). When the actual solution is better than the average quality of the last $k > 0$ solutions then it increases its corresponding pheromone values and otherwise it decreases them.

- *Minimum pheromone values.* The use of minimum pheromone values was proposed in order to guarantee that each possible choice always has a minimum probability to be chosen (Stützle and Hoos, 2000).

Other ACO Variants   Several variants of ACO algorithms which do not regard the pheromone update have also been proposed (see Dorigo and Di Caro (1999) for an overview):

- *Candidate lists.* A candidate list defines for each decision a set of preferable choices (Dorigo and Gambardella, 1997). For the TSP a candidate list can be defined for each city to determine the set of preferred successor cities. An ant then chooses, if possible, the next city only from cities that are in the selection set S and also in the candidate list. Only if no city in the candidate list is selectable is one of the other cities from the selection set S chosen.

- *Lower bounds.* The use of lower bounds on the cost of completing a partial solution was proposed in Maniezzo (1999). The lower bounds give additional heuristic information about the possible choices.

- *Lookahead.* A lookahead strategy was proposed in Michels and Middendorf (1999) where for each possible choice of an ant the maximum $\tau_{ij}^{\alpha} \cdot \eta_{ij}^{\beta}$ value that would result from this choice is evaluated and taken into account when actually making a decision.

- *Stagnation recovery.* For longer runs of an ACO algorithm there is the danger that after some time the search concentrates too much on a small search region. Several authors have proposed methods for modification of the pheromone information to counteract such stagnation behavior of ACO algorithms. When stagnation is detected the approach of Gambardella et al. (1999) is to reset all elements of the pheromone matrix to their initial values. In Stützle and Hoos (1997a) it was suggested to increase the pheromone values proportionately to their difference to the maximum pheromone value. A temporary reduction of α to negative values was proposed in Randall and Tonkes (2002).

- *Changing α, β values.* In Merkle et al. (2002) it was proposed to reduce the value of β during a run to increase the influence of the pheromone at later stages of the algorithm. See "Stagnation recovery" for changing α values.

- *Repelling pheromone.* Some experiments with pheromone that let the ants avoid choosing an edge have been conducted in Kawamura et al. (2000) and Montgomery and Randall (2002) in order to enforce ants (or different colonies of ants) to search in different regions of the search space. A similar idea has also been applied for an ant-based network routing algorithm (Amin et al., to appear; Schoonderwoerd et al., 1996).

- *Moving direction.* The use of ants that "move in different directions" can improve the optimization behavior (Michels and Middendorf, 1999). For an example for a permutation problem this could mean that some ants decide first which item is on place one of the permutation and other ants decide first which item is on the last place. One aspect is that the ants should make important decisions early (Merkle and Middendorf, 2001a). For some permutation problems where an unwanted bias in the decision of the ants can occur it can be advantageous to let the ants decide randomly about the sequence in which the places of the permutation are fixed (for details see Merkle and Middendorf, 2001a).

- *Local improvement of solutions.* The use of local optimization strategies to improve the solutions that have been found by the ants has been applied quite successfully for many ACO algorithms (e.g., Dorigo and Gambardella, 1997; Stützle et al., 2000; Stützle and Dorigo, 1999). Most state-of-the-art ACO algorithms use local improvement strategies. Two variants of the use of local improvement strategies exist: (i) to determine how much pheromone is updated for a solution, the quality or rank of the solution is computed after the local improvement has been applied but the actual pheromone update is done according to the original solution before the local improvement; (ii) as (i), but the pheromone update is done according to the solution after local improvement.

14.2.5   Some Promising Areas for Future Application of ACO

An important area of research for ACO that is often underestimated in its practical importance is to gain a deeper understanding how the use of different pheromone models influences the optimization behavior. Other promising fields, like multi-objective optimization, dynamic and probabilistic optimization, hybrid techniques algorithms, and theoretical aspects, cannot be covered in this introductory tutorial.

Pheromones and Optimization Behavior of ACO   A few works which consider this aspect have already been mentioned. Some recent papers have focused on the investigation of pheromone models. In Dorigo et al. (2002b) pheromone update rules for ACO are systematically derived based on the stochastic gradient ascent algorithm and cross-entropy method. A deterministic model for ACO is proposed in Merkle and Middendorf (2002a) and used to explain the dynamic change of pheromone values based on fixed-point analysis. In Blum and Sampels (2002b) and Merkle and Middendorf (2001a, 2001b) it is investigated how the pheromone model can introduce a strong bias to some regions of the search space.

14.3   PARTICLE SWARM OPTIMIZATION

The roots of the metaheuristic that is described in this section lie in computing models that have been created by scientists in the last two decades to simulate bird flocking and fish schooling. The coordinated search for food which lets a swarm of birds land at a certain place where food can be found was modeled with simple rules for information sharing between the individuals of the swarm. These studies inspired Kennedy and Eberhart to develop a method for function optimization that they called particle swarm optimization (PSO) (Kennedy and Eberhart, 1995). A PSO algorithm maintains a population of particles (the swarm), where each particle represents a location in a multidimensional search space (also called problem space). The particles start at random locations and search for the minimum (or maximum) of a given objective function by moving through the search space. The analogy to reality (in the case of search for a maximum) is that the function measures the quality or amount of the food at each place and the particle swarm searches for the place with the best or most food. The movements of a particle depend only on its velocity and the locations where good solutions have already been found by the particle itself or other (neighboring) particles in the swarm. This is again in analogy to bird flocking where each individual makes its decisions based on cognitive aspects (modeled by the influence of good solutions found by the particle itself) and social aspects (modeled by the influence of good solutions found by other particles). Note that, unlike many deterministic methods for continuous function optimization, PSO uses no gradient information.

In a typical PSO algorithm each particle keeps track of the coordinates in the search space which are associated with the best solution it has found so far. The corresponding value of the objective function (fitness value) is also stored. Another "best" value that is tracked by each particle is the best value obtained so far by any particle in its topological neighborhood. When a particle takes the whole population as its neighbors, the best value is a global best. At each iteration of the PSO algorithm the velocity of each particle is changed towards the personal and global best (or neighborhood best) locations. But also some random component is incorporated into the velocity update. A scheme for a PSO algorithm is given below.

> *PSO scheme*:
> Initialize location and velocity of each particle
> **repeat**
>   **for** each particle
>   evaluate objective function f at the particles location
>   **endfor**
>   **for** each particle
>   update the personal best position
>   **endfor**
>   update the global best position
>   **for** each particle
>   update the velocity
>   compute the new location of the particle
>   **endfor**
> **until** stopping criterion is met

An active field of research on PSO has developed, with the main use of PSO being for continuous function optimization. An increasing number of works have begun to investigate the use of PSO algorithms as function optimizers embedded into more complex application contexts. Examples are the use of PSO for neural network training (van den Bergh and Engelbrecht, 2000; Conradie et al., 2002), gene clustering (Xiao et al., 2003), power systems (Yoshida et al., 2000), and multimodal biometric systems (Veeramachaneni and Osadciw, 2003). Some works apply PSO also to discrete problems, such as subset problems (Kennedy and Eberhart, 1997; Ko and Lin, 2004) and permutation problems (see Hu et al., 2003).

In the following we describe in more detail how the PSO scheme can be applied to optimization problems. The first example considers the typical use of PSO for continuous function optimization. A subset problem is addressed in the second example to illustrate how PSO can be applied to other types of optimization problems.

14.3.1   Example 1: Basic PSO and Continuous Function Optimization

In order to describe the PSO algorithm for function optimization we need some notation. Let f be a given objective function over a D-dimensional problem space. The location of a particle $i \in \{1, \ldots, m\}$ is represented by a vector $\vec{x}_i = (x_{i1}, \ldots, x_{iD})$ and the velocity of the particle by the vector $\vec{v}_i = (v_{i1}, \ldots, v_{iD})$. Let l_d and u_d be lower and upper bounds for the particles coordinates in the dth dimension, $d \in [1 : D]$. The best previous position of a particle is recorded as $\vec{p}_i = (p_{i1}, \ldots p_{iD})$ and is called *pBest*. The index of the particle with the so far best found position in the swarm is denoted by g and $\vec{p}_g$ is called *gBest*.

At each iteration of a PSO algorithm after the evaluation of function f the personal best position of each particle i is updated, i.e. if $f(\vec{x}_i) < f(\vec{p}_i)$ then set $\vec{p}_i = \vec{x}_i$. If $f(\vec{p}_i) < f(\vec{p}_g)$ then i becomes the new global best solution, i.e. set $g = i$. Then the new velocity of each particle i is determined during the update of velocity in every dimension $d \in [1 : D]$ as follows:

$$v_{id} = w \cdot v_{id} + c_1 \cdot r_1 \cdot (p_{id} - x_{id}) + c_2 \cdot r_2 \cdot (p_{gd} - x_{id}) \qquad (14.6)$$

where

- parameter w is called the *inertia weight*, it determines the influence of the old velocity; the higher the value of w the more the individuals tend to search in new areas; typical values for W are near 1.0;

- c_1 and c_2 are the *acceleration coefficients*, which are also called the cognitive and the social parameter respectively, because they are used to determine the influence of the local best position and the global best position respectively; typical values are $c_1 = c_2 = 2$;

- r_1 and r_2 are random values uniformly drawn from $[0, 1]$.

After velocity update the new position of the particle i is then determined by

$$x_{id} = x_{id} + v_{id}$$

Sphere	$f_1(x) = \sum_{i=1}^{D} x_i^2$
Rastrigin	$f_2(x) = \sum_{i=1}^{D} (x_i^2 - 10\cos(2\pi x_i) + 10)$
Rosenbrock	$f_3(x) = \sum_{i=1}^{D-1} (100(x_{i+1} - x_i^2)^2 + (x_i - 1)^2)$
Schaffer's f6	$f_4(x) = 0.5 - \dfrac{(\sin\sqrt{x_1^2+x_2^2})^2 - 0.5}{(1+0.001(x_1^2+x_2^2))^2}$
Griewank	$f_5(x) = \frac{1}{4000} \sum_{i=1}^{D} x_i^2 - \prod_{i=1}^{D} \cos(\frac{x_i}{\sqrt{i}}) + 1$

Table 14.3. Test functions.

If there is a maximum range for the location in dimension d, i.e. $x_g \in [l_d, u_d]$, then the particle is reflected.

The behavior of PSO algorithms is usually studied and compared on a set of standard test functions. Examples of the most prominent test functions are given in Table 14.3. These functions represent different types of functions, e.g. the variables in Sphere and Rastrigin are uncorrelated which is not the case for the other functions in the table. Most of these functions are typically used for dimensions D of 10–100.

As an example we consider a test run of the standard PSO with a swarm of size $m = 10$ on the two-dimensional Sphere function (the PSO parameters used are $w = 0.729$, $c1 = c2 = 1.494$). It can be seen from Figure 14.3 (left) that the swarm proceeds from initial random positions at iteration $t = 0$ towards the single minimum value of the Sphere function. The velocity vectors of the particles at iteration $t = 10$ are shown in Figure 14.3 (right).

14.3.2 Example 2: Discrete Binary PSO for Subset Problems

Subset problems are a broad class of optimization problems where the aim is to find a good subset of a given set of items. For many practical problems additional restrictions will be given so that not all subsets of the given set are valid. Unlike many permutation problems like the TSP, subset problems allow solutions of different sizes. As an example subset problem, we consider a problem from the financial sector where the earnings of a company have to be forecast. The forecast is based on financial ratios that are generated from the company's results and other economic indicators from the last quarters. We assume that a forecast method is given that computes for each financial ratio a forecast and the final forecast is the average of the forecasts for all given values. Since many different financial ratios are in use, for example the book value per share or the total growth rate of a company, the problem is to select

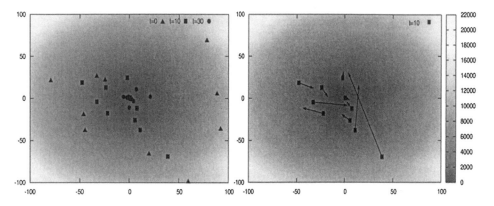

Figure 14.3. Swarm on the two-dimensional Sphere function; particles positions at iterations $t \in \{0, 10, 30\}$ (left); particles positions and velocity vectors at iteration $t = 10$ (right).

a not too large subset of these so that the forecast method gives good forecasts when applied to the selected financial ratios.

To solve a subset problem with PSO an individual of the swarm can be encoded by a D-dimensional vector where $D = |M|$ equals the size of the given set M (in the example M is the set of all considered financial ratios). Each dimension represents one binary bit that determines whether the corresponding item respectively the corresponding financial ratio is selected to be member of the subset. The crucial part in the design of the PSO algorithm is to connect the continuous movement of the particles to the discrete solution space.

In the so-called discrete binary PSO algorithm this is done as follows (Kennedy and Eberhart, 1997). As for the continuous PSO, the position of a particle corresponds to a solution and the velocity has an influence on the new position. But how the position is computed is different. Since the solution space is discrete and a particle should not stay at the same place a random component is used in the computation of the new position. The idea is to let a high velocity in one dimension give a high probability that the corresponding bit of the position vector is one.

Formally, the velocity of a particle is determined exactly as in (14.6). In order to determine the probabilities for the computation of the position vector a function is used that maps a velocity value onto the interval [0, 1]. A function often used is

$$\text{sig}(v_{id}) = \frac{1}{1 + \exp(-v_{id})} \tag{14.7}$$

To determine then the ith bit of the position vector of particle d a random number r_{id} is drawn from the interval $[0, 1]$ and the ith bit is set to one if $r_{id} < \text{sig}(v_{id})$ and otherwise it is set to zero.

The results of a comparative study between the discrete binary PSO and a genetic algorithm for the financial ratio selection problem are presented in Ko and Lin (2004). It was shown for a problem of dimension $D = 64$ that PSO is faster and gives better results than the genetic algorithm. Another application of discrete binary PSO to data mining can be found in Souza et al. (2003). The problem was to select a good subset of a given set of classification rules so that certain data can be classified accordingly.

14.3.3   Advanced Features of PSO

Several variations of the velocity update in PSO and extensions of the standard PSO for a better control of the behavior of the swarm have been proposed in the literature. Some of them are reviewed in this section.

- *Adaptive PSO.* In Clerc (2002) a version of PSO has been proposed where most values of the algorithms parameters are adapted automatically at run time. One example is the swarm size that varied during execution. A particle is removed when it is the worst (with respect to the best solution found so far) of a neighborhood of particles and the best particle in its neighborhood has improved significantly since its creation. Other rules have been implemented for creating new particles. Experimental results reported in Parsopoulos and Vrahatis (2002) have shown that an approach to use a second PSO during run time for determining the best parameter values for the first PSO were not successful. A certain evolutionary algorithm (Differential Evolution algorithm) performed better for this purpose.

- *Neighborhood best velocity update.* Several PSO algorithms establish a neighborhood relation between particles. In that case instead of using the global best position *gBest* for velocity update for each particle the best position of the particles in its neighborhood is used. This position is called neighborhood best and is denoted by *lBest*. A PSO variant where all particles in the neighborhood of a particle have an influence on its velocity is proposed in Kennedy and Mendes (2003). The following formula describes such an all-neighborhood-velocity-update of particle i with neighborhood N_i in dimension $d \in [1 : D]$ (note that a particle is included in its own neighborhood):

$$v_{id} = w \cdot v_{id} + \sum_{j \in N_i} \frac{c \cdot r_j \cdot (p_{jd} - x_{id})}{|N_i|} \qquad (14.8)$$

- *Maximum velocity.* A parameter $v_{\max}$ is introduced for some PSO algorithms to restrict the size of the elements v_{ij} of the velocity vector so that $v_{ij} \in [-v_{\max}, v_{\max}]$. Hence, when the velocity of a particle becomes larger than $v_{\max}$ during velocity update it is set to $v_{\max}$. A typical range for values of $v_{\max}$ is $[0.1 \cdot x_{max}, 1.0 \cdot x_{\max}]$. Observe that such values for $v_{\max}$ do not restrict the possible locations of a particle to $[-x_{\max}, x_{\max}]$.

- *Queen particle.* The addition of a queen particle, which is always located at the swarm's actual gravity center was proposed in Clerc (1999). Since the gravity center of the swarm might often be near a possible optimum it is reasonable to evaluate the objective function at this location. Experimental results have shown that it depends on the type of function whether the introduction of a queen particle is advantageous.

- *Convergence enforcement.* Several authors have considered the problem of how to improve the rate of convergence of PSO.

  - A constriction coefficient K was introduced in Clerc and Kennedy (2002) to reduce undesirable explosive feedback effects where the average distance between the particles grows during an execution. With the constriction coefficient as computed in Kennedy and Eberhart (1999) and Kennedy et al. (2001) the formula for velocity update becomes

$$v_{id} = K \cdot (v_{id} + c_1 \cdot r_1 \cdot (p_{id} - x_{id}) + c_2 \cdot r_2 \cdot (p_{gd} - x_{id})) \quad (14.9)$$

    Note that the constriction factor is just another way of choosing parameters w, c_1, and c_2. It can be shown that the swarm converges when parameter K is determined as (see Kennedy and Eberhart, 1999)

$$K = \frac{2}{|2 - c - \sqrt{c^2 - 4c}|} \quad (14.10)$$

    with $c = c_1 + c_2$, $c > 4$.

  - Parameter w can be decreased over time during execution to diminish the diversity of the swarm and to more quickly reach a state of equilibrium. A linear decrease of w from a maximum value $w_{\max}$ to a minimum value $w_{\min}$ is used by several authors (e.g. Kennedy et al., 2001). Typical values are $w_{\max} = 0.9$ and $w_{\min} = 0.4$.

  - In Vesterstrøm et al. (2002) the concept of division of labor and specialization was applied to PSO. Specialization to a task in this approach means for a particle to search near the global best position. A particle that has not found a better solution for a longer time span is replaced to the global best solution in order to start searching around the global best solution *gBest*. To prevent too many particles searching around *gBest* it was suggested to use a maximum number of particles that can switch to *gBest* or to make it more difficult to switch to *gBest* for those particles that are far away from *gBest*. Note that a similar approach to let particles that have not found good solutions jump to the place of good particles is realized in the hybrid techniques PSO, described below.

- *Controlling diversity.* To prevent the swarm from too early convergence to a small area so that the particles become too similar some methods have been proposed to keep the diversity of the swarm high enough. A common measure for the diversity of the swarm S in PSO is the "distance-to-average-point"

$$\text{diversity}(S) := \frac{1}{|S|} \cdot \sum_{i=1}^{|S|} \sqrt{\sum_{j=1}^{D}(p_{ij} - \bar{p}_j)} \qquad (14.11)$$

where $\bar{p}$ is the average vector of all vectors $\vec{p}_i$. In order to make the diversity measure independent of the range of the search space some authors use the measure $\text{diversity}(S)/|L|$ where $|L|$ is the length of the longest diagonal in the search space. Some methods to keep the diversity of the swarm high enough are described in the following.

- Xie et al. (2002) proposed adding an additional random element to the movement of the particles. In the new algorithm called dissipative PSO (DPSO) immediately after velocity update and determination of the new position of the particles the following computations are performed to introduce additional "chaos" to the system:

$$\textbf{if } rand() < c_v \textbf{ then}$$
$$v_{id} = rand() \cdot v_{\max,d} \ \{\text{chaos for velocity}\}$$
$$\textbf{if } rand() < c_l \textbf{ then}$$
$$x_{id} = rand(l_d, u_d) \ \{\text{chaos for location}\}$$

where $c_v, c_l \in [0, 1]$ are parameters which control the probability to add chaos, $rand(a, b)$ is random number that is uniformly distributed in (a, b) ($rand()$ is a shortcut for $rand(0, 1)$) and l_d, u_d are lower and upper bounds for the location in dimension d. Observe, that if $rand() < c_l$ the dth dimension of the new position is a random location within the search area.

- The idea of using particles that have a spatial extension in the search space was introduced in Krink et al. (2002) to hinder particles from coming too close to each other and forming too dense clusters. In this variation of PSO, particles that come too close to each other bounce away. A problem is to choose the direction in which the bouncing particles move away. Three strategies for defining the bouncing direction have been studied: (i) random bouncing, where the particles move in a random direction, (ii) physical bouncing, in which the particles bounce like physical objects, and (iii) velocity line-bouncing, in which the particles do not change their direction but move with increased speed. Preliminary experimental results show that bouncing can be advantageous for complex objective functions. For objective functions with a single optimum clustering is not a problem and bouncing is not advantageous. A similar approach to hinder the swarm to collapse uses an analogy to electrostatic energy. In this approach so-called charged particles experience a repulsive force when they come too close to each other (Blackwell and Bentley, 2002). Swarms with different ratios of charged particles have been studied.

- A strategy to explicitly control the diversity is to have two different phases of the PSO algorithm that can increase (repulsion phase) or reduce (attraction phase) the diversity of the swarm (Riget and Vesterstrøm, 2002). Two threshold values have been introduced that are used two determine when an exchange between the two phases should take place. When the diversity becomes lower than the threshold d_{low} the algorithm switches to the repulsion phase and when the diversity becomes larger than threshold $d_{high} > d_{low}$ the algorithm changes to the attraction phase. The only thing that happens when the phase of the algorithm changes is that every velocity vector is changed so that it points in the opposite direction. The authors have shown by experiments that nearly all improvements of the global best solutions were found during the attraction phases. Therefore, they propose to do no function evaluations during the repulsion phase to reduce the run time of the algorithm.

■ *Stagnation recovery.* For multi-modal functions there is the danger of premature convergence of standard PSO which results in suboptimal solutions. Stagnation recovery means to detect such a situation and then to react accordingly.

- Re-initialization of the swarm is proposed in Clerc (1999) when the diameter of the area that is actively searched by the swarm has become too small. The new swarm is initialized around the previous best position.

14.3.4   Some Promising Areas for Future Application of PSO

Complex multimodal functions that possess multiple, possibly similarly good, local optimal solutions occur in many applications. Often in such applications it is not enough to know just a single of these local optimal solutions but several or all of them are needed. Two areas of future PSO research that are relevant for optimizing complex multimodal functions are: (i) to find good neighborhood relations between the particles that might lead to an increased optimization efficiency and (ii) to investigate how several swarms can work co-operatively. We briefly review some works in both areas. Other interesting application and research areas, such as multi-objective and dynamic optimization, hybrid techniques algorithms, and theoretical aspects of PSO, cannot be covered by this introductory tutorial.

Neighborhood Relations   One possible advantage of neighborhood relations between the particles is an increased efficiency of PSO because the particles have to react only with respect to their neighbors. Another advantage is that the introduction of neighborhood relations can support the specialization of subsets of particles to different parts of the search space. A neighborhood scheme is explored in Suganthan (1999) that is defined by a particle's actual position so that a certain number of the closest other particles are considered to be neighbors. A different approach is to define the neighborhood independently from the particles' positions. In Kennedy and Mendes (2003) several such fixed neighborhoods are examined. For example, the particles are mapped onto a grid that defines the neighborhood. A hierarchical PSO where the particles of the swarm are arranged in a dynamic hierarchy that depends on the quality of the actual positions of the particles and defines the neighborhood structure was proposed in Janson and Middendorf (2003).

Co-operative Swarms   Co-operative swarms have been introduced to divide the work between several swarms. One motivation is that it can be very difficult for a single swarm to solve problems with large dimension D. An example is the co-operative swarm optimizer (CPSO) or split swarm that uses a set of swarms and splits the work equally between them in the following way (van den Bergh and Engelbrecht, 2000). The vector to be optimized is split across the swarms so that each swarm optimizes a different part of the vector, i.e. the swarms optimize with respect to different dimensions of the search space. Co-operation between the swarms is established in that for every evaluation of a new position of some particle its partial vector is combined with one partial

vector from each of the others swarms so that the quality of the resulting position is best. Experiments with CPSO for function minimization and neural
network learning have shown that it is good for problems where the depen-
dences between the component vectors are not too strong. Another motivation
to use co-operative swarms is for solving multi-objective problems where several functions have to be optimized so that the swarms optimize with respect
to different functions. A problem is then to find good methods for exchanging
information about the best positions between the swarms (see e.g. Parsopoulos
et al., 2004).

A PSO method that intends to form subswarms of particles searching for
the same local minimum is proposed in Kennedy (2000). A standard k-means
cluster method was used to divide the swarm into several clusters of individuals. For velocity update then each particle i uses the center of its cluster instead
of its personal best position p_{id}, see (14.6). Test results have shown that this
velocity update modification can be advantageous, especially for multimodal
functions like the Rastrigin function (see Table 14.3).

Niching techniques can also be used to promote the formation of subswarms
around different local optima. Niching is a concept that is inspired by the well
known observation from ecology that coexisting species can survive because
they occupy different niches, which roughly means that they have different
tasks. Various niching techniques have been developed for genetic algorithms
but meanwhile some authors have used niching also for PSO. A niching technique for PSO that aims to find all good local minima was proposed in Parsopoulos and Vrahatis (2001). It uses a function "stretching" method that
changes the objective function during execution as follows. Assume that a
position x has been found where the objective function f to be minimized has
a small value. Then f is transformed with the aim to remove local minima
that are larger than $f(x)$ and a subswarm is created that searches for a local
minimum near x on the transformed function. In addition, a second transformation is applied to f which increases the function values in the neighborhood
of x. This function is then used by the main swarm which will be repelled from
the area around x and searches for a different local minimum. Another niching approach for PSO was proposed in Brits et al. (2002). The niching PSO
starts with particles that move according to the so-called cognition-only model
where velocity update is done only according to the personal best position of
an individual, i.e. $c_2 = 0$ in (14.6). The particles then basically perform local search. When the quality of a particle has not changed significantly for
several iterations it is assumed that is has reached the region of a local minimum. To search for this minimum a subswarm is formed. At creation time
the subswarm consists only of two particles, the founding particle and its closest neighbor in the search space. Each subswarm is assigned a search region
(initially all positions that are not further away from the founding particle as

its closest neighbor). Several rules are used to define how other particles can enter a subswarm or how subswarms with intersecting regions are merged.

14.4   TRICKS OF THE TRADE

For newcomers to the field of Swarm Intelligence it is an advantage that ACO and PSO algorithms are relatively easy to implement so that one can develop practical experience without too much effort. Often even the standard form of ACO and PSO algorithms that do not use many problem-specific features work reasonably well for different types of optimization problems. This is especially true for certain types of problems: for example, scheduling problems in the case of ACO, and continuous indexcontinuous!function function indexcontinuous!optimization optimization in the case of PSO. Clearly, such early success should not lead to the illusion that Swarm Intelligence is a field where good algorithms can be obtained more or less for free because principles are used that have been inspired by successful strategies which occur in nature. The following hints may help the newcomer arrive at a deeper understanding of Swarm Intelligence.

- Read papers which apply Swarm Intelligence methods to problems that are similar to the problem you want to solve. No less important is to also study other good papers to learn about specific aspects of Swarm Intelligence methods or where carefully designed state of the art algorithms are described.

- Preferably do not start with too complicated an algorithm that you do not understand. Critically evaluate every step of your algorithm.

- Investigate how your algorithm behaves on different types of problem instances and try to verify your explanations. Test your algorithm on benchmark instances if available to make comparisons with the works of other researchers easier. Ideally, use random instances and real-world instances for the tests. Random instances have the advantage that their properties can be characterized by their generation method. A disadvantage is that they are often too artificial to reflect important characteristics of real-world problem. In addition, carefully designed artificial problem instances can sometimes help to study special aspects of the behavior of algorithms.

- Investigate how robust your algorithm is with respect to changes of the parameters (e.g. the α, β, and ρ parameters for ACO and the w, c_2, and c_2 parameters for PSO).

- Consider the optimization behavior of your algorithm at different numbers of iterations. Then you can discover, for example, whether the algorithm converges too early.

For ACO, the following hints should be considered:

- It is important to use pheromone information so that the ants are guided to good solutions. Two connected aspects are important here: (i) the pheromone should be used to encode properties of a solution that are most relevant in the sense that they can be used to characterize the good solutions, and (ii) the pheromone information should be interpreted by the ants in the best possible way.

- Find a solution construction process so that the ants can use a good (deterministic) heuristic. Such heuristics can be found in the literature for many problems.

For PSO, the following hint should be considered:

- For function optimization it is important to understand the characteristics of the search landscape (see Chapter 19) of the application functions. When there is basically a single valley in the search space a single swarm where convergence is enforced might work. But for search landscapes with many valleys a more sophisticated approach might be necessary where the diversity of the swarm is controlled, stagnation recovery mechanisms are introduced, or several co-operative swarm are used.

14.5   CONCLUSIONS

The field of Swarm Intelligence with the vision to learn from the behavior of natural swarms for the development of new methods in optimization has produced with ACO and PSO two successful metaheuristics that have found an increasing number of applications in the last few years. The basic principles of swarm intelligence methods and a selection of example applications have been explained in this tutorial. A number of new application areas is emerging in which Swarm Intelligence will play its part. One promising concept are hybrid techniques methods where swarm intelligence algorithms work in line with other metaheuristics. Might this tutorial also be a starting point for the reader to further explore the field of Swarm Intelligence.

SOURCES OF ADDITIONAL INFORMATION

- Good introductory books that cover various aspects of Swarm Intelligence are

- E. Bonabeau, M. Dorigo and G. Theraulaz, *Swarm Intelligence: From Natural to Artificial Systems,* 1999, Oxford University Press, New York.

- Kennedy, J., R. C. Eberhart and Y. Shi, *Swarm Intelligence,* 2001, Morgan Kaufmann, San Mateo, CA.

Recent overview papers are Cordón et al. (2002), Maniezzo et at. (2001) and Stützle and Dorigo (2002b) on ACO and van den Bergh (2002) on PSO.

- The following book will become the ultimate reference book for ACO: M. Dorigo and T. Stützle, *Ant Colony Optimization,* 2004, MIT Press, Boston, MA.

- A valid source of information for optimization techniques in general that contains four chapters on ACO and one chapter on PSO is *New Ideas in Optimisation,* D. Corne, M. Dorigo, and F. Glover (eds), 1999, McGraw-Hill, New York.

- Special issues of journals that are devoted to Swarm Intelligence are

  - Special section on Ant Algorithms and Swarm Intelligence in *IEEE Transactions on Evolutionary Computation* 6(4), M. Dorigo, L. Gambardella, M. Middendorf and T. Stützle, guest editors, 2002.

  - Special issue on Ant Colony Optimization, *Mathware & Soft Computing* 9, O. Cordón, F. Herrera and T. Stützle, guest editors, 2002.

  - Special issue on Ant Algorithms, *Future Generation Computer Systems Journal* 16(8), M. Dorigo, G. Di Caro, and T. Stützle, guest editors, 2000.

- A valuable source of recent research papers are the proceedings of the following workshop series that focus on Swarm Intelligence: International Workshop on Ant Colony Optimization and Swarm Intelligence (ANTS); IEEE Swarm Intelligence Symposium (the latest proceedings are Dorigo et al. (2002a) and Eberhart et al. (2003), respectively).

References

Agassounoun W., Martinoli, A. and Goodman, R., 2001, A scalable, distributed algorithm for allocating workers in embedded systems, in: *Proc. 2001 IEEE Systems, Man and Cybernetics Conf.,* pp. 3367–3373.

Amin, K. A., Mikler, A. R. and Iyengar, P. V., Dynamic agent population in agent-based distance vector routing, *J. Neural Parallel Sci. Comput.* **11**:127–142.

Bauer, A., Bullnheimer, B., Hartl, R. F. and Strauss, C., 1999, An ant colony optimization approach for the single machine total tardiness problem, in: *Proc. CEC'99*, IEEE, Piscataway, NJ, pp. 1445–1450.

Bautista J. and Pereira, J., 2002, Ant algorithms for assembly line balancing, in: *Proc. 3rd Int. Workshop on Ant Algorithms*, M. Dorigo et al., eds, Lecture Notes in Computer Science, Vol. 2463, Springer, Berlin, pp. 65–75.

van den Bergh, F., 2002, An analysis of particle swarm optimizers, *Ph.D. Thesis*, Department of Computer Science, University of Pretoria, South Africa.

van den Bergh, F. and Engelbrecht, A. P., 2000, Cooperative learning in neural networks using particle swarm optimizers, *S. African Comput. J.* **26**:84–90.

Blackwell, T. M. and Bentley, P. J., 2002, Dynamic search with charged swarms, *Proc. Genetic and Evolutionary Computation Conf. 2002 (GECCO 2002)*, Morgan Kaufmann, San Mateo, CA, pp. 19–26.

Blum C. and Sampels, M., 2002a, Ant colony optimization for FOP shop scheduling: a case study on different pheromone representations, in: *Proc. 2002 Congress on Evolutionary Computation (CEC'02)*, pp. 1558–1563.

Blum C. and Sampels, M., 2002b, When model bias is stronger than selection pressure. *Proc. 7th Int. Conf. on Parallel Problem Solving from Nature (PPSN VII)*, Lecture Notes in Computer Science, Vol. 2439, Springer, Berlin, pp. 893–902.

Bonabeau, E., Dorigo, M. and Theraulaz, G., 1999, *Swarm Intelligence: From Natural to Artificial Systems*, Oxford University Press, Oxford.

Brits, R., Engelbrecht, A. P. and van den Bergh, F., 2002, A niching particle swarm optimizer, in: *Proc. 4th Asia-Pacific Conf. on Simulated Evolution and Learning (SEAL 2002)*, pp. 692–696.

Bullnheimer, B., Hartl, R. F. and Strau, C. A., 1999, New rank based version of the Ant System: a computational study, *Central Eur. J. Oper. Res. Econ.* **7**:25–38.

Camazine, S., Franks, N. R. and Deneubourg, J.-L., 2001, *Self-Organization in Biological Systems*, Princeton Studies in Complexity, Princeton University Press, Princeton, NJ.

Clerc, M., 1999, The swarm and the queen: towards a deterministic and adaptive particle swarm optimization, in: *Proc. Congress of Evolutionary Computation*, IEEE, Piscataway, NJ, pp. 1951–1957.

Clerc, M., 2002, Think locally, act locally—a framework for adaptive particle swarm optimizers, *IEEE J. Evol. Comput.*, submitted.

Clerc, M. and Kennedy, J., 2002, The particle swarm—explosion, stability, and convergence in a multidimensional complex space, *IEEE Trans. Evol. Comput.* **6**:58–73.

Conradie, A., Miikkulainen, R., and Aldrich, C., 2002, Adaptive control utilizing neural swarming, *Proc. Genetic and Evolutionary Computation Conf. (GECCO 2002)*, Morgan Kaufmann, San Mateo, CA, pp. 60–67.

Cordón, O., Fernandez, I., Herrera, F. and Moreno, L., 2000, A new ACO model integrating evolutionary computation concepts: the best–worst ant system, in: *Proc. 2nd Int. Workshop on Ant Algorithms,* pp. 22–29.

Cordon, O., Herrera, F. and Stützle, T., 2002, A review on the ant colony optimization metaheuristic: basis, models and new trends, *Mathware Soft Comput.* **9**:141–175.

Deneubourg, J.-L., Aron, S., Goss, S. and Pasteels, J. M., 1990, The self-organizing exploratory pattern of the Argentine ant, *J. Insect Behav.* **32**:159–168.

Dorigo, M., 1992, Optimization, learning and natural algorithms (in Italian), *Ph.D. Thesis,* Dipartimento di Elettronica, Politecnico di Milano, Italy.

Dorigo, M., and Di Caro, G., 1999, The ant colony optimization meta-heuristic, in: *New Ideas in Optimization,* D. Corne, M. Dorigo, and F. Glover, eds, McGraw-Hill, New York, pp. 11–32.

Dorigo, M., Di Caro, G. and Sampels, M., eds, 2002a, *Proc. 3rd Int. Workshop on Ant Algorithms (ANTS 2002),* Lecture Notes in Computer Science, Vol. 2463, Springer, Berlin.

Dorigo, M. and Gambardella, L. M., 1997, Ant colony system: a cooperative learning approach to the traveling salesman problem, *IEEE Trans. Evol. Comput.* **1**:53–66.

Dorigo, M., Maniezzo, V. and Colorni, A., 1991, Positive feedback as a search strategy, *Technical Report* 91-016, Politecnico di Milano, Italy.

Dorigo, M., Maniezzo, V., and Colorni, A., 1996, The ant system: optimization by a colony of cooperating agents, *IEEE Trans. Syst., Man Cybernet.* B **26**:29–41.

Dorigo, M., Zlochin, M., Meuleau, N. and Birattari, M., 2002b, Updating ACO pheromones using stochastic gradient ascent and cross-entropy methods, in: *Proc. EvoWorkshops 2002,* S. Cagnoni et al., eds, Lecture Notes in Computer Science, Vol. 2279, Springer, Berlin, pp. 21–30.

Eberhart, R. C., Kennedy, J. and Shi, Y., eds, 2003, *Proc. Swarm Intelligence Symp.* (Indianapolis, IN).

Gambardella, L. M., Taillard, E. and Dorigo, M., 1999, Ant colonies for the quadratic assignment problem, *J. Oper. Res. Soc.* **50**:167–76.

Goldberg, D. and Mataric, M. J., 2000, Robust behavior-based control for distributed multi-robot collection tasks, *USC Institute for Robotics and Intelligent Systems Technical Report* IRIS-00-387.

Goss, S., Aron, S., Deneubourg, J. L. and Pasteels, J. M., 1989, Self-organized shortcuts in the Argentine ant, *Naturwissenschaften* **76**:579–581.

Guntsch M. and Middendorf, M., 2002a, Applying population based ACO to dynamic optimization problems, in: *Proc. 3rd Int. Workshop (ANTS 2002),* Lecture Notes in Computer Science, Vol. 2463, Springer, Berlin, pp. 111–122.

Guntsch M., and Middendorf, M., 2002b, A population based approach for ACO, in: *Applications of Evolutionary Computing—Proc. EvoWorkshops 2002*, Lecture Notes in Computer Science, Vol. 2279, Springer, Berlin, pp. 72–81.

Gutjahr, W., 2000, A graph-based Ant System and its convergence, *Future Generation Comput. Syst.* **16**:873–888.

Gutjahr, W., 2002, ACO algorithms with guaranteed convergence to the optimal solution. *Inform. Process. Lett.* **82**:145–153.

Handl, J. and Meyer, B., 2002, Improved ant-based clustering and sorting in a document retrieval interface, in: *Proc. 7th Int. Conf. Parallel Problem Solving from Nature—PPSN VII*, J. J. Merelo Guervos et al., eds, Lecture Notes in Computer Science, Vol. 2439, Springer, Berlin, pp. 913–923.

Hu, X., Eberhart, R. and Shi, Y., 2003, Swarm intelligence for permutation optimization: a case study on n-queens problem, in: *Proc. IEEE Swarm Intelligence Symp. 2003* (Indianapolis, IN).

Janson, S. and Middendorf, M., 2003, A hierarchical particle swarm optimizer, in: *Proc. Congress on Evolutionary Computation (CEC 2003)*, IEEE, Piscataway, NJ, pp. 770–776.

Kawamura, H., Yamamoto, M., Suzuki, K., and Ohucke, A., 2000, Multiple ant colonies algorithm based on colony level interactions, *IEICE Trans. Fundamentals* A **83**:371-379.

Kennedy, J. 2000. Stereotyping: improving particle swarm performance with cluster analysis, in: *Proc. IEEE Int. Conf. on Evolutionary Computation*, pp. 1507–1512.

Kennedy J. and Eberhart, R. C., 1995, Particle swarm optimization, in: *Proc. IEEE Int. Conf. on Neural Networks,* pp. 1942–1948.

Kennedy, J. and Eberhart, R. C., 1997, A discrete binary version of the particle swarm algorithm. *Proc. 1997 Conf. on Systems, Man, and Cybernetics*, Piscataway, NJ: IEEE Service Center, pp. 4104-4109.

Kennedy, J. and Eberhart, R. C., 1999, The particle swarm: social adaption in information processing systems, in: D. Corne et al., eds, *New Ideas in Optimization,* McGraw-Hill, New York, pp. 379–387.

Kennedy, J., Eberhart, R. C., and Shi, Y., 2001, *Swarm Intelligence,* Morgan Kaufmann, San Francisco, CA.

Kennedy, J. and Mendes, R., 2003, Neighborhood topologies in fully-informed and best-of-neighborhood particle swarms, in: *Proc. IEEE Int. Workshop on Soft Computing in Industrial Applications.*

Ko, P.-C. and Lin, P.-C., 2004, A hybrid swarm intelligence based mechanism for earning forecast, in: *Proc. 2nd Int. Conf. on Information Technology and Applications (ICITA 2004).*

Krink, T., Vesterstrøm, J. S. and Riget, J., 2002, Particle swarm optimisation with spatial particle extension, in: *Proc. 4th Congress on Evolutionary Computation (CEC-2002)*, pp. 1474–1479.

Kube, C. R. and Bonabeau, E., 2000, Cooperative transport by ants and robots, *Robot. Auton. Syst.* **30**:85–101.

Lambrinos, D., Möller, R., Labhart, T., Pfeifer, R., and Wehner, R., 2000, A mobile robot employing insect strategies for navigation, *Robot. Auton. Syst.* **30**:39–64.

Lumer, E. D. and Faieta, B., 1994, Diversity and adaptation in populations of clustering ants, in: *Proc. 3rd Int. Conf. on Simulation of Adaptive Behavior: From Animals to Animats 3 (SAB 94)*, MIT Press, Cambridge, MA, pp. 501–508.

Maniezzo, V., 1999, Exact and approximate nondeterministic tree-search procedures for the quadratic assignment problem, *INFORMS J. Comput.* **11**:358–369.

Maniezzo, V., and Carbonaro, A., 2001, Ant colony optimization: an overview, in: *Essays and Surveys in Metaheuristics,* C. Ribeiro, ed., Kluwer, Dordrecht, pp. 21–44.

Merkle, D. and Middendorf, M., 2001a, A new approach to solve permutation scheduling problems with ant colony optimization, in: *Applications of Evolutionary Computing: Proc. EvoWorkshops 2001,* Lecture Notes in Computer Science, Vol. 2037, E. J. W. Boers et al., eds, Springer, Berlin, pp. 484–493.

Merkle, D. and Middendorf, M., 2001b, On solving permutation scheduling problems with ant colony optimization, *Int. J. Syst. Sci.*, submitted.

Merkle, D. and Middendorf, M., 2002a, Modelling the dynamics of ant colony optimization algorithms, *Evol. Comput.* **10**:235–262.

Merkle D. and Middendorf, M., 2002b, Ant colony optimization with the relative pheromone evaluation method, in: *Applications of Evolutionary Computing: Proc. EvoWorkshops 2001*, Lecture Notes in Computer Science, Vol. 2279, Springer, Berlin, pp. 325–333.

Merkle, D., and Middendorf, M., 2003a, On the behavior of ACO algorithms: studies on simple problems, in: *Metaheuristics: Computer Decision-Making,* M. G. C. Resende and J. Pinho de Sousa, eds, Kluwer, Dordrecht, pp. 465–480.

Merkle, D., and Middendorf, M., 2003b, An ant algorithm with global pheromone evaluation for scheduling a single machine, *Appl. Intell.* **18**:105–111.

Merkle, D., Middendorf, M. and Schmeck, H., 2002, Ant colony optimization for resource-constrained project scheduling, *IEEE Trans. Evol. Comput.* **6**:333–346.

Michels, R. and Middendorf, M., 1999, An ant system for the shortest common supersequence problem, in: *New Ideas in Optimization,* D. Corne, M. Dorigo and F. Glover, eds, McGraw-Hill, New York, pp. 51–61.

Middendorf, M., 2002. Ant colony optimization, in: *Tutorial Proc. Genetic and Evolutionary Computation Conf. (GECCO-2002).*

Montgomery, J., and Randall, M., 2002, Anti-pheromone as a tool for better exploration of search space, in: *Proc. 3rd Int. Workshop ANTS 2002,* Lecture Notes in Computer Science, Vol. 2463, Springer, Berlin, pp. 100–110.

Parsopoulos, K. E. and Vrahatis, M. N., 2001, Modification of the particle swarm optimizer for locating all the global minima, in: *Artificial Neural Networks and Genetic Algorithms,* V. Kurkova et al., eds, Springer, Berlin, pp. 324–327.

Parsopoulos, K. E. and Vrahatis, M. N., 2002, Recent approaches to global optimization problems through particle swarm optimization, *Nat. Comput.* **1**:235–306.

Parsopoulos, K. E., Tasoulis, D. K. and Vrahatis, M. N., 2004, Multiobjective optimization using parallel vector evaluated particle swarm optimization, in: *Proc. IASTED Int. Conf. on Artificial Intelligence and Applications.*

Randall, M. and Tonkes, E., 2002, Intensification and diversification strategies in ant colony optimisation, *Complexity Int.* **9**:randal01.

Riget, J. and Vesterstrøm, J. S., 2002, A diversity-guided particle swarm optimizer—the ARPSO, *Technical Report* 2002-02, Department of Computer Science, University of Aarhus.

Schoonderwoerd, R., Holland, O., Bruten, J. and Rothkrantz, L., 1996, Ant-based load balancing in telecommunications networks, *Adapt. Behav.* **5**:169–207.

Stützle, T., den Besten, M. and Dorigo, M., 2000, Ant colony optimization for the total weighted tardiness problem, in: *Proc. 6th Int. Conf. on Parallel Problem Solving from Nature (PPSN-VI),* Lecture Notes in Computer Science, Vol. 1917, Deb et al., eds, Springer, Berlin, pp. 611–620.

Stützle, T. and Dorigo, M., 1999, ACO algorithms for the traveling salesman problem, in: *Evolutionary Algorithms in Engineering and Computer Science,* K. Miettinen et al., eds, Wiley, New York, pp. 163–183.

Stützle, T. and Dorigo, M., 2001, An experimental study of the simple ant colony optimization algorithm, in: *Proc. 2001 WSES Int. Conf. on Evolutionary Computation (EC'01).*

Stützle, T. and Dorigo, M., 2002a, A short convergence proof for a class of ACO algorithms. *IEEE Trans. Evol. Comput.* **6**:358–365.

Stützle, T. and Dorigo, M., 2002b, The ant colony optimization metaheuristic: algorithms, applications, and advances, in: *Handbook of Metaheuristics,* F. Glover and G. Kochenberger, eds, Kluwer, Dordrecht.

Stützle, T. and Hoos, H., 1997, Improvements on the ant system: Introducing MAX(MIN) ant system, in: *Proc. Int. Conf. on Artificial Neutral Networks and Genetic Algorithms,* Springer, Berlin, pp. 245–249.

Stützle, T., and Hoos, H., 2000, MAX–MIN ant system, *Future Gener. Comput. Syst. J.* **16**:889–914.

Suganthan, P. N., 1999, Particle swarm optimizer with neighborhood optimizer, in: *Proc. Congress on Evolutionary Computation (CEC 1999),* pp. 1958–1961.

Sousa, T., Neves, A., and Silva, A., 2003, Swarm optimization as a new tool for data mining, in: *Proc. Workshop on Nature Inspired Distributed Computing (NIDISC)—at IPDPS'03,* p. 144.

Veeramachaneni, K., and Osadciw, L., 2003, Adaptive multimodal biometric fusion algorithm using particle swarm, in: *Proc. SPIE Aerosense 2003* (Orlando, FL), pp. 211–222.

Vesterstrøm, J. S., Riget, J. and Krink, T., 2002, Division of labor in particle swarm optimisation, in: *Proc. 4th IEEE Congress on Evolutionary Computation (CEC-2002),* pp. 1570–1575.

Xiao, X., Dow, E. R., Eberhart, R., Ben Miled, Z. and Oppelt, R. J., 2003, Gene clustering using self-organizing maps and particle swarm optimization, in: *Online Proc. 2nd IEEE Int. Workshop on High Performance Computational Biology (HICOMB 2003).*

Xie, X.-F., Zhang, W.-J. and Yang, Z.-L., 2002, A dissipative particle swarm optimization, in: *Proc. 4th IEEE Congress on Evolutionary Computation (CEC-2002)* (Honolulu, HI).

Yoshida, H., Kawata, K., Fukuyama, Y., Takayama, S., and Nakanishi, Y., 2000, A particle swarm optimization for reactive power and voltage control considering voltage security assessment, *IEEE Trans. Power Syst.* **15**:1232–1239.

OR-Library, 2004, http://mscmga.ms.ic.ac.uk/jeb/orlib/wtinfo.html

Chapter 15

FUZZY REASONING

Costas P. Pappis
University of Piraeus
Piraeus, Greece

Constantinos I. Siettos
National Technical University of Athens
School of Applied Mathematics and Physics
Athens, Greece

15.1   INTRODUCTION

The derivation of mathematical models that can efficiently describe real-world problems is most of the time an overwhelming or even impossible task due to the complexity and the inherent ambiguity of characteristics that these problems may possess. As Zadeh (1973), the founder of the theory of fuzzy sets, puts it,

> ...as the complexity of a system increases, our ability to make precise and yet significant statements about its behavior diminishes until a threshold is reached beyond which precision and significance (or relevance) become almost mutually exclusive characteristics.

Fuzzy Reasoning is based on the theory of fuzzy sets and it encompasses Artificial Intelligence, information processing and theories from logic to pure and applied mathematics, like graph theory, topology and optimization. The theory of fuzzy sets was introduced in 1965. In his introductory paper, Zadeh, while stating his intention ("to explore in a preliminary way some of the basic properties and implications" of fuzzy sets) he noted that

> ...the notion of a fuzzy set provides a convenient point of departure for the construction of a conceptual framework which parallels in many respects the framework used in the case of ordinary sets, but is more general than the latter and, potentially, may prove to have a much wider scope of applicability, particularly in the fields of pattern classification and information processing.

Table 15.1. A chronology of critical points in the development of fuzzy reasoning.

First paper on fuzzy systems	Zadeh, 1965
Linguistic approach	Zadeh, 1973
Fuzzy logic controller	Assilian and Mamdani, 1974
Heat exchanger control based on fuzzy logic	Ostergaard, 1977
First industrial application of fuzzy logic:	
cement kiln control	Homblad and Ostergaard, 1982
Self-organizing fuzzy controller	Procyk and Mamdani, 1979;
Fuzzy pattern recognition	Bezdek, 1981
Fuzzy controllers on Tokyo subway shuttles	Hitachi, 1984
Fuzzy chip	Togai and Watanabe, 1986
Takagi–Sugeno fuzzy modeling	Takagi and Sugeno, 1985
Hybrid neural–fuzzy systems	Kosko, 1992

Indeed, in subsequent years, the theory of fuzzy sets was more decisively established as a new approach to complex systems theory and decision processes. The application of fuzzy logic has dramatically increased since 1990, ranging from production, finance, marketing and other decision-making problems to micro-controller-based systems in home appliances and large-scale process control systems (Sugeno and Yasukawa, 1993; Karr and Gentry, 1993; Lee, 1990). For systems involving nonlinearities and lack of a reliable analytical model, fuzzy logic control has emerged as one of the most promising approaches. Without doubt, fuzzy inference is a step towards the simulation of human thinking.

The main advantage of fuzzy logic techniques, i.e. techniques based on the theory of fuzzy sets, over more conventional approaches in solving complex, nonlinear and/or ill-defined problems lies in their capability of incorporating a priori qualitative knowledge and expertise about system behavior and dynamics. This renders fuzzy logic systems almost indispensable for obtaining a more transparent and tactile qualitative insight for systems whose representation with exact mathematical models is poor and inadequate. Besides, fuzzy schemes can be used either as enabling to other approaches or as self-reliant methodologies providing thereby a plethora of alternative structures and schemes.

In fact, fuzzy control theory generates nonlinear functions according to a representation theorem by Wang (1992), who stated that any continuous nonlinear function can be approximated as exactly as needed with a finite set of fuzzy variables, values and rules. Therefore, by applying appropriate design procedures, it is always possible to design a fuzzy controller that is suitable for the nonlinear system under control. Table 15.1 depicts some benchmarks in the history of fuzzy logic, particularly in the domain of fuzzy control.

This chapter is intended to present an overview of the basic notions of the theory of fuzzy sets and fuzzy logic. The chapter is organized as follows: in the next section, an introduction to the theory of fuzzy sets is presented, covering topics of the most commonly used types of membership functions, logical and transformation operators, fuzzy relations, implication and inference rules, and fuzzy similarity measures. Section 15.3 introduces the basic structure of a fuzzy inference system and its elements are described. Section 15.4 presents the topic of fuzzy control system and an example is demonstrated. In particular a fuzzy controller is proposed for the control of a plug flow tubular reactor, which is a typical nonlinear distributed parameter system. The proposed fuzzy controller is compared with a conventional proportional–integral (PI) controller. In the same section an introduction to the field of fuzzy adaptive control systems is given and the self-organizing scheme is presented. In Section 15.5 reviews are given on the topics of model identification and stability of fuzzy systems, respectively. Conclusions and perspectives of fuzzy reasoning are given in Section 15.6.

15.2 BASIC DEFINITIONS OF FUZZY SET THEORY

15.2.1 Fuzzy Sets and the Notion of Membership

A classical set A is defined as a collection of elements or objects. Any element or object x either belongs or does not belong to A. The membership $\mu_A(x)$ of x in A is a mapping:

$$\mu_A : X \rightarrow \{0, 1\}$$

that is, it may take the value 1 or 0, which represent the truth value of x in A. It follows that, if is the complement set of A and $\cap$ represents intersection of sets, then

$$A \cap \bar{A} = \emptyset$$

Fuzzy logic is a logic based on fuzzy sets, i.e. sets of elements or objects characterized by truth-values in the [0,1] interval rather than crisp 0 and 1, as in the conventional set theory. The function that assigns a number in [0,1] to each element of the universe of discourse of a fuzzy set is called the membership function.

15.2.2 Membership Functions

Let X denote the universe of discourse of a fuzzy set A. A is completely characterized by its membership function μ_A:

$$\mu_A : X \rightarrow [0, 1]$$

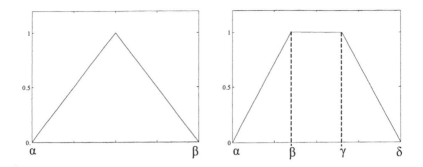

Figure 15.1. (a) Triangular, (b) trapezoid membership function.

and is defined as a set of pairs:

$$A = \{(x, \mu_A(x))\}$$

The most commonly used membership functions are the following (Dubois and Prade, 1980; Zimmermann, 1996):

- triangular membership function
- trapezoid membership function
- linear membership function
- sigmoidal membership function
- Π-type membership function
- Gaussian membership function.

The *triangular* membership function, see Figure 15.1(a), is defined as

$$\mathrm{Tri}(x; \alpha, \beta, \gamma) = \begin{cases} 0 & x < \alpha \\ \dfrac{x - a}{\beta - \alpha} & \alpha \leq x \leq \beta \\ -\dfrac{x - \gamma}{\gamma - \beta} & \beta \leq x < \gamma \\ 0 & x \geq \gamma \end{cases}$$

The *trapezoid* membership function (Figure 15.1(b)) is defined as

$$\mathrm{Tra}(x; \alpha, \beta, \gamma, \delta) = \begin{bmatrix} 0 & x < \alpha \\ \dfrac{x - \alpha}{\beta - \alpha} & \alpha \leq x < \beta \\ 1 & \beta \leq x < \gamma \\ -\dfrac{x - \delta}{\delta - \gamma} & \gamma \leq x < \delta \\ 0 & x \geq \delta \end{bmatrix}$$

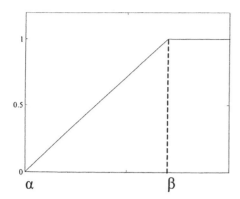 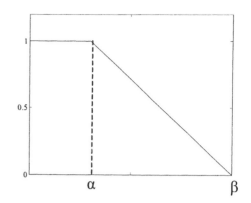

Figure 15.2. (a) Monotonically increasing linear, (b) monotonically decreasing linear membership function.

The *monotonically increasing linear* membership function (Figure 15.2(a)) is given by

$$
L(x; \alpha, \beta) = \begin{cases} 0 & x < \alpha \\ \dfrac{x - \alpha}{\beta - \alpha} & \alpha \leq x \leq \beta \\ 1 & x > \beta \end{cases}
$$

The *monotonically decreasing linear* membership function (Figure 15.2(b)) is given by

$$
L(x; \alpha, \beta) = \begin{cases} 1 & x < \alpha \\ -\dfrac{x - \alpha}{\beta - \alpha} & \alpha \leq x \leq \beta \\ 0 & x > \beta \end{cases}
$$

The *monotonically increasing sigmoidal* membership function (Figure 15.3(a)) is given by

$$
S(x; \alpha, \beta, \gamma) = \begin{cases} 0 & x < \alpha \\ 2\left(\dfrac{x - \alpha}{\gamma - \alpha}\right)^2 & \alpha \leq x \leq \beta \\ 1 - 2\left(\dfrac{x - \alpha}{\gamma - \alpha}\right)^2 & \beta \leq x \leq \gamma \\ 1 & x > \gamma \end{cases}
$$

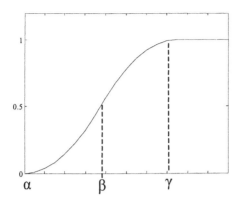 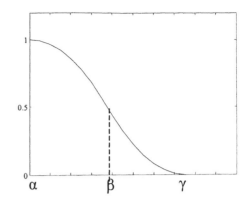

Figure 15.3. (a) Monotonically increasing sigmoidal, (b) monotonically decreasing sigmoidal membership function.

The *monotonically decreasing sigmoidal* membership function (Figure 15.3(b)) reads as

$$S(x; \alpha, \beta, \gamma) = \begin{cases} 1 & x < \alpha \\ 1 - 2\left(\dfrac{x - \alpha}{\gamma - \alpha}\right)^2 & \alpha \le x \le \beta \\ 2\left(\dfrac{x - \alpha}{\gamma - \alpha}\right)^2 & \beta \le x \le \gamma \\ 0 & x > \gamma \end{cases}$$

The Π-membership function (Figure 15.4(a)) is defined as

$$\Pi(x; \beta, \gamma) = \begin{cases} S\left(x; \gamma - \beta, \dfrac{\gamma - \beta}{2}, \gamma\right) & x \le \gamma \\ 1 - S\left(x; \gamma, \dfrac{\gamma + \beta}{2}, \gamma + \beta\right) & x > \gamma \end{cases}$$

The *Gaussian* membership function (Figure 15.4(b)) is given by

$$G(x; k, \sigma) = \exp\left(-\frac{(\gamma - x)^2}{2\sigma^2}\right)$$

where σ is the standard deviation.

Examples on Fuzzy Sets   Maintaining a "comfortable" room temperature is of great importance for the work productivity. Fuzzy logic climate control is one of the many successive commercial applications of the theory. For example, the room temperature for low-level activities could be described by the following five fuzzy sets, where a temperature around 18° C is a comfortable

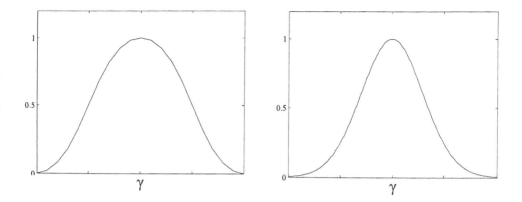

Figure 15.4. (a) Π, (b) Gaussian membership functions.

one, around 26° C a warm one (though not during summer!), while above 40° C is definitely too warm, and around 12° C can be characterized as cold and below that too cold.

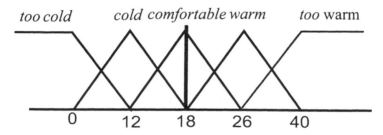

"Fast" cars can be described by the horsepower (HP) using the following membership function:

$$\mu(x) = \left\{ \begin{array}{ll} 0 & 0 \leq x \leq 75 \\ \dfrac{x-120}{25} & 75 \leq x \leq 120 \\ 1 & 120 \leq x \leq 150 \\ 0 & x \geq 150 \end{array} \right\}$$

Notice that in the above characterization, for a horsepower above 150 HP "fast" cars have a zero membership. A nonzero membership could have been assigned in another fuzzy set (e.g. the fuzzy set of "very fast" cars).

15.2.3   Fuzzy Set Operations

Knowledge and understanding of the operations of the theory of fuzzy sets is important for the design of fuzzy systems. The fuzzy set operations are defined with respect to the sets' membership functions.

Two fuzzy sets A and B on the universe of discourse X are **equal** if their membership functions are equal for each $x \in X$:

$$\forall x \in X : \mu_A(x) = \mu_B(x)$$

A fuzzy set A is a *subset* of B $(A \subseteq B)$ if

$$\forall x \in X : \mu_A(x) \leq \mu_B(x)$$

For the operation of *intersection* $\cap$ of two fuzzy sets A and B, there is a plethora of definitions in the bibliography. The choice is application dependent:

$$\forall x \in X : \mu_{A \cap B} = \left\{ \begin{array}{c} \min(\mu_A(x), \mu_B(x)) \\ \dfrac{\mu_A(x) + \mu_B(y)}{2} \\ \mu_A(x)\mu_B(y) \\ \cdots \end{array} \right\}$$

The *union* $\cup$ of two fuzzy sets A and B is also defined in several ways:

$$\forall x \in X : \mu_{A \cup B} = \left\{ \begin{array}{c} \max(\mu_A(x), \mu_B(y)) \\ \dfrac{2\min(\mu_A(x), \mu_B(y)) + 4\max(\mu_A(x), \mu_B(y))}{6} \\ \mu_A(x) + \mu_B(y) - \mu_A(x)\mu_B(y) \\ \cdots \end{array} \right.$$

The *complement* A' of a fuzzy set A is defined as

$$\forall x \in X : \mu_{A'}(x) = 1 - \mu_A(x)$$

Examples on Fuzzy Set Operations Let us consider the fuzzy sets A and B:

$$A = \{0/1 + 0.2/2 + 0.8/3 + 1/4 + 1/5)$$
$$B = \{0.1/1 + 0.4/2 + 0.5/3 + 0.7/4 + 0.3/5)$$

Table 15.2. Examples of transformation operators.

Very	$\mu_{\tilde{A}}(x) = (\mu_A(x))^n \quad n > 1$
More/less	$\mu_{\tilde{A}}(x) = (\mu_A(x))^n \quad 0 < n < 1$
More than (lt)	
$= 1 - \mu_A(x)$ for $x < x_0$	$M_{lt(A)}(x) = 0$ for $x \geq x_0, x_0 : \mu_A(x_0) = \max \mu_A(x)$
More/less (mt)	
$= 1 - \mu_A(x)$ for $x > x_0$	$\mu_{mt(A)}(x) = 0$ for $x \leq x_0, x_0 : \mu_A(x_0) = \max \mu_A(x)$

Then

$$A \cap B = \{0/1 + 0.2/2 + 0.5/3 + 0.7/4 + 0.3/5)$$
$$\text{using the min operator}$$
$$= \{0/1 + 0.08/2 + 0.4/3 + 0.7/4 + 0.3/5)$$
$$\text{using the product operator}$$
$$A \cup B = \{0.1/1 + 0.4/2 + 0.8/3 + 1/4 + 1/5)$$
$$\text{using the max operator}$$
$$A' \text{ (complement of A)} = \{1/1 + 0.8/2 + 0.2/3 + 0/4 + 0/5\}$$
$$B' = \{0.9/1 + 0.6/2 + 0.5/3 + 0.3/4 + 0.7/5)$$
$$(A \cap B)' = A' \cup B' = \{1/1 + 0.8/2 + 0.5/3 + 0.3/4 + 0.7/5)$$
$$\text{using the max operator}$$

15.2.4   Transformation Operators

The transformation operator (or hedge or modifier) acts on a membership function to modify the concept of the linguistic term that describes the fuzzy set. For example, in the clause "number *very* close to 10", the transformation operator *very* acts on the linguistic term "close to 10" which corresponds to a fuzzy set. Examples of such operators are given in Table 15.2 (Ross, 1995; Zimmermann, 1996; Pappis and Mamdani, 1977).

Example on Transformation Operators   Let us consider the fuzzy set "young" on the discrete set $U = \{0, 20, 40, 60, 80\}$:

$$F_{\text{young}} = \{(0, 1), (20, 0.75), (40, 0.52), (60, 0.23), (80, 0)\}$$

Then we can derive the fuzzy set $F' = $ "very young" by using the relevant transformation operator. Choosing $\nu = 1.5$ we obtain

$$F' = \{(0, 1), (20, 0.6495), (40, 0.0.375), (60, 0.1103), (80, 0)\}$$

15.2.5   Cartesian Inner Product of Fuzzy Sets

If $A_1, A_2, \ldots, A_\nu$ are fuzzy sets defined in $U_1, U_2, \ldots, U_\nu$, their Cartesian inner product is a fuzzy set $F = A_1 \times A_2 \times \cdots \times A_\nu$ in $U_1 \times U_2 \times \cdots \times U_\nu$ with membership function $\mu_F(u_1, u_2, \ldots, u_\nu) = \cap_{i=1,\nu} \mu_{A_i}(u_i)$, e.g.

$$\mu_F(u_1, u_2, \ldots, u_\nu) = \min\{\mu_{A1}(u_1), \mu_A 2(u_2), \ldots, \mu_A \nu(u_\nu)\}$$

or

$$\mu_F(u_1, u_2, \ldots, u_\nu) = \mu_{A1}(u_1), \mu_A 2(u_2), \ldots, \mu_A \nu(u_\nu)$$

Example   The objective in climate control is to find the optimum conditions in terms of both temperature T and humidity H. Suppose that the discrete sets of temperature and humidity are given by $T = \{T_1, T_2, T_3, T_4\}$ and $H = \{H_1, H_2, H_3\}$ respectively, and that of the desired temperature by the discrete fuzzy set

$$A = \frac{0.12}{T_1} + \frac{0.65}{T_2} + \frac{1}{T_3} + \frac{0.25}{T_4}$$

while that for the desired level of humidity by

$$B = \frac{0.5}{H_1} + \frac{0.9}{H_2} + \frac{0.1}{H_3}$$

Then the Cartesian product $A \times B$ reads as

$$\begin{aligned}
A \times B = {} & \frac{0.12}{T_1, H_1} + \frac{0.12}{T_1, H_2} + \frac{0.1}{T_1, H_3} + \frac{0.5}{T_2, H_1} + \frac{0.65}{T_2, H_2} + \frac{0.1}{T_2, H_3} \\
& + \frac{0.5}{T_3, H_1} + \frac{0.9}{T_3, H_2} + \frac{0.1}{T_3, H_3} + \frac{0.25}{T_4, H_1} + \frac{0.25}{T_4, H_2} + \frac{0.1}{T_4, H_3}
\end{aligned}$$

Then the optimum conditions are those for $T = T_3$ and $H = H_2$.

15.2.6   Fuzzy Relations

Let U_1 and U_2 be two universes of discourse and the membership function $\mu_R : U_1 \times U_2 \to [0, 1]$. Then a fuzzy relation R on $U_1 \times U_2$ is defined as (Zimmermann, 1996)

$$R = \int_{U \times B} \mu_R \frac{(u_1, u_2)}{(u_1, u_2)} \quad \text{if } U_1, U_2 \quad \text{are continuous}$$

or

$$R_d = \sum_{U \times V} \mu_R \frac{(u_1, u_2)}{(u_1, u_2)} \quad \text{if } U_1, U_2 \quad \text{are discrete}$$

Example Consider the coordinates of three atoms, denoted by i, j, k in a cubic crystal with a lattice constant of 3 A and their corresponding x, y, z coordinates $U = \{(0, 0, 0), (0.5, 0.5, 0.5), (1.2, 1.2, 1.2)\}$ (in A). Then the fuzzy relation "near neighbors" can be described by the following fuzzy relation:

$$R = \frac{1.0}{i, i} + \frac{1}{j, j} + \frac{1}{k, k} + \frac{0.9}{i, j} + \frac{0.1}{i, k} + \frac{0.9}{j, i} + \frac{0.6}{j, k} + \frac{0.1}{k, i} + \frac{0.6}{k, j}$$

(note that for the particular problem we should have excluded the pairs (i, i), (j, j) and (k, k), but we have kept them for the completeness of the example).

15.2.7 Fuzzy Set Composition

Let R_1 and R_2 be two fuzzy relations on $U_1 \times U_2$ and $U_2 \times U_3$ respectively, then the composition C of R_1 and R_2 is a fuzzy relation defined as follows:

$$C = R_1 \cdot R_2 = \{(u_1, u_3) \cup (\mu_{R_1}(u_1, u_2) \cap \mu_{R_2}(u_1, u_2))\}$$
$$u_1 \in U_1, u_2 \in U_2, u_3 \in U_3$$

Example Consider the following fuzzy relations (in matrix form):

$$R = \begin{bmatrix} 0.2 & 0.6 \\ 0.9 & 0.4 \end{bmatrix} \quad \text{and} \quad S = \begin{bmatrix} 1 & 0.4 & 0.3 \\ 0.8 & 0.5 & 0.1 \end{bmatrix}$$

Then using a min operator for $\cap$ and a max operator for $\cup$ their composition gives

$$T = R \cdot S = \begin{bmatrix} 0.2 & 0.6 \\ 0.9 & 0.4 \end{bmatrix} \cdot \begin{bmatrix} 1 & 0.4 & 0.3 \\ 0.8 & 0.5 & 0.1 \end{bmatrix} = \begin{bmatrix} 0.6 & 0.5 & 0.2 \\ 0.9 & 0.4 & 0.3 \end{bmatrix}$$

15.2.8 Fuzzy Implication

Let A and B be two fuzzy sets in U_1, U_2 respectively. The implication $I : A \rightarrow B \in U_1 \times U_2$ is defined as (Ross, 1994; Zimmermann, 1996):

$$I = A \times B = \int_{U_1 \times U_2} \mu_A(u_1) \cap \mu_B(u_2)/(u_1, u_2)$$

The rule "*If* the error is negative big *then* control output is positive big" is an implication: error x implies control action y.

Let there be two discrete fuzzy sets $A = \{(u_i, \mu_A(u_i)), i = 1, \ldots, n\}$ defined on U and $B = \{(v_j, \mu_B(v_j)), j = 1, \ldots, m\}$ defined on V. Then the implication $A \rightarrow B$ is a fuzzy relation R:

$$R = \left\{((u_i, v_j), \mu_R(u_i, v_j)), i = 1, \ldots, n, j = 1, \ldots, m\right\}$$

defined on $U \times V$, whose membership function $\mu_R(u_i, v_j)$ is given by

$$\begin{bmatrix} \mu_A(u_1) \\ \mu_A(u_2) \\ \ldots \\ \mu_A(u_n) \end{bmatrix} \times [\mu_B(v_1)\mu_B(v_2)\ldots\mu_B(v_m)]$$

$$= \begin{bmatrix} \mu_A(u_1) \wedge \mu_B(v_1) & \mu_A(u_1) \wedge \mu_B(v_2) & \ldots & \mu_A(u_1) \wedge \mu_B(v_m) \\ \mu_A(u_2) \wedge \mu_B(v_1) & \mu_A(u_2) \wedge \mu_B(v_2) & \ldots & \mu_A(u_2) \wedge \mu_B(v_m) \\ \ldots & \ldots & \ldots & \ldots \\ \mu_A(u_n) \wedge \mu_B(v_1) & \mu_A(u_n) \wedge \mu_B(v_2) & \ldots & \mu_A(u_n) \wedge \mu_B(v_m) \end{bmatrix}$$

15.2.9   Inference Rules

Let R be a fuzzy relation on $U_1 \times U_2$ and A be a fuzzy set in U_1. The composition $A \cdot R = B$ is a fuzzy set in U_2, which represents the conclusion made from the fuzzy set A (fact) based on the implication R (rule). Let there be a multiple-input–single-output (MISO) rule base with N rules. The ith rule is given by

$$\textit{If } A_{i1} \textit{ and } A_{i2} \textit{ and } \ldots \textit{ and } A_{in} \textit{ then } B_i$$

where n is the number of input variables x_i, A_{ij} is the fuzzy set of input variable x_j in the ith rule, and B_i is the fuzzy set of output variable y_j in the ith rule. The ith rule is the implication

$$I_i = A_i \to B_i, \, A_i = A_{i1} \cap A_{i2} \cap \ldots \cap A_{in} = \cap_{i=1}^{n} A_{ij}$$

Then the implication I_{tot} of N rules is given by

$$I_{tot} = R_1 \cup R_2 \cup \ldots \cup R_N = \cup_{i=1}^{N} R_i = \cup_{i=1}^{N} A_i \to B_i$$

15.2.10   The Inverse Problem

The inverse problem is defined as follows: given two fuzzy relations S and T find R such that $R \cdot S = T$. In application terms, the problem may be defined as follows: let S be the input–output relation describing a system and T a desired output of the system. Find input R, which produces T.

Sanchez (1976) showed an existence condition of the solutions associated with their least upper bound and presented a method for obtaining it analytically. Pappis (1976) and Pappis and Sugeno (1985) presented a method to obtain the whole set of solutions.

15.2.11   Fuzzy Similarity Measures

The fuzzy similarity measures introduce the notion of approximate equality (or similarity) between fuzzy sets. The most commonly used fuzzy similarity measures are the following (Pappis, 1991; Pappis and Karacapilidis, 1993, 1995; Wang, 1997; Wang et al., 1995; Cross and Sudkamp, 2002):

L-fuzzy Similarity Measure The $L(A, B)$ similarity measure of two fuzzy sets A, B is defined as

$$L(A, B) = 1 - \max_{x \in X} |A(x) - B(x)|$$

M-fuzzy Similarity Measure The $M(A, B)$ similarity measure of two fuzzy sets A, $B \in X$ is defined as

$$M(A, B) = \begin{cases} 1 & \text{if } A = B = \emptyset \\ \dfrac{\sum\limits_{x \in X} \min(A(x), B(x))}{\sum\limits_{x \in X} \max(A(x), B(x))} & \text{in every other case} \end{cases}$$

Two fuzzy sets are ε-"almost" equal ($A \sim B$) if and only if $M(A, B) \leq \varepsilon$, where $\varepsilon \in [0, 1]$.

S-fuzzy Similarity Measure The $S(A, B)$ similarity measure of two fuzzy sets A, $B \in X$ is defined as

$$S(A, B) = \begin{cases} 1 & \text{if } A = B = \emptyset \\ 1 - \dfrac{\sum\limits_{x \in X} |A(x) - B(x)|}{\sum\limits_{x \in X} (A(x) + B(x))} & \text{in every other case} \end{cases}$$

W-fuzzy Similarity Measure The $W(A, B)$ similarity measure of two fuzzy sets A, $B \in X$ is defined as

$$W(A, B) = 1 - \sum_{x \in X} |A(x) - B(x)|$$

P-fuzzy Similarity Measure The $P(A, B)$ similarity measure of two fuzzy sets A, $B \in X$ is defined as

$$P(A, B) = \frac{\sum\limits_{x \in X} A(x) B(x)}{\max\left(\sum\limits_{x \in X} A(x) A(x), \ \sum\limits_{x \in X} B(x) B(x)\right)}$$

15.3 BASIC STRUCTURE OF A FUZZY INFERENCE SYSTEM

The basic structure of a fuzzy inference system consists of a fuzzification unit, a fuzzy logic reasoning unit (process logic), a knowledge base, and a

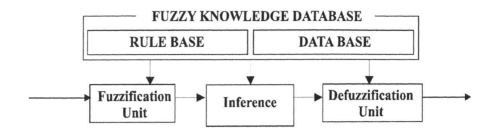

Figure 15.5. Basic structure of a fuzzy inference system.

defuzzification unit (Figure 15.5). The key element of the system is the fuzzy logic-reasoning unit that contains two main types of information:

1 A data base defining the number, labels and types of the membership functions the fuzzy sets used as values for each system variable. These are of two types: the input and the output variables. For each one of them the designer has to define the corresponding fuzzy sets. The proper selection of these is one of the most critical steps in the design process and can dramatically affect the performance of the system. The fuzzy sets of each variable form the universe of discourse of the variable.

2 A rule base, which essentially maps fuzzy values of the inputs to fuzzy values of the outputs. This actually reflects the decision-making policy. The control strategy is stored in the rule base, which in fact is a collection of fuzzy control rules and typically involves weighting and combining a number of fuzzy sets resulting from the fuzzy inference process in a calculation, which gives a single crisp value for each output. The fuzzy rules incorporated in the rule base express the control relationships usually in an IF-THEN format. For instance, for a two-input-one-output fuzzy logic controller, that is the case in this work, a control rule has the general form

$$\text{Rule } i: \text{ IF } x \text{ is } A_i \text{ and } y \text{ is } B_i \text{ THEN } z \text{ is } C_i$$

where x and y are input variables, z is the output variable; A_i, B_i and C_i are linguistic terms (fuzzy sets) such as "negative", "positive" or "zero". The *if-part* of the rule is called condition or premise or antecedent, and the *then-part* is called the consequence or action.

Usually the actual values acquired from or sent to the system of concern are crisp, and therefore fuzzification and defuzzification operations are needed to map them to and from the fuzzy values used internally by the fuzzy inference system.

The fuzzy reasoning unit performs various fuzzy logic operations to infer the output (decision) from the given fuzzy inputs. During fuzzy inference, the following operations are involved for each fuzzy rule:

1 Determination of the degree of match between the fuzzy input data and the defined fuzzy sets for each system input variable.

2 Calculation of the fire strength (degree of relevance or applicability) for each rule based on the degree of match and the connectives (e.g. AND, OR) used with input variables in the antecedent part of the rule.

3 Derivation of the control outputs based on the calculated fire strength and the defined fuzzy sets for each output variable in the consequent part of each rule.

Several techniques have been proposed for the inference of the fuzzy output based on the rule base. The most common used are the following:

- the Max–Min fuzzy inference method

- the Max-product fuzzy inference method.

Assume that there are two input variables, e (error) and ce (change of error), one output variable, cu (change of output), and two rules:

Rule 1 If e is A_1 AND ce is B_1 THEN cu is C_1

Rule 2 If e is A_2 AND ce is B_2 THEN cu is C_2

In the Max–Min inference method, the fuzzy operator AND (intersection) means that the minimum value of the antecedents is taken:

$$\mu_A \text{ AND } \mu_B = \min\{\mu_A, \mu_B\}$$

while for the Max-product one the product of the antecedents is taken:

$$\mu_A \text{ AND } \mu_B = \mu_A\mu_B$$

for any two membership values μ_A and μ_B of the fuzzy subsets A, B, respectively. All the contributions of the rules are aggregated using the union operator, thus generating the output fuzzy space C.

15.3.1 Defuzzification Unit

Defuzzification typically involves weighting and combining a number of fuzzy sets resulting from the fuzzy inference process in a calculation, which gives a single crisp value for each output.

The most commonly used defuzzification methods are those of mean of maximum, centroid, and center of sum of areas (Lee, 1990; Ross, 1995; Driankov et al., 1993).

Mean of Maximum Defuzzification Technique   The technique of the mean value of maximum is given by the following equation (Yan et al., 1994):

$$x = \frac{\sum\limits_{i=1}^{n} \alpha_i H_i x_i}{\alpha_i H_i}$$

where x is the control (output) value to be applied, n is the number of rules in a MISO system, H_i is the maximum value of the membership function of the output fuzzy set, which corresponds to rule I_i, x_i is the corresponding control (output) value, and α_i is the degree that the rule i is fired.

Centroid Defuzzification Technique   This is the most prevalent and intuitively appealing among the defuzzification methods (Lee, 1990; Ross, 1995). This method takes the center of gravity of the final fuzzy space in order to produce a result (the value u of the control variable) sensitive to all rules; it is described by the following equation (Ross, 1995):

$$u = \frac{\sum\limits_{i=1}^{n} \alpha_i M_i}{\sum\limits_{i=1}^{n} \alpha_i A_i}$$

where M_i is the value of the membership function of the output fuzzy set of rule i, A_i is the corresponding area and α_i is the degree that the rule i is fired. Note that the overlapping areas are merged (Figure 15.6(a)).

In the case of a continuous space (universe of discourse), the output value is given by (Ross, 1994; Taprantzis et al., 1997)

$$u = \frac{\int_U u \mu_U(u) \, du}{\int_U \mu_U(u) \, du}$$

Center of Sums Defuzzification Technique   A similar technique to the centroid technique, but computationally more efficient, in terms of speed, is that of the center of sums. The difference is that the overlapping (between the output fuzzy sets) areas are not merged (Figure 15.6(b)). The discrete value of the output is given by (Lee, 1990; Driankov et al., 1993)

$$u = \frac{\sum\limits_{i=1}^{l} u_i \cdot \sum\limits_{k=1}^{n} \mu_k(u_i)}{\sum\limits_{i=1}^{l} \sum\limits_{k=1}^{n} \mu_k(u_i)}$$

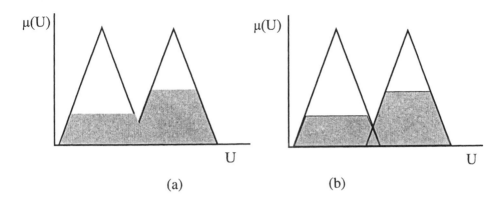

Figure 15.6. Defuzzification techniques: (a) centroid, (b) center of sums.

15.3.2 Design of the Rule Base

Two are the main approaches in the design of rule bases (Yan et al., 1994):

- the heuristic approach,

- the systematic approach.

Heuristic approaches (Yan et al., 1994; King and Mamdani, 1977; Pappis and Mamdani, 1977) provide a convenient way to build fuzzy control rules in order to achieve the desired output response, requiring only qualitative knowledge for the behavior of the system under study. For a two-input (e and ce) one-output (cu) system these rules are of the form

IF e is P (positive) AND ce is N (negative) THEN cu is P (positive)
IF e is N (negative) AND ce is P (positive) THEN cu is N (negative).

The reasoning for the construction of the fuzzy control rules can be summarized as follows:

- If the system output has the desired value and the change of the error (ce) is zero then keep the control action constant.

- If the system output diverges from the desired value then the control action changes with respect to the sign and the magnitude of the error e and the change of error ce. Table 15.3 compresses the design of a rule base for the linguistic term sets NB (negative big), NM (negative medium), NS (negative small), ZE (zero), PS (positive small), PM (positive medium) and PB (positive big) of the fuzzy variables e, ce, cu. The input variables are laid out along the axes, and each matrix element represents the output variable.

Table 15.3. A fuzzy rule base with two inputs and one output.

		ce						
		NB	NM	NS	Z	PS	PM	PB
e	PB	ZE	PS	NM	NB	NB	NB	NB
	PM	PS	ZE	NS	NM	NM	NB	NB
	PS	PM	PS	ZE	NS	NS	NM	NB
	Z	PB	PM	PS	ZE	NS	NM	NB
	NS	PB	PM	PS	PS	ZE	NS	NM
	NM	PB	PB	PM	PM	PS	ZE	ZE
	NB	PB	PB	PB	PB	PM	ZE	ZE

Systematic approaches provide the decision-making strategy (rule base) with the aid of system identification and pattern recognition techniques from input–output data.

15.4   A CASE STUDY: A FUZZY CONTROL SYSTEM

15.4.1   The Fuzzy Logic Control Closed Loop

Over the last 20 years, a large number of conventional modeling and control methods have been proposed to cope with nonlinear and/or time-varying systems including input-state linearization (Isidori, 1995), input–output linearization (Cravaris and Chung, 1987; Henson and Seborg, 1990), model predictive schemes (Patwardhan et al., 1992) and various direct and indirect adaptive control schemes (Isermann, 1989; Batur and Kasparian, 1991).

However, the poor modeling of system uncertainties and the inherent difficulty of incorporating a priori qualitative information about the system dynamics limit the efficiency and the applicability of the classical approaches. The fuzzy logic approach to process control provides a convenient way to build the control strategy, by requiring only qualitative knowledge for the behaviour of the control system. The heuristics employed offer a very attractive way of handling imprecision in the data and/or complex systems, where the derivation of an accurate model is difficult or even impossible. On the other hand, modeling and control techniques based on fuzzy logic comprise very powerful approaches of handling imprecision and nonlinearity in complex systems. The basic structure of a fuzzy logic controller is given in Figure 15.7. Usually the input and output variables are normalized through scaling factors G_{in} and G_{out} in the interval $[-1, 1]$.

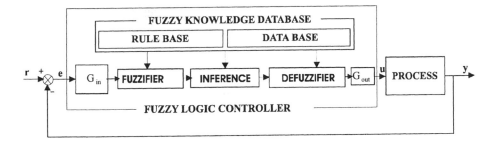

Figure 15.7. The fuzzy logic control closed loop.

15.4.2   Fuzzy Logic Controllers in Proportional–Integral (PI) and Proportional–Differential (PD) Forms

In what follows, in order to enable a comparison basis, the fuzzy logic controller (FLC) with two input variables, the error and the change of error, is represented in PI- and PD-like forms.

PI-like Fuzzy Controller   The PI controller in the z-domain has the following form (Stephanopoulos, 1984):

$$C(z) = \frac{u(z)}{e(z)} = K_c \left(1 + K \frac{1}{1 - z^{-1}} \right)$$

In the time domain the above equation can be rewritten as

$$cu = K_c ce + (K_c K)e$$

where e is the error between a predefined set point and the process output, ce is the change in error, and u is the control output signal. In order to generate an equivalent fuzzy controller, the same inputs e, ce and the same output, cu, will be used in its design.

Based on the above, a two-input–single-output FLC is derived with the following variables:

- input variables: $e(t) = r(t) - y(t)$
  $$ce(t) = e(t) - e(t - 1)$$

- output variable: $cu(t) = u(t) - u(t - 1)$

where $r(t)$ is the set point at time t (set point moisture), $y(t)$ is the process output at time t (output moisture), $e(t)$, $ce(t)$ are the error and the change of error at time t, respectively, and $cu(t)$ is the change in the control variable at time t.

In a general form the control action cu can be represented as a nonlinear function of the input variables $e(t)$, $ce(t)$:

$$cu = f(e', ce', t) = f(GEe, GCEce, t)$$

For small perturbations δe, δce around equilibrium, the above equation is approximated by the linearized equation

$$cu = \left[\frac{\partial f}{\partial e}\right]_{ce} \delta e + \left[\frac{\partial f}{\partial ce}\right]_{e} \delta ce$$

By substitution one finally obtains the simplified discretized equation (Mizumoto, 1995)

$$cu(t) = GEe(t) + GCEce(t)$$

which gives the incremental control output at time t. GE and GCE are the scaling factors for the error and change of error, respectively.

PD-like Fuzzy Controller   In an analogous manner the PD-like fuzzy controller is of the form

$$u(t) = GEe(t) + GCEce(t)$$

Note that the above expressions are derived using the max-product inference technique.

15.4.3   An Illustrative Example

The design procedure of a fuzzy controller (FLC) is demonstrated through an illustrative example: the system under study is a plug flow tubular reactor, which is a nonlinear distributed parameter with time lag system. The design of the FLC is based on a heuristic approach. The proposed controller is compared with a conventional PI controller, which is tuned with two methods: the process reaction curve tuning method and by using time integral performance criteria such as integral of absolute error (IAE). Based on dynamic performance criteria, such as IAE, ISE, ITAE, it is shown that the proposed fuzzy controller exhibits a better performance compared to the PI controller tuned by the process reaction curve tuning method and an equivalent, if not better, dynamic behavior, compared to the optimal tuned via the time performance criteria PI controller, for a wide range of disturbances.

Case Study: Fuzzy Control of a Plug Flow Tubular Reactor   The process of concern is shown in Figure 15.8. It is the problem of the control of a jacketed tubular reactor in which a simple exothermic reaction $A \rightarrow B$ with first-order kinetics takes place. Assuming plug flow conditions, constant temperature for

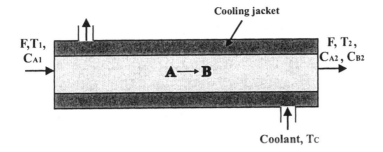

Figure 15.8. The process under study: control of a plug flow tubular reactor.

the coolant, which flows around the tube of the reactor, the governing equations consists of a set of nonlinear time-dependent partial differential equations listed below. The system is a nonlinear distributed parameter with time delay system:

$$\frac{\partial C_A}{\partial t} + u \frac{\partial C_A}{\partial z} = -k C_A$$

$$c_p \rho A \frac{\partial T}{\partial t} + c_p \rho u A \frac{\partial T}{\partial t} = h A_t (T_C - T) + (-D H_R) k A C_A$$

$$k = k_O \exp\left(-\frac{E}{RT}\right)$$

The nominal values of the tubular reactor parameters are as follows:

$C_{A1} = 1.6 \text{ kmol m}^{-3}$

$T_1 = 440 \text{ K}$

$k_o = 3.34 \times 10^8 \text{ min}^{-1}$

$D H_R = -44\,000 \text{ kcal kmol}^{-1}$

$c_p = 25 \text{ kcal kmol}^{-1} \text{ K}^{-1}$

$T_{co} = 293 \text{ K}$

$A = 0.002 \text{ m}^2$

$C_{A2} = 0.11 \text{ kmol m}^{-3}$

$T_2 = 423 \text{ K}$

$E/R = 8600 \text{ K}$

$U = 25 \text{ kcal m}^2 \text{ min}^{-1} \text{ grad}^{-1}$

$\rho = 47 \text{ kmol m}^{-3}$

$A_t = 0.01 \text{ m}^2$

$U = 2 \text{ m min}^{-1}$

The solution of nonlinear, time-dependent, partial differential equations is possible only by means of modern computer-aided methods. The choice here is the combination of Galerkin's method of weighted residuals and finite-element basis functions (Zienkiewicz and Morgan, 1983).

The control objective is to maintain the control variable, which is the composition of the reacting mixture at the output of the reactor, within the desired operational settings and, particularly, to keep the A reactant concentration at the output below its nominal steady-state value, eliminating mostly input concentration disturbances. The manipulated variable is taken to be the coolant temperature. The incremental fuzzy controller, a two-input–single-output FLC, is derived with the following variables: $e(t) = r(t) - y(t)$,

$ce(t) = e(t) - e(t-1)$, $cu(t) = u(t) - u(t-1)$, where $r(t)$ is the set point at time t (set point moisture), $y(t)$ is the process output at time t (output moisture), and $e(t)$, $ce(t)$ are the error and the change of error at time t.

For the fuzzification of the input–output variables, seven fuzzy sets are defined for each variable, $e(t)$, $ce(t)$ and $cu(t)$ with fixed triangular shaped membership functions normalized in the same universe of discourse, as it is shown in Figure 15.9. For the development of the rule base a heuristic approach was employed.

Given the fact that a reduction in the coolant temperature decreases the output concentration, and inversely, the reasoning for the construction of the fuzzy control rules is outlined as follows:

- Keep the output of the FLC constant if the output has the desired value and the change of error is zero.

- Change the control action of the FLC according to the values and signs of the error, e, and the change of error, ce:

  1 *If* the error is negative (the process output is above the set point) *and* the change of error is negative (at the previous step the controller was driving the system output upwards), *then* the controller should turn its output downwards. Hence, considering negative feedback, the change in control action should be positive, i.e. $cu > 0$, since $u(t) = u(t-1) + cu$.

  2 *If* the error is positive (the process output is below the set point) *and* the change of error is positive (at the previous step the controller was driving the system output downwards), *then* the controller should turn its output upwards. Hence, considering negative feedback, the change in control action should be negative, i.e. $cu < 0$, since $u(t) = u(t-1) + cu$.

  3 *If* the error is positive (the process output is below the set point) *and* the change of error is negative, implying that at the previous step the controller was driving the system output upwards, trying to correct the control deviation, *then* the controller need not take any further action.

  4 *If* the error is negative (the process output is above the set point) *and* the change of error is positive, implying that at the previous step the controller was driving the system output downwards, *then* the controller need not take any further action.

Table 15.4 compresses the design of the control rules for the term sets (nb: negative big, nm: negative medium, ns: negative small, ze: zero, ps: positive small, pm: positive medium, pb: positive big) of the fuzzy variables e, ce, cu.

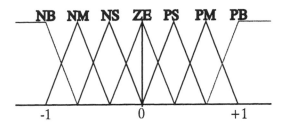

Figure 15.9. Input–output fuzzy sets.

Table 15.4. Fuzzy control rules.

		ce						
		nb	nm	ns	ze	ps	pm	pb
	pb	ze	ze	nm	nb	nb	nb	nb
	pm	ps	ze	ns	nm	nm	nb	nb
e	ps	pm	ps	ze	ns	ns	nm	nb
	ze	pb	pm	ps	ze	ns	nm	nb
	ns	pb	pm	ps	ps	ze	ns	nm
	nm	pb	pb	pm	pm	ps	ze	ze
	nb	pb	pb	pb	pb	pm	ze	ze

The input variables are laid out along the axes, and each matrix element represents the output variable. This structure of the rule base provides negative feedback control in order to maintain stability under any condition. For the evaluation of the rules, the fuzzy reasoning unit of the FLC has been developed using the Max–Min fuzzy inference method (Lee, 1990; Driankov et al., 1993). In the particular FLC, the centroid defuzzification method (Zimmermann, 1996; Driankov et al., 1993) is used. Finally, for the projection of the input and output variable values to the normalized universe of discourse, the following values of scaling factors have been chosen: $G_{e(t)} = 5$, $G_{ce(t)} = 45$, $G_{cu(t)} = 2.5$.

Performance Analysis: Results and Discussion   To study the performance of the FLC controller, a comparison with a conventional PI controller is made. The parameters of the PI controller are adjusted using two methods of tuning. First it is assumed that the dynamics of the process are poorly known and the tuning of the PI controller is based on the process reaction curve, an empirical tuning method, which provides an experimental model of the process near the operating point. The results of this analysis are: $Gain_1 = 350$, *Integral time constant*$_1 = 1.5$ min.

In the second approach, the optimal values of the PI controller are determined by minimizing the integral of absolute value of error (IAE) of the control variable for a predetermined disturbance in input concentration. Here the optimal parameters of the PI controller are adjusted by minimizing the IAE at the +20% step disturbance in input reactant concentration. The resulting tuning parameters are $Gain_{II} = 155$, *integral time constant*$_{II} = 1.0$ min. The relatively large deviation between the parameters obtained by minimizing the IAE and those obtained by the process reaction curve method is results from the fact that the process reaction curve method is based on the approximation of the open loop process response by a first-order system plus dead time.

In the case under study, this approximation seems to be rather poor. In order to objectively compare the FLC controller with the conventional PI controller, in addition to the IAE criterion, the integral of time multiplied by the absolute value of error (ITAE) and the integral of the square of the error (ISE) performance criteria are used for both control and manipulated variables.

Simulation results are presented for step change disturbances ranging from −5% up to −20% in input reactant concentration. Figure 15.11 depicts in histograms the calculated three dynamic performance criteria IAE, ISE, ITAE for the fuzzy and the PI controller tuned with the two different methods.

The performance criteria are determined for both control and manipulated variables. Based on Figure 15.10, it is apparent that the overall performance of the FLC seems better than the conventional PI controller tuned by the empirical process reaction curve method (controller PI-1) and equivalent, if not better, than the optimal PI controller tuned by minimizing the IAE (controller PI-2). The PI-1 controller has the highest values of IAE (Figure 15.10(a, b)), ISE (Figure 15.10(c, d)) and ITAE (Figure 15.10(e, f)) criteria. As is shown, the fuzzy controller exhibits up to 60% lower IAE (Figure 15.10(b)), up to 30% lower ISE (Figure 15.10(c)) and up to 200% lower ITAE (Figure 15.10(f)) compared to the PI controller tuned by the process reaction curve method. In comparison to the PI controller, whose parameters are optimally adjusted by minimizing the IAE criterion, the fuzzy controller shows an equivalent, if not better, performance, based on IAE, ISE and ITAE criteria for all the range of step disturbances (from 5% up to 20%).

However, the approach of optimally adjusting the parameters of the PI controller to some dynamic performance criterion, such as IAE, requires an exact mathematical model of the process, which in real-world processes is very difficult, if not impossible, to derive. In contrast, the design of the fuzzy logic controller is based on a heuristic approach and a mathematical model of the process is not vital.

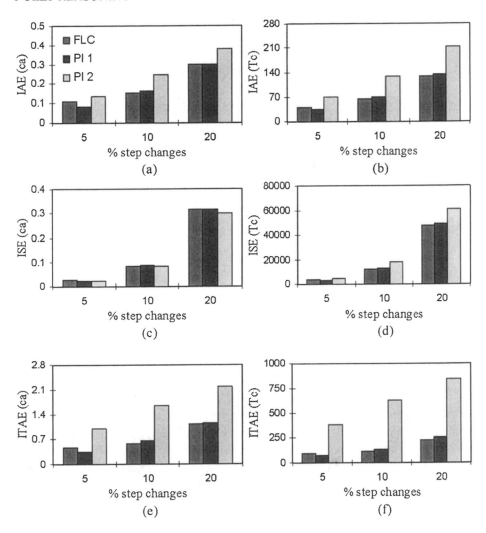

Figure 15.10. Performance comparison of the fuzzy and the PI controller tuned by the process reaction method (PI-1), by minimizing the IAE criterion (PI-2): (a) IAE of the control variable, (b) IAE of the manipulated variable, (c) ISE of the control variable, (d) ISE of the manipulated variable, (e) ITAE of the control variable, (f) ITAE of the manipulated variable.

15.4.4 Fuzzy Adaptive Control Schemes

A major problem encountered in nonlinear and/or time-dependent systems is the degradation of the closed-loop performance as the system shifts away from the initial operational settings. This drawback imposes the need of using adaptive controllers, i.e. controllers, which adjust their parameters optimally, according to some objective criteria (Astrom, 1983).

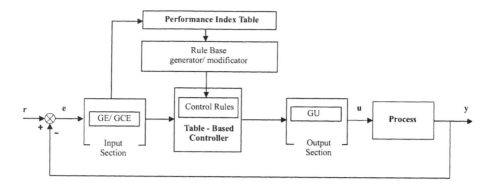

Figure 15.11. The self-organizing fuzzy logic controller.

To date, many schemes have been proposed for fuzzy adaptive control, including self-organizing control (Procyk and Mamdani, 1979; Siettos et al., 1999), membership functions adjustment (Batur and Kasparian, 1991; Zheng, 1992) and scaling factors adjustment (Maeda and Murakami, 1992; Daugherity et al., 1992; Palm, 1993; Jung et al., 1995; Chou and Lu, 1994; Chou, 1998; Sagias et al., 2001). Maeda and Murakami and Daugherity et al. proposed adjustment mechanisms for the tuning of scaling factors by evaluating the control result based on system performance indices such as overshoot, rising time, amplitude and settling time. Palm (1993) addressed the method of adjusting optimally the scaling factors by measuring on-line the linear dependence between each input and output signal of the fuzzy controller. According to the above method the scaling factors are expressed in terms of input–output cross-correlation functions. Jung et al. proposed a real-time tuning of the scaling factors, based on a variable reference tuning index and an instantaneous system fuzzy performance according to system response characteristics. Chou and Lu presented an algorithm for the adjustment of the scaling factors using tuning rules, which are based on heuristics. Sagias et al. (2001) presented a model-identification fuzzy adaptive controller for real-time scaling factor adjustment.

Among the first attempts to apply a fuzzy adaptive system for the control of dynamic systems was that of Procyk and Mamdani (1979), who introduced the self-organizing controller (SOC). The configuration of the proposed controller is shown in Figure 15.11.

It has a two-level structure, in which the lower level consists of a table-based controller with two inputs and one output. The upper level consists of a performance index table, which relates the state of the process to the deviation from the desired overall response, and defines the corrections required in the table-based controller to bring the system to the desired state. From this point of view, the SOC performs two tasks simultaneously: namely, (a) performance evaluation of the system and (b) system performance improvement by creation and/or modification of the control actions based on experience gained from past system states. Hence, the controller accomplishes its learning through repetition over a sequence of operations. The elements of the table are the control actions as they are calculated from a conventional or a fuzzy controller for a fixed operational range of input variables. Here, the (i, j) element of the table contains the changes in control action inferred for the ith value of error and jth value of change of error.

15.5 MODEL IDENTIFICATION AND STABILITY OF FUZZY SYSTEMS

15.5.1 Fuzzy Systems Modeling

Mathematical models, which can describe efficiently the dynamics of the system under study, play an essential role in process analysis and control. However, most of the real-world processes are complicated and nonlinear in nature, making the derivation of mathematical models and/or subsequent analysis rather formidable tasks. In practice, such models are not available. For these cases, models need to be developed based solely on input–output data. Many approaches based on nonlinear time series (Hernadez and Arkun, 1993; Ljung, 1987), several nonlinear approaches (Henon, 1982; Wolf et al., 1985) and normal form theory (Read and Ray, 1998a,b,c) have been applied in nonlinear system modeling and analysis. During the last decade, a considerable amount of work has been published on the dynamic modeling of nonlinear systems using neural networks (Narendra and Parthasarathy, 1990; Chen and Billings, 1992; Shaw et al., 1997; Haykin, 1999) and/or fuzzy logic methodologies (Sugeno and Yasukawa, 1993; Laukoven and Pasino, 1995; Babuska and Verbruggen, 1996). Most of them, excluding normal forms and fuzzy-based approaches, are numerical in nature providing therefore only black-box representations. On the other hand, fuzzy logic methodologies (Laukoven and Pasino, 1995; Park et al., 1999; Sugeno and Kang, 1988; Sugeno and Yasukawa, 1993; Takagi and Sugeno, 1985) can incorporate a priori qualitative knowledge of the system dynamics. In Siettos et al. (2001) and Alexandridis et al. (2002) fuzzy logic and Kohonen's neural networks are combined for the derivation of truncated time series models.

Fuzzy logic can incorporate expertise and a priori qualitative knowledge of the system. In the last 20 years, strikingly results have been obtained by using various fuzzy design methods. In many cases the fuzzy control systems outperform other more traditional approaches. However, the extensive applicability of the former is limited due to the deficiency of formal and systematic design techniques, which can fulfill the two essential requirements of a control system: the requirement for robust stability and that of satisfactory performance. As a consequence, due to the complexity of nonlinear processes, it is rather difficult to construct a proper fuzzy rule base based only on observation. Moreover, the lack of a mathematical model, which characterizes fuzzy systems, often limits their applicability, since various vital tasks, such as stability analysis, are difficult to accomplish.

15.5.2   Stability of Fuzzy Systems

The problem of designing reliable fuzzy control systems in terms of stability and performance has found a remarkable resonance among engineers and scientists. So far, various approaches to this problem have been presented. One of the first contributions to this topic was that of Braae and Ratherford (1979), where they utilized the phase plane method for analyzing the stability of a fuzzy system. Kickert and Mamdani (1978) proposed the use of describing functions for the stability analysis of unforced fuzzy control systems. In Kiszka et al. (1985) the notion of the energy of fuzzy relations to investigate the local stability of a free fuzzy dynamic system is introduced. Motivated by the work of Tanaka and Sugeno (1992), many schemes have been proposed for analyzing the stability of fuzzy systems (Feng, et al., 1997; Kiriakidis et al., 1998; Leung et al., 1998; Kim et al., 1995; Thathachar and Viswanath, 1997; Wang et al., 1996). The main idea behind this approach lies in the decomposition of a global fuzzy model into simpler linear fuzzy models, which locally represent the dynamics of the whole system. In Kiendl and Ruger (1995) and Michels (1997) the authors proposed numerical methods for the stability analysis for fuzzy controllers in the sense of Lyapunov's direct method. In Fuh and Tung (1997) and Kandel et al. (1999) the stability analysis of fuzzy systems using Popov–Lyapunov techniques is proposed. In recent years, the problem of designing stable robust and adaptive fuzzy controllers with satisfactory performance based on the sliding mode approach has attracted much attention (Chen and Chang, 1998; Chen and Chen, 1998; Chen and Fukuda, 1998; Palm, 1992; Tang, et al., 1999; Wang, 1994; Yi and Chung, 1995; Yu et al., 1998). The design of such schemes is based on the Lyapunov direct method. The proposed schemes take advantage of both sliding and fuzzy features. A systematic practical way of deriving analytical expressions for fuzzy systems for use in control, system identification and stability using well-established classical theory methods is presented in Siettos et al. (2001). Finally, in Siettos and Bafas (2001) singular perturbation methods (Kokotovic et al., 1976) based on a Lyapunov approach are implemented for the derivation of sufficient conditions for the semiglobal stabilization with output tracking of nonlinear systems having internal dynamics, incorporating fuzzy controllers.

15.6   TRICKS OF THE TRADE

Newcomers to the field of fuzzy reasoning often ask themselves (and/or other more experienced fuzzy researchers) questions such as "What is the best way to get started with fuzzy reasoning?", or "Which papers should I read?"

A very helpful tutorial, a step by step introduction to the basic ideas and definitions of fuzzy set theory, with simple and well designed illustrative examples, is available at http://www.mathworks.com/access/helpdesk/help/toolbox/fuzzy/fuzzytu2.html

A variety of other sources of information are available, including the first publication on fuzzy reasoning by L. A. Zadeh, the founder of fuzzy logic, which appeared in 1965, as well as his subsequent publications (notably "Outline of a new approach to the analysis of complex systems and decision processes" *IEEE Trans. Syst., Man. Cybernet.* **3**:28–44, 1973), which inspired so many researchers in this new and fascinating field of research.

Another question often asked is "How should I be acquainted with the world of fuzzy systems and fuzzy reasoning?". This question is best answered by consulting information available on the Web. For example, in http://www.cse.dmu.ac.uk/~rij/tools.html, information useful to practitioners is given about fuzzy logic tools and companies. Information may also be found about books and journals as well as research groups and national and international associations and networks, whose members are researchers and practitioners working on fuzzy sets and systems. For this and other relevant information see the next section.

15.7   CONCLUSIONS AND PERSPECTIVES

In this chapter an overview of the basics of fuzzy reasoning has been presented. The theory of fuzzy sets has been introduced and definitions concerning the membership function, logical and transformation operators, fuzzy relations, implication and inference rules, and fuzzy similarity measures have been stated. The basic structure of a fuzzy inference system and its elements have been described. Fuzzy control is introduced and an example of a fuzzy logic controller has been demonstrated, which applies to the control of a plug flow tubular reactor. The issue of fuzzy adaptive control systems has been discussed and the self-organizing scheme has been presented. Subsequently, the topics of stability and model identification of fuzzy systems have been outlined and the presentation has concluded with an introduction of fuzzy classification and clustering systems in pattern recognition.

The above are only an elementary attempt to outline only a small part of the introductory concepts and areas of interest of fuzzy reasoning, whose theory and applications are fast developing. Indeed, during recent years, the literature on fuzzy logic theory and applications has exploded. Areas of current research include an enormous set of topics, from basic fuzzy set-theoretic concepts and fuzzy mathematics to fuzzy methodology and fuzzy logic in practice. The statement by H.-J. Zimmermann that "theoretical publications are already so specialized and assume such a background in fuzzy set theory that they are hard to understand" (Zimmermann, 1985) holds much more today than 20 years ago, when it was first stated.

In particular, research is continuing on the various basic fuzzy set-theoretic concepts, including possibility theory (Ben Amor et al., 2002), fuzzy operators (Pradera et al., 2002; Yager, 2002; Ying, 2002; Wang et al., 2003), fuzzy relations (Wang et al., 1995; Naessens et al., 2002; Pedrycz and Vasilakos, 2002), measures of information and comparison (Hung, 2002; Yager, 2002), etc.

In the area of fuzzy mathematics, research focuses on various issues of non-classical logics (Biacino and Gerla, 2002; Novak, 2002), algebra (Di Nola et al., 2002), topology (Albrecht, 2003), etc.

The research on fuzzy methodology is very extensive. It encompasses issues related to inference systems (del Amo et al, 2002; Marin-Blazquez and Shen; 2002), computational linguistics and knowledge representation (Intan and Mukaidono, 2002), production scheduling (Adamopoulos et al., 2000, Karacapilidis et al., 2000), neural networks (Alpaydin et al., 2002; Oh et al., 2002), genetic algorithms (Spiegel and Sudkamp, 2002), information processing (Liu et al., 2002; Hong et al., 2002; Nikravesh et al., 2002), pattern analysis and classification (Gabrys and Bargiela, 2002; de Moraes et al., 2002; Pedrycz and Gacek, 2002), fuzzy systems modeling and control (Mastrocostas and Theocharis, 2002; Pomares et al., 2002; Tong et al., 2002; Yi and Heng, 2002), decision making (Yager, 2002c; Wang, 2000; Zimmermann et al., 2000; Wang, 2003), etc.

Finally, extensive research is also reported on various applications of fuzzy logic, including process control (Tamhane et al., 2002), robotics (Lin and Wang, 1998; Ruan et al., 2003), scheduling (Muthusamy et al., 2003), transportation (Chou and Teng, 2002), nuclear engineering (Kunsch and Fortemps, 2002), medicine (Blanco et al., 2002; Kilic et al., 2002), economics (Kahraman et al., 2002), etc.

It is this last area and the reported applications of fuzzy reasoning, which proves the relevance and vigor of this new approach to understanding, modeling and solving many problems of modern society.

SOURCES OF ADDITIONAL INFORMATION

A most valuable source of additional information about fuzzy reasoning is the site http://www.abo.fi/~rfuller/fuzs.html. It includes information on almost everything one might like to know about the world of fuzzy systems and fuzzy reasoning, from L. A. Zadeh, the founder of fuzzy logic, fuzzy national and international associations and networks, personal home pages of fuzzy researchers, and fuzzy-mail archives, to fuzzy logic tools and companies, Conferences and Workshops on fuzzy systems, fuzzy logic journals and books and research groups. An excellent Internet course on fuzzy logic control and fuzzy clustering can be found in the site http://fuzzy.iau.dtu.dk/ from Jan Jantzen, Professor at the Technical University of Denmark, Oersted-DTU.

References

Adamopoulos, G. I., Pappis, C. P. and Karacapilidis, N. I., 2000, A methodology for solving a range of sequencing problems with uncertain data, in: *Advances in Scheduling and Sequencing under Fuzziness,* R. Slowinski and M. Hapke, eds, Physica, Heidelberg, pp. 147–164.

Albrecht, R. F., 2003, Interfaces between fuzzy topological interpretation of fuzzy sets and intervals, *Fuzzy Sets Syst.* **135**:11–20.

Alexandridis, A., Siettos, C. I., Sarimveis, H., Boudouvis, A. G. and Bafas, G. V., 2002, Modeling of nonlinear process dynamics using Kohonen's Neural Networks, *Comput. Chem. Engng.* **26**:479–486.

Alpaydin, G., Dündar, G. and Balkir, S., 2002, Evolution-based design of neural fuzzy networks using self-adapting genetic parameters, *IEEE Trans. Fuzzy Syst.* **10**:211–221.

Assilian, S. and Mamdani, E. H., 1974, An experiment in linguistic synthesis with a fuzzy logic controller, *Int. J. Man Machine Studies* **1**:1–13.

Astrom, K. J., 1983, Theory and applications of adaptive control—a survey, *Automatica* **19**:471–486.

Batur, C. and Kasparian, V., 1991, Adaptive expert control, *Int. J. Control* **54**:867–881.

Ben Amor, N., Melloyli, K., Benfeshat, S., Dubios, D. and Prade, H., 2002, A theoretical framework for possibilistic independence in a weakly ordered setting, *Int. J. Uncertainty Fuzziness Knowledge-Based Syst.* **10**:117–155.

Bezdek, J. C., 1981, *Pattern Recognition with Objective Function Algorithms,* Plenum, London.

Biacino, L. and Gerla, G., 2002, Fuzzy logic, continuity and effectiveness, *Arch. Math. Logic* **41**:643–667.

Blanco, A., Pelta, D. A. and Verdegay, J. L., 2002, Applying a fuzzy sets-based heuristic to the protein structure prediction problem, *Int. J. Intell. Syst.* **17**:629–643.

Braae, M. and Rutherford, D. A., 1979, Selection of parameters for a fuzzy logic controller, *Fuzzy Sets Syst.* **2**:185–199.

Chen, C. L. and Chang, M. H., 1998, Optimal design of fuzzy sliding-mode control: a comparative study, *Fuzzy Sets Syst.* **93**:37–48.

Chen, C. S. and Chen, W. L., 1998, Analysis and design of a stable fuzzy control system, *Fuzzy Sets Syst.* **96**:21–35.

Chen, S. and Billings, S., 1992, Neural networks for nonlinear dynamic system modelling and identification, *Int. J. Control* **56**:319–346.

Chen, X. and Fukuda, T., 1998, Robust adaptive quasi-sliding mode controller for discrete-time systems, *Syst. Control Lett.* **35**:165–173.

Chou, C. H. and Lu, H. C., 1994, A heuristic self-tuning fuzzy controller, *Fuzzy Sets Syst.* **61**:249–264.

Chou, C. H., 1998, Model reference adaptive fuzzy control: a linguistic approach, *Fuzzy Sets Syst.* **96**:1–20.

Chou, C. H. and Teng, J. C., 2002, A fuzzy logic controller for traffic junction signals, *Inform. Sci.—Informatics Comput. Sci.* **143**:73–97.

Cravaris, C. and Chung, C., 1987, Nonlinear state feedback synthesis by global input/output linearization, *AIChE J.* **33**:592–603.

Cross, V. V. and Sudkamp, T.,A., 2002, *Similarity and Compatibility in Fuzzy Set Theory—Assessment and Applications,* Studies in Fuzziness and Soft Computing, Vol.93, Springer, Berlin.

Daugherity, W., Rathakrishnan, B. and Yen, J., 1992, Performance evaluation of a self-tuning fuzzy controller, in: *Proc. IEEE Int. Conf. on Fuzzy Systems* (San Diego,CA), pp. 389–397.

De Moraes, R. M., Banon, G. J. F. and Sandri, S. A., 2002, Fuzzy expert systems architecture for image classification using mathematical morphology operators, *Inform. Sci.* **142**:7–21.

del Amo, A., Comez, D., Montero, J. and Biging, G., 2001, Relevance and redundancy in fuzzy classification systems, *Mathware Soft Comput.* **VIII**:203–216.

Di Nola, A., Esteva, F., Garcia, P., Godo, L. and Sessa, S., 2002, Subvarieties of BL-algebras generated by single component chains, *Arch. Math. Logic* **41**:673–685.

Driankov, D., Hellendoorn, H. and Reinfrank, M., 1993, *An Introduction to Fuzzy Control,* Springer, Berlin.

Dubois, D. and Prade, H., 1980, Fuzzy Sets and Systems: Theory and Applications, Academic, New York.

Feng, G., Cao, S. G., Rees, N. W. and Chak, C. K., 1997, Design of fuzzy control systems with guaranteed stability, *Fuzzy Sets Syst.* **85**:1–10.

Fuh, C. C. and Tung, P. C., 1997, Robust stability analysis of fuzzy control systems, *Fuzzy Sets Syst.* **88**:289–298.

Gabrys, B. and Bargiela, A., 2002, General fuzzy min–max neural network for clustering and classification, *IEEE Trans. Neural Networks* **11**:769–783.

Haykin, S., 1999, *Neural Networks*, 2nd edn, Prentice-Hall, Englewood Cliffs, NJ.

Henon, M., 1982, On the numerical computation of Poincaré maps, *Physica* D **5**:412–414.

Henson, M. and Seborg, D., 1990, Input–output linearization of general non-linear processes, *AIChE J.* **36**:1753–1895.

Hernadez, E. and Arkun, Y., 1993, Control of nonlinear systems using polynomial ARMA models, *AIChE J.* **39**:446–460.

Homblad, P. and Ostergaard, J.-J., 1982, Control of a cement kiln by fuzzy logic, in: *Fuzzy Information and Decision Processes*, M. M. Gupta and E. Sanchez, eds, North-Holland, Amsterdam, pp. 398–399.

Hong, T. P., Lin, K. Y. and Wang, S. L., 2002, Mining linguistic browsing patterns in the world wide web, *Soft Comput.* **6**:329–336.

Hung, W. L., 2002, Partial correlation coefficients of intuitionist fuzzy sets, *Int. J. Uncertainty Fuzziness Knowledge-Based Syst.* **10**:105–112.

Intan, R. and Mukaidono, M., 2002, On knowledge-based fuzzy sets, *Int. J. Fuzzy Syst.* **4**:655–664.

Isermann, R., 1989, *Digital Control Systems II*, Springer, Berlin.

Isidori, A., 1995, *Nonlinear Control Systems,* 3rd edn, Springer, Berlin.

Jung, C. H., Ham, C. S. and Lee, K. I., 1995, A real-time self-tuning controller through scaling factor adjustment for the steam generator of NPP, *Fuzzy Sets Syst.* **74**:53–60.

Kahraman, C., Ruan, D. and Tolga, E., 2002, Capital budgeting techniques using discounted fuzzy versus probabilistic cash flows, *Inform. Sci.* **142**:57–56.

Kandel, A., Luo, Y. and Zhang, Y. Q., 1999, Stability analysis of fuzzy control systems, *Fuzzy Sets Syst.* **105**:33–48.

Karacapilidis, N. I., Pappis, C. P. and Adamopoulos, G.,I., 2000, Fuzzy set approaches to lot sizing, in: *Advances in Scheduling and Sequencing under Fuzziness*, R. Slowinski and M. Hapke, eds, Physica, Heidelberg, pp. 291–304.

Karr, C. L. and Gentry, E. J., 1993, Fuzzy control of pH using genetic algorithms, *IEEE Trans. Fuzzy Syst.* **1**:46–53.

Kickert, W. M. and Mamdani, E. H., 1978, Analysis of a fuzzy logic controller, *Fuzzy Sets Syst.* **1**:29–44.

Kiendl, H. and Ruger, J. J., 1995, Stability analysis of fuzzy control systems using facet functions, *Fuzzy Sets Syst.* **70**:275–285.

Kilic, K. Sproule, B. A., Türksen, I. B. and Naranjo, C. A., 2002, Fuzzy system modeling in pharmacology: an improved algorithm, *Fuzzy Sets Syst.* **130**:253–264.

Kim, W. C., Ahn, S. C. and Kwon, W. H., 1995, Stability analysis and stabilization of fuzzy state space models, *Fuzzy Sets Syst.* **71**:131–142.

King, P. J. and Mamdani, E. H., 1977, Analysis of fuzzy control systems to industrial processes, *Automatica* **13**:235–242.

Kiriakidis, K., Grivas, A. and Tzes, A., 1998, Quadratic stability analysis of the Takagi–Sugeno fuzzy model, *Fuzzy Sets Syst.* **98**:1–14.

Kiszka, J. B., Gupta, M. M. and Nikiforuk, P. N., 1985, Energetistic stability of fuzzy dynamic systems, *IEEE Trans. Syst. Man and Cybernet.* **15**:783–791.

Kokotovic, P. V., O'Malley, R. E. and Sannuti, P., 1976, Singular perturbation and order reduction in control theory—an overview, *Automatica* **12**:123–132.

Kosko, B., 1992, *Neural Networks and Fuzzy Systems: A Dynamical System Approach*, Prentice-Hall, Englewood Cliffs, NJ.

Kunsch, P. L. and Fortemps, P., 2002, A fuzzy decision support system for the economic calculus in radioactive waste management, *Inform. Sci.* **142**:103–116.

Laukoven, E. G. and Pasino, K. M., 1995, Training fuzzy systems to perform estimation and identification, *Engng. Appl. Artif. Intell.* **8**:499–514.

Lee, C. C., 1990, Fuzzy logic in control systems: fuzzy logic controllers—parts I and II, *IEEE Trans. Syst. Man. Cybernet.* **20**:404–435.

Leung, F. H. F., Lam, H. K. and Tam, P. K. S., 1998, Design of fuzzy controllers for uncertain nonlinear systems using stability and robustness analyses, *Syst. Control Lett.* **35**:237–243.

Lin, C. K. and Wang, S. D., 1998, A self-organizing fuzzy control approach for bank-to-turn missiles, *Fuzzy Sets Syst.* **96**:281–306.

Liu, M., Wan, C. and Wang, L., 2002, Content-based audio classification and retrieval using a fuzzy logic system: towards multimedia search engines, *Soft Comput.* **6**:357–364.

Ljung, L., 1987, *System Identification: Theory for the User*, Prentice-Hall, Englewood Cliffs, NJ.

Maeda, M. and Murakami, S., 1992, A self–tuning fuzzy controller, *Fuzzy Sets Syst.* **51**:29–40.

Marin-Blazquez, J. G. and Shen, Q., 2002, From approximative to descriptive fuzzy classifiers, *IEEE Trans. Fuzzy Systems* **10**:484–497.

Mastrokostas, P. A. and Theocharis, J. B., 2002, A recurrent fuzzy–neural model for dynamic system identification, *IEEE Trans. Syst., Man Cybernet.* B **32**:176–190.

Michels, K., 1997, Numerical stability analysis for a fuzzy or neural network controller, *Fuzzy Sets Syst.* **89**:335–350.

Mizumoto, M., 1995, Realization of PID controls by fuzzy control methods, *Fuzzy Sets Syst.* **70**:171–182.

Muthusamy, K., Sung, S. C., Vlach, M. and Ishii, H., 2003, Scheduling with fuzzy delays and fuzzy precedences, *Fuzzy Sets Syst.* **134**:387–395.

Naessens, H., De Meyer, H. and De Baets, B., 2002, Algorithms for the computation of T-transitive closures, *IEEE Trans. Fuzzy Syst.* **10**:541–551.

Narendra, K. S. and Parthasarathy, K., 1990, Identification and control of dynamical systems using neural networks, *IEEE Trans. Neural Networks* 1:4–27.

Nikravesh, M., Loia, V. and Azvine, B., 2002, Fuzzy logic and the Internet (FLINT): Internet, World Wide Web and search engines, *Soft Comput.* 6:287–299.

Novak, V., 2002, Joint consistency of fuzzy theories, *Math. Log. Quart.* 48:563–573.

Oh, S. K., Kim, D. W. and Pedrycz, W., 2002, Hybrid fuzzy polynomial neural networks, *Int. J. Uncertainty Fuzziness Knowledge-Based Syst.* 10:257–280.

Ostergaard, J. J., 1977, Fuzzy logic control of a heat process, in: *Fuzzy Automata and Design Processes*, M. M. Gupta, G. N. Gains and B. R. Saridis, eds, Elsevier, New York.

Palm, R., 1992, Sliding mode fuzzy control, in: *Proc. IEEE Int. Conf. on Fuzzy Systems* (San Diego), pp. 519–526.

Palm, R., 1993, Tuning of scaling factors in fuzzy controllers using correlation functions, in: *Proc. IEEE Int. Conf. on Fuzzy Systems* (San Francisco), pp. 691–696.

Pappis, C. P. and Karacapilidis N. I., 1995, Application of a similarity measure of fuzzy sets to fuzzy relational equations, *Fuzzy Sets Syst.* 75:35–142.

Pappis, C. P. and Karacapilidis, N. I., 1993, A comparative assessment of measures of similarity of fuzzy values, *Fuzzy Sets Syst.* 56:171–174.

Pappis, C. P., 1976, On a fuzzy set theoretic approach to aspects of decision making in ill-defined systems, *Ph.D. Thesis,* University of London.

Pappis, C. P., 1991, Value Approximation of fuzzy systems variables, *Fuzzy Sets Syst.* 39:111–115.

Pappis, C. P. and Mamdani, E. H., 1977, A fuzzy logic controller for a traffic junction, *IEEE Systems, Man Cybernet. (SMC-7)* 10:707–717.

Pappis, C. P. and Sugeno, M., 1985, Fuzzy relational equations and the inverse problem, *Fuzzy Sets Syst.* 15:79–90.

Park, M., Ji, S., Kim, E. and Park, M., 1999, A new approach to the identification of a fuzzy model, *Fuzzy Sets Syst.* 104:169–181.

Pedrycz, W. and Gacek, A., 2002, Temporal granulation and its application to signal analysis, *Inform. Sci.* 143:47–71.

Pedrycz, W. and Vasilakos, A. V., 2002, Modularization of fuzzy relational equations, *Soft Comput.* 6:33–37.

Pomares, H., Rojas, I., Gonzalez, J. and Prieto, A., 2002, Structure identification in complete rule-Based fuzzy systems, *IEEE Trans. Fuzzy Syst.* 10:349–359.

Pradera, A., Trillas, E. and Calvo, T., 2002, A general class of triangular norm-based aggregation operators: quasilinear T-S operators, *Int. J. Approx. Reason.* 30:57–72.

Procyk, T. J. and Mamdani, E. H., 1979, A linguistic self-organizing process controller, *Automatica* **15**:15–30.

Read, N. K. and Ray, W. H., 1998a, Application of nonlinear dynamic analysis in the identification and control of nonlinear systems: I. Simple dynamics, *J. Process Control* **8**:1–15.

Read, N. K. and Ray, W. H., 1998b, Application of nonlinear dynamic analysis in the identification and control of nonlinear systems: II. More complex dynamics, *J. Process Control* **8**:17–34.

Read, N. K. and Ray, W. H., 1998c, Application of nonlinear dynamic analysis in the identification and control of nonlinear systems: III. *n*-dimensional systems, *J. Process Control* **8**:35–46.

Ross, T. J., *Fuzzy Logic with Engineering Applications,* McGraw-Hill, New York, 1995.

Ruan, D., Zhou, C. and Gupta, M. M., 2003, Fuzzy set techniques for intelligent robotic systems, *Fuzzy Sets Syst.* **134**:1–4.

Sagias, D. I., Sarafis, E. N., Siettos, C. I. and Bafas, G. V., 2001, Design of a model identification fuzzy adaptive controller and stability analysis of nonlinear processes, *Fuzzy Sets Syst.* **121**:169–179.

Sanchez, E., 1976, Resolution of composite fuzzy relational equations, *Inform. Control* **30**:38–48.

Shaw, A. M., Doyle III, F. J. and Schwaber, J. S., 1997, A dynamic neural approach to nonlinear process modeling, *Comput. Chem. Eng.* **21**:371–385.

Siettos, C. I. and Bafas, G. V., 2001, Semiglobal stabilization of nonlinear systems using fuzzy control and singular perturbation methods, *Fuzzy Sets Syst.* **129**:275–294.

Siettos, C. I., Boudouvis, A. G. and Bafas, G. V., 1999, Implementation and performance of a fuzzy adaptive controller for a tubular reactor with limit points, *Syst. Anal. Model. Simul.* **38**:725–739.

Siettos, C. I., Boudouvis, A. G. and Bafas, G. V., 2001, Approximation of fuzzy control systems using truncated Chebyshev series, *Fuzzy Sets Syst.* **126**:89–104.

Siettos, C. I., Kiranoudis, C. T. and Bafas, G. V., 1999, Advanced control strategies for fluidized bed dryers, *Drying Technol.* **17** 2271–2292.

Spiegel, D. and Sudkamp, T., 2002, Employing locality in the evolutionary generation of fuzzy rule bases, *IEEE Trans. Syst. Man Cybernet.* B **32**:296–305.

Stephanopoulos, G., 1984, *Chemical Process Control: an Introduction to Theory and Practice,* Prentice-Hall, New York.

Sugeno, M. and Yasukawa, T., 1993, A fuzzy-logic-based approach to qualitative modelling, *IEEE Trans. Fuzzy Syst.* **1**:7–31.

Sugeno, M. and Kang, G. T., 1988, Structure identification of fuzzy model, *Fuzzy Sets Syst.* **28**:15–23.

Takagi, T. and Sugeno, M., 1985, Fuzzy identification of systems and its application to modelling and control, *IEEE Trans. Syst. Man Cybernet.* **15**:116–132.

Tamhane, D., Wong, P. M. and Aminzadeh, F., 2002, Integrating linguistic descriptions and digital signals in petroleum reservoirs, *Int. J. Fuzzy Syst.* **4**:586–591.

Tanaka, K. and Sugeno, M., 1992, Stability analysis and design of fuzzy control systems, *Fuzzy Sets Syst.* **45**:135–156.

Tang, Y., Zhang, N. and Li, Y., 1999, Stable fuzzy adaptive control for a class of nonlinear systems, *Fuzzy Sets Syst.* **104**:279–288.

Taprantzis, A. V., Siettos, C. I. and Bafas, G. V., 1997, Fuzzy control of a fluidized bed dryer, *Drying Technol.* **15**:511–537.

Thathachar, M. A. and Viswanath, P., 1997, On the stability of fuzzy systems, *IEEE Trans. Fuzzy Syst.* **5**:145–151.

Togai, M. and Watanabe, 1986, Expert systems on a chip: an engine for real-time approximate reasoning, *IEEE Expert Magazine* **1**:55–62.

Tong, S., Wang, T. and Li, H. X., 2002, Fuzzy robust tracking control for uncertain nonlinear systems, *Int. J. Approx. Reason* **30**:73–90.

Wang, H. O., Tanaka, K. and Griffin, M. F., 1996, An approach to fuzzy control of nonlinear systems: stability and design issues, *IEEE Trans. Fuzzy Syst.* **4**:14–23.

Wang, H. F., 2000, Fuzzy multicriteria decision making—an overview, *J. Intell. Fuzzy Syst.* **9**:61–84.

Wang, J. and Lin, Y. I., 2003, A fuzzy multicriteria group decision making approach to select configuration items for software development, *Fuzzy Sets Syst.* **134**:43–363.

Wang, L. X., 1994, Adaptive Fuzzy Systems and Control: Design and Stability Analysis, Prentice-Hall, Englewood Cliffs, NJ.

Wang, L. X., (1992) Fuzzy systems are universal approximators, in: *Proc. IEEE Int. Conf. on Fuzzy Systems* (San Diego), pp. 1163–1170.

Wang, S. M., Wang, B. S. and Wang, G. J., 2003, A triangular-norm-based propositional fuzzy logic, *Fuzzy Sets Syst.* **136**:55–70.

Wang, W., 1997, New similarity measures on fuzzy sets and on elements, *Fuzzy Sets Syst.* **85**:305–309.

Wang, W., De Baets, B. and Kerre, E., 1995, A comparative study of similarity measures, *Fuzzy Sets Syst.* **73**:259–268.

Wolf, A., Swift, J. B., Swinney, H. L. and Vastano, J. A., 1985, Determining Lyapunov exponents from a time series, *Physica* D, **16**:285–317.

Yager, R. R., 2002, On the cardinality index and attitudinal character of fuzzy measures, *Int. J. Gen. Syst.* **31**:303–329.

Yager, R. R., 2002, The power average operator, *IEEE Trans. Syst. Man Cybernet.* A **31**:724–730.

Yager, R. R. (2002) On the valuation of alternatives for decision-making under uncertainty, *Int. J. Intell. Syst.* **17**:687–707.

Yan, J., Ryan, M. and Power, J., 1994, Using Fuzzy Logic, Prentice-Hall, Englewood Cliffs, NJ.

Yi, S. Y. and Chung, M. J., 1995, Systematic design and stability analysis of a fuzzy logic controller, *Fuzzy Sets Syst.* **72**:271–298.

Yi, Z. and Heng, P. A., 2002, Stability of fuzzy control systems with bounded uncertain delays, *IEEE Trans. Fuzzy Syst.* **10**:92–97.

Ying, M., 2002, Implication operators in fuzzy logic, *IEEE Trans. Fuzzy Syst.* **10**:88–91.

Yu, X., Man, Z. and Wu, B., 1998, Design of fuzzy sliding-mode control systems, *Fuzzy Sets Syst.* **95**:295–306.

Zadeh, L. A., 1973, Outline of a new approach to the analysis complex systems and decision processes, *IEEE Trans. Syst. Man. Cybernet.* **3**:28–44.

Zheng, L., 1992, A practical guide to tune of proportional and integral (PI) like fuzzy controllers, in: *Proc. 1st IEEE Int. Conf. on Fuzzy Systems* (San Diego), pp. 633–640.

Zienkiewicz, O. C. and Morgan, K., 1983, *Finite Elements and Approximation,* Wiley, New York.

Zimmermann, H. J., 1996, *Fuzzy Set Theory and its Applications,* 3rd edn, Kluwer, Dordrecht.

Zimmermann, H. J., Ruan D. and Huang C., eds, 2000, *Fuzzy Sets and Operations Research for Decision Support: Key Selected Papers,* Beijing Normal University Press, Beijing.

Chapter 16

ROUGH SET BASED DECISION SUPPORT

Roman Slowinski
Institute of Computing Science
Poznan University of Technology
and
Institute for Systems Research
Polish Academy of Sciences
Warsaw, Poland

Salvatore Greco, Benedetto Matarazzo
Faculty of Economics
University of Catania, Italy

16.1   INTRODUCTION

In this chapter, we are concerned with discovering knowledge from data. The aim is to find concise classification patterns that agree with *situations* that are described by the data. Such patterns are useful for explanation of the data and for the prediction of future situations. They are particularly useful in such decision problems as technical diagnostics, performance evaluation and risk assessment. The situations are described by a set of *attributes*, which we might also call properties, features, characteristics, etc. Such attributes may be concerned with either the *input* or *output* of a situation. These situations may refer to states, examples, etc. Within this chapter, we will refer to them as *objects*. The goal of the chapter is to present a knowledge discovery paradigm for multi-attribute and multicriteria decision making, which is based upon the concept of rough sets. Rough set theory was introduced by Pawlak (1982, 1991). Since then, it has often proved to be an excellent mathematical tool for the analysis of a *vague* description of objects. The adjective vague (referring to the quality of information) is concerned with inconsistency or ambiguity. The rough set philosophy is based on the assumption that with every object of the universe U there is associated a certain amount of information (data, knowledge). This information can be expressed by means of a number of attributes. The attributes describe the object. Objects which have the same description are said to be indiscernible (similar) with respect to the available information. The *indiscernibility relation* thus generated constitutes the mathematical basis of rough set theory. It induces a partition of the universe into blocks of indiscernible objects, called elementary sets, which can be used to build knowledge about a real or abstract world. The use of the indiscernibility relation results in information *granulation*.

Any subset X of the universe may be expressed in terms of these blocks either precisely (as a union of elementary sets) or approximately. In the latter case, the subset X may be characterized by two ordinary sets, called the *lower* and *upper approximations*. A rough set is defined by means of these two approximations, which coincide in the case of an ordinary set. The lower approximation of X is composed of all the elementary sets included in X (whose elements, therefore, certainly belong to X), while the upper approximation of X consists of all the elementary sets which have a non-empty intersection with X (whose elements, therefore, may belong to X). The difference between the upper and lower approximation constitutes the boundary region of the rough set, whose elements cannot be characterized with certainty as belonging or not to X (by using the available information). The information about objects from the boundary region is, therefore, inconsistent or ambiguous. The cardinality of the boundary region states, moreover, the extent to which it is possible to express X in exact terms, on the basis of the available information. For this reason, this cardinality may be used as a measure of vagueness of the information about X.

Some important characteristics of the rough set approach makes it a particularly interesting tool in a variety of problems and concrete applications. For example, it is possible to deal with both quantitative and qualitative input data and inconsistencies need not to be removed prior to the analysis. In terms of the output information, it is possible to acquire a posteriori information regarding the relevance of particular attributes and their subsets to the quality of approximation considered within the problem at hand. Moreover, the lower and upper approximations of a partition of U into decision classes, prepare the ground for handling *certain* and *possible* classification patterns in the form of "*if...*, *then...*" decision rules.

Several attempts have been made to employ rough set theory for decision support (Pawlak and Slowinski, 1994; Slowinski 1993). The Classical Rough Set Approach is not able, however, to deal with preference-ordered attribute domains and preference-ordered decision classes. In decision analysis, an attribute with a preference-ordered domain (scale) is called a *criterion*.

In the late 1990s, adapting the classical rough set approach to knowledge discovery from preference-ordered data became a particularly challenging problem within the field of multicriteria decision analysis. Why might it be so important? The answer is connected with the nature of the input preferential information available in multicriteria decision analysis and of the output of that analysis. As to the input, the rough set approach requires a set of decision examples which is also convenient for the acquisition of preferential information from decision makers. Very often in multicriteria decision analysis, this information has to be given in terms of preference model parameters, such as importance weights, substitution ratios and various thresholds. Presenting such information requires significant effort on the part of the decision maker. It is generally acknowledged that people often prefer to make exemplary decisions and cannot always explain them in terms of specific parameters. For this reason, the idea of inferring preference models from exemplary decisions provided by the decision maker is very attractive. Furthermore, the exemplary decisions may be inconsistent because of some additional aspects which are not included in the considered family of criteria and because of hesitation on the part of the decision maker (see e.g. Roy 1996). These inconsistencies cannot be considered as a simple error or as noise. They can convey important information that should be taken into account in the construction of the decision maker's preference model. The rough set approach is intended to deal with inconsistency and this is a major argument to support its application to multicriteria decision analysis. Note also that the output of the analysis, i.e. the model of preferences in terms of decision rules is very convenient for decision support because it is intelligible and *speaks the same language* as the decision maker.

An extension of the classical rough set approach which enables the analysis of preference-ordered data was proposed by Greco et al. (1998a, 1999a,b). This extension, called the Dominance-Based Rough Set Approach is mainly based on the substitution of the indiscernibility relation by a dominance relation in the rough approximation of decision classes. An important consequence of this fact is the possibility of inferring (from exemplary decisions) the preference model in terms of decision rules which are logical statements of the type "*if...*, *then...*". The separation of *certain* and *uncertain* knowledge about the decision maker's preferences is carried out by the distinction of different kinds of decision rules, depending upon whether they are induced from lower approximations of decision classes or from the difference between upper and lower approximations (composed of inconsistent examples). Such a preference model is more general than the classical functional models considered within multi-attribute utility theory or the relational models considered, for example, in outranking methods.

In the next section, we begin this tutorial by presenting the basic version of the classical rough set approach by way of an example.

Table 16.1. Examples of traffic signs described by S and PC.

Traffic sign		Shape (S)	Primary color (PC)	Class
(a)		triangle	yellow	*W*
(b)		circle	white	*I*
(c)		circle	blue	*I*
(d)		circle	blue	*O*

16.2   ROUGH SET FUNDAMENTALS

16.2.1   Explanation by an Example

Let us assume that we want to describe the classification of basic traffic signs to a novice. We start by saying that there are three main classes of traffic signs corresponding to

- Warning (W),

- Interdiction (I),

- Order (O).

Then, we say that these classes may be distinguished by such attributes as the shape (S) and the principal color (PC) of the sign. Finally, we give a few examples of traffic signs, like those shown in Table 16.1.
    These are

(a) sharp right turn,

(b) speed limit of 50 km h^{-1},

(c) no parking,

(d) go ahead.

One can remark that the sets of signs indiscernible by "Class" are

$$W = \{a\}_{\text{Class}} \quad I = \{b, c\}_{\text{Class}} \quad O = \{d\}_{\text{Class}}$$

Table 16.2. Examples of traffic signs described by S, PC and SC.

Traffic sign		Shape (S)	Primary color (PC)	Secondary color (SC)	Class
(a)		triangle	yellow	red	*W*
(b)		circle	white	red	*I*
(c)		circle	blue	red	*I*
(d)		circle	blue	white	*O*

and the sets of signs indiscernible by S and PC are as follows:

$$\{a\}_{\text{S,PC}} \quad \{b\}_{\text{S,PC}} \quad \{c,d\}_{\text{S,PC}}$$

The above sets are *granules of knowledge* generated by, on the one hand, the classification of traffic signs by "Class" and, on the other hand, their description by S and PC. The sets of signs indiscernible by "Class" are denoted by $\{\cdot\}_{\text{Class}}$ and those by S and PC are denoted by $\{\cdot\}_{\text{S,PC}}$. We can see that granule $W = \{a\}_{\text{Class}}$ is characterized precisely by granule $\{a\}_{\text{S,PC}}$. In order to characterize granules $I = \{b,c\}_{\text{Class}}$ and $O = \{d\}_{\text{Class}}$, one needs granules $\{b\}_{\text{S,PC}}$ and $\{c,d\}_{\text{S,PC}}$; however, only granule $\{b\}_{\text{S,PC}}$ is included in $I = \{b,c\}_{\text{Class}}$ while $\{c,d\}_{\text{S,PC}}$ has a non-empty intersection with both $I = \{b,c\}_{\text{Class}}$ and $O = \{d\}_{\text{Class}}$. It follows, from this characterization, that by using attributes S and PC, one can characterize class W precisely, while classes I and O can only be characterized approximately:

- class W includes sign (a) certainly and possibly no other sign,

- class I includes sign (b) certainly and possibly signs (b), (c) and (d),

- class O includes no sign certainly and possibly signs (c) and (d).

The terms *certainly* and *possibly* refer to the absence or presence of ambiguity between the description of signs by S and PC from the one side, and by "Class", from the other side. In other words, using knowledge about the description of signs by S and PC, one can say that all signs from granules $\{\cdot\}_{\text{S,PC}}$

included in granules $\{\,\cdot\,\}_{\text{Class}}$ belong certainly to the corresponding class, while all signs from $\{\,\cdot\,\}_{\text{S,PC}}$ having a non-empty intersection with granules $\{\,\cdot\,\}_{\text{Class}}$ belong to the corresponding class only possibly. The two sets of certain and possible signs are, respectively, the *lower* and *upper approximations* of the corresponding class by attributes S and PC:

$$\text{lower\_approx.}_{\text{S,PC}}(W) = \{a\} \qquad \text{upper\_approx.}_{\text{S,PC}}(W) = \{a\}$$
$$\text{lower\_approx.}_{\text{S,PC}}(I) = \{b\} \qquad \text{upper\_approx.}_{\text{S,PC}}(I) = \{b, c, d\}$$
$$\text{lower\_approx.}_{\text{S,PC}}(O) = \{\emptyset\} \qquad \text{upper\_approx.}_{\text{S,PC}}(O) = \{c, d\}$$

The *quality of approximation* of the classification by attributes S and PC is equal to the number of all the signs in the lower approximations divided by the number of all the signs in the table, i.e. $1/2$.

One way to increase the quality of the approximation is to add a new attribute so as to decrease the ambiguity. Let us introduce the secondary color (SC) as a new attribute. The new situation is shown in Table 16.2.

As one can see, the sets of signs indiscernible by S, PC and SC, i.e. the granules $\{\,\cdot\,\}_{\text{S,PC,SC}}$, are now

$$\{a\}_{\text{S,PC,SC}} \quad \{b\}_{\text{S,PC,SC}} \quad \{c\}_{\text{S,PC,SC}} \quad \{d\}_{\text{S,PC,SC}}.$$

It is worth noting that the granularity is finer than before and it enables the ambiguity to be eliminated. Consequently, the quality of approximation of the classification by attributes S, PC and SC is now equal to 1.

A natural question occurring here is to ask if, indeed, all three attributes are necessary to characterize precisely the classes W, I and O. When we eliminate attribute S or attribute PC from the description of the signs, we obtain the granules $\{\,\cdot\,\}_{\text{PC,SC}}$ or $\{\,\cdot\,\}_{\text{S,SC}}$, respectively, as follows:

$$\{a\}_{\text{PC,SC}} \quad \{b\}_{\text{PC,SC}} \quad \{c\}_{\text{PC,SC}} \quad \{d\}_{\text{PC,SC}}$$
$$\{a\}_{\text{S,SC}} \quad \{b, c\}_{\text{S,SC}} \quad \{d\}_{\text{S,SC}}$$

Using any one of the above sets of granules, it is possible to characterize (approximate) classes W, I and O with the same quality (equal to 1) as it is when using the granules $\{\,\cdot\,\}_{\text{S,PC,SC}}$ (i.e. those generated by the complete set of three attributes). Thus, the answer to the above question is that the three attributes are not necessary to characterize precisely the classes W, I and O. It is, in fact, sufficient to use either PC and SC or S and SC. The subsets of attributes {PC, SC} and {S, SC}) are called *reducts* of {S, PC, SC} because they have this property. Note that the identification of reducts enables us to *reduce* knowledge about the signs from the table to that which is *relevant*.

Other useful information can be generated from the identification of reducts by taking their intersection. This is called the *core*. In our example, the core contains attribute SC. This tells us that it is clearly an *indispensable* attribute i.e. it cannot be eliminated from the description of the signs without decreasing the quality of the approximation. Note that other attributes from the reducts (i.e. S and PC) are *exchangeable*. If there happened to be some other attributes which were not included in any reduct, then they would be *superfluous* i.e. they would not be useful at all in the characterization of the classes W, I and O.

If, however, we eliminate column S or PC from Table 16.2 then we do not yet have a minimal representation of knowledge about the classification of the four traffic signs. Note that, in order to characterize class W in Table 16.2, it is sufficient to use the descriptor "S = triangle". Moreover, class I is characterized by two descriptors ("S = circle" and "SC = red") and class O is characterized by the descriptor "SC = white". Thus, the minimal representation of this knowledge requires only four descriptors (rather than the eight descriptors that are presented in Table 16.2 with either column S or PC eliminated). This representation corresponds to the following set of *decision rules* which may be seen as classification patterns discovered in the dataset contained in Table 16.2 (in braces are the symbols of signs covered by the corresponding rule):

rule #1 : *if* $S =$ triangle, *then* Class $= W$ $\{a\}$

rule #2 : *if* $S =$ circle *and* $SC =$ red, *then* Class $= I$ $\{b, c\}$

rule #3 : *if* $SC =$ white, *then* Class $= O$ $\{d\}$

This is not the only representation, because an alternative set of rules is

rule #1$'$: *if* $PC =$ yellow, *then* Class $= W$ $\{a\}$

rule #2$'$: *if* $PC =$ white, *then* Class $= I$ $\{b\}$

rule #3$'$: *if* $PC =$ blue *and* $SC =$ red, *then* Class $= I$ $\{c\}$

rule #4$'$: *if* $SC =$ white, *then* Class $= O$ $\{d\}$

It is interesting to return to Table 16.1 and ask what decision rules represent the knowledge contained in this dataset. As the description of the four signs by S and PC is not sufficient to characterize precisely all the classes, it is not surprising that not all the rules will have a non-ambiguous decision. Indeed, we have

rule #1$''$: *if* $S =$ triangle, *then* Class $= W$ $\{a\}$

rule #2$''$: *if* $PC =$ white, *then* Class $= I$ $\{b\}$

rule #3$''$: *if* $PC =$ blue, *then* Class $= I$ *or* O $\{c, d\}$

Note that these rules can be induced from the lower and upper approximations of classes W, I and O defined above. Indeed, for rule #1$''$, the supporting example is in lower_approx.$_{S,PC}(W) = \{a\}$; for rule #2$''$ it is in lower_approx.$_{S,PC}(I) = \{b\}$; and the supporting examples for rule #3$''$ are in the set called the *boundary* of both I and O:

$$
\begin{aligned}
\text{boundary}_{S,PC}(I) \;&=\; \text{upper\_approx.}_{S,PC}(I) \\
&\quad -\text{lower\_approx.}_{S,PC}(I) = \{c, d\} \\
\text{boundary}_{S,PC}(O) \;&=\; \text{upper\_approx.}_{S,PC}(O) \\
&\quad -\text{lower\_approx.}_{S,PC}(O) = \{c, d\}
\end{aligned}
$$

As a result of the approximate characterization of classes W, I and O by S and PC, we can thus obtain an approximate representation in terms of decision rules. Since the quality of the approximation is 1/2, *certain rules* (#1″ and #2″) cover one half of the examples and the other half is covered by the *approximate rule* (#3″). Now, the quality of approximation by S and SC or by PC and SC was equal to 1, so all examples were covered by certain rules (#1–3 or #1′–4′, respectively).

We can see, from this simple example, that the rough set analysis of a dataset provides some useful information. In particular, we can determine

- a characterization of decision classes in terms of chosen attributes through lower and upper approximation;

- a measure of the quality of approximation which indicates how good the chosen set of attributes is for approximation of the classification;

- a reduction of the knowledge contained in the table to a description by relevant attributes, i.e. those belonging to reducts; at the same time, superfluous attributes are also identified;

- the core which indicates indispensable attributes;

- a set of decision rules which is induced from the lower and upper approximations of the decision classes; this shows classification patterns which exist in the dataset.

Other important information can also be induced but it cannot be illustrated by such a simple example. In the next section, we will present a more formal treatment. For more details, the reader is referred to Pawlak (1991), Polkowski (2002), Slowinski (1992b) and many others (see Sources of Additional Information at the end of the chapter).

16.2.2   A Formal Description of the Classical Rough Set Approach

For algorithmic reasons, we supply the information regarding the objects in the form of a *data table*, whose separate rows refer to distinct *objects* and whose columns refer to the different *attributes* considered. Each cell of this table indicates an *evaluation* (quantitative or qualitative) of the object placed in that row by means of the attribute in the corresponding column.

Formally, a *data table* is the 4-tuple $S = \langle U, Q, V, f \rangle$, where U is a finite set of *objects* (universe), $Q = \{q_1, q_2, \ldots, q_m\}$ is a finite set of *attributes*, V_q is the domain of the attribute q, $V = \bigcup_{q \in Q} V_q$ and $f : U \times Q \to V$ is a total function such that $f(x, q) \in V_q$ for each $q \in Q, x \in U$, called the *information function*.

Each object x of U is described by a vector (string) $\text{Des}_Q(x) = [f(x, q_1), f(x, q_2), \ldots, f(x, q_m)]$, called the *description* of x in terms of the evaluations of the attributes from Q. It represents the available information about x.

To every (non-empty) subset of attributes P we associate an *indiscernibility relation* on U, denoted by I_P and defined as follows:

$$I_P = \{(x, y) \in U \times U : f(x, q) = f(y, q), \text{ for each } q \in P\}.$$

If $(x, y) \in I_P$, we say that the objects x and y are P-indiscernible. Clearly, the indiscernibility relation thus defined is an equivalence relation (reflexive, symmetric and transitive). The family of all the equivalence classes of the relation I_P is denoted by $U \mid I_P$ and the equivalence class containing an element $x \in U$ is denoted by $I_P(x)$. The equivalence classes of the relation I_P are called the *P-elementary sets* or *granules of knowledge* encoded by P.

Let S be a data table, X be a non-empty subset of U and $\emptyset \neq P \subseteq Q$. The set X may be characterized by two ordinary sets, called the *P-lower approximation* of X (denoted by $\underline{P}(X)$) and the *P-upper approximation* of X (denoted by $\overline{P}(X)$) in S. They can be defined, respectively, as

$$\begin{aligned} \underline{P}(X) &= \{x \in U : I_P(x) \subseteq X\} \\ \overline{P}(X) &= \bigcup_{x \in X} I_P(x) \end{aligned}$$

The family of all the sets $X \subseteq U$ having the same P-lower and P-upper approximations is called a *P-rough set*. The elements of $\underline{P}(X)$ are all and only those objects $x \in U$ which belong to the equivalence classes generated by the indiscernibility relation I_P *contained* in X. The elements of $\overline{P}(X)$ are all and only those objects $x \in U$ which belong to the equivalence classes generated by the indiscernibility relation I_P *containing at least one* object x belonging to X. In other words, $\underline{P}(X)$ is the largest union of the P-elementary sets included in X, while $\overline{P}(X)$ is the smallest union of the P-elementary sets containing X.

The *P-boundary* of X in S, denoted by $Bn_P(X)$, is defined as

$$Bn_P(X) = \overline{P}(X) - \underline{P}(X)$$

The term *rough approximation* is a general term used to express the operation of the P-lower and P-upper approximation of a set or of a union of sets. The rough approximations obey the following basic laws (see Pawlak, 1991):

- the *inclusion property*: $\underline{P}(X) \subseteq X \subseteq \overline{P}(X)$

- the *complementarity property*: $\underline{P}(X) = U - \overline{P}(U - X)$

Therefore, if an object x belongs to $\underline{P}(X)$, it is also *certainly* contained in X, while if x belongs to $\overline{P}(X)$, it is only *possibly* contained in X. $Bn_P(X)$ constitutes the *doubtful region* of X: using the knowledge encoded by P nothing can be said with certainty about the inclusion of its elements in set X.

If the P-boundary of X is empty (i.e. $Bn_P(X) = \emptyset$) then the set X is an ordinary set, called the P-exact set. By this, we mean that it may be expressed as the union of a certain number of P-elementary sets. Otherwise, if $Bn_P(X) \neq \emptyset$, then the set X is a P-rough set and may be characterized by means of $\underline{P}(X)$ and $\overline{P}(X)$.

The following ratio defines an *accuracy* measure of the approximation of X ($X \neq \emptyset$) by means of the attributes from P: $\alpha_P(X) = \frac{|\underline{P}(X)|}{|\overline{P}(X)|}$, where $|Y|$ denotes the cardinality of a (finite) set Y.

Obviously, $0 \leq \alpha_P(X) \leq 1$. If $\alpha_P(X) = 1$, then X is a P-exact set. If $\alpha_P(X) < 1$, then X is a P-rough set.

Another ratio defines a *quality* measure of the approximation of X by means of the attributes from P: $\gamma_P(X) = \frac{|\underline{P}(X)|}{|X|}$. The quality $\gamma_P(X)$ represents the relative frequency of the objects correctly assigned by means of the attributes from P. Moreover, $0 \leq \alpha_P(X) \leq \gamma_P(X) \leq 1$, and $\gamma_P(X) = 0$ iff $\alpha_P(X) = 0$, while $\gamma_P(X) = 1$ iff $\alpha_P(X) = 1$.

The definition of approximations of a subset $X \subseteq U$ can be extended to a classification, i.e. a partition $Y = \{Y_1, \ldots, Y_n\}$ of U. The subsets Y_i, $i = 1, \ldots, n$, are disjunctive classes of Y. By the P-lower and P-upper approximations of Y in S we mean the sets $\underline{P}Y = \{\underline{P}Y_1, \ldots, \underline{P}Y_n\}$ and $\overline{P}Y = \{\overline{P}Y_1, \ldots, \overline{P}Y_n\}$, respectively. The coefficient

$$\gamma_P(Y) = \frac{\sum_{i=1}^{n} |\underline{P}Y_i|}{|U|}$$

is called the *quality of approximation of classification* Y by the set of attributes P, or in short, the *quality of classification*. It expresses the ratio of all P-correctly classified objects to all objects in the data table.

The main issue in rough set theory is the approximation of subsets or partitions of U, representing *knowledge* about U, with other sets or partitions that have been built up using available information about U. From the perspective of a particular object $x \in U$, it may be interesting, however, to use the available information to assess the degree of its membership to a subset X of U. The subset X can be identified with the knowledge to be approximated. Using the rough set approach one can calculate the membership function $\mu_X^P(x)$ (*rough membership function*) as

$$\mu_X^P(x) = \frac{|X \cap I_P(x)|}{|I_P(x)|}$$

The value of $\mu_X^P(x)$ may be interpreted analogously as conditional probability and may be understood as the *degree of certainty* (credibility) to which x belongs to X. Observe that the value of the membership function is calculated from the available data, and not subjectively assumed, as it is in the case of membership functions of fuzzy sets.

Between the rough membership function and the rough approximations of X the following relationships hold:

$$
\begin{aligned}
\underline{P}(X) &= \left\{ x \in U : \mu_X^P(x) = 1 \right\} \\
\overline{P}(X) &= \left\{ x \in U : \mu_X^P(x) > 0 \right\} \\
\underline{P}(U - X) &= \left\{ x \in U : \mu_X^P(x) = 0 \right\} \\
\overline{P}(U - X) &= \left\{ x \in U : \mu_X^P(x) < 1 \right\} \\
Bn_P(X) &= Bn_P(U - X) = \left\{ x \in U : 0 < \mu_X^P(x) < 1 \right\}
\end{aligned}
$$

In rough set theory there is, therefore, a close link between the granularity connected with the rough approximation of sets and the uncertainty connected with the rough membership of objects to sets.

A very important concept for concrete applications is that of the dependence of attributes. Intuitively, a set of attributes $T \subseteq Q$ *totally depends* upon a set of attributes $P \subseteq Q$ if all the values of the attributes from T are uniquely determined by the values of the attributes from P. In other words, this is the case if a functional dependence exists between evaluations by the attributes from P and by the attributes from T. This means that the partition (granularity) generated by the attributes from P is at least as "fine" as that generated by the attributes from T, so that it is sufficient to use the attributes from P to build the partition $U|I_T$. Formally, T totally depends on P iff $I_P \subseteq I_T$.

Therefore, T is totally (partially) dependent on P if all (some) objects of the universe U may be univocally assigned to granules of the partition $U|I_T$, using only the attributes from P.

Another issue of great practical importance is that of *knowledge reduction*. This concerns the elimination of superfluous data from the data table, without deteriorating the information contained in the original table.

Let $P \subseteq Q$ and $p \in P$. It is said that attribute p is *superfluous* in P if $I_P = I_{P-\{p\}}$; otherwise, p is *indispensable* in P.

The set P is *independent* if all its attributes are indispensable. The subset P' of P is a *reduct* of P (denoted by RED(P)) if P' is independent and $I_{P'} = I_P$.

A reduct of P may also be defined with respect to an approximation of the classification Y of objects from U. It is then called a *Y-reduct* of P (denoted by $\text{RED}_Y(P)$) and it specifies a minimal subset P' of P which keeps the quality of the classification unchanged, i.e. $\gamma_{P'}(Y) = \gamma_P(Y)$. In other words, the attributes that do not belong to a Y-reduct of P are superfluous with respect to the classification Y of objects from U.

More than one Y-reduct (or reduct) of P may exist in a data table. The set containing all the indispensable attributes of P is known as the *Y-core* (denoted by $\text{CORE}_Y(P)$). In formal terms, $\text{CORE}_Y(P) = \bigcap \text{RED}_Y(P)$. Obviously, since the Y-core is the intersection of all the Y-reducts of P, it is included in every Y-reduct of P. It is the most important subset of attributes of Q, because none of its elements can be removed without deteriorating the quality of the classification.

16.2.3   Decision Rules Induced From Rough Approximations

If in a data table the attributes of the set Q are divided into *condition* attributes (set $C \neq \emptyset$) and *decision* attributes (set $D \neq \emptyset$), then such a table is called a *decision table*. Note that $C \cup D = Q$ and $C \cap D = \emptyset$. The decision attributes induce a partition of U deduced from the indiscernibility relation I_D in a way that is independent of the condition attributes. D-elementary sets are called *decision classes*. There is a tendency to reduce the set C while keeping all important relationships between C and D, in order to make decisions on the basis of a smaller amount of information. When the set of condition attributes is replaced by one of its reducts, the quality of approximation of the classification induced by the decision attributes does not deteriorate.

Since the tendency is to underline the functional dependencies between condition and decision attributes, a decision table may also be seen as a set of *decision rules*. These are logical statements (consequence relations) of the type "*if..., then...*", where the antecedent (condition part) specifies values assumed by one or more condition attributes (describing C-elementary sets) and the consequence (decision part) specifies an assignment to one or more decision classes (describing D-elementary sets). Therefore, the syntax of a rule can be outlined as follows:

> *if* $f(x,q_1)$ is equal to r_{q1}
> *and* $f(x,q_2)$ is equal to r_{q2}
> *and* ... *and* $f(x,q_p)$ is equal to r_{qp},

then x belongs to Y_{j1} *or* Y_{j2} *or* $\ldots Y_{jk}$

where $\{q_1, q_2, \ldots, q_p\} \subseteq C$, $(r_{q1}, r_{q2}, \ldots, r_{qp}) \in V_{q1} \times V_{q2} \times \cdots \times V_{qp}$ and $Y_{j1}, Y_{j2}, \ldots, Y_{jk}$ are some decision classes of the considered classification (*D*-elementary sets). If there is only one possible consequence, i.e. $k = 1$, then the rule is said to be *certain*, otherwise it is said to be *approximate* or *ambiguous*.

An object $x \in U$ *supports* decision rule r if its description is matching both the condition part and the decision part of the rule. We also say that decision rule r *covers* object x if it matches at least the condition part of the rule. Each decision rule is characterized by its *strength* defined as the number of objects supporting the rule. In the case of approximate rules, the strength is calculated for each possible decision class separately.

Let us observe that certain rules are supported only by objects from the lower approximation of the corresponding decision class. Approximate rules are supported, in turn, only by objects from the boundaries of the corresponding decision classes.

Procedures for the generation of decision rules from a decision table use an *inductive learning* principle. The objects are considered as examples of decisions. In order to induce decision rules with a unique consequent assignment to a *D*-elementary set, the examples belonging to the *D*-elementary set are called *positive* and all the others *negative*. A decision rule is *discriminant* if it is consistent (i.e. if it distinguishes positive examples from negative ones) and *minimal* (i.e. if removing any attribute from a condition part gives a rule covering negative objects). It may be also interesting to look for *partly discriminant* rules. These are rules that, besides positive examples, could cover a limited number of negative ones. They are characterized by a coefficient, called the *level of confidence*, which is the ratio of the number of positive examples (supporting the rule) to the number of all examples covered by the rule.

The generation of decision rules from decision tables is a complex task and a number of procedures have been proposed to solve it (see e.g. Grzymala-Busse, 1992, 1997; Skowron, 1993; Skowron and Polkowski, 1997; Slowinski et al., 2000; Stefanowski, 1998; Ziarko and Shan, 1994). The existing induction algorithms use one of the following strategies:

- The generation of a minimal set of rules covering all objects from a decision table.

- The generation of an exhaustive set of rules consisting of all possible rules for a decision table.

- The generation of a set of "strong" decision rules, even partly discriminant, covering relatively many objects from the decision table (but not necessarily all of them).

16.2.4   From Indiscernibility to Similarity

As mentioned above, the classical definitions of lower and upper approximations are based on the use of the binary indiscernibility relation which is an equivalence relation. The indiscernibility implies the impossibility of distinguishing between two objects of U having the *same* description in terms of the attributes from Q. This relation induces equivalence classes on U, which constitute the basic granules of knowledge. In reality, due to the imprecision of data describing the objects, small differences are often not considered significant for the purpose of discrimination. This situation may be formally modeled by considering similarity or tolerance relations (see e.g. Nieminen, 1988; Marcus, 1994; Slowinski, 1992a; Polkowski et al., 1995; Skowron and Stepaniuk, 1995; Slowinski and Vanderpooten, 1997, 2000; Stepaniuk, 2000; Yao and Wong, 1995).

Replacing the indiscernibility relation by a weaker binary *similarity* relation has considerably extended the capacity of the rough set approach. This is because, in the least demanding case, the similarity relation requires reflexivity only, relaxing the assumptions of symmetry and transitivity.

In general, the similarity relations R do not generate partitions on U. The information regarding similarity may be represented using *similarity classes* for each object $x \in U$. More precisely, the similarity class of x, denoted by $R(x)$, consists of the set of objects which are similar to x:

$$R(x) = \{y \in U : yRx\}$$

It is obvious that an object y may be similar to both x and z, while z is not similar to x, i.e. $y \in R(x)$ and $y \in R(z)$, but $z \notin R(x)$, $x, y, z \in U$. The similarity relation is of course reflexive (each object is similar to itself). Slowinski and Vanderpooten (1995, 2000) have proposed a *similarity* relation which is only *reflexive*. The abandonment of the transitivity requirement is easily justifiable. For example, see Luce's (1956) paradox of the cups of tea:

- one cannot distinguish a cup of tea without sugar from a cup of tea with one grain of sugar, and

- one cannot distinguish a cup of tea with n grains of sugar from a cup of tea with $n + 1$ grains of sugar, however

- one can distinguish a cup of tea without sugar from a cup of tea with two spoons of sugar.

As for the symmetry, one should notice that yRx, which means "y is similar to x", is directional. There is a *subject* y and a *referent* x, and in general this is not equivalent to the proposition "x is similar to y", as maintained by Tversky (1977). This is immediate when the similarity relation is defined in terms of a percentage difference between evaluations of the objects compared on the attribute in hand, calculated with respect to a numerical evaluation of the referent object. Therefore, the symmetry of the similarity relation should not be imposed. It then makes sense to consider the inverse relation of R, denoted by R^{-1}, where $x R^{-1} y$ means again "y is similar to x". Thus $R^{-1}(x)$, $x \in U$, is the class of referent objects to which x is similar:

$$R^{-1}(x) = \{y \in U : x R y\}$$

Given a subset $X \subseteq U$ and a similarity relation R on U, an object $x \in U$ is said to be *non-ambiguous* in each of the two following cases:

- x belongs to X without ambiguity, that is $x \in X$ and $R^{-1}(x) \subseteq X$; such objects are also called *positive*;

- x does not belong to X without ambiguity (x clearly does not belong to X), that is $x \in U - X$ and $R^{-1}(x) \subseteq U - X$ (or $R^{-1}(x) \cap X = \emptyset$); such objects are also called *negative*.

The objects which are neither positive nor negative are said to be *ambiguous*. A more general definition of lower and upper approximation may thus be offered (see Slowinski and Vanderpooten, 2000). Let $X \subseteq U$ and let R be a reflexive binary relation defined on U. The lower approximation of X, denoted by $\underline{R}(X)$, and the upper approximation of X, denoted by $\overline{R}(X)$, are defined, respectively, as

$$\begin{aligned} \underline{R}(X) &= \left\{ x \in U : R^{-1}(x) \subseteq X \right\} \\ \overline{R}(X) &= \bigcup_{x \in X} R(x) \end{aligned}$$

It may be demonstrated that the key properties, inclusion and complementarity, still hold and that

$$\overline{R}(X) = \left\{ x \in U : R^{-1}(x) \cap X \neq \emptyset \right\}$$

Moreover, the above definition of rough approximation is the *only one* that correctly characterizes the set of positive objects (lower approximation) and the set of positive or ambiguous objects (upper approximation) when a similarity relation is reflexive, but not necessarily symmetric nor transitive.

Using a similarity relation, we are able to induce decision rules from a decision table. The syntax of a rule is represented as follows:

if $f(x, q_1)$ is similar to r_{q1} and $f(x, q_2)$ is similar to r_{q2} and ... $f(x, q_p)$ is similar to r_{qp}, then x belongs to Y_{j1} or Y_{j2} or ... Y_{jk}

where $\{q_1, q_2, \ldots, q_p\} \subseteq C$, $(r_{q1}, r_{q2}, \ldots, r_{qp}) \in V_{q1} \times V_{q2} \times \cdots \times V_{qp}$ and $Y_{j1}, Y_{j2}, \ldots, Y_{jk}$ are some classes of the considered classification (*D*-elementary sets). As mentioned above, if $k = 1$ then the rule is *certain*, otherwise it is *approximate* or *ambiguous*. Procedures for generation of decision rules follow the induction principle described in Section 16.2.3. One such procedure has been proposed by Krawiec et al. (1998)—it involves a similarity relation that is learned from data. We would also like to point out that Greco et al. (1998b, 2000a) proposed a fuzzy extension of the similarity, that is, rough approximation of fuzzy sets (decision classes) by means of fuzzy similarity relations (reflexive only).

16.3   THE KNOWLEDGE DISCOVERY PARADIGM AND PRIOR KNOWLEDGE

The data set in which classification patterns are searched for is called the *learning sample*. The learning of patterns from this sample should take into account available *prior knowledge* that may include the following items (see Slowinski et al., 2002a):

(i) Domains of attributes, i.e. sets of values that an attribute may take while being meaningful to the user.

(ii) A division of attributes into condition and decision attributes, which restricts the range of patterns to functional relations between condition and decision attributes.

(iii) A preference order in the domains of some attributes and a semantic correlation between pairs of these attributes, requiring the patterns to observe the dominance principle.

In fact, item (i) is usually taken into account in knowledge discovery. With this prior knowledge only, one can discover patterns called *association rules* (Agrawal et al., 1996) which show strong relationships between values of some attributes, without fixing which attributes will be on the condition and which ones on the decision side in all rules.

If item (i) is combined with item (ii) in the prior knowledge, then one can consider a partition of the learning sample into decision classes defined by decision attributes. The patterns to be discovered have then the form of *decision trees* or *decision rules* representing functional relations between condition and decision attributes. These patterns are typically discovered by machine learning (see Chapter 12) and data mining methods (Michalski et al., 1998). As there is a direct correspondence between a decision tree and rules, we will concentrate our attention on decision rules only.

As item (iii) is crucial for decision support, let us explain it in more detail. Consider an example of a data set concerning pupils' achievements in a high school. Suppose that among the attributes describing the pupils there are results in *Mathematics* (*Math*) and *Physics* (*Ph*). There is also a *General Achievement* (*GA*) result. The domains of these attributes are composed of three values: *bad, medium* and *good*. This information constitutes item (i) of prior knowledge. Item (ii) is also available because, clearly, *Math* and *Ph* are condition attributes while *GA* is a decision attribute. The preference order of the attribute values is obvious: *good* is better than *medium* and *bad*, and *medium* is better than *bad*. It is known, moreover, that both *Math* and *Ph* are semantically correlated with *GA*. This is, precisely, item (iii) of the prior knowledge.

Attributes with preference-ordered domains are called *criteria* because they involve an evaluation. We will use the name of *regular attributes* for those attributes whose domains are not preference-ordered. *Semantic correlation between two criteria* (condition and decision) means that an improvement on one criterion should not worsen the evaluation on the second criterion, while other attributes and criteria are unchanged. In our example, an improvement of a pupil's score in *Math* or *Ph*, with other attribute values unchanged, should not worsen the pupil's general achievement (*GA*), but rather improve it. In general, semantic correlation between condition criteria and decision criteria requires that an object x dominating object y on all condition criteria (i.e. x having evaluations at least as good as y on all condition criteria) should also dominate y on all decision criteria (i.e. x should have evaluations at least as good as y on all decision criteria). This principle is called the *dominance principle* (or Pareto principle) and it is the only objective principle that is widely agreed upon in the multicriteria comparisons of objects.

Let us consider two questions:

- What classification patterns can be drawn from the pupils' data set?

- How does item (iii) influence the classification patterns?

The answer to the first question is "*if...*, *then...*" decision rules. Each decision rule is characterized by a *condition profile* and a *decision profile*, corresponding to vectors of threshold values of regular attributes and criteria in the condition and decision parts of the rule, respectively. The answer to the second question is that condition and decision profiles of a decision rule should observe the dominance principle if the rule has at least one pair of semantically correlated criteria spanned over the condition and decision part. We say that one profile *dominates* another if they both involve the same values of regular attributes and the values of criteria of the first profile are not worse than the values of criteria of the second profile.

Let us explain the dominance principle with respect to decision rules on the pupils' example. Suppose that two rules induced from the pupils' data set relate *Math* and *Ph* on the condition side, with *GA* on the decision side:

rule #1: if *Math* = *medium* and *Ph* = *medium*, then *GA* = *good*

rule #2: if *Math* = *good* and *Ph* = *medium*, then *GA* = *medium*

The two rules do not observe the dominance principle because the condition profile of rule #2 dominates the condition profile of rule #1, while the decision profile of rule #2 is dominated by the decision profile of rule #1. Thus, in the sense of the dominance principle, the two rules are inconsistent, i.e. they are wrong.

One could say that the above rules are true because they are supported by examples of pupils from the learning sample, but this would mean that the examples are also inconsistent. The *inconsistency* may come from many sources. Examples include:

- Missing attributes (regular ones or criteria) in the description of objects. Maybe the data set does not include such attributes as the *opinion of the pupil's tutor* expressed only verbally during an assessment of the pupil's GA by a school assessment committee.

- Unstable preferences of decision makers. Maybe the members of the school assessment committee changed their view on the influence of *Math* on GA during the assessment.

Handling these inconsistencies is of crucial importance for knowledge discovery. They cannot be simply considered as noise or error to be eliminated from data, or amalgamated with consistent data by some averaging operators. They should be identified and presented as uncertain patterns.

If item (iii) were ignored in prior knowledge, then the handling of the above mentioned inconsistencies would be impossible. Indeed, there would be nothing wrong with rules #1 and #2. They would be supported by different examples discerned by considered attributes.

It has been acknowledged by many authors that *rough set theory* provides an excellent framework for dealing with inconsistency in knowledge discovery (Grzymala-Busse, 1992; Pawlak, 1991; Pawlak et al., 1995; Polkowski, 2002; Polkowski and Skowron, 1999; Slowinski, 1992b; Slowinski and Zopounidis, 1995; Ziarko, 1998). As we have shown in Section 16.2, the paradigm of rough set theory is that of *granular computing*, because the main concept of the theory (rough approximation of a set) is built up of blocks of objects which are indiscernible by a given set of attributes, called *granules of knowledge*. In the space of regular attributes, the granules are bounded sets. Decision rules induced from rough approximation of a classification are also built up of such granules. While taking into account prior knowledge of type (i) and (ii), the rough approximation and the inherent rule induction ignore, however, prior knowledge of type (iii). In consequence, the resulting decision rules may be inconsistent with the dominance principle.

The authors have proposed an extension of the granular computing paradigm that enables us to take into account prior knowledge of type (iii), in addition to either (i) only (Greco et al., 2002d), or (i) and (ii) together (Greco et al., 1998a, 1999b, 2000d, 2001a, 2002a, 2002b; Slowinski et al., 2000, 2002a). The combination of the new granules with the idea of rough approximation is called the Dominance-Based Rough Set Approach.

In the following, we present the concept of granules which permit us to handle prior knowledge of type (iii) when inducing decision rules.

Let U be a finite set of objects (universe) and let Q be a finite set of attributes divided into a set C of *condition attributes* and a set D of *decision attributes* where $C \cap D = \emptyset$. Also, let

$$X_C = \prod_{q=1}^{|C|} X_q \quad \text{and} \quad X_D = \prod_{q=1}^{|D|} X_q$$

be attribute spaces corresponding to sets of condition and decision attributes, respectively. The elements of X_C and X_D can be interpreted as possible evaluation of objects on attributes from set $C = \{1, \ldots, |C|\}$ and from set $D = \{1, \ldots, |D|\}$, respectively. Therefore, X_q is the set of possible evaluations of considered objects with respect to attribute q. The value of object x on attribute $q \in Q$ is denoted by x_q. Objects x and y are *indiscernible* by $P \subseteq C$ if $x_q = y_q$ for all $q \in P$ and, analogously, objects x and y are indiscernible by $R \subseteq D$ if $x_q = y_q$ for all $q \in R$. The sets of indiscernible objects are equivalence classes of the corresponding *indiscernibility relation* I_P or I_R. Moreover, $I_P(x)$ and $I_R(x)$ denote equivalence classes including object x. I_D generates a partition of U into a finite number of decision classes $Cl = \{Cl_t, t = 1, \ldots, n\}$. Each $x \in U$ belongs to one and only one class $Cl_t \in Cl$.

The above definitions take into account prior knowledge of type (i) and (ii) only. In this case, the granules of knowledge are bounded sets in X_P and X_R ($P \subseteq C$ and $R \subseteq D$), defined by partitions of U induced by the indiscernibility relations I_P and I_R, respectively. Then, classification patterns to be discovered are functions representing granules $I_R(x)$ by granules $I_P(x)$ in the condition attribute space X_P, for any $P \subseteq C$ and for any $x \in U$.

If prior knowledge includes item (iii) in addition to (i) and (ii), then the indiscernibility relation is unable to produce granules in X_C and X_D that would take into account the preference order. To do so, the indiscernibility relation has to be substituted by a dominance relation in X_P and X_R ($P \subseteq C$ and $R \subseteq D$). Suppose, for simplicity, that all condition attributes in C and all decision attributes in D are criteria, and that C and D are semantically correlated.

Let $\succeq_q$ be a *weak preference relation* on U (often called *outranking*) representing a preference on the set of objects with respect to criterion $q \in \{C \cup D\}$. Now, $x_q \succeq_q y_q$ means "x_q is at least as good as y_q with respect to criterion q".

On the one hand, we say that x *dominates* y with respect to $P \subseteq C$ (shortly, x *P-dominates* y) in the condition attribute space X_P (denoted by $x D_P y$) if $x_q \succeq_q y_q$ for all $q \in P$. Assuming, without loss of generality, that the domains of the criteria are numerical (i.e. $X_q \subseteq R$ for any $q \in C$) and that they are ordered so that the preference increases with the value, we can say that $x D_P y$ is equivalent to $x_q \geq y_q$ for all $q \in P$, $P \subseteq C$. Observe that for each $x \in X_P$, $x D_P x$, i.e. P-dominance is reflexive. On the other hand, the analogous definition holds in the decision attribute space X_R (denoted by $x D_R y$), where $R \subseteq D$.

The dominance relations $x D_P y$ and $x D_R y$ ($P \subseteq C$ and $R \subseteq D$) are directional statements where x is a subject and y is a referent.

If $x \in X_P$ is the referent, then one can define a set of objects $y \in X_P$ dominating x, called the *P-dominating set* (denoted by $D_P^+(x)$) and defined as $D_P^+(x) = \{y \in U : y D_P x\}$.

If $x \in X_P$ is the subject, then one can define a set of objects $y \in X_P$ dominated by x, called the *P-dominated set* (denoted by $D_P^-(x)$) and defined as $D_P^-(x) = \{y \in U : x D_P y\}$.

P-dominating sets $D_P^+(x)$ and P-dominated sets $D_P^-(x)$ correspond to *positive* and *negative dominance cones* in X_P, with the origin x.

With respect to the decision attribute space X_R (where $R \subseteq D$), the R-dominance relation enables us to define the following sets:

$$Cl_R^{\geq x} = \{y \in U : y D_R x\} \qquad Cl_R^{\leq x} = \{y \in U : x D_R y\}$$

$Cl_{t_q} = \{x \in X_D : x_q = t_q\}$ is a decision class with respect to $q \in D$; $Cl_R^{\geq x}$ is called the *upward union* of classes, and $Cl_R^{\leq x}$ is the *downward union* of classes. If $x \in Cl_R^{\geq x}$, then x belongs to class Cl_{t_q}, $x_q = t_q$, or better, on each decision attribute $q \in R$. On the other hand, if $x \in Cl_R^{\leq x}$, then x belongs to class Cl_{t_q}, $x_q = t_q$, or worse, on each decision attribute $q \in R$. The downward and upward unions of classes correspond to the *positive* and *negative dominance cones* in X_R, respectively.

In this case, the granules of knowledge are open sets in X_P and X_R defined by dominance cones $D_P^+(x)$, $D_P^-(x)$ ($P \subseteq C$) and $Cl_R^{\geq x}$, $Cl_R^{\leq x}$ ($R \subseteq D$), respectively. Then, classification patterns to be discovered are functions representing granules $Cl_R^{\geq x}$, $Cl_R^{\leq x}$ by granules $D_P^+(x)$, $D_P^-(x)$, respectively, in the condition attribute space X_P, for any $P \subseteq C$ and $R \subseteq D$ and for any $x \in X_P$.

In both cases above, the functions are sets of decision rules.

16.4 THE DOMINANCE-BASED ROUGH SET APPROACH

16.4.1 Granular Computing with Dominance Cones

When discovering classification patterns, a set D of decision attributes is, usually, a singleton, $D = \{d\}$. Let us take this assumption for further presentation, although it is not necessary for the Dominance-Based Rough Set Approach. The decision attribute d makes a partition of U into a finite number of classes, $Cl = \{Cl_t, t = 1, \ldots, n\}$. Each $x \in U$ belongs to one and only one class, $Cl_t \in Cl$. The upward and downward unions of classes boil down, respectively, to

$$Cl_t^{\geq} = \bigcup_{s \geq t} Cl_s$$

$$Cl_t^{\leq} = \bigcup_{s \leq t} Cl_s$$

where $t = 1, \ldots, n$. Notice that for $t = 2, \ldots, n$ we have $Cl_n^{\leq} = U - Cl_{t-1}^{\leq}$, i.e. all the objects not belonging to class Cl_t or better, belong to class Cl_{t-1} or worse.

Let us explain how the rough set concept has been generalized to the Dominance-Based Rough Set Approach in order to enable granular computing with dominance cones (for more details, see Greco et al. (1998a, 1999b, 2000d, 2001a, 2002a) and Slowinski et al. (2000)).

Given a set of criteria, $P \subseteq C$, the inclusion of an object $x \in U$ to the upward union of classes $Cl_t^{\geq}, t = 2, \ldots, n$, is *inconsistent with the dominance principle* if one of the following conditions holds:

- x belongs to class Cl_t or better but it is P-dominated by an object y belonging to a class worse than Cl_t, i.e.

$$x \in Cl_t^{\geq} \quad \text{but} \quad D_P^+(x) \cap Cl_{t-1}^{\leq} \neq \emptyset$$

- x belongs to a worse class than Cl_t but it P-dominates an object y belonging to class Cl_t or better, i.e.

$$x \notin Cl_t^{\geq} \quad \text{but} \quad D_P^-(x) \cap Cl_t^{\geq} \neq \emptyset$$

If, given a set of criteria $P \subseteq C$, the inclusion of $x \in U$ to $Cl_t^{\geq}$, where $t = 2, \ldots, n$, is inconsistent with the dominance principle, we say that x belongs to $Cl_t^{\geq}$ *with some ambiguity*. Thus, x belongs to $Cl_t^{\geq}$ *without any ambiguity* with respect to $P \subseteq C$, if $x \in Cl_t^{\geq}$ and there is no inconsistency with the dominance principle. This means that all objects P-dominating x belong to

$Cl_t^\geq$, i.e. $D_P^+(x) \subseteq Cl_t^\geq$. Geometrically, this corresponds to the inclusion of the complete set of objects contained in the positive dominance cone originating in x, in the positive dominance cone $Cl_t^\geq$ originating in Cl_t.

Furthermore, x *possibly belongs to* $Cl_t^\geq$ with respect to $P \subseteq C$ if one of the following conditions holds:

- According to decision attribute d, x belongs to $Cl_t^\geq$.

- According to decision attribute d, x does not belong to $Cl_t^\geq$, but it is inconsistent in the sense of the dominance principle with an object y belonging to $Cl_t^\geq$.

In terms of ambiguity, x possibly belongs to $Cl_t^\geq$ with respect to $P \subseteq C$, if x belongs to $Cl_t^\geq$ with or without any ambiguity. Due to the reflexivity of the dominance relation D_P, the above conditions can be summarized as follows: x *possibly belongs* to class Cl_t or better, with respect to $P \subseteq C$, if among the objects P-dominated by x there is an object y belonging to class Cl_t or better, i.e. $D_P^-(x) \cap Cl_t^\geq \neq \emptyset$. Geometrically, this corresponds to the non-empty intersection of the set of objects contained in the negative dominance cone originating in x, with the positive dominance cone $Cl_t^\geq$ originating in Cl_t.

For $P \subseteq C$, the set of all objects belonging to $Cl_t^\geq$ without any ambiguity constitutes the *P-lower approximation* of $Cl_t^\geq$, denoted by $\underline{P}(Cl_t^\geq)$, and the set of all objects that possibly belong to $Cl_t^\geq$ constitutes the *P-upper approximation* of $Cl_t^\geq$, denoted by $\overline{P}(Cl_t^\geq)$. More formally, we can say

$$\underline{P}(Cl_t^\geq) = \{x \in U : D_P^+(x) \subseteq Cl_t^\geq\}$$
$$\overline{P}(Cl_t^\geq) = \{x \in U : D_P^-(x) \cap Cl_t^\geq \neq \emptyset\}$$

where $t = 1, \ldots, n$. Analogously, one can define the *P-lower approximation* and the *P-upper approximation* of $Cl_t^\leq$ as follows:

$$\underline{P}(Cl_t^\leq) = \{x \in U : D_P^-(x) \subseteq Cl_t^\leq\}$$
$$\overline{P}(Cl_t^\leq) = \{x \in U : D_P^+(x) \cap Cl_t^\leq \neq \emptyset\}$$

where $t = 1, \ldots, n$. The P-lower and P-upper approximations so defined satisfy the following *inclusion properties* for each $t \in \{1, \ldots, n\}$ and for all $P \subseteq C$:

$$\underline{P}(Cl_t^\geq) \subseteq Cl_t^\geq \subseteq \overline{P}(Cl_t^\geq)$$
$$\underline{P}(Cl_t^\leq) \subseteq Cl_t^\leq \subseteq \overline{P}(Cl_t^\leq)$$

All the objects belonging to $Cl_t^\geq$ and $Cl_t^\leq$ with some ambiguity constitute the *P-boundary* of $Cl_t^\geq$ and $Cl_t^\leq$, denoted by $Bn_P(Cl_t^\geq)$ and $Bn_P(Cl_t^\leq)$, respectively. They can be represented, in terms of upper and lower approximations, as follows:

$$Bn_P(Cl_t^\geq) = \overline{P}(Cl_t^\geq) - \underline{P}(Cl_t^\geq)$$
$$Bn_P(Cl_t^\leq) = \overline{P}(Cl_t^\leq) - \underline{P}(Cl_t^\leq)$$

where $t = 1, \ldots, n$. The P-lower and P-upper approximations of the unions of classes $Cl_t^\geq$ and $Cl_t^\leq$ have an important *complementarity property*. It says that if object x belongs without any ambiguity to class Cl_t or better, then it is impossible that it could belong to class Cl_{t-1} or worse, i.e. $\underline{P}(Cl_t^\geq) = U - \overline{P}(Cl_{t-1}^\leq)$, $t = 2, \ldots, n$.

Due to the complementarity property, $Bn_P(Cl_t^\geq) = Bn_P(Cl_{t-1}^\leq)$, for $t = 2, \ldots, n$, which means that if x belongs with ambiguity to class Cl_t or better, then it also belongs with ambiguity to class Cl_{t-1} or worse.

From the knowledge discovery point of view, P-lower approximations of unions of classes represent *certain knowledge* given by criteria from $P \subseteq C$, while P-upper approximations represent *possible knowledge* and the P-boundaries contain *doubtful knowledge* given by the criteria from $P \subseteq C$.

The above definitions of rough approximations are based on a strict application of the dominance principle. However, when defining non-ambiguous objects, it is reasonable to accept a limited proportion of negative examples, particularly for large data tables. This extended version of the Dominance-Based Rough Set Approach is called the Variable-Consistency Dominance-Based Rough Set Approach model (Greco et al., 2001f).

For any $P \subseteq C$, we say that $x \in U$ belongs to $Cl_t^\geq$ *with no ambiguity at consistency level* $l \in (0, 1]$, if $x \in Cl_t^\geq$ and at least $l \times 100\%$ of all objects $y \in U$ dominating x with respect to P also belong to $Cl_t^\geq$, i.e.

$$\frac{|D_P^+(x) \cap Cl_t^\geq|}{|D_P^+(x)|} \geq l$$

The level l is called the *consistency level* because it controls the degree of consistency between objects qualified as belonging to $Cl_t^\geq$ without any ambiguity. In other words, if $l < 1$, then at most $(1 - l) \times 100\%$ of all objects $y \in U$ dominating x with respect to P do not belong to $Cl_t^\geq$ and thus contradict the inclusion of x in $Cl_t^\geq$.

Analogously, for any $P \subseteq C$ we say that $x \in U$ belongs to $Cl_t^\leq$ *with no ambiguity at consistency level* $l \in (0, 1]$, if $x \in Cl_t^\leq$ and at least $l \times 100\%$ of all the objects $y \in U$ dominated by x with respect to P also belong to $Cl_t^\leq$, i.e.

$$\frac{|D_P^-(x) \cap Cl_t^\leq|}{|D_P^-(x)|} \geq l$$

Thus, for any $P \subseteq C$, each object $x \in U$ is either ambiguous or non-ambiguous at consistency level l with respect to the upward union $Cl_t^\geq$ ($t = 2, \ldots, n$) or with respect to the downward union $Cl_t^\leq$ ($t = 1, \ldots, n - 1$).

The concept of non-ambiguous objects at some consistency level l leads naturally to the definition of P-lower approximations of the unions of classes $Cl_t^{\geq}$ and $Cl_t^{\leq}$ which can be formally presented as follows:

$$\underline{P}^l\left(Cl_t^{\geq}\right) = \left\{x \in Cl_t^{\geq} : \frac{|D_P^+(x) \cap Cl_t^{\geq}|}{|D_P^+(x)|} \geq l\right\}$$

$$\underline{P}^l\left(Cl_t^{\leq}\right) = \left\{x \in Cl_t^{\leq} : \frac{|D_P^-(x) \cap Cl_t^{\leq}|}{|D_P^-(x)|} \geq l\right\}$$

Given $P \subseteq C$ and consistency level l, we can define the *P-upper approximations* of $Cl_t^{\geq}$ and $Cl_t^{\leq}$, denoted by $\overline{P}^l\left(Cl_t^{\geq}\right)$ and $\overline{P}^l\left(Cl_t^{\leq}\right)$, respectively, by complementation of $\underline{P}^l\left(Cl_{t-1}^{\leq}\right)$ and $\underline{P}^l\left(Cl_{t+1}^{\geq}\right)$ with respect to U as follows:

$$\overline{P}^l\left(Cl_t^{\geq}\right) = U - \underline{P}^l\left(Cl_{t-1}^{\leq}\right)$$

$$\overline{P}^l\left(Cl_t^{\leq}\right) = U - \underline{P}^l\left(Cl_{t+1}^{\geq}\right)$$

$\overline{P}^l\left(Cl_t^{\geq}\right)$ can be interpreted as the set of all the objects belonging to $Cl_t^{\geq}$, which are *possibly ambiguous* at consistency level l. Analogously, $\overline{P}^l\left(Cl_t^{\leq}\right)$ can be interpreted as the set of all the objects belonging to $Cl_t^{\leq}$, which are *possibly ambiguous* at consistency level l. The *P-boundaries* (*P*-doubtful regions) of $Cl_t^{\geq}$ and $Cl_t^{\leq}$ are defined as

$$Bn_P(Cl_t^{\geq}) = \overline{P}^l(Cl_t^{\geq}) - \underline{P}^l(Cl_t^{\geq})$$

$$Bn_P(Cl_t^{\leq}) = \overline{P}^l(Cl_t^{\leq}) - \underline{P}^l(Cl_t^{\leq})$$

where $t = 1, \ldots, n$. The *variable consistency* model of the Dominance-Based Rough Set Approach provides some degree of flexibility in assigning objects to lower and upper approximations of the unions of decision classes. It can easily be demonstrated that for $0 < l' < l \leq 1$ and $t = 2, \ldots, n$,

$$\underline{P}^l\left(Cl_t^{\geq}\right) \subseteq \underline{P}^{l'}\left(Cl_t^{\geq}\right) \quad \text{and} \quad \overline{P}^{l'}\left(Cl_t^{\geq}\right) \subseteq \overline{P}^l\left(Cl_t^{\geq}\right)$$

The *variable consistency* model is inspired by Ziarko's model of the *variable precision* rough set approach (Ziarko 1993, 1998). However, there is a significant difference in the definition of rough approximations because $\underline{P}^l\left(Cl_t^{\geq}\right)$ and $\overline{P}^l\left(Cl_t^{\geq}\right)$ are composed of non-ambiguous and ambiguous objects at the consistency level l, respectively, while Ziarko's $\underline{P}^l\left(Cl_t\right)$ and $\overline{P}^l\left(Cl_t\right)$ are composed of P-indiscernibility sets such that at least $l \times 100\%$ of these sets are included in Cl_t or have an non-empty intersection with Cl_t, respectively. If one would like to use Ziarko's definition of variable precision rough approximations in the context of multiple-criteria classification, then the P-

indiscernibility sets should be substituted by P-dominating sets $D_P^+(x)$. However, then the notion of ambiguity that naturally leads to the general definition of rough approximations (see Slowinski and Vanderpooten, 2000) loses its meaning. Moreover, a bad side effect of the direct use of Ziarko's definition is that a lower approximation $\underline{P}^l\left(Cl_t^{\geq}\right)$ may include objects y assigned to Cl_h, where h is much less than t, if y belongs to $D_P^+(x)$, which was included in $\underline{P}^l\left(Cl_t^{\geq}\right)$. When the decision classes are preference ordered, it is reasonable to expect that objects assigned to far worse classes than the considered union are not counted to the lower approximation of this union.

For every $P \subseteq C$, the objects being consistent in the sense of the dominance principle with all upward and downward unions of classes are *P-correctly classified*. For every $P \subseteq C$, the *quality of approximation of classification* Cl by the set of criteria P is defined as the ratio between the number of P-correctly classified objects and the number of all the objects in the data sample set. Since the objects which are P-correctly classified are those that do not belong to any P-boundary of unions $Cl_t^{\geq}$ and $Cl_t^{\leq}$, $t = 1, \ldots, n$, the quality of approximation of classification Cl by set of criteria P, can be written as

$$
\gamma_P(Cl) = \frac{\left|\left(U - \left(\bigcup_{t\in\{1,\ldots,n\}} Bn_P\left(Cl_t^{\leq}\right)\right) \cup \left(\bigcup_{t\in\{1,\ldots,n\}} Bn_P\left(Cl_t^{\geq}\right)\right)\right)\right|}{|U|}
$$

$$
= \frac{\left|\left(U - \left(\bigcup_{t\in\{1,\ldots,n\}} Bn_P\left(Cl_t^{\geq}\right)\right)\right)\right|}{|U|}
$$

$\gamma_P(Cl)$ can be seen as a measure of the quality of knowledge that can be extracted from the data table, where P is the set of criteria and Cl is the considered classification.

Each minimal subset $P \subseteq C$ such that $\gamma_P(Cl) = \gamma_C(Cl)$ is called a *reduct* of Cl and is denoted by RED_{Cl}. Note that a decision table can have more than one reduct. The intersection of all reducts is called the *core* and is denoted by CORE_{Cl}. Criteria from CORE_{Cl} cannot be removed from the data sample set without deteriorating the knowledge to be discovered. This means that in set C there are three categories of criteria:

- *indispensable* criteria included in the core

- *exchangeable* criteria included in some reducts but not in the core

- *redundant* criteria being neither indispensable nor exchangeable, thus not included in any reduct.

Note that reducts are minimal subsets of attributes and criteria conveying the relevant knowledge contained in the learning sample. This knowledge is relevant for the explanation of patterns in a given decision table but not necessarily for prediction.

It has been shown in Greco et al. (2001d) that the quality of classification satisfies properties of set functions which are called *fuzzy measures*. For this reason, we can use the quality of classification for the calculation of indices which measure the relevance of particular attributes and/or criteria, in addition to the strength of interactions between them. The useful indices are: the value index and interaction indices of Shapley and Banzhaf; the interaction indices of Murofushi-Soneda and Roubens; and the Möbius representation. All these indices can help to assess the interdependence of the considered attributes and criteria, and can help to choose the best reduct.

16.4.2   Induction of Decision Rules

The dominance-based rough approximations of upward and downward unions of classes can serve to induce a generalized description of the objects contained in the decision table, in terms of "*if..., then...*" decision rules. For a given upward or downward union of classes, $Cl_t^{\geq}$ or $Cl_s^{\leq}$, the decision rules induced under a hypothesis that objects belonging to $\underline{P}(Cl_t^{\geq})$ or $\underline{P}(Cl_s^{\leq})$ are positive and all the others are negative, suggests an assignment to "class Cl_t or better", or to "class Cl_s or worse", respectively. On the other hand, the decision rules induced under a hypothesis that objects belonging to the intersection $\overline{P}(Cl_s^{\leq}) \cap \overline{P}(Cl_t^{\geq})$ are positive and all the others are negative, are suggesting an assignment to some classes between Cl_s and Cl_t ($s < t$).

In the case of preference-ordered data it is meaningful to consider the following five types of decision rules:

1 Certain $D_{\geq}$-decision rules. These provide lower profile descriptions for objects belonging to $Cl_t^{\geq}$ without ambiguity: *if* $x_{q1} \succeq_{q1} r_{q1}$ *and* $x_{q2} \succeq_{q2} r_{q2}$ *and* $\ldots x_{qp} \succeq_{qp} r_{qp}$, *then* $x \in Cl_t^{\geq}$, where for each $w_q, z_q \in X_q$, "$w_q \succeq_q z_q$" means "w_q is <u>at least</u> as good as z_q".

2 Possible $D_{\geq}$-decision rules. Such rules provide lower profile descriptions for objects belonging to $Cl_t^{\geq}$ with or without any ambiguity: *if* $x_{q1} \succeq_{q1} r_{q1}$ *and* $x_{q2} \succeq_{q2} r_{q2}$ *and* $\ldots x_{qp} \succeq_{qp} r_{qp}$, *then* x possibly belongs to $Cl_t^{\geq}$.

3 Certain $D_{\leq}$-decision rules. These give upper profile descriptions for objects belonging to $Cl_t^{\leq}$ without ambiguity: *if* $x_{q1} \preceq_{q1} r_{q1}$ *and* $x_{q2} \preceq_{q2} r_{q2}$ *and* $\ldots x_{qp} \preceq_{qp} r_{qp}$, *then* $x \in Cl_t^{\leq}$, where for each $w_q, z_q \in X_q$, "$w_q \preceq_q z_q$" means "w_q is <u>at most</u> as good as z_q".

4 Possible $D_{\leq}$-decision rules. These provide upper profile descriptions for objects belonging to $Cl_t^{\leq}$ with or without any ambiguity: *if* $x_{q1} \preceq_{q1} r_{q1}$ *and* $x_{q2} \preceq_{q2} r_{q2}$ *and* $\ldots x_{qp} \preceq_{qp} r_{qp}$, *then* x possibly belongs to $Cl_t^{\leq}$.

5 Approximate $D_{\geq\leq}$-decision rules. These represent simultaneously lower and upper profile descriptions for objects belonging to $Cl_s \cup Cl_{s+1} \cup \ldots \cup Cl_t$ without the possibility of discerning the actual class: *if* $x_{q1} \succeq_{q1} r_{q1}$ *and* $\ldots x_{qk} \succeq_{qk} r_{qk}$ *and* $x_{qk+1} \preceq_{qk+1} r_{qk+1}$ *and* $\ldots x_{qp} \preceq_{qp} r_{qp}$, *then* $x \in Cl_s \cup Cl_{s+1} \cup \ldots \cup Cl_t$.

In the left-hand side of a $D_{\geq\leq}$-decision rule we can have $x_q \succeq_q r_q$ and $x_q \preceq_q r_q'$, where $r_q \leq r_q'$, for the same $q \in C$. Moreover, if $r_q = r_q'$, the two conditions boil down to $x_q \sim_q r_q$, where for each $w_q, z_q \in X_q$, "$w_q \sim_q z_q$" means "w_q is indifferent to z_q".

A *minimal* rule is a consequence relation where we understand that there is no other consequence relation with a left-hand side which has at least the same weakness (which means that it uses a subset of elementary conditions and/or weaker elementary conditions) and which has a right-hand side that has at least the same strength (which means a $D_{\geq}$- or a $D_{\leq}$-decision rule assigning objects to the same union or sub-union of classes, or a $D_{\geq\leq}$-decision rule assigning objects to the same or larger set of classes).

Rules of type 1 and 3 represent certain knowledge extracted from the data table, while rules of type 2 and 4 represent possible knowledge. Rules of type 5 represent doubtful knowledge.

The rules of type 1 and 3 are *exact* if they do not cover negative examples; they are *probabilistic*, otherwise. In the latter case, each rule is characterized by a confidence ratio, representing the probability that an object matching the left-hand side of the rule matches also its right-hand side. Probabilistic rules concord to the Variable-Consistency Dominance-Based Rough Set Approach model mentioned above.

We will now comment upon the application of decision rules to some objects described by criteria from C. When applying $D_{\geq}$-decision rules to an object x, it is possible that x either matches the left-hand side of at least one decision rule or it does not. In the case of at least one such match, it is reasonable to conclude that x belongs to class Cl_t, that is the lowest class of the upward union $Cl_t^{\geq}$ which results from intersection of all the right-hand sides of the rules covering x. More precisely, if x matches the left-hand side of rules $\rho_1, \rho_2, \ldots, \rho_m$, having right-hand sides $x \in Cl_{t1}^{\geq}, x \in Cl_{t2}^{\geq}, \ldots, x \in Cl_{tm}^{\geq}$, then x is assigned to class Cl_t, where $t = \max\{t1, t2, \ldots, tm\}$. In the case of no matching, we can conclude that x belongs to Cl_1, i.e. to the worst class, since no rule with a right-hand side suggesting a better classification of x is covering this object.

Analogously, when applying $D_{\leq}$-decision rules to the object x, we can conclude that x belongs either to class Cl_z, (that is the highest class of the downward union $Cl_t^\leq$ resulting from the intersection of all the right-hand sides of the rules covering x) or to class Cl_n, i.e. to the best class, when x is not covered by any rule. More precisely, if x matches the left-hand side of rules $\rho_1, \rho_2, \ldots, \rho_m$, having right-hand sides $x \in Cl_{t1}^\leq$, $x \in Cl_{t2}^\leq, \ldots, x \in Cl_{tm}^\leq$, then x is assigned to class Cl_t, where $t = \min\{t1, t2, \ldots, tm\}$. In the case of no matching, it is concluded that x belongs to the best class Cl_n because no rule with a right-hand side suggesting a worse classification of x is covering this object.

Finally, when applying $D_{\geq\leq}$-decision rules to x, it is possible to conclude that x belongs to the union of all the classes suggested in the right hand side of the rules covering x.

A set of decision rules is *complete* if it is able to cover all objects from the decision table in such a way that consistent objects are re-classified to their original classes and inconsistent objects are classified to clusters of classes which refer to this inconsistency. Each set of decision rules that is complete and non-redundant is called *minimal*. Note that an exclusion of any rule from this set makes it non-complete.

In the case of the Variable-Consistency Dominance-Based Rough Set Approach, the decision rules are induced from the P-lower approximations whose composition is controlled by the user-specified consistency level l. Consequently, the value of confidence α for the rule should be constrained from the bottom. It is reasonable to require that the smallest accepted confidence level of the rule should not be lower than the currently used consistency level l. Indeed, in the worst case, some objects from the P-lower approximation may create a rule using all the criteria from P thus giving a confidence $\alpha \geq l$.

Observe that the syntax of decision rules induced from dominance-based rough approximations uses the concept of dominance cones: each condition profile is a dominance cone in X_C, and each decision profile is a dominance cone in X_D. In both cases the cone is positive for $D_\geq$-rules and negative for $D_\leq$-rules.

Also note that dominance cones which correspond to condition profiles can originate in any point of X_C, without the risk of being too specific. Thus, in contrast to traditional granular computing, the condition attribute space X_C need not be discretized.

Some procedures for rule induction from rough approximations have been proposed in Greco et al. (2001g). In Giove et al. (2002), a new methodology for the induction of monotonic decision trees from dominance-based rough approximations of preference-ordered decision classes has been proposed.

16.4.3 An Illustrative Example

To illustrate the application of the Dominance-Based Rough Set Approach to multicriteria classification, we will use a part of some data provided by a Greek industrial bank ETEVA which finances industrial and commercial firms in Greece (Slowinski and Zopounidis, 1995). A sample composed of 39 firms has been chosen for the study in co-operation with the ETEVA financial manager. The manager has classified the selected firms into three classes of bankruptcy risk. The sorting decision is represented by decision attribute d making a trichotomic partition of the 39 firms:

$d = $ A means "acceptable"

$d = $ U means "uncertain"

$d = $ NA means "non-acceptable".

The partition is denoted by $Cl = \{Cl_A, Cl_U, Cl_{NA}\}$ and, obviously, class Cl_A is better than Cl_U which is better than Cl_{NA}.

The firms were evaluated using the following 12 criteria (↑ means *preference increasing with value* and ↓ means *preference decreasing with value*):

- $A_1 = $ earnings before interests and taxes/total assets, ↑

- $A_2 = $ net income/net worth, ↑

- $A_3 = $ total liabilities/total assets, ↓

- $A_4 = $ total liabilities/cash flow, ↓

- $A_5 = $ interest expenses/sales, ↓

- $A_6 = $ general and administrative expenses/sales, ↓

- $A_7 = $ managers' work experience, ↑ (very low $= 1$, low $= 2$, medium $= 3$, high $= 4$, very high $= 5$)

- $A_8 = $ firm's market niche/position, ↑ (bad $= 1$, rather bad $= 2$, medium $= 3$, good $= 4$, very good $= 5$)

- $A_9 = $ technical structure-facilities, ↑ (bad $= 1$, rather bad $= 2$, medium $= 3$, good $= 4$, very good $= 5$)

- $A_{10} = $ organization-personnel, ↑ (bad $= 1$, rather bad $= 2$, medium $= 3$, good $= 4$, very good $= 5$)

- $A_{11} = $ special competitive advantage of firms, ↑ (low $= 1$, medium $= 2$, high $= 3$, very high $= 4$)

- A_{12} = market flexibility, ↑ (very low = 1, low = 2, medium = 3, high = 4, very high = 5).

The first six criteria are cardinal (financial ratios) and the last six are ordinal. The data table is presented in Table 16.3.

The main questions to be answered by the knowledge discovery process were the following:

- Is the information contained in Table 16.3 consistent?

- What are the reducts of criteria ensuring the same quality of approximation of the multicriteria classification as the whole set of criteria?

- What decision rules can be extracted from Table 16.3?

- What are the minimal sets of decision rules?

We will answer these questions using the Dominance-Based Rough Set Approach. The *first result* from this approach is a discovery that the financial data matrix is **consistent** for the complete set of criteria C. Therefore, the C-lower and C-upper approximations of $Cl_{NA}^{\leq}$, $Cl_U^{\leq}$ and $Cl_U^{\geq}$, $Cl_A^{\geq}$ are the same. In other words, the quality of approximation of all upward and downward unions of classes, as well as the quality of classification, is equal to 1.

The *second discovery* is a set of 18 *reducts* of criteria ensuring the same quality of classification as the whole set of 12 criteria:

$$\text{RED}_{Cl}^1 = \{A_1, A_4, A_5, A_7\} \qquad \text{RED}_{Cl}^2 = \{A_2, A_4, A_5, A_7\}$$
$$\text{RED}_{Cl}^3 = \{A_3, A_4, A_6, A_7\} \qquad \text{RED}_{Cl}^4 = \{A_4, A_5, A_6, A_7\}$$
$$\text{RED}_{Cl}^5 = \{A_4, A_5, A_7, A_8\} \qquad \text{RED}_{Cl}^6 = \{A_2, A_3, A_7, A_9\}$$
$$\text{RED}_{Cl}^7 = \{A_1, A_3, A_4, A_7, A_9\} \qquad \text{RED}_{Cl}^8 = \{A_1, A_5, A_7, A_9\}$$
$$\text{RED}_{Cl}^9 = \{A_2, A_5, A_7, A_9\} \qquad \text{RED}_{Cl}^{10} = \{A_4, A_5, A_7, A_9\}$$
$$\text{RED}_{Cl}^{11} = \{A_5, A_6, A_7, A_9\} \qquad \text{RED}_{Cl}^{12} = \{A_4, A_5, A_7, A_{10}\}$$
$$\text{RED}_{Cl}^{13} = \{A_1, A_3, A_4, A_7, A_{11}\} \qquad \text{RED}_{Cl}^{14} = \{A_2, A_3, A_4, A_7, A_{11}\}$$
$$\text{RED}_{Cl}^{15} = \{A_4, A_5, A_6, A_{12}\} \qquad \text{RED}_{Cl}^{16} = \{A_1, A_3, A_5, A_6, A_9, A_{12}\}$$
$$\text{RED}_{Cl}^{17} = \{A_3, A_4, A_6, A_{11}, A_{12}\}$$
$$\text{RED}_{Cl}^{18} = \{A_1, A_2, A_3, A_6, A_9, A_{11}, A_{12}\}$$

All the 18 subsets of criteria are equally good and sufficient for the perfect approximation of the classification performed by ETEVA's financial manager on the 39 firms. The core of *Cl* is empty ($\text{CORE}_{Cl} = \emptyset$) which means that no criterion is indispensable for the approximation. Moreover, all the criteria are exchangeable and no criterion is redundant.

The *third discovery* is the set of **all** decision rules. We obtained 74 rules describing $Cl_{NA}^{\leq}$, 51 rules describing $Cl_U^{\leq}$, 75 rules describing $Cl_U^{\geq}$ and 79 rules describing $Cl_A^{\geq}$.

Table 16.3. Financial data matrix.

Firm	A_1	A_2	A_3	A_4	A_5	A_6	A_7	A_8	A_9	A_{10}	A_{11}	A_{12}	d
F1	16.4	14.5	59.82	2.5	7.5	5.2	5	3	5	4	2	4	A
F2	35.8	67.0	64.92	1.7	2.1	4.5	5	4	5	5	4	5	A
F3	20.6	61.75	75.71	3.6	3.6	8.0	5	3	5	5	3	5	A
F4	11.5	17.1	57.1	3.8	4.2	3.7	5	2	5	4	3	4	A
F5	22.4	25.1	49.8	2.1	5.0	7.9	5	3	5	5	3	5	A
F6	23.9	34.5	48.9	1.7	2.5	8.0	5	3	4	4	3	4	A
F7	29.9	44.0	57.8	1.8	1.7	2.5	5	4	4	5	3	5	A
F8	8.7	5.4	27.4	3.3	4.5	4.5	5	2	4	4	1	4	A
F9	25.7	29.7	46.8	1.7	4.6	3.7	4	2	4	3	1	3	A
F10	21.2	24.6	64.8	3.7	3.6	8.0	4	2	4	4	1	4	A
F11	18.32	31.6	69.3	4.4	2.8	3.0	4	3	4	4	3	4	A
F12	20.7	19.3	19.7	0.7	2.2	4.0	4	2	4	4	1	3	A
F13	9.9	3.5	53.1	4.5	8.5	5.3	4	2	4	4	1	4	A
F14	10.4	9.3	80.9	9.4	1.4	4.1	4	2	4	4	3	3	A
F15	17.7	19.8	52.8	3.2	7.9	6.1	4	4	4	4	2	5	A
F16	14.8	15.9	27.94	1.3	5.4	1.8	4	2	4	3	2	3	A
F17	16.0	14.7	53.5	3.9	6.8	3.8	4	4	4	4	2	4	A
F18	11.7	10.01	42.1	3.9	12.2	4.3	5	2	4	2	1	3	A
F19	11.0	4.2	60.8	5.8	6.2	4.8	4	2	4	4	2	4	A
F20	15.5	8.5	56.2	6.5	5.5	1.8	4	2	4	4	2	4	A
F21	13.2	9.1	74.1	11.21	6.4	5.0	2	2	4	4	2	3	U
F22	9.1	4.1	44.8	4.2	3.3	10.4	3	4	4	4	3	4	U
F23	12.9	1.9	65.02	6.9	14.01	7.5	4	3	3	2	1	2	U
F24	5.9	−27.7	77.4	−32.2	16.6	12.7	3	2	4	4	2	3	U
F25	16.9	12.4	60.1	5.2	5.6	5.6	3	2	4	4	2	3	U
F26	16.7	13.1	73.5	7.1	11.9	4.1	2	2	4	4	2	3	U
F27	14.6	9.7	59.5	5.8	6.7	5.6	2	2	4	4	2	4	U
F28	5.1	4.9	28.9	4.3	2.5	46.0	2	2	3	3	1	2	U
F29	24.4	22.3	32.8	1.4	3.3	5.0	2	3	4	4	2	3	U
F30	29.7	8.6	41.8	1.6	5.2	6.4	2	3	4	4	2	3	U
F31	7.3	−64.5	67.5	−2.2	30.1	8.7	3	3	4	4	2	3	NA
F32	23.7	31.9	63.6	3.5	12.1	10.2	3	2	3	4	1	3	NA
F33	18.9	13.5	74.5	10.0	12.0	8.4	3	3	3	4	3	4	NA
F34	13.9	3.3	78.7	25.5	14.7	10.1	2	2	3	4	3	4	NA
F35	−13.3	−31.1	63.0	−10.0	21.2	23.1	2	1	4	3	1	2	NA
F36	6.2	−3.2	46.1	5.1	4.8	10.5	2	1	3	3	2	3	NA
F37	4.8	−3.3	71.9	34.6	8.6	11.6	2	2	4	4	2	3	NA
F38	0.1	−9.6	42.5	−20.0	12.9	12.4	1	1	4	3	1	3	NA
F39	13.6	9.1	76.0	11.4	17.1	10.3	1	1	2	1	1	2	NA

The *fourth discovery* is the finding of **minimal sets** of decision rules. Several minimal sets were found. One of them is shown below. The number in parentheses indicates the number of objects which support the corresponding rule, i.e. the rule strength:

1 If $f(x, A_3) \geq 67.5$ and $f(x, A_4) \geq -2.2$ and $f(x, A_6) \geq 8.7$, then $x \in Cl_{NA}^{\leq}$, (4);

2 If $f(x, A_2) \leq 3.3$ and $f(x, A_7) \leq 2$, then $x \in Cl_{NA}^{\leq}$, (5);

3 If $f(x, A_3) \geq 63.6$ and $f(x, A_7) \leq 3$ and $f(x, A_9) \leq 3$, then $x \in Cl_{NA}^{\leq}$, (4);

4 If $f(x, A_2) \leq 12.4$ and $f(x, A_6) \geq 5.6$, then $x \in Cl_{U}^{\leq}$, (14);

5 If $f(x, A_7) \leq 3$, then $x \in Cl_{U}^{\leq}$, (18);

6 If $f(x, A_2) \geq 3.5$ and $f(x, A_5) \leq 8.5$, then $x \in Cl_{U}^{\geq}$, (26);

7 If $f(x, A_7) \geq 4$, then $x \in Cl_{U}^{\geq}$, (21);

8 If $f(x, A_1) \geq 8.7$ and $f(x, A_9) \geq 4$, then $x \in Cl_{U}^{\geq}$, (27);

9 If $f(x, A_2) \geq 3.5$ and $f(x, A_7) \geq 4$, then $x \in Cl_{A}^{\geq}$, (20).

As the minimal set of rules is complete and composed of $D_{\geq}$-decision rules and $D_{\leq}$-decision rules only, application of these rules to the 39 firms will result in their exact re-classification to classes of risk.

Minimal sets of decision rules represent the most concise and non-redundant knowledge representations. The above minimal set of nine decision rules uses eight criteria and 18 elementary conditions, i.e. 3.85% of descriptors from the data matrix.

The well-known machine discovery methods cannot deal with multicriteria classification because they do not consider preference orders in the domains of attributes and among the classes. There are multicriteria decision analysis methods for such classification. However, they are not discovering classification patterns from data. They simply apply a preference model, like the utility function in scoring methods (see e.g. Thomas et al. 1992), to a set of objects to be classified. In this sense, they are not knowledge discovery methods at all.

Comparing the Dominance-Based Rough Set Approach to the Classical Rough Set Approach, we can notice the following differences between the two approaches. The Classical Rough Set Approach extracts knowledge about a partition of U into classes which are not preference-ordered. The granules used for knowledge representation are sets of objects which are indiscernible by a set of condition attributes.

In the case of the Dominance-Based Rough Set Approach and multicriteria classification, the condition attributes are criteria and the classes are preference-ordered. The extracted knowledge concerns a collection of upward and downward unions of classes and the granules used for knowledge representation are sets of objects defined using the dominance relation. This is the main difference between the Classical Rough Set Approach and the Dominance-Based Rough Set Approach.

There are three notable advantages of the Dominance-Based Rough Set Approach over the Classical Rough Set Approach. The first one is the ability to handle criteria, preference-ordered classes and inconsistencies in the set of decision examples that the Classical Rough Set Approach is simply not able to discover. Consequently, the rough approximations separate the certain information from the doubtful, which is taken into account in rule induction. The second advantage is the ability to analyze a data table without any preprocessing of data. The third advantage lies in the richer syntax of decision rules that are induced from rough approximations. The elementary conditions of decision rules resulting from Dominance-Based Rough Set Approach use relations from $\{\leq, =, \geq\}$, while those resulting from the Classical Rough Set Approach only use $=$. The Dominance-Based Rough Set Approach syntax is more understandable to practitioners. The minimal sets of decision rules are smaller than the minimal sets which result from the Classical Rough Set Approach.

16.5   THE DOMINANCE-BASED ROUGH SET APPROACH FOR MULTICRITERIA CHOICE AND RANKING

One of the very first extensions of the Dominance-Based Rough Set Approach concerned preference-ordered data representing pairwise comparisons (i.e. binary relations) between objects on both, condition and decision attributes (Greco et al., 1999a, 1999b, 2000d, 2001c). Note that while classification is based on the absolute evaluation of objects, choice and ranking refer to pairwise comparisons of objects. In this case, the patterns (i.e. decision rules) to be discovered from the data characterize a comprehensive binary relation on the set of objects. If this relation is a preference relation and if, from among the condition attributes, there are some criteria which are semantically correlated with the comprehensive preference relation, then the data set (serving as the learning sample) can be considered to be preferential information given by a decision maker in a multicriteria choice or ranking problem. In consequence, the comprehensive preference relation characterized by the decision rules discovered from this data set can be considered as a *preference model* for the decision maker. It may be used to explain the decision policy of the decision maker and to recommend a good choice or preference ranking with respect to new objects.

Let us consider a finite set A of objects evaluated by a finite set of criteria C. The best choice (or the preference ranking) in set A is semantically correlated with the criteria from set C. The preferential information concerning the multicriteria choice or ranking problem is a data set in the form of a pairwise comparison table which includes pairs of some *reference objects* from a subset $B \subseteq A \times A$. This is described by preference relations on particular criteria from C and a comprehensive preference relation. One such example is a weak preference relation called the *outranking relation*. By using the Dominance-Based Rough Set Approach for the analysis of the pairwise comparison table, we can obtain a rough approximation of the outranking relation by a dominance relation. The decision rules induced from the rough approximation are then applied to the complete set A of the objects associated with the choice or ranking. As a result, one obtains a four-valued outranking relation on this set. In order to obtain a recommendation, it is advisable to use an exploitation procedure based on the net flow score of the objects. We present this methodology in more detail below.

16.5.1 The Pairwise Comparison Table as Preferential Information and as a Learning Sample

A set of reference objects represent a decision problem and a decision maker can express the preferences by pairwise comparisons. In the following, $x S y$ denotes the presence, while $x S^c y$ denotes the absence of the outranking relation for a pair of objects $(x, y) \in A \times A$.

For each pair of reference objects $(x, y) \in B \subseteq A \times A$, the decision maker can select one of the three following possibilities:

1 object x is as good as y, i.e. $x S y$,

2 object x is worse then y, i.e. $x S^c y$,

3 the two objects are incomparable at the present stage.

An $m \times (n + 1)$ pairwise comparison table, denoted by S_{PCT}, is then created on the basis of this information. The first n columns correspond to the criteria from set C. The last, i.e. the $(n + 1)$th, column represents the comprehensive binary preference relation S or S^c. The m rows are pairs from B. For each pair in S_{PCT}, a difference between criterion values is put in the corresponding column. If the decision maker judges that two objects are incomparable, then the corresponding pair does not appear in S_{PCT}.

We will define S_{PCT} more formally. For any criterion $g_i \in C$, let T_i be a finite set of binary relations defined on A on the basis of the evaluations of objects from A with respect to the considered criterion g_i, such that for every $(x, y) \in A \times A$ exactly one binary relation $t \in T_i$ is verified. More precisely, given the domain V_i of $g_i \in C$, if $v_i', v_i'' \in V_i$ are the respective evaluations of $x, y \in A$ by means of g_i and $(x, y) \in t$, with $t \in T_i$, then for each $w, z \in A$

having the same evaluations v_i', v_i'' by means of g_i, $(w,z) \in t$. Furthermore, let T_d be a set of binary relations defined on set A (comprehensive pairwise comparisons) such that at most one binary relation $t \in T_d$ is verified for every $(x, y) \in A \times A$.

The *pairwise comparison table* is defined as data table $S_{PCT} = \langle B, C \cup \{d\}, T_C \cup T_d, f \rangle$, where $B \subseteq A \times A$ is a non-empty *set of exemplary pairwise comparisons of reference objects*, $T_C = \bigcup_{g_i \in C} T_i$, d is a decision corresponding to the comprehensive pairwise comparison (comprehensive preference relation), and $f : B \times (C \cup \{d\}) \rightarrow T_C \cup T_d$ is a total function such that $f[(x, y), q] \in T_i$ for every $(x, y) \in A \times A$ and for each $g_i \in C$, and $f[(x, y), q] \in T_d$ for every $(x, y) \in B$. It follows that for any pair of reference objects $(x, y) \in B$ there is verified one and only one binary relation $t \in T_d$. Thus, T_d induces a partition of B. In fact, the data table S_{PCT} can be seen as a decision table, since the set of considered criteria C and the decision d are distinguished.

We assume that the exemplary pairwise comparisons made by the decision maker can be represented in terms of *graded preference relations* (for example "very large preference", "large preference", "strict preference", "strong preference" and "very strong preference"), denoted by P_q^h: for each $q \in C$ and for every $(x, y) \in A \times A$,

$$T_i = \left\{ P_i^h, h \in H_i \right\}$$

where H_i is a particular subset of the relative integers and

- $x P_i^h y$, $h > 0$, means that object x is preferred to object y by degree h with respect to criterion g_i,

- $x P_i^h y$, $h < 0$, means that object x is not preferred to object y by degree h with respect to criterion g_i,

- $x P_i^0 y$ means that object x is similar (asymmetrically indifferent) to object y with respect to criterion g_i.

Within the preference context, the similarity relation P_i^0, even if not symmetric, resembles the indifference relation. Thus, in this case, we call this similarity relation "asymmetric indifference". Of course, for each $g_i \in C$ and for every $(x, y) \in A \times A$,

$$\left[x P_i^h y, h > 0\right] \Rightarrow \left[y P_i^k x, k \leq 0\right], \quad \left[x P_i^h y, h < 0\right] \Rightarrow \left[y P_i^k x, k \geq 0\right]$$

The set of binary relations T_d may be defined in a similar way, but $x P_d^h y$ means that object x is comprehensively preferred to object y by degree h. We are considering a pairwise comparison table where the set T_d is composed of two binary relations defined on A:

- x outranks y (denoted by xSy or $(x, y) \in S$), where $(x, y) \in B$,

- x does not outrank y (denoted by $xS^c y$ or $(x, y) \in S^c$), where $(x, y) \in B$, and $S \cup S^c = B$.

Observe that the binary relation S is reflexive, but not necessarily transitive or complete.

16.5.2   Rough Approximation of the Outranking and Non-outranking Relations Specified in the Pairwise Comparison Table

In the following we will distinguish between two types of evaluation scales of criteria: *cardinal* and *ordinal*. Let C^N be the set of criteria expressing preferences on a cardinal scale, and let C^O, be the set of criteria expressing preferences on an ordinal scale, such that $C^N \cup C^O = C$ and $C^N \cap C^O = \emptyset$. Moreover, for each $P \subseteq C$, we denote by P^O the subset of P composed of criteria expressing preferences on an ordinal scale, i.e. $P^O = P \cap C^O$, and by P^N we denote the subset of P composed of criteria expressing preferences on a cardinal scale, i.e. $P^N = P \cap C^N$. Of course, for each $P \subseteq C$, we have $P = P^N \cup P^O$ and $P^N \cap P^O = \emptyset$.

The meaning of the two scales is such that in the case of the cardinal scale we can specify the intensity of preference for a given difference of evaluations, while in the case of the ordinal scale, this is not possible and we can only establish an order of evaluations.

Multigraded Dominance   Let $P = P^N$ and $P^O = \emptyset$. Given $P \subseteq C$ ($P \neq \emptyset$), $(x, y),(w, z) \in A \times A$, the pair of objects (x, y) is said to dominate (w, z) with respect to criteria from P (denoted by $(x, y)D_P(w, z)$), if x is preferred to y at least as strongly as w is preferred to z with respect to each $g_i \in P$. More precisely, "at least as strongly as" means "by at least the same degree", i.e. $h_i \geq k_i$, where $h_i, k_i \in H_i$, $x P_i^{hi} y$ and $w P_i^{ki} z$, for each $g_i \in P$.

Let $D_{\{i\}}$ be the dominance relation confined to the single criterion $g_i \in P$. The binary relation $D_{\{i\}}$ is reflexive $((x, y)D_{\{i\}}(x, y)$, for every $(x, y) \in A \times A)$, transitive $((x, y)D_{\{i\}}(w, z)$ and $(w, z)D_{\{i\}}(u, v)$ imply $(x, y)D_{\{i\}}(u, v)$, for every $(x, y),(w, z),(u, v) \in A \times A)$, and complete $((x, y)D_{\{i\}}(w, z)$ and/or $(w, z)D_{\{i\}}(x, y)$, for all $(x, y),(w, z) \in A \times A)$. Therefore, $D_{\{i\}}$ is a complete preorder on $A \times A$. Since the intersection of complete preorders is a partial preorder and $D_P = \bigcap_{g_i \in P} D_{\{i\}}$, $P \subseteq C$, then the dominance relation D_P is a partial preorder on $A \times A$.

Let $R \subseteq P \subseteq C$ and $(x, y),(u, v) \in A \times A$; then the following implication holds:

$$(x, y)D_P(u, v) \Rightarrow (x, y)D_R(u, v)$$

Given $P \subseteq C$ and $(x, y) \in A \times A$, we define the following:

- A set of pairs of objects dominating (x, y), called the *P-dominating set*, denoted by $D_P^+(x, y)$ and defined to be

$$\{(w, z) \in A \times A : (w, z)D_P(x, y)\}$$

- A set of pairs of objects dominated by (x, y), called the *P-dominated set*, denoted by $D_P^-(x, y)$ and defined as

$$\{(w, z) \in A \times A : (x, y)D_P(w, z)\}$$

The P-dominating sets and the P-dominated sets defined on B for all pairs of reference objects from B are "granules of knowledge" that can be used to express P-lower and P-upper approximations of the comprehensive outranking relations S and S^c, respectively:

$$
\begin{aligned}
\underline{P}(S) &= \{(x, y) \in B : D_P^+(x, y) \subseteq S\} \\
\overline{P}(S) &= \bigcup_{(x,y)\in S} D_P^+(x, y) \\
\underline{P}\left(S^c\right) &= \{(x, y) \in B : D_P^-(x, y) \subseteq S^c\} \\
\overline{P}\left(S^c\right) &= \bigcup_{(x,y)\in S^c} D_P^-(x, y)
\end{aligned}
$$

It has been proved in Greco et al. (1999a) that

$$\underline{P}(S) \subseteq S \subseteq \overline{P}(S) \qquad \underline{P}\left(S^c\right) \subseteq S^c \subseteq \overline{P}\left(S^c\right)$$

Furthermore, the following complementarity properties hold:

$$
\begin{aligned}
\underline{P}(S) &= B - \overline{P}\left(S^c\right) & \overline{P}(S) &= B - \underline{P}\left(S^c\right) \\
\underline{P}\left(S^c\right) &= B - \overline{P}(S) & \overline{P}\left(S^c\right) &= B - \underline{P}(S)
\end{aligned}
$$

The P-boundaries (P-doubtful regions) of S and S^c are defined as

$$Bn_P(S) = \overline{P}(S) - \underline{P}(S) \qquad Bn_P(S^c) = \overline{P}\left(S^c\right) - \underline{P}\left(S^c\right)$$

From the above it follows that $Bn_P(S) = Bn_P(S^c)$.

The concepts of the quality of approximation, reducts and core can be extended also to the approximation of the outranking relation by multigraded dominance relations.

In particular, the coefficient

$$\gamma_P = \frac{\left|\underline{P}(S) \cup \underline{P}(S^c)\right|}{|B|}$$

defines the *quality of approximation of S* and S^c by $P \subseteq C$. It expresses the
ratio of all pairs of reference objects $(x, y) \in B$ correctly assigned to S and
S^c by the set P of criteria to all the pairs of objects contained in B. Each
minimal subset $P \subseteq C$, such that $\gamma_P = \gamma_C$, is called a *reduct* of C (denoted by
$\text{RED}_{S_{\text{PCT}}}$). Note that S_{PCT} can have more than one reduct. The intersection of
all B-reducts is called the *core* (denoted by $\text{CORE}_{S_{\text{PCT}}}$).

It is also possible to use the Variable Consistency Model on S_{PCT} (Slowinski
et al., 2002b) but being aware that some of the pairs in the positive or negative
dominance sets belong to the opposite relation although at least $l \times 100\%$ of
pairs belong to the correct one. Then the definition of the lower approximations
of S and S^c boils down to

$$\underline{P}(S) = \left\{ (x, y) \in B : \frac{\left| D_P^+(x, y) \cap S \right|}{\left| D_P^+(x, y) \right|} \geq l \right\}$$

$$\underline{P}(S^c) = \left\{ (x, y) \in B : \frac{\left| D_P^-(x, y) \cap S^c \right|}{\left| D_P^-(x, y) \right|} \geq l \right\}$$

Dominance Without Degrees of Preference   The degree of graded preference considered above is defined on a cardinal scale of the strength of preference. However, in many real world problems, the existence of such a quantitative scale is rather questionable. This is the case with ordinal scales of criteria.
In this case, the dominance relation is defined directly on evaluations $g_i(x)$ for
all objects $x \in A$. Let us explain this latter case in more detail.

Let $P = P^O$ and $P^N = \emptyset$, then, given $(x, y), (w, z) \in A \times A$, the pair (x, y)
is said to dominate the pair (w, z) with respect to criteria from P (denoted by
$(x, y) D_P(w, z)$), if for each $g_i \in P$, $g_i(x) \geq g_i(w)$ and $g_i(z) \geq g_i(y)$.

Let $D_{\{i\}}$ be the dominance relation confined to the single criterion $g_i \in P^O$.
The binary relation $D_{\{i\}}$ is reflexive, transitive, but non-complete (it is possible that *not* $(x, y) D_{\{i\}}(w, z)$ and *not* $(w, z) D_{\{i\}}(x, y)$ for some $(x, y), (w, z) \in$
$A \times A$). Therefore, $D_{\{i\}}$ is a partial preorder. Since the intersection of partial preorders is also a partial preorder and $D_P = \bigcap_{g_i \in P} D_{\{i\}}$, $P = P^O$, then the
dominance relation D_P is a partial preorder.

If some criteria from $P \subseteq C$ express preferences on a quantitative or a numerical non-quantitative scale and others on an ordinal scale, i.e. if $P^N \neq \emptyset$
and $P^O \neq \emptyset$, then, given $(x, y), (w, z) \in A \times A$, the pair (x, y) is said to dominate the pair (w, z) with respect to criteria from P, if (x, y) dominates (w, z)
with respect to both P^N and P^O. Since the dominance relation with respect
to P^N is a partial preorder on $A \times A$ (because it is a multigraded dominance)
and the dominance with respect to P^O is also a partial preorder on $A \times A$ (as
explained above), then the dominance D_P, being the intersection of these two
dominance relations, is a partial preorder. In consequence, all the concepts in-

troduced in the previous section can be restored using this specific definition of dominance.

16.5.3 Induction of Decision Rules From Rough Approximations of Outranking and Non-outranking Relations

Using the rough approximations of S and S^c defined in Section 16.5.2, it is possible to induce a generalized description of the preferential information contained in a given S_{PCT} in terms of suitable decision rules. The syntax of these rules is based on the concept of *upward cumulated preferences* (denoted by $P_i^{\geq h}$) and *downward cumulated preferences* (denoted by $P_i^{\leq h}$), having the following interpretation:

- $x P_i^{\geq h} y$ means "x is preferred to y with respect to g_i by at least degree h"

- $x P_i^{\leq h} y$ means "x is preferred to y with respect to g_i by at most degree h".

Exact definition of the cumulated preferences, for each $(x,y) \in A \times A$, $g_i \in C$ and $h \in H_i$, can be represented as follows:

- $x P_i^{\geq h} y$ if $x P_i^k y$, where $k \in H_i$ and $k \geq h$

- $x P_i^{\leq h} y$ if $x P_i^k y$, where $k \in H_i$ and $k \leq h$.

Let also $G_i = \{g_i(x), x \in A\}$, $g_i \in C^O$. The decision rules have then the following syntax.

1 *Certain $D_\geq$-decision rules.* If $x P_{i1}^{\geq h(i1)} y$ *and...* $x P_{ie}^{\geq h(ie)} y$ *and* $g_{ie+1}(x) \geq r_{ie+1}$ *and* $g_{ie+1}(y) \leq s_{ie+1}$ *and* ... $g_{ip}(x) \geq r_{ip}$ *and* $g_{ip}(y) \leq s_{ip}$, *then* xSy, where $P = \{g_{i1}, \ldots, g_{ip}\} \subseteq C$, $P^N = \{g_{i1}, \ldots, g_{ie}\}$, $P^O = \{g_{ie+1}, \ldots, g_{ip}\}$, $(h(i1), \ldots, h(ie)) \in H_{i1} \times \cdots \times H_{ie}$ *and* $(r_{ie+1}, \ldots, r_{ip}),(s_{ie+1}, \ldots, s_{ip}) \in G_{ie+1} \times \cdots \times G_{ip}$. These rules are supported by pairs of objects from the P-lower approximation of S only.

2 *Certain $D_\leq$-decision rules.* If $x P_{i1}^{\leq h(i1)} y$ *and...* $x P_{ie}^{\leq h(ie)} y$ *and* $g_{ie+1}(x) \leq r_{ie+1}$ *and* $g_{ie+1}(y) \geq s_{ie+1}$ *and* ... $g_{ip}(x) \leq r_{ip}$ *and* $g_{ip}(y) \geq s_{ip}$, *then* $xS^c y$, where $P = \{g_{i1}, \ldots, g_{ip}\} \subseteq C$, $P^N = \{g_{i1}, \ldots, g_{ie}\}$, $P^O = \{g_{ie+1}, \ldots, g_{ip}\}$, $(h(i1), \ldots, h(ie)) \in H_{i1} \times \cdots \times H_{ie}$ *and* $(r_{ie+1}, \ldots, r_{ip}), (s_{ie+1}, \ldots, s_{ip}) \in G_{ie+1} \times \cdots \times G_{ip}$. These rules are supported by pairs of objects from the P-lower approximation of S^c only.

3 *Approximate $D_{\geq\leq}$-decision rules.* If $x P_{i1}^{\geq h(i1)} y$ *and...* $x P_{ie}^{\geq h(ie)} y$ *and* $x P_{ie+1}^{\leq h(ie+1)} y \ldots x P_{if}^{\leq h(if)} y$ *and* $g_{if+1}(x) \geq r_{if+1}$ *and* $g_{if+1}(y) \leq s_{if+1}$

and...$g_{ig}(x) \geq r_{ig}$ and $g_{ig}(y) \leq s_{ig}$ and $g_{ig+1}(x) \leq r_{ig+1}$ and $g_{ig+1}(y) \geq s_{ig+1}$ and...$g_{ip}(x) \leq r_{ip}$ and $g_{ip}(y) \geq s_{ip}$, then xSy or xS^cy, where $O' = \{g_{i1}, \ldots, g_{ie}\} \subseteq C$, $O''=\{g_{ie+1}, \ldots, g_{if}\}\} \subseteq C$, $P^N = O' \cup O''$, O' and O'' are not necessarily disjoint, $P^O = \{g_{if+1}, \ldots, g_{ip}\}$, $(h(i1), \ldots, h(if)) \in H_{i1} \times \cdots \times H_{if}$, $(r_{if+1}, \ldots, r_{ip})$, $(s_{if+1}, \ldots, s_{ip}) \in G_{if+1} \times \cdots \times G_{ip}$. These rules are supported by pairs of objects from the P-boundary of S and S^c only.

16.5.4   Use of Decision Rules for Decision Support

The decision rules induced from a given S_{PCT} describe the comprehensive preference relations S and S^c either exactly ($D_{\geq}$- and $D_{\leq}$-decision rules) or approximately ($D_{\geq\leq}$-decision rules). A set of these rules covering all pairs of S_{PCT} represents a preference model of the decision maker who gave the pairwise comparison of reference objects. The application of these decision rules on a new subset $M \subseteq A$ of objects induces a specific preference structure on M.

In fact, any pair of objects $(u, v) \in M \times M$ can match the decision rules in one of four ways:

- at least one $D_{\geq}$-decision rule and neither $D_{\leq}$- nor $D_{\geq\leq}$-decision rules,

- at least one $D_{\leq}$-decision rule and neither $D_{\geq}$- nor $D_{\geq\leq}$-decision rules,

- at least one $D_{\geq}$-decision rule and at least one $D_{\leq}$-decision rule, or at least one $D_{\geq\leq}$-decision rule, or at least one $D_{\geq\leq}$-decision rule and at least one $D_{\geq}$- and/or at least one $D_{\leq}$-decision rule,

- no decision rule.

These four ways correspond to the following four situations of indexoutranking outranking, respectively:

- uSv and *not* uS^cv, i.e. *true* outranking (denoted by uS^Tv),

- uS^cv and *not* uSv, i.e. *false* outranking (denoted by uS^Fv),

- uSv and uS^cv, i.e. *contradictory* outranking (denoted by uS^Kv),

- *not* uSv and *not* uS^cv, i.e. *unknown* outranking (denoted by uS^Uv).

The four above situations, which together constitute the so-called *four-valued outranking* (Greco et al., 1998c), have been introduced to underline the presence and absence of *positive* and *negative* reasons for the outranking. Moreover, they make it possible to distinguish contradictory situations from unknown ones.

A final *recommendation* (choice or ranking) can be obtained upon a suitable exploitation of this structure, i.e. of the presence and the absence of outranking S and S^c on M. A possible exploitation procedure consists of calculating a specific score, called the Net Flow Score, for each object $x \in M$:

$$S_{nf}(x) = S^{++}(x) - S^{+-}(x) + S^{-+}(x) - S^{--}(x)$$

where

$$S^{++}(x) = |\{y \in M: \text{ at least one decision rule affirms } xSy\}|$$

$$S^{+-}(x) = |\{y \in M: \text{ at least one decision rule affirms } ySx\}|$$

$$S^{-+}(x) = |\{y \in M: \text{ at least one decision rule affirms } yS^cx\}|$$

$$S^{--}(x) = |\{y \in M: \text{ at least one decision rule affirms } xS^cy\}|$$

The recommendation in ranking problems consists of the total preorder determined by $S_{nf}(x)$ on M. In choice problems, it consists of the object(s) $x^* \in M$ such that $S_{nf}(x^*) = \max_{x \in M} \{S_{nf}(x)\}$.

The above procedure has been characterized with reference to a number of desirable properties in Greco et al. (1998c).

16.5.5   An Illustrative Example

Let us suppose that a company managing a chain of warehouses wants to buy some new warehouses. To choose the best proposals or to rank them all, the managers of the company decide to analyze first the characteristics of eight warehouses already owned by the company (reference objects). This analysis should give some indications for the choice and ranking of the new proposals. Eight warehouses belonging to the company have been evaluated by the following three criteria: capacity of the sales staff (A_1), perceived quality of goods (A_2) and high traffic location (A_3). The domains (scales) of these attributes are presently composed of three preference-ordered echelons: $V_1 = V_2 = V_3 = \{\text{sufficient, medium, good}\}$. The decision attribute (d) indicates the profitability of warehouses, expressed by the *return on equity* (ROE) ratio (in %). Table 16.4 presents a decision table which represents this situation.

With respect to the set of criteria $C = C^N = \{A_1, A_2, A_3\}$, the following multigraded preference relations P_i^h, $i = 1, 2, 3$, are defined:

- $x P_i^0 y$ (and $y P_i^0 x$), meaning that x is *indifferent* to y with respect to A_i, if $f(x, A_i) = f(y, A_i)$,

- $x P_i^1 y$ (and $y P_i^{-1} x$), meaning that x is *preferred* to y with respect to A_i, if $f(x, A_i) = \text{good}$ and $f(y, A_i) = \text{medium}$, or if $f(x, A_i) = \text{medium}$ and $f(y, A_i) = \text{sufficient}$,

Table 16.4. Decision table with reference objects.

Warehouse	A_1	A_2	A_3	d (ROE %)
1	good	medium	good	10.35
2	good	sufficient	good	4.58
3	medium	medium	good	5.15
4	sufficient	medium	medium	−5
5	sufficient	medium	medium	2.42
6	sufficient	sufficient	good	2.98
7	good	medium	good	15
8	good	sufficient	good	−1.55

- $x P_i^2 y$ (and $y P_i^{-2} x$), meaning that x is *strongly preferred* to y with respect to A_i, if $f(x, A_i) =$ good and $f(y, A_i) =$ sufficient.

Using the decision attribute, the comprehensive outranking relation was built as follows: warehouse x is at least as good as warehouse y with respect to profitability $(x S y)$ if

$$\text{ROE}(x) \geq \text{ROE}(y) - 2\%$$

Otherwise, i.e. if $\text{ROE}(x) < \text{ROE}(y) - 2\%$, warehouse x is *not* at least as good as warehouse y with respect to profitability $(x S^c y)$.

The pairwise comparisons of the reference objects result in S_{PCT}. The rough set analysis of the S_{PCT} leads to the conclusion that the set of decision examples on the reference objects is inconsistent. The quality of approximation of S and S^c by all criteria from set C is equal to 0.44. Moreover, $\text{RED}_{S_{\text{PCT}}} = \text{CORE}_{S_{\text{PCT}}} = \{A_1, A_2, A_3\}$. This means that no criterion is superfluous.

The C-lower approximations of S and S^c, obtained by means of multigraded dominance relations, are

$$\underline{C}(S) = \{(1, 2), (1, 4), (1, 5), (1, 6), (1, 8), (3, 2), (3, 4), (3, 5), (3, 6),$$
$$(3, 8), (7, 2), (7, 4), (7, 5), (7, 6), (7, 8)\}$$
$$\underline{C}(S^c) = \{(2, 1), (2, 7), (4, 1), (4, 3), (4, 7), (5, 1), (5, 3), (5, 7), (6, 1),$$
$$(6, 3), (6, 7), (8, 1), (8, 7)\}$$

All the remaining 36 pairs of reference objects belong to the C-boundaries of S and S^c, i.e. $Bn_C(S) = Bn_C(S^c)$.

The following minimal $D_\geq$-decision rules and $D_\leq$-decision rules can be induced from lower approximations of S and S^c, respectively (the figures within parentheses represent the pairs of objects supporting the corresponding rules):

If $x P_1^{\geq 1} y$ *and* $x P_2^{\geq 1} y$, *then* $x S y$;
  $((1, 6), (3, 6), (7, 6))$
If $x P_2^{\geq 1} y$ *and* $x P_3^{\geq 0} y$, *then* $x S y$;
  $((1, 2), (1, 6), (1, 8), (3, 2), (3, 6), (3, 8), (7, 2), (7, 6), (7, 8))$
If $x P_2^{\geq 0} y$ *and* $x P_3^{\geq 1} y$, *then* $x S y$;
  $((1, 4), (1, 5), (3, 4), (3, 5), (7, 4), (7, 5))$
If $x P_1^{\leq -1} y$ *and* $x P_2^{\leq -1} y$, *then* $x S^c y$;
  $((6, 1), (6, 3), (6, 7))$
If $x P_2^{\leq 0} y$ *and* $x P_3^{\leq -1} y$, *then* $x S^c y$;
  $((4, 1), (4, 3), (4, 7), (5, 1), (5, 3), (5, 7))$
If $x P_1^{\leq 0} y$ *and* $x P_2^{\leq -1} y$ *and* $x P_3^{\leq 0} y$, *then* $x S^c y$;
  $((2, 1), (2, 7), (6, 1), (6, 3), (6, 7), (8, 1), (8, 7))$

Moreover, it is possible to induce five minimal $D_{\geq \leq}$-decision rules from the boundary of approximation of S and S^c:

If $x P_1^{\leq 0} y$ *and* $x P_2^{\geq 0} y$ *and* $x P_3^{\leq 0} y$ *and* $x P_3^{\geq 0} y$, *then* $x S y$ *or* $x S^c y$;
  $((1, 1), (1, 3), (1, 7), (2, 2), (2, 6), (2, 8), (3, 1), (3, 3), (3, 7), (4, 4), (4, 5),$
  $(5, 4), (5, 5), (6, 2), (6, 6), (6, 8), (7, 1), (7, 3), (7, 7), (8, 2), (8, 6), (8, 8))$
If $x P_2^{\leq -1} y$ *and* $x P_3^{\geq 1} y$, *then* $x S y$ *or* $x S^c y$;
  $((2, 4), (2, 5), (6, 4), (6, 5), (8, 4), (8, 5))$
If $x P_2^{\geq 1} y$ *and* $x P_3^{\leq -1} y$, *then* $x S y$ *or* $x S^c y$;
  $((4, 2), (4, 6), (4, 8), (5, 2), (5, 6), (5, 8))$
If $x P_1^{\geq 1} y$ *and* $x P_2^{\leq 0} y$ *and* $x P_3^{\leq 0} y$, *then* $x S y$ *or* $x S^c y$;
  $((1, 3), (2, 3), (2, 6), (7, 3), (8, 3), (8, 6))$,
If $x P_1^{\geq 1} y$ *and* $x P_2^{\leq -1} y$, *then* $x S y$ *or* $x S^c y$;
  $((2, 3), (2, 4), (2, 5), (8, 3), (8, 4), (8, 5))$.

Using all the above decision rules and the Net Flow Score exploitation procedure on ten other warehouses proposed for purchase, the managers can obtain the result presented in Table 16.5. The dominance-based rough set approach gives a clear recommendation:

- For the *choice problem* it suggests the *selection of warehouse* 2' *and* 6', having maximum score (11);

- For the *ranking problem* it suggests the *ranking* presented in the last column of Table 16.5, as follows:

$$(2', 6') \rightarrow (8') \rightarrow (9') \rightarrow (1') \rightarrow (4') \rightarrow (5') \rightarrow (3') \rightarrow (7', 10')$$

Table 16.5. Ranking of warehouses for sale by decision rules and the Net Flow Score procedure.

Warehouse for sale	A_1	A_2	A_3	Net flow score	Ranking
$1'$	good	sufficient	medium	1	5
$2'$	sufficient	good	good	11	1
$3'$	sufficient	medium	sufficient	-8	8
$4'$	sufficient	good	sufficient	0	6
$5'$	sufficient	sufficient	medium	-4	7
$6'$	sufficient	good	good	11	1
$7'$	medium	sufficient	sufficient	-11	9
$8'$	medium	medium	medium	7	3
$9'$	medium	good	sufficient	4	4
$10'$	medium	sufficient	sufficient	-11	9

16.5.6   Summary

We have briefly presented the contribution of the Dominance-Based Rough Set Approach to multicriteria choice and ranking problems. Let us point out the main features of the described methodology:

- The decision maker is asked for the preference information necessary to deal with a multicriteria decision problem in terms of exemplary decisions.

- The rough set analysis of preferential information supplies some useful elements of knowledge about the decision situation. These are: the relevance of particular attributes and/or criteria, information about their interaction, minimal subsets of attributes or criteria (reducts) conveying important knowledge contained in the exemplary decisions and the set of the non-reducible attributes or criteria (core).

- The preference model induced from the preferential information is expressed in a natural and comprehensible language of "*if...*, *then...*" decision rules. The decision rules concern pairs of objects and from them we can determine either the presence or the absence of a comprehensive preference relation. The conditions for the presence are expressed in "at least" terms, and for the absence in "at most" terms, on particular criteria.

- The decision rules do not convert ordinal information into numeric but keep the ordinal character of input data due to the syntax proposed.

- Heterogeneous information (qualitative and quantitative, ordered and non-ordered) and scales of preference (ordinal, cardinal) can be processed within the Dominance-Based Rough Set Approach, while classical methods consider only quantitative ordered evaluations (with rare exceptions).

- No prior discretization of the quantitative domains of criteria is necessary.

16.6 TRICKS OF THE TRADE

Below we give some hints about how to start a typical session of rough set analysis of a multi-attribute or multicriteria classification problem.

1. First, prepare the data set so it is composed of objects (examples) described by a set of attributes. In the set of attributes, distinguish the decision attribute from other (condition) attributes. For example, in Section 16.4.3, we considered a set of firms evaluated by financial and managerial criteria, assigned to three classes of bankruptcy risk. In terms of the size of the data set: in the case of, say, five condition attributes and three decision classes, the number of objects should not be less than a dozen per class.

2. Check if the decision classes labeled by the decision attribute are preference-ordered. Check also whether or not, among the condition attributes, there is at least one whose domain is also preference ordered such that there is a semantic correlation between this condition attribute and the decision attribute (e.g. the bankruptcy risk of a firm and its "net income/net worth" ratio). If the check is positive, then the Dominance-Based Rough Set Approach should be used, otherwise, the Classical Rough Set Approach is sufficient. In the latter case, in order to avoid getting decision rules which are too specific, you may need to group some values of particular attributes (say, to at most seven values per attribute)—this step is called discretization.

3. Choose the appropriate software (web addresses for free download are given in the next section) and proceed with your calculations.

4. Calculate the quality of approximation of the classification for the complete set of condition attributes/criteria. A quality value above 0.75 is usually satisfactory. In the case of a lower quality value, there are too many inconsistencies in the data. So try to get data about the evaluation of the objects on additional attributes/criteria, or eliminate some extremely inconsistent objects from the doubtful region of the classification, or add some new and consistent objects. For example, in Sec-

tion 16.2.1 (the traffic signs example), we have added one additional attribute—secondary color (SC). Of course, you may continue the analysis even if the quality is low but then you will get weaker decision rules.

5 Calculate the minimal subsets of attributes/criteria conveying the relevant knowledge contained in the data (reducts) and the set of non-reducible attributes/criteria (core). You may continue the analysis with a data set confined to a chosen reduct—then, the decision rules induced from the reduced data set will represent knowledge contained in the data in terms of attributes/criteria from the reduct only. For example, with the traffic signs, one could eliminate from the data table the column of either shape (S) or primary color (PC), without decreasing the quality of knowledge representation.

6 Using the lower and upper approximations of either decision classes (Classical Rough Set Approach) or unions of preference-ordered decision classes (Dominance-Based Rough Set Approach), induce decision rules from the reduced or original decision table. You may either induce a minimal set of rules covering all the objects from the decision table or choose from all induced decision rules a subset of the most interesting rules. For example, this might be the rules with a minimal support of 50% of objects per class or per union of classes, or rules with no more than three elementary conditions in the premise (see the example of traffic signs in Section 16.2.1 and the example of bankruptcy risk in Section 16.4.3). Usually, the "minimal cover" set of rules is chosen in the perspective of prediction and the "most interesting" set of rules is chosen in the perspective of explanation. At this stage, an expert may disagree with some rules but they say nothing apart from the truth hidden in the decision table, so you can show what objects from the decision table support the rules in question and the expert may want to eliminate at least some of them from the data. It is also possible that decision rules seem strange for the expert because there are not enough examples in the decision table.

7 If the expert finds your decision rules too specific and/or too numerous, you may use the variable-precision (Classical Rough Set Approach) model or the variable-consistency (Dominance-Based Rough Set Approach) model. Then, you have to specify the required precision or consistency level, say 80%, and you will finally get fewer decision rules. However, their confidence will vary between 80% and 100%.

16.7   CONCLUSIONS AND PROMISING AREAS OF FUTURE WORK

We introduced a knowledge discovery paradigm for multi-attribute and multicriteria decision support, based on the concept of rough sets. Rough set theory provides mathematical tools for dealing with granularity of information and possible inconsistencies in the description of objects. Considering this description as an input data about a decision problem, the knowledge discovery paradigm consists of searching for patterns in the data that facilitate an understanding of the decision maker's preferences and that enable us to recommend a decision which is in line with these preferences. An original component of this paradigm is that it takes into account prior knowledge about preference semantics in the patterns to be discovered.

Knowledge discovery from preference-ordered data differs from usual knowledge discovery, since the former involves preference orders in domains of attributes and in the set of decision classes. This requires that a knowledge discovery method applied to preference-ordered data respects the dominance principle. As this is not the case for the well-known methods of data mining and knowledge discovery, they are not able to discover all relevant knowledge contained in the analyzed data sample and, even worse, they may yield unreasonable discoveries, because of inconsistency with the dominance principle. These deficiencies are addressed in the Dominance-Based Rough Set Approach. Moreover, this approach enables us to apply a rough set approach to multicriteria decision making. We showed how the approach could be used for multicriteria classification, choice and ranking. In more advanced papers, we have presented many extensions of the approach that make it a useful tool for other practical applications. These are:

- The Dominance-Based Rough Set Approach for decision under risk and uncertainty (Greco et al., 2001e).

- The Dominance-Based Rough Set Approach for incomplete decision tables (Greco et al., 1999c, 2000a).

- Fuzzy set extensions of the approach (Greco et al., 1999b, 2000b, 2000c, 2002e, 2003).

- The Dominance-Based Rough Set Approach for hierarchical decision making (Dembczynski et al., 2002).

- A dominance-based approach to induction of association rules (Greco et al., 2002d).

The Dominance-Based Rough Set Approach leads to a preference model of a decision maker in terms of decision rules. The decision rules have a special syntax which involves partial evaluation profiles and dominance relations on these profiles. The clarity of the rule representation of preferences enables us to see the limits of other traditional aggregation functions: the utility function and the outranking relation. In several studies (Greco et al., 2001b, 2002c, 2004; Slowinski et al., 2002c), we proposed an axiomatic characterization of these aggregation functions in terms of conjoint measurement theory and in terms of a set of decision rules. In comparison to other studies on the characterization of aggregation functions, our axioms do not require any preliminary assumptions about the scales of criteria. A side-result of these investigations is that the decision rule aggregation (preference model) is the most general among the known aggregation functions. The decision rule preference model fulfils, moreover, the postulate of transparency and interpretability of preference models in decision support.

SOURCES OF ADDITIONAL INFORMATION

A list of rough set references can be found at the web site of the International Rough Set Society: http://www.roughsets.org. This page includes information about rough set conferences and about Transactions on Rough Sets that Springer has started to publish as a sub-series of the Lecture Notes in Computer Science. This page also includes tutorial presentations on rough sets and links to the following available software:

- ROSE2 (http://idss.cs.put.poznan.pl/site/software.html; Rough Set Data Explorer),

- 4eMka, JAMM (http://idss.cs.put.poznan.pl/site/software.html; Dominance-Based Rough Set Approach to Multicriteria Classification),

- RSES (http://logic.mimuw.edu.pl; Rough Set Exploration System),

- ROSETTA (http://www.idi.ntnu.no/ aleks/rosetta; Rough Set Toolkit for Analysis of Data).

Acknowledgments. The first author wishes to acknowledge financial support from the State Committee for Scientific Research, KBN research grant no. 4T11F 002 22, and from the Foundation for Polish Science, subsidy no. 11/2001. The other two authors have been supported by the Italian Ministry of Education, University and Scientific Research (MIUR).

References

Agrawal, R., Mannila, H., Srikant, R., Toivinen, H. and Verkamo, I., 1996, Fast discovery of association rules, in: *Advances in Knowledge Discovery and Data Mining,* U. M. Fayyad et al., eds, AAAI Press, Menlo Park, CA, pp. 307–328.

Dembczynski, K., Greco, S. and Slowinski, R., 2002, Methodology of rough-set-based classification and sorting with hierarchical structure of attributes and criteria, *Control Cybernet.* **31**:891–920.

Giove, S., Greco, S., Matarazzo, B. and Slowinski, R., 2002, Variable consistency monotonic decision trees, in: *Rough Sets and Current Trends in Computing,* Lecture Notes in Artificial Intelligence, Vol. 2475, J. J. Alpigini, J. F. Peters, A. Skowron, and N. Zhong, eds, Springer, Berlin, pp. 247–254.

Greco, S., Inuiguchi, M. and Slowinski, R., 2002e, Dominance-based rough set approach using possibility and necessity measures, in: *Rough Sets and Current Trends in Computing,* Lecture Notes in Artificial Intelligence, Vol. 2475, J. J. Alpigini, J. F. Peters, A. Skowron and N. Zhong, eds, Springer, Berlin, pp. 85–92.

Greco, S., Inuiguchi, M., and Slowinski, R., 2003, A new proposal for fuzzy rough approximations and gradual decision rule representation, in: *Rough Fuzzy and Fuzzy Rough Sets,* D. Dubois, J. Grzymala-Busse, M. Inuiguchi and L. Polkowski, eds, Springer, Berlin.

Greco, S., Matarazzo, B. and Slowinski, R., 1998a, A new rough set approach to evaluation of bankruptcy risk, in: *Operational Tools in the Management of Financial Risk,* C. Zopounidis, ed., Kluwer, Dordrecht, pp. 121–136.

Greco, S., Matarazzo, B. and Slowinski, R., 1998b, Fuzzy similarity relation as a basis for rough approximation, in: *Rough Sets and Current Trends in Computing,* L. Polkowski and A. Skowron, eds, Lecture Notes in Artificial Intelligence, Vol. 1424, Springer, Berlin, pp. 283–289.

Greco, S., Matarazzo, B. and Slowinski, R., 1999a, Rough approximation of a preference relation by dominance relations, *Eur. J. Oper. Res.* **117**:63–83.

Greco, S., Matarazzo, B. and Slowinski, R., 1999b, The use of rough sets and fuzzy sets in MCDM, in: *Advances in Multiple Criteria Decision Making,* T. Gal, T. Stewart and T. Hanne, eds, Kluwer, Dordrecht, pp. 14.1–14.59.

Greco, S., Matarazzo, B. and Slowinski, R., 1999c, Handling missing values in rough set analysis of multi-attribute and multi-criteria decision problems, in: N. Zhong, A. Skowron and S. Ohsuga, eds, *New Directions in Rough Sets, Data Mining and Granular-Soft Computing,* Lecture Notes in Artificial Intelligence, Vol. 1711, Springer, Berlin, pp. 146–157.

Greco, S., Matarazzo, B. and Slowinski, R., 2000a, Dealing with missing data in rough set analysis of multi-attribute and multi-criteria decision problems, in: *Decision Making: Recent Developments and Worldwide Applications,* S. H. Zanakis, G. Doukidis and C. Zopounidis, eds, Kluwer, Dordrecht, pp. 295–316.

Greco, S., Matarazzo, B. and Slowinski, R., 2000b, Rough set processing of vague information using fuzzy similarity relations, in: *Finite Versus*

Infinite—Contributions to an Eternal Dilemma, C. S. Calude and G. Paun, eds, Springer, Berlin, pp. 149–173.

Greco, S., Matarazzo, B. and Slowinski, R., 2000c, Fuzzy extension of the rough set approach to multicriteria and multiattribute sorting, in: *Preferences and Decisions under Incomplete Knowledge,* J. Fodor, B. De Baets and P. Perny, eds, Physica, Heidelberg, pp. 131–151.

Greco, S., Matarazzo, B. and Slowinski, R., 2000d, Extension of the rough set approach to multicriteria decision support, *INFOR* **38**:161–196.

Greco, S., Matarazzo, B. and Slowinski, R., 2001a, Rough sets theory for multicriteria decision analysis, *Eur. J. Oper. Res.* **129**:1–47.

Greco, S., Matarazzo, B. and Slowinski, R., 2001b, Conjoint measurement and rough set approach for multicriteria sorting problems in presence of ordinal criteria, in: *A-MCD-A: Aide Multi-Critère à la Décision—Multiple Criteria Decision Aiding,* A. Colorni, M. Paruccini and B. Roy, eds, European Commission Report, EUR 19808 EN, Ispra, pp. 117–144.

Greco, S., Matarazzo, B. and Slowinski, R., 2001c, Rule-based decision support in multicriteria choice and ranking, in: *Symbolic and Quantitative Approaches to Reasoning with Uncertainty,* S. Benferhat and P. Besnard, eds, Lecture Notes in Artificial Intelligence, Vol. 2143, Springer, Berlin, pp. 29–47.

Greco, S., Matarazzo, B. and Slowinski, R., 2001d, Assessment of a value of information using rough sets and fuzzy measures, in: J. Chojcan and J. Leski, eds, *Fuzzy Sets and their Applications,* Silesian University of Technology Press, Gliwice, pp. 185–193.

Greco, S., Matarazzo, B. and Slowinski, R., 2001e, Rough set approach to decisions under risk, in: *Rough Sets and Current Trends in Computing,* W. Ziarko and Y. Yao, eds, Lecture Notes in Artificial Intelligence, Vol. 2005, Springer, Berlin, pp. 160–169.

Greco, S., Matarazzo, B. and Slowinski, R., 2002a, Rough sets methodology for sorting problems in presence of multiple attributes and criteria. *Eur. J. Oper. Res.* **138**:247–259.

Greco, S., Matarazzo, B. and Slowinski, R., 2002b, Multicriteria classification, in: *Handbook of Data Mining and Knowledge Discovery,* W. Kloesgen and J. Zytkow, eds, Oxford University Press, Oxford, pp. 318–328.

Greco, S., Matarazzo, B. and Slowinski, R, 2002c, Preference representation by means of conjoint measurement & decision rule model, in: *Aiding Decisions with Multiple Criteria—Essays in Honor of Bernard Roy,* D. Bouyssou, E. Jacquet-Lagrèze, P. Perny, R. Slowinski, D. Vanderpooten and P. Vincke, eds, Kluwer, Dordrecht, pp. 263–313.

Greco, S., Matarazzo, B. and Slowinski, R., 2004, Axiomatic characterization of a general utility function and its particular cases in terms of conjoint measurement and rough-set decision rules. *Eur. J. Oper. Res.* **158**:271–292.

Greco, S., Matarazzo, B., Slowinski, R. and Stefanowski, J., 2001f, Variable consistency model of dominance-based rough set approach, in: *Rough Sets and Current Trends in Computing*, Lecture Notes in Artificial Intelligence, Vol. 2005, W. Ziarko and Y. Yao, eds, Springer, Berlin, pp. 170–181.

Greco, S., Matarazzo, B., Slowinski, R. and Stefanowski, J., 2001g, An algorithm for induction of decision rules consistent with dominance principle, in: *Rough Sets and Current Trends in Computing*, Lecture Notes in Artificial Intelligence, Vol. 2005, W. Ziarko and Y. Yao, eds, Springer, Berlin, pp. 304–313.

Greco, S., Matarazzo, B., Slowinski, R. and Stefanowski, J., 2002d, Mining association rules in preference-ordered data, in: *Foundations of Intelligent Systems*, Lecture Notes in Artificial Intelligence, Vol. 2366, M.-S. Hacid, Z. W. Ras, D. A. Zighed and Y. Kodratoff, eds, Springer, Berlin, pp. 442–450.

Greco, S., Matarazzo, B., Slowinski, R. and Tsoukias, A., 1998c, Exploitation of a rough approximation of the outranking relation in multicriteria choice and ranking, in: *Trends in Multicriteria Decision Making*, Lecture Notes in Economics and Mathematical Systems, Vol. 465, T. J. Stewart and R. C. van den Honert, eds, Springer, Berlin, pp. 45–60.

Grzymala-Busse, J. W., 1992, LERS—a system for learning from examples based on rough sets, in: *Intelligent Decision Support. Handbook of Applications and Advances of the Rough Sets Theory*, R. Slowinski, ed., Kluwer, Dordrecht, pp. 3–18.

Grzymala-Busse, J. W., 1997, A new version of the rule induction system LERS, *Fund. Inform.* **31**:27–39.

Krawiec, K., Slowinski, R. and Vanderpooten, D., 1998, Learning of decision rules from similarity based rough approximations, in: *Rough Sets in Knowledge Discovery*, Vol. 2, L. Polkowski and A. Skowron, eds, Physica, Heidelberg, pp. 37–54.

Luce, R. D., 1956, Semi-orders and a theory of utility discrimination, *Econometrica* **24**:178–191.

Marcus, S., 1994, Tolerance rough sets, Cech topologies, learning processes, *Bull. Polish Acad. Sci., Tech. Sci.* **42**:471–487.

Michalski, R. S., Bratko, I. and Kubat, M., eds, 1998, *Machine Learning and Data Mining—Methods and Applications*, Wiley, New York.

Nieminen, J., 1988, Rough tolerance equality, *Fund. Inform.* **11**:289–296.

Pawlak, Z., 1982, Rough sets, *Int. J. Inform. Comput. Sci.* **11**:341–356.

Pawlak, Z., 1991, *Rough Sets. Theoretical Aspects of Reasoning about Data*, Kluwer, Dordrecht.

Pawlak, Z., Grzymala-Busse, J. W., Slowinski, R. and Ziarko, W., 1995, Rough sets, *Commun. ACM* **38**:89–95.

Pawlak, Z. and Slowinski, R., 1994, Rough set approach to multi-attribute decision analysis, *Eur. J. Oper. Res.* **72**:443–459.

Polkowski, L., 2002, *Rough Sets: Mathematical Foundations,* Physica, Heidelberg.

Polkowski, L. and Skowron, A., 1999, Calculi of granules based on rough set theory: approximate distributed synthesis and granular semantics for computing with words, in: *New Directions in Rough sets, Data Mining and Soft-Granular Computing,* Lecture Notes in Artificial Intelligence, Vol. 1711, N. Zhong, A. Skowron and S. Ohsuga, eds, Springer, Berlin, pp. 20–28.

Polkowski, L., Skowron, A. and Zytkow, J., 1995, Rough foundations for rough sets, in: *Soft Computing,* T. Y. Lin and A. Wildberger, eds, Simulation Councils, Inc., San Diego, CA, pp. 142–149.

Roy, B., 1996, *Multicriteria Methodology for Decision Aiding,* Kluwer, Dordrecht.

Skowron, A, Boolean reasoning for decision rules generation, 1993, in: *Methodologies for Intelligent Systems,* J. Komorowski and Z. W. Ras, eds, Lecture Notes in Artificial Intelligence, Vol. 689, Springer, Berlin, pp. 295–305.

Skowron, A. and Polkowski, L., 1997, Decision algorithms: a survey of rough set-theoretic methods, *Fund. Inform.* **27**:345–358.

Skowron, A. and Stepaniuk, J., 1995, Generalized approximation spaces, in: *Soft Computing,* T. Y. Lin and A. Wildberger, eds, Simulation Councils, Inc., San Diego, CA, pp. 18–21.

Slowinski, R., 1992, A generalization of the indiscernibility relation for rough set analysis of quantitative information, *Riv. Mat. Sci. Econ. Soc.* **15**:65–78.

Slowinski, R., ed., 1992b, *Intelligent Decision Support. Handbook of Applications and Advances of the Rough Sets Theory,* Kluwer, Dordrecht.

Slowinski, R., 1993, Rough set learning of preferential attitude in multi-criteria decision making, in: *Methodologies for Intelligent Systems,* J. Komorowski and Z. W. Ras, eds, Lecture Notes in Artificial Intelligence, Vol. 689, Springer, Berlin, pp. 642–651

Slowinski, R., Greco, S. and Matarazzo, B., 2002a, Rough set analysis of preference-ordered data, in: *Rough Sets and Current Trends in Computing,* Lecture Notes in Artificial Intelligence, Vol. 2475, J. J. Alpigini, J. F. Peters, A. Skowron and N. Zhong, eds, Springer, Berlin, pp. 44–59.

Slowinski, R., Greco, S. and Matarazzo, B., 2002b, Mining decision-rule preference model from rough approximation of preference relation, in: *Proc. 26th IEEE Annual Int. Conf. on Computer Software and Applications* (Oxford), pp. 1129–1134.

Slowinski, R., Greco, S. and Matarazzo, B., 2002c, Axiomatization of utility, outranking and decision-rule preference models for multiple-criteria classification problems under partial inconsistency with the dominance principle, *Control Cybernet.* **31**:1005–1035.

Slowinski, R., Stefanowski, J., Greco, S. and Matarazzo, B., 2000, Rough sets based processing of inconsistent information in decision analysis, *Control Cybernet.* **29**:379–404.

Slowinski, R. and Vanderpooten, D., 1997, Similarity relation as a basis for rough approximations, in: *Advances in Machine Intelligence and Soft-Computing,* Vol. 4, P. P. Wang, ed., Duke University Press, Durham, NC, pp. 17–33.

Slowinski, R. and Vanderpooten, D., 2000, A generalised definition of rough approximations, *IEEE Trans. Data Knowledge Engng* **12**:331–336.

Slowinski, R. and Zopounidis, C., 1995, Application of the rough set approach to evaluation of bankruptcy risk, *Int. J. Intell. Syst. Acc. Fin. Mgmt* **4**:27–41.

Stefanowski J., 1998, On rough set based approaches to induction of decision rules, in: *Rough Sets in Data Mining and Knowledge Discovery,* Vol. 1, L. Polkowski and A. Skowron, eds, Physica, Heidelberg, pp. 500–529.

Stepaniuk, J., 2000, Knowledge discovery by application of rough set models, in: *Rough Set Methods and Application,* L. Polkowski, S. Tsumoto, and T. Y. Lin, eds, Physica, Heidelberg, pp. 137–231.

Thomas, L. C., Crook, J. N. and Edelman, D. B., eds, 1992, *Credit Scoring and Credit Control,* Clarendon, Oxford.

Tversky, A., 1977, Features of similarity, *Psychol. Rev.* **84**:327–352.

Yao, Y. and Wong, S., 1995, Generalization of rough sets using relationships between attribute values, in: *Proc. 2nd Annual Joint Conf. on Information Sciences* (Wrightsville Beach, NC), pp. 30–33.

Ziarko, W., 1993, Variable precision rough sets model, *J. Comput. Syst. Sci.* **46**:39–59.

Ziarko, W., 1998, Rough sets as a methodology for data mining, in: *Rough Sets in Knowledge Discovery,.* Vol. 1, L. Polkowski and A. Skowron, eds, Physica, Heidelberg, pp. 554–576.

Ziarko, W. and Shan, N., 1994, An incremental learning algorithm for constructing decision rules. in: *Rough Sets, Fuzzy Sets and Knowledge Discovery,* W. P. Ziarko, ed., Springer, Berlin, pp. 326–334.

Chapter 17

HYPER-HEURISTICS

Peter Ross
School of Computing
Napier University, Edinburgh, UK

17.1   THE CONCEPT OF HYPER-HEURISTICS

The term "hyper-heuristics" is fairly new, although the notion has been hinted at in papers from time to time since the 1960s (e.g. Crowston et al., 1963). The key idea is to devise new algorithms for solving problems by combining known heuristics in ways that allow each to compensate, to some extent, for the weaknesses of others. They might be thought of as *heuristics to choose heuristics*. They are methods which work with a search space of heuristics. In this sense, they differ from most applications of metaheuristics (see Glover and Kochenberger, 2003) which usually work with search spaces of solutions. One of the main goals of research in this area is to devise algorithms that are fast and exhibit good performance across a whole family of problems, presumably because the algorithms address some shared features of the whole set of problems.

Many practical problems cannot be tackled by exhaustive search, either to find an optimal solution or even to find a very good quality solution. For such problems people often resort to *heuristic methods*: incomplete search methods that offer no guarantees of success and perhaps also involving some random elements. There are many varieties of heuristic search methods in regular use. Examples include:

Local Search Methods come in many flavors. They start from some chosen place and seek improvements by searching in some kind of neighborhood of that place. When an improvement is found, the process restarts from that improved position. Often, the order in which candidate improvements are considered is determined by some heuristic selection process. The many categories of local search methods include metaheuristic ideas such as variable neighborhood search (see Chapter 8) which adjusts the size of the neighborhood, and guided local search

(Voudouris, 1997) which adjusts the notion of the value, or fitness, of a solution as the search progresses.

Simulated Annealing resembles local search but allows the search to accept even a worse solution than the current one, particularly when early on in the search process. A probabilistic decision is made about whether to accept a worsening step; the probability depends not only on how much worse it is but also on how long the search has been running. Simulated annealing thus permits a local search to get off a local hilltop. See Chapter 7 for an introduction.

Evolutionary Algorithms typically manage a population of candidate solutions and apply some ideas from the theory of evolution. The simplest is that of selection. Imagine a search that starts from 100 different places and does local search in each. Rather than keeping 100 local searches running (as it were) in parallel, selection chooses the hundred best points found and the search restarts from those points, but those hundred best points may not be one from the neighborhood of each of the previous hundred points. Thus, under the influence of selection alone, the number of regions of the search space that are being considered may decrease. The idea of crossover, or recombination, can counteract this; it forms new candidate points by recombining fragments of existing solutions, thus potentially generating points that are nowhere near the existing regions being sampled. There are many varieties of evolutionary algorithm. For an introduction to Genetic Algorithms and Genetic Programming see Chapters 4 and 5, respectively.

Ant-Colony Algorithms borrow ideas from studies of the collective behavior of ants, who communicate by individually laying down chemical information and react to chemical signals already laid by others. This has much in common with many other reinforcement learning schemes that seek to discover what are the good decisions to make at each step in a search, by some balance between exploration of possibilities and exploitation of what seems good in terms of past experience. See Chapter 14 for an introduction.

The term *metaheuristics* often appears in the literature as a general term for such varieties. Some authors reserve the term *heuristic* for the decision procedure applied at a single step in the search and apply the term *metaheuristics* to overall control strategies; but there is no great consensus about these terms. See Glover and Kochenberger (2003) for a comprehensive treatment of metaheuristics.

Despite the fact that all these search techniques are often very effective, there can often be some reluctance to use them for money-critical problems. In practice, experience suggests that people often prefer to use very simple and readily understandable search methods even if those methods deliver relatively inferior results. Reasons might include:

1 the above kinds of search techniques involve making a significant range of parameter or algorithm choices and it is not clear to inexperienced users what is best and whether certain choices are actively bad;

2 the state of the art in the above methods for real world problems tends to represent bespoke problem-specific methods which are particularly resource intensive to develop and implement;

3 such algorithms often involve making some probabilistic choices, so that two seemingly identical runs may produce significantly different answers to the very same problem;

4 there is little knowledge or understanding of the average- or worst-case behavior of some of these techniques; where such results do exist (see Ausiello et al. (1999) for a compendium), it is usually for very straight-forward algorithms that can be analysed mathematically;

5 some of these techniques can be relatively slow;

6 even if the technique generates a good-looking solution, it can be hard to understand how the solution was arrived at; this matters because it is often important to a group of people to feel that a proposed solution is intuitively acceptable; people are often unwilling to trust a computer's results implicitly, and often that is for good reasons.

Research on *hyper-heuristics* is an attempt to respond to such legitimate criticisms. The broad aim is to discover some algorithm for solving a whole range of problems that is fast, reasonably comprehensible, trustable in terms of quality and repeatability and with good worst-case behavior across that range of problems. The goal is to develop algorithms that are more generally applicable than many of the implementations of the approaches outlined above. The space of possible algorithms for any given sort of problem is of course vast and there are many ways to search certain parts of such a space. For example, *genetic programming* (described in Chapter 5) uses evolutionary techniques to explore a chosen space composed of problem-specific functions and variables and also algorithmic control structures such as for-loops and if-statements. One of the difficulties facing research in genetic programming is that it can be hard to offer the search process much guidance about how to fit the available control structures together. For example, suppose you wish to sort a two-dimensional array of numbers so that each row is in decreasing order and each column is in decreasing order. Even if a sorting function is provided as a primitive, it is true but not obvious that it is sufficient to sort each row exactly once and each column exactly once; you do not need to continually sort the rows and then the columns and then check for violations, repeating until none remain.

Although hyper-heuristics might be regarded as a special form of genetic programming, the key intuition underlying research in this area is that, for a given type of problem, there are often a number of straightforward heuristics already in existence that can work well (but perhaps not optimally) for certain sorts of instances of that type of problem. Perhaps it is possible to combine those existing heuristics into some more elaborate algorithm that will work well across a range of problems.

17.2   A STRAIGHTFORWARD EXAMPLE: BIN PACKING

Bin packing is an easily understandable problem that appears as a factor in many other practical problems. The basic scenario is as follows. You have an unlimited supply of identical bins, each capable of containing one or more items totalling at most W in weight. You have n items to pack, of weights $w_1, w_2, \ldots, w_n$; the weights may not all be different but they are all individually packable: $0 < w_i \leq W$. The task is to put items into bins in such a way as to minimize the total number of bins used.

At first glance this may seem easy, but it is not. For example, if you are considering distributing 50 items among 10 bins then there are 9.484×10^{49} ways of doing it without leaving any bin empty, but ignoring that maximum-weight constraint. It can be tempting to adopt a strategy that (besides anything else it does) tries to fill each bin as close to capacity as possible. Before giving in to any such temptation it is a sensible research strategy to devote a little energy to trying to construct a counter-example. For example, suppose bins have capacity 20 and there are six items of weights 12, 11, 11, 7, 7 and 6. One bin can be completely filled $(7 + 7 + 6)$ but then the remaining items need a bin each, for a grand total of four bins, whereas if large and small items were paired $(12 + 7, 11 + 7, 11 + 6)$ then only three bins would be needed. With a little thought it is not hard to see how to construct such a counter-example; in bin packing, small (or smallish) items are useful as a filling material to fill up those odd bits of wasted space, so why not adjust the numbers so that all the smallish stuff gets used up to completely fill one bin leaving no items to fill up wasted space in others?

There is a simple lower bound on the number M of bins that are necessary:

$$ M \geq \left\lceil \left(\sum_{i=1}^{n} w_i \right) / W \right\rceil $$

(note: this notation rounds M up to the nearest integer) and if the packing algorithm never starts a new bin unnecessarily (that is, when the item it is considering could be placed in a partially-filled bin), then $M \leq$ bins used $< 2M$ because if $2M$ or more bins were used then there would be at least two bins whose combined contents weighed at most W (because the average load per bin would be $W/2$ or less) and why did the first item placed into the second-started of these two bins not get placed in the first-started one instead?

A good introduction to bin packing can be found in Martello and Toth (1990) and a survey of results about the performance of various algorithms can be found in Coffman et al. (1996). One very popular heuristic is the so-called first-fit-decreasing algorithm: pack items in order of weight, largest first; never start a new bin unnecessarily; given an item to pack, look at the bins in the order in which they were started and put that item in the first bin that is capable of holding it. This is popular because, although it often fails to get the optimal answer, it has good worst-case behavior; it has been proved (Johnson, 1973) that it will never use more than $11M/9+4$ bins, that is, more than about 22% too many. Because of this, it is an ingredient in many commercial container-packing algorithms. But as a heuristic it still has some behavioral quirks, as the example below due to R. L. Graham shows. Suppose the bins have capacity 524 and the following items (given in order of size, reading down the columns) are to be packed:

442	252	127	106	37	10	10
252	252	127	106	37	10	9
252	252	127	85	12	10	9
252	127	106	84	12	10	
252	127	106	46	12	10	

Then the algorithm uses seven bins. But if you *delete* the item of weight 46, the same algorithm now requires eight bins—there is an apparent discontinuity of performance. It is not hard to figure out how to construct such an example once you observe that the numbers in the table add up to exactly 7×524 so that in a seven-bin packing every bin is completely full. The numbers merely have to be chosen so that each item gets put in its "right" place by the first-fit-decreasing algorithm. Then if you delete an item that is not equal in weight to the sum of two or three smaller items, the algorithm will quite probably fail to get the packing just right. People have suggested many other heuristic ideas for bin packing. For example, Djang and Finch (1998) suggested this: given a newly-started bin, pack items in it (taking them largest-first) until the bin is at least one-third full; it could be much fuller than one-third full after this, of course. Then look for one single item that exactly fills the bin; or else

```
Given: a small set S of heuristics,
       and initial problem state P
Repeat until no items remain:
       - choose a heuristic H from S,
         in a way that depends on P;
       - apply H to pack the next bin;
       - update P accordingly.
```

Figure 17.1. The general form of algorithm sought.

look for two items that exactly fill the bin; or else look for three items that exactly fill the bin. If none of those are possible, then look for one item that fills the bin to $W - 1$; or else two items; or else three items. If that still is not possible, let $W - 2$ be the target load; and so on. This heuristic works excellently well on many benchmark bin-packing problems that are known to be hard. However, on easy problems it can work abysmally badly. Consider a problem in which the bins have capacity 1000 and there are 10 000 items each of weight 1. This needs only 10 bins. However, the above algorithm will first fill a bin until it contains 334 (just over one-third) and then put just three more items into the bin, so the bin contains 337. Thus 30 bins will be needed ($337 \times 29 = 9773$). This illustrates one of the difficulties of designing good heuristics: problems known to be hard have certain characteristics. In bin packing, the hard benchmark problems involve items whose weights are typically a significant fraction of the bin capacity, for example at least 20% of bin capacity, so that there will be no more than five items per bin but there will be a very large number of items so that the difficulty arises when trying to find the subsets of items that are to reside in each bin. If there were many very small items, those items could be used essentially as "sand" to fill up space wasted when large items were packed.

Because there is a range of available heuristics with different strengths and weaknesses, it makes sense to try to combine them in some way that permits one heuristic to compensate for the weakness of another. See Ross et al. (2002, 2003) for two hyper-heuristic ways of doing this, applied to large sets of benchmark problems and with very good worst-case performance. Both happen to use forms of evolutionary algorithm to conduct a search for a good combination of heuristics. The aim in both is to discover an algorithm that builds a solution to any given bin-packing problem incrementally, as shown in Figure 17.1.

In Ross et al. (2002), Wilson's XCS classifier system (Wilson, 1995) is used to try to discover a set of rules, each of which associate a short description of the current state of the problem with a heuristic to apply. The description uses just 11 bits. Two bits are used to describe the proportion of items still to pack in each of the four size ranges shown in Table 17.1, thus accounting for eight bits.

Table 17.1. Size ranges.

Huge:	items of weight $w > W/2$
Large:	items with $W/3 < w \le W/2$
Medium:	items with $W/4 < w \le W/3$
Small:	items with $w \le W/4$

Table 17.2. Encoding proportions.

Bits	Proportion of items
0 0	0–10%
0 1	10–20%
1 0	20–50%
1 1	50–100%

These are rationally chosen: at most one huge item will fit in a bin; at most two large items will fit, and so on. For each range the two-bit encoding of proportion is shown in Table 17.2.

The final three bits encode the proportion of items that remain to be packed, dividing the range from 0% to 100% into eight equal-sized intervals so that if (say) 27% of items remain to be packed this is encoded as 010. The rationale is that this information is useful, one heuristic might be best if few items remain but another might be best if there are many still to pack, even if the relative proportions of huge, large etc items are the same. Although this encoding was chosen on the basis of intuition and experience, it is ultimately justified by the results. XCS was able to discover a collection of rules that was very good at bin packing, finding the optimal result in nearly 80% of a collection of 890 publicly-available benchmark problems. See the cited paper for further details, including details of the heuristics used.

In Ross et al. (2003), a messy genetic algorithm (Goldberg et al., 1989) is used instead and rather than using bits, real numbers are used to encode the proportions that describe the state. The task of the messy genetic algorithm is to place a number of control points, each labeled with a given heuristic; the labels do not need to be all different. Figure 17.2 shows a simplified, three-dimensional rather than five-dimensional representation of the idea. Given a set of labeled control points, any bin-packing problem is tackled by finding the point that represents the problem's current state, identifying the control point that is closest and applying the heuristic that labels that point to pack the

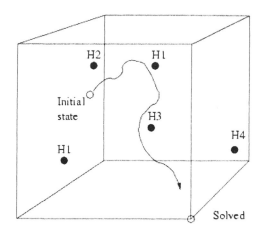

Figure 17.2. A messy genetic algorithm approach: the basic idea.

next bin. This changes the problem's state. The process is repeated until the problem is solved.

Note that it can be beneficial to allow control points to be located just outside the cube, even though the actual problem state cannot be outside the cube. Overall the results are a little better than those obtained with XCS, and on a larger set of benchmark problems. The enlarged problem set included a modest number of easy problems; as noted above, the hard bin-packing problems available in benchmark collections tend to have a certain character, but in practice real-life problems are sometimes relatively easy and it is therefore interesting to see if a discovered algorithm can do a good job of solving more than just the hard problems!

This discussion of bin-packing problems has illustrated the concept of hyper-heuristics, and some general points should be apparent:

- The algorithm being sought, which selects and applies heuristics, ought to have a clear and simple general form. Algorithms with an over-elaborate general form can be open to the same kinds of criticism listed at the start of this chapter.

- It is important to have a substantial set of problems to work with. What is being sought is an algorithm that is fast, reliable and with good overall and worst-case performance, and you cannot judge worst-case performance on the basis of a small set of problems. For real-world use best-case performance statistics, although highly valued in many academic papers, is often not so important.

- Much may depend on the particular set of heuristics used and on the choice of representation and other factors that affect the nature of the space to be searched.

- Although evolutionary algorithms were used in the examples above, it is not clear that they are necessary. Some simpler search process may suit the particular search space better.

Below, various research issues will be discussed but first it will be useful to survey past work briefly.

17.3   A BRIEF SURVEY

An early example of what might be called a hyper-heuristic approach was the scheduler used within the COMPOSER system (Gratch et al., 1993). COMPOSER's task was to plan communication schedules between earth-orbiting satellites and ground stations, with a maximum interval between communications with a given satellite. The ground stations that could be used were constrained by satellite orbits. The scheduler used heuristic methods to try to build a schedule, deciding which unsatisfied constraints to focus on next and how to try to satisfy them. Because there were several possible heuristics to use in each case, the system used a simple hill-climbing approach to investigate combinations of them, testing each on 50 different problems, and was able to discover an effective combination.

Hyper-heuristic ideas have been applied to various other scheduling problems. In job-shop and open-shop problems, there are m machines and j jobs. Each job consists of a number of tasks, which in job-shop problems occur in a job-specific order and in open-shop problems can occur in any order, and each task involves visiting a given machine for a given length of time. A task may also have a ready time, before which it cannot start, and a due date, by which it ought to have been completed. Various criteria have been studied, such as minimizing the makespan (total time to complete all jobs) or minimizing the worst delay beyond due date of any task, perhaps weighted by some measure of task importance. Fang et al. (1994) used a genetic algorithm to solve open-shop problems. In their work a chromosome consisted of a sequence of pairs of integers (j_i, h_i), each pair being interpreted in turn to build a complete schedule and meaning "consider the j_ith uncompleted job (modulo the total number of uncompleted jobs) and use heuristic h_i to select a task from that job to insert next into the schedule". At the time, this produced good results but the genetic algorithm was used to produce a separate solution for each problem rather than to find a single algorithm that could be applied to many problems. Hart and Ross (1998) applied a variant of this idea to dynamic job-shop problems (in which not all jobs can start immediately): chromosomes again contained a sequence of pairs of integers (a_i, h_i) but now was 0 or 1 to indicate which of two heuristically-guided task selection algorithms to employ, and indicated which heuristic was to be plugged into the chosen algorithm. As in the work of Fang et al., they used a separate genetic algorithm run for each scheduling problem rather than seeking one generally-powerful algorithm. It is also worth noting that in later work, Vázquez and Whitley (2000) were able to obtain even better results with a genetic algorithm that used a direct encoding and a special-purpose crossover operator. Terashima-Marin et al. (1999) applied hyper-heuristic to solve a range of large university exam timetabling problems, but once again the approach was used to find a different solution procedure for each problem.

A recent general introduction to hyper-heuristic ideas can be found in Burke et al. (2003a). More recently, hyper-heuristic ideas have been applied in various ways to a range of other problems. For example, Kendall et al. (2002) use a performance-rating function to rank heuristics, in trying to solve three personnel scheduling problems. Burke et al. (2003b) combine hyper-heuristic ideas with those of tabu search on some variants of a nurse scheduling problem and eleven university course timetabling problems. Burke et al. (2003c, 2005) and Petrovic and Qu (2002) have also demonstrated that case based reasoning can be employed as an effective heuristic selection methodology for timetabling problems. The interested reader can find further information on the ASAP (2004) website.

17.4   SOME RESEARCH ISSUES

17.4.1   No Free Lunch

The No Free Lunch theorem of Wolpert and Macready (1995) showed that, when considering the set of all finite search problems on a given finite domain, all search algorithms have exactly the same average performance. See Chapter 10 for more details. Although it caused a stir at the time of its first publication, this result should be unsurprising to you. The vast majority of problems have no exploitable structure to them; the only way they can be described is by a full-sized lookup table, rather than by (say) a closed-form mathematical function. Imagine, for example, a search problem $f()$ defined on the integers $1, \ldots, 100$ in which each $f(i)$ was a randomly chosen positive integer. No search algorithm can do better than to look at every $f(i)$ if it aims to find the maximum or minimum of the function. However, in practice we are not interested in random or random-seeming functions; the problems we humans are interested in have some internal structure and some degree of predictability about them. We therefore want to find algorithms that can perform well on the subset of problems that arise naturally in some context; the difficulty lies in characterizing that subset properly in such a way that we can design a really effective algorithm for the members of that subset. Hyper-heuristics

```
superAlg(Problem P) {
  switch(P) {
    case P1 : solution = s1; break;
    case P2 : solution = s2; break;
    ....
  }
  return(solution);
}
```

Figure 17.3. A possible super-algorithm?

sidesteps these issues of characterization and design to some degree by instead conducting a search for an adequately effective algorithm.

Before leaving the theoretically interesting but practically rather limited topic of the No Free Lunch theorem, it is worth considering why some super-algorithm of the form shown in Figure 17.3 does not violate the theorem. After all, for a given finite range and domain there is a finite but normally huge number of possible problems.

The difficulty lies in determining which problem P is. As discussed, most problems can only be described by a full-sized lookup table and so the question of whether $P = P1$ (say) might only be resolved by examining every entry in such a table. This observation also has certain consequences for hyper-heuristic algorithms. For example, it is tempting to wonder whether the control points in the kind of algorithm illustrated in Figure 17.2 could be moved about a bit, say by some amount $\delta > 0$, without affecting performance. If this were so then it might be possible to construct an algorithm akin to the super-algorithm but in terms of control points rather than problems; and if the control points only had to be specified to a given minimal accuracy then there would only be a modest number of branches to consider and each branch test would be of modest complexity, thus contradicting the theorem. This highly informal argument suggests that even for hyper-heuristically-generated algorithms there must be certain problems for which the sequence of heuristic choices made will be absolutely critical.

17.4.2   What is a Problem Family?

It is therefore natural to wonder, for what particular families of problems might hyper-heuristic methods be acceptably effective? One way to address this question is to generate whole families of problems that have particular characteristics and investigate how performance varies across members of that family. In recent years work on binary constraint satisfaction in particular has thrown up a number of interesting phenomena in relation to performance of algorithms across a spectrum of problems. The kind of problems studied have a standard form: there are n variables $x_1, \ldots, x_n$, each can potentially take values from a finite domain $v_1, \ldots, v_n$. However, there are constraints between pairs of variables of this form: the pair (x_3, x_7) cannot take any of the values (v_6, v_2), (v_2, v_6), (v_9, v_8), (v_{11}, v_4). In early work, problems were generated probabilistically by picking a probability p_1 that there were any constraints at all between a pair of variables and a probability p_2 that if there was a constraint between a given pair of variables, then a particular pair of values was included in that constraint. It was possible to derive an exact mathematical formula for the expected number of solutions of a random problem generated with given p_1 and p_2, and experiments suggested that problems for which this expected number was very close to 1 were particularly hard to solve by any of the known (and exhaustive) methods. There is nothing intrinsically significant about the number 1 here; rather, if the expected number is much less than 1 then most problems are unsolvable and it is easy to locate an unsatisfiable sub-problem, and if the expected number is much larger than 1 then most problems are solvable with many solutions and it is fairly easy to find one. More recently people have looked at generating constraint problems in a less uniformly-random way, imposing more internal structure on them. Further details can be found at web pages such as Hewlett-Packard (2004) and APES (2004). However, outside the constraint satisfaction community, this kind of study of performance across large families of problems is still regrettably rare.

17.4.3   What Heuristics Should Be Chosen?

It is also natural to wonder how sensitive the results might be to the particular choice of heuristics involved, and the best way forward is probably to conduct systematic experiments. Clearly, in the interests of generating an algorithm that terminates, every heuristic used should make some discernible progress toward a solution; if not, the algorithm might repeatedly choose it and therefore loop forever. Also, although some authors have employed a random heuristic that simply makes a random choice, it would be sensible to exclude the use of such random heuristics from the final algorithm because they lay the algorithm open to the same criticisms, of non-repeatability and incomprehensibility, that originally motivated the idea of hyper-heuristics. However, there *is* a place for using random heuristics during the development stage: if the inclusion of a random heuristic improves the results then it suggests that the other heuristics between them are not so capable of generating good performance and therefore that set of heuristics perhaps needs to be extended or changed. But beware of including too large a set of heuristics and thereby conducting a kind of "buckshot attack" on your chosen problem set. Bear in mind also that sometimes an individual heuristic is excellent and no hyper-heuristically generated combination might be able to beat it. For example, Kruskal's greedy algorithm (Kruskal, 1956) for generating minimum spanning trees in weighted graphs always finds optimal solutions (and such, perhaps does not count as a heuristic; the point is, it can sometimes be hard to tell!).

Heuristics often take the form of optimally-determine-then-act rules (such as "find the most constrained item and insert it next") but they need not always do so. For example, a heuristic might prescribe or control the amount of back-tracking that a search can do, or otherwise influence search parameters, rather than dictating the search step directly. See Terashima-Marin et al. (1999) for an example.

17.4.4   What Search Algorithm Should Be Used?

It is still unclear what kind of search method should be employed to try to generate an algorithm. Much depends on the nature of the search space, and not simply on its size. Many authors, not specifically concerned with hyper-heuristics, advocate evolutionary algorithms as a good way to search huge spaces, but no sensible search is unbiased; they focus on certain regions of the space and quickly exclude vast parts of the space from the search, so raw size is no guide. Remember also that in any given situation, when considering which heuristic to apply, it may well be the case that several heuristics would generate the same choice. Thus there may be many solutions to be found even in a large-seeming space, and local search methods such as hill climbing may sometimes be the best choice in terms of cost to find an acceptable algorithm. Further comparative studies are needed.

17.4.5   During the Search, How Should Performance Be Evaluated?

As noted earlier, some authors have used hyper-heuristic methods to solve single problems and for them, the question of how to evaluate a generated hyper-heuristic algorithm is easy: simply apply it to the problem. On the other hand, if you are trying to find an algorithm that exhibits good performance across a large set of problems, then it can be very costly to evaluate a generated algorithm's performance on every member of the set. Therefore, don't do it. A simple alternative was first proposed in Gathercole and Ross (1997): choose a few problems initially for evaluation purposes, but also keep track of when each problem was last used for evaluation and how well it was solved when it was used. At each subsequent cycle, choose a small subset for evaluation purposes in a way that takes account of both apparent problem difficulty and problem recency. The size of the chosen subset might also be made to grow with time so that as the search narrows toward a good algorithm, candidates are tested on a growing subset.

17.4.6   What Kind of Algorithm Should Be Sought?

In the section on bin packing, two examples of hyper-heuristic algorithm were described but both were essentially constructive, building the solution of a given problem incrementally and without backtracking. This might not be the best option even for bin packing; it might be better to find a reasonable initial packing by some straightforward means and then use heuristics to shift and swap items between bins with the aim of reducing the number of bins used.

For many operations-research-type problems, the best existing methods for solving individual problems start by constructing a complete initial solution, or perhaps a collection of them, and then search for improvements by heuristically-guided local search methods. To date, hyper-heuristic research has perhaps tended to focus on finding incremental solution-constructing algorithms but those which modify full candidate solutions might turn out to be better.

The next sections discuss some possible places to start research in hyper-heuristics.

17.5   SOME PROMISING AREAS FOR FUTURE APPLICATIONS

17.5.1   Timetabling

Timetabling problems have been intensively studied for many years; see, for example, Burke and Ross (1996), Burke and Carter (1998), Burke and Erben (2001) and Burke and De Causemacker (2003) for collections of relevant papers that apply a variety of methods to solve a wide variety of timetabling problems. However, many authors confine themselves to solving only the problems that arise at their own home institution without considering whether those problems are intrinsically difficult or simply are so packed with detail that a computer is needed to keep track of them. Timetabling problems remain interesting to study because there is such a wide variety of them; every school and university seems to have its own peculiarly idiosyncratic constraints. But consider, say, one reasonably generic kind of problem, that of university exam timetabling. This arises because many universities wish to fit all their exams into a predetermined period; shortly after that period all the exams must have been marked and the results entered into university databases so that degree-class decisions can be announced in good time for degree-award ceremonies. In scheduling exams a university must obey some hard constraints: no student should have to be in two places at once, or should find it impossible to get from one exam to the next in time; exam halls have finite capacity, even though different exams of identical duration may perhaps be scheduled to take place in the same exam hall if there is enough space; and so on. There are also soft constraints, such as trying to ensure that no student has to take too many exams in a day and to ensure that as many students as possible get some rest and revision time between exams, or to ensure that very large exams happen early on in order to ease the marking burden.

Exam timetabling therefore provides an ideal context for studying different families of problems. Some universities are short of space for exams, and the nature of their problem is largely a matter of bin packing first and then shuffling exams about to try to improve the satisfaction of soft constraints. Other universities may have plenty of exam hall space but may allow students a relatively unfettered choice of what subjects they can study, so that there are many constraints between different exams. Yet other universities may have problems that are almost decomposable into separate sub-problems, for example because science students almost always take courses only from the science faculty and arts students nearly always stick within the arts faculty for their education, and so on. Some universities have a few huge courses and many small courses, so that the essence of the problems is to deal somehow with those huge courses first and then pack the rest in around them. Some benchmark problems can be obtained from OR (2004a, 2004b) and Carter (2004) but this area is ripe for generating many more kinds of problem in a systematic way.

17.5.2 Vehicle Routing with Time Windows

This topic has also attracted much attention over the years. The typical problem is as follows. There is a single depot; think of it as a warehouse. There are customers at specified locations, to whom deliveries are to be made, and you know all distances involved; in benchmark problems the distance is simply the Euclidean distance, although in real life this is unrealistic. Each customer C needs to take delivery of a quantity g_C of goods from the depot and can only take delivery within a specific time-window $[s_C, e_C]$, and the actual delivery consumes a time t_C. This does correspond to real life; for example many large supermarkets plan to hire temporary staff for just two hours in order to unload an expected delivery, and impose contractual penalties on the transport company for failing to arrive in time, because the temporary staff will have to be paid even for waiting.

The transport company uses vehicles of capacity V, so it may be possible to put the wanted goods for two or more customers onto the one vehicle and get it to deliver to those customers before returning to the depot. The task is a multi-objective one: minimize both the number of vehicles required and the total distance traveled by all the vehicles. Vehicles start from the depot and all must return to the depot. There is also a global time-window $[0, T]$ (think of it as the working day): all vehicles must start from the depot and finally return to the depot within this window. Strictly speaking, it is not the number of vehicles but the number of depot-to-depot vehicle trips that is to be minimized; if a vehicle can return to the depot, pick up another load and complete another delivery run in the time available, that counts as though an extra vehicle were being used.

It might not be possible to minimize both the number of vehicles and the distance simultaneously, because of the time windows or vehicle capacity constraints. If one solution beats another in terms of both objectives it is said to dominate that other solution, and solutions which cannot be dominated are said to be *Pareto-optimal* (after the Italian economist Vilfredo Pareto, 1848–1923, who extensively discussed the issue of how to deal with incompatible objectives in his 1906 *Manual of Political Economy*).

Many authors work with some standard sets of benchmark problems and strive to minimize the average distance and average number of vehicles required for the problems in a whole set, as well as to find new Pareto-optimal solutions for individual problems. The most widely used benchmark problems are Solomon's six problem sets (Soloman, 1987) named R1, C1, RC1, R2, C2, and RC2. Each problem involves 100 customers. In the two R* sets customers were located randomly (and within any one set, always at the same place). In the two C* sets the customers were located in distinct clusters. In the two RC* sets a combination of randomly-placed and clustered customers were used. The three *1 sets have fairly short time windows and low vehicle capacity; the three *2 sets have longer time windows and much larger vehicle capacity. Within any one set it is the time-window information for each customer that varies. The data can be obtained from Gambardella (2004). Problems generated in a similar way but involving up to 1000 customers can be obtained from Gehring and Homberger (2004).

The aim of getting good performance across a set of problems is very much in keeping with the motivations of hyper-heuristics but, as yet, hyper-heuristic ideas have not been applied to these problems. However, in vehicle routing it seems essential to start by generating some candidate routes and then to search for improved solutions by shifting customers, or sequences of customers, between routes so as to reduce distance and to try to empty some routes and thus reduce the number of vehicles. This is where it would make sense to try to apply heuristics. For example, if there is a current route that is short and with few customers it might make sense to focus on eliminating that route. Or, if route X is incapable of being extended because it almost fills the global time-window or its vehicle is close to capacity, but there is a customer on route X whose time-window is compatible with nearby route Y that can be extended, then consider shifting that customer onto route Y; and so on. A purely constructive approach that tries to "grow" routes by choosing the next customer to add to a route by some heuristic means seems very unlikely to work; the decision to allocate a customer to one route can have a very significant effect on what happens to all the other routes.

A large number of papers have been published over the years on vehicle routing with time-windows, using a wide variety of techniques including simulated annealing, genetic algorithms and ant colony methods. In many such papers the focus has been mainly on describing the use of a biologically-motivated technique but it appears that much of the productive work of obtaining good solutions has actually been due to the inclusion of some local search that tries to improve a set of routes generated by the biologically-motivated technique. For example, some authors pre-compute a short list of customers that are close in space and time-window to a given customer and use such lists to guide a limited local search that seeks to shift customers between routes. Despite the plethora of biologically-inspired approaches, perhaps the best results so far obtained have been generated by a deterministic approach based on variable-neighborhood search due to Bräsy (2003), although Berger et al. (2001) obtain almost as good results using a parallel hybrid techniques genetic algorithm. These papers provide a good entry point into the extensive literature.

17.5.3   Other Promising Areas

The Operations Research library (OR, 2004a) maintained by John Beasley contains problem data for many different sorts of combinatorial optimization problem, although it is usually necessary to do your own literature search to discover the best currently available results. Examples of areas include:

- *Job scheduling:* in which there is a set of jobs to be processed, and each job consists of a number of separate tasks, and each task requires the full use of certain resources for a period of time. There are many variants: the order of tasks for any given job may or may not be fixed; there may be limits or uncertainties about when some tasks can be done; certain resources may be replicated so that only one of a set has to be committed to a task; the set of jobs may or may not be fully known at the start; the stream of jobs may be endless; and so on. The aim may be to minimize total time taken, or each job may have a "due date" by which it should be completed and the aim is to ensure that no job is particularly late; there are many variants.

- *Staff scheduling:* this resembles job scheduling in many ways. Staff with differing sets of skills are to be allocated to tasks, ensuring that all work gets done but no staff member is overloaded and all contractual commitments are met.

- *Cutting and packing:* in which boxes of different shapes and sizes are to be efficiently packed into containers, or some given set of shapes is to be cut from given supplies of raw material so as to minimize waste. There are many variants; for example, the two sides of a length of cloth may differ, so that shapes to be cut from the cloth to make up suits cannot be turned over to achieve a better packing, but can be rotated.

Sometimes there are web sites that specialize in one particular kind of problem. For example, there is a website (KZZI, 2004) at the Konrad-Zuse-Zentrum für Informationstechnik, Berlin, devoted to so-called *frequency assignment problems* that arise in telecommunications. Typically each link in a wireless network has to have two frequencies assigned to it (one for each direction of communication), but the choices are limited by the potential for interference between similar frequencies used on geographically close links, and by legal and other factors. The website describes the various flavors of frequency assignment problems and has links to sets of test problems.

17.6   TRICKS OF THE TRADE

17.6.1   A Ski-Lodge Problem

If you wish to start doing research in the area of hyper-heuristics, it can be daunting to have to learn about a well-studied application area, its past work and its successes, failures and current best techniques. This section describes a modest problem that seems suitable for an initial exploration of hyper-heuristic ideas, and gives some target results and details of resources. The problem is of a scale suitable for a short-term exploratory project. It was suggested by Mark Bucci (2001), who used simulated annealing to tackle it—see the reference for access to his C++ source code.

The problem concerns a time-shared ski-lodge containing four identical apartments, each capable of sleeping up to eight people. However, fire regulations require that there be at most 22 people resident in the building during any week. There is a 16-week ski-ing season and the first five weeks (numbered 0–4) are somewhat more popular than the other 11. Owners do not buy the right to use a particular apartment during a particular week; instead, what they get for their money is the right to give a first, second and third choice of week when they would like to use one of the apartments. They also have to state the total size of the party including themselves (up to the maximum of eight) that they propose to bring. The management of the ski-lodge must try to meet everyone's requests as far as possible, but are contractually committed to paying compensation to those who do not get their first choice of week. In particular, if an owner is offered his second choice then he is entitled to compensation of two free one-day ski-lift passes per member of his party. If he can only be offered his third choice, the compensation is four passes per member. If he can only be offered a week he did not list at all, the compensation is free ski-ing (seven passes per member) plus a cash sum equivalent to 50 passes. If an owner cannot be offered any week at all, the compensation is a cash sum equivalent to 1000 passes.

Because there are four apartments and a 16-week season, there are 64 owners. Given the preference lists and party sizes of all the owners, the management's problem is to allocate owners to weeks in a way that obeys the constraints and also minimizes the total compensation to be paid out. An example problem (problem-10.txt) is shown in Table 17.3. There are various kinds of observation about this data that might guide the development of some suitable heuristics. For example, in this particular problem there are a total of 351 people, very close to the maximum possible of $22 \times 16 = 352$, so a major aspect of the allocation task for this particular problems is going to be a form of bin packing, because every week except one must have maximum occupancy if that 1000-pass payout is to be avoided. Just one owner has each listed three of the popular weeks, ten owners have listed two of them and one of the less popular weeks, 42 owners have chosen one popular week and two unpopular weeks and just 11 owners have avoided the popular weeks entirely. Only five owners have listed week 12. Only seven owners have listed week 11, only seven have listed week 13; and so on.

Eleven such problems are available for immediate download as plain-text files from http: /www.dcs.napier.ac.uk/~peter/ski-lodge/. The web page also includes the following:

- The source code, in C, of a problem generator so that you can generate many more problems. Use $ski - prob - gen - s13$ to seed the random number generator with 13, use $ski - prob - gen - c$ if you want the output files to use Windows CRLF end-of-line termination rather than Unix/Linux LF. Note that the generator does not guarantee that the number of people involved will be 352 or less, but many generated problems will satisfy this and therefore potentially be solvable without any 1000-pass payout. All of the eleven problems can be solved without a single 1000-pass payout.

- Source code in C++ (almost C) of an interactive program that enables you to load a given problem and try to solve it yourself. The program merely enforces the constraints and displays the compensation payout; enter -1 to de-assign an owner. Linux and Windows executables are also provided. The program uses the Fast Light Toolkit, available free from www.fltk.org, for the GUI aspects. The code will compile under either Linux or Windows, and executables for both are provided.

- Source code in Java of a genetic algorithm to solve such problems. The Java program has no graphical user interface, it needs to be run from a command prompt: java EA problem-10.txt 13 50 000 runs the program

Table 17.3. A ski-lodge problem.

Owner	Size	Choices	Owner	Size	Choices
0	4	2 0 11	32	6	3 0 8
1	5	0 4 6	33	4	1 3 14
2	4	4 15 5	34	4	9 11 5
3	3	3 7 9	35	5	3 15 6
4	7	0 5 11	36	6	3 10 5
5	3	6 15 10	37	3	2 10 9
6	3	3 5 10	38	7	0 10 14
7	4	4 7 13	39	8	1 6 12
8	5	4 5 14	40	4	3 9 6
9	7	2 6 13	41	7	2 4 15
10	4	0 14 5	42	6	1 2 10
11	8	0 15 6	43	4	2 15 8
12	8	14 10 13	44	5	0 9 11
13	5	3 15 14	45	6	4 6 15
14	3	11 14 4	46	7	3 1 2
15	3	10 7 9	47	6	14 10 7
16	8	6 8 4	48	8	14 10 6
17	3	2 0 8	49	3	0 6 15
18	6	13 9 6	50	5	1 10 14
19	4	14 11 7	51	3	1 8 14
20	7	3 15 5	52	8	0 1 9
21	8	4 14 15	53	6	1 6 7
22	4	14 5 3	54	5	6 10 9
23	5	1 14 13	55	8	2 0 7
24	5	2 10 9	56	7	3 14 8
25	5	3 9 8	57	7	4 13 9
26	7	3 9 5	58	8	3 12 6
27	7	0 1 9	59	5	0 15 10
28	7	10 11 4	60	6	13 14 6
29	6	15 0 8	61	5	8 6 12
30	4	3 15 12	62	5	2 9 7
31	6	1 9 6	63	6	0 12 8

on the problem in file problem-10.txt, using seed 13, for up to 50 000 iterations of the main loop. On any recent PC this should be very fast. Output is to standard output, redirect it to a file as you wish.

- Source code in C++ for a simulated annealing program to solve such programs. It too has no graphical user interface.

- A summary of the results of 25 runs of the Java genetic algorithm and the C++ simulated annealing program on each of the problems.

The genetic algorithm uses a simple representation. A chromosome consists of 64 integers, and $c[i]$ is simply the week offered to owner i. Initialization is done by first setting $c[i]$ to a random one of owner i's three choices, and then (because this may produce constraint violations) repairing it by the following process. First, for each owner in turn, check if the suggested week is possible. If not, mark that owner as unassigned using the special flag value of -1. This generates a number of unassigned owners, call them "losers". Second, for each week from 0 to 15 in turn, check if there are available apartments. If there is just one free apartment, find the loser who would best fit (in terms of party size) and assign him to that week; break ties by choosing the first such suitable loser. If there are two free apartments, find the pair of losers who would best fit (in terms of combined party sizes) and assign them to that week. What if there are three or four free apartments in a week? We simply hope that the genetic algorithm's crossover and mutation will deal with them! Finally, we do some local search: for 1000 tries, pick two owners at random and see if swapping their assignments would improve the payout figure; if so, do that swap.

The genetic algorithm uses a population size of 100 and tournament selection of size 2. In each iteration, two parents are chosen, one child is produced by single-point crossover and the child is then mutated by altering two randomly-chosen genes to be a randomly-chosen one of the relevant owner's choices. This mutated child is then repaired as above. Finally, the child displaces the worse parent, but only if the child has an equal or better payout level than that parent.

Table 17.4 summarizes the results of 25 runs for each problem, using random number seed values 1..25. Note that in the best solutions (and in nearly all others) the genetic algorithm manages to find a week for every owner, no owner remains unassigned.

Note that on the two problems which involve 351 people (numbers 04 and 10) the range of values found is wide, because these are difficult problems. In problems with relatively few people the genetic algorithm produces consistently good results, with only small variation in final payouts. As the number of people approaches 352, the consistency gets worse.

Is the genetic algorithm properly configured? Here are some experimental observations, with details omitted:

- crossover matters: turning it off produces worse results;

- two-point crossover is worse than one-point crossover;

- larger tournament sizes produce worse results;

- producing two children, and overwriting both parents (assuming the child has a cost that is $\leq$ parent's cost) produces somewhat worse results;

Table 17.4. Genetic algorithm results.

Problem	Size	Min	Max	Average over 25 runs
01	344	641	707	667.48
02	337	404	457	415.88
03	338	450	502	479.92
04	351	732	1616	1362.68
05	315	304	308	305.76
06	328	360	392	373.84
07	347	730	842	787.76
08	326	481	493	484.12
09	316	404	412	406.00
10	351	684	1604	1164.20
11	320	386	408	393.04

- getting the child to overwrite the parent only if the child has strictly lower cost (that is, not ≤ but <) also produces worse results;

- mutating just one gene, or no genes, or three genes, produces somewhat worse results;

- population sizes of 50 or 150 produce somewhat worse results;

- biasing either (or both) the initialization and mutation steps so that the owner's first choice is more likely than the second and the second is more likely than the third, makes very little difference.

However, with these variants, the worse results are not dramatically worse. In general, this genetic algorithm seems pretty good; for most problems, the range of values produced is not particularly wide, and the cost of doing 25 runs in order to get a good low score is very reasonable. And in practice, the management company might want to use a genetic algorithm such as this to help them determine suitable levels of compensation in the first place.

But, although this genetic algorithm seems reasonable its results are not always as good as they might be. For example, its best score on problem-10 is 684. This is considerably better than a human can typically do; some informal experiments using the interactive program mentioned above suggest that humans often have great difficulty in pushing the payout below 2000 for this problem. But a solution of 653 is possible, obtained by a different genetic algorithm (due to Henry Liang) that performs better on the largest of these problems but not quite as well as the above results on others.

The simulated annealing code available from the website has also been tuned to suit these sorts of problems, but the results it obtains are a little worse than the results shown in Table 16-4. The web page gives more details.

Your challenge is to develop some decent heuristics for such ski-lodge problems, perhaps using the interactive program, and then investigate whether a hyper-heuristic approach can do better than the results listed here. For each heuristic that you consider, work out what characterizes those problems that it works reasonably well on and see if you can pick a set that in some way covers the spectrum of problems. You will need many more than the 11 problems provided.

The problem suits either a constructive approach, in which heuristics are used to build a solution incrementally by repeatedly picking an owner and picking a week to which to allocate that owner, or a search-based approach in which an initial solution is built (perhaps by one of the downloadable programs) and then heuristics are used to search for improving alterations.

17.6.2 A Simple Framework for a Constructive Approach

In line with the ideas outlined earlier, here are some suggestions for a program that implements a "messy evolutionary search" for a set of labeled points, as suggested in Figure 17.2.

First, choose a problem-state representation—Section 17.2 above should give you some ideas. Devise a small set of heuristics (perhaps 5–10) that choose an owner and another small set that choose a week for the owner (or leave the owner unassigned, in the worst case). This gives you a set H of, say, 10–20 heuristics, half of which are owner-choosers and half of which are week-choosers.

The next step is to create an initial population, each of which represents a complete set of labeled points as in Figure 17.2. A member of the initial population might perhaps contain anywhere from three to six labeled points. The label of each point is randomly chosen from the set H, and the implicit algorithm is repeatedly to find the nearest owner-choosing point and also the nearest week-choosing point, and use them to place one more owner into the growing solution. Clearly, there needs to be some provision to handle the case in which there are no owner-choosing points at all, or no week-choosing points.

Generate a set of problems (say, 100 of them) each involving at most 352 people. Any member of the population can be evaluated by testing it on all the problems and noting the total compensation required, but this method has the disadvantage that it can conceal a few very bad performances if it performs generally well on most of them. It would be better to devise a performance measure that penalizes poor worst-case performance, say by picking a subset S of the problems, evaluating all members of the population on that subset and keeping track, over time, of the best and worst result that any member of any population has ever achieved on that problem. The subset S can be altered as the search proceeds.

The messy evolutionary search is based on a steady-state genetic algorithm. In any one step of the search:

- two parents are chosen from the current population, using tournament selection with a tournament of size 2 (or if not 2, it should be a small number to avoid having a heavy bias in favor of the better-scoring members of the population);

- they are recombined, by choosing at random between two recombination operators:

- the first applies a form of uniform crossover, copying labeled points from the first parent into the first child with 90% probability and into the second child with 10% probability; labeled points are similarly copied from the second parent into the children, with reversed probabilities; therefore, the children may contain different numbers of points than the parents do;

- the second applies a form of two-point crossover; two points are chosen in the first parent; two points are also chosen in the second parent, with the aim of swapping the segments between parents to form the children; so the points in the second parent must occur at the same *type* of location as in the first parent: for example, if the first point in the first parent occurs just before a point's label (that is, a number identifying a heuristic in the set H) then the first point in the second parent is constrained to be just before some point's label too;

- mutation is applied to the children; mutation may add a new labeled point, delete an existing labeled point or modify an existing labeled point by moving it a bit and/or altering its choice of heuristic;

- the children are evaluated, and overwrite the parents if they are fitter; essentially, the places occupied by the parents are reserved for the best two of the set of four members consisting of the two parents and the two children.

Some results obtained by applying this kind of algorithm to class and exam timetabling problems can be found in Ross et al. (2004).

SOURCES OF ADDITIONAL INFORMATION

As yet there are still relatively few resources available about hyper-heuristics. This section lists some places to look:

- the *Handbook of Metaheuristics* is a recent book (Glover and Kochenberg, 2003) that contains a great deal of information about different kinds of heuristics, and includes a chapter about hyper-heuristics (Burke et al., 2003a);

- the ASAP group at Nottingham University has a good website (ASAP, 2004) that includes research publications about hyper-heuristics and links to timetabling problems and other resources;

- the *Journal of Heuristics*, published by Kluwer, contains many papers about heuristic methods generally; the tables of contents and the abstracts of papers are available online; full papers are available to subscribers to Kluwer Online;

- the *European Journal of Operational Research* also contains many papers relating to heuristics and to problems that might be tackled by hyper-heuristic methods; again, abstracts are freely available online;

- the Metaheuristics Network site at www.metaheuristics.org provides information about various metaheuristic techniques, references to papers and links to sets of problems in several areas: quadratic assignment, maximum-satisfiability, timetabling, scheduling, vehicle routing and an industrial hose-optimization problem; the aim of the Metaheuristics Network is to conduct scientific comparisons of performance between various metaheuristic techniques in different problem areas; although hyper-heuristic methods are not explicitly considered, the site is valuable because the problems have been generated or contributed by the members and performance results are being made available.

Acknowledgments. This work was supported by UK EPSRC research grant GR/N36660.

References

APES, 2004, APES research group page. http://www.dcs.st-and.ac.uk/ ~apes/
ASAP, 2004, ASAP Research Group, http://www.asap.cs.nott.ac.uk/.
Ausiello, G., Crescenzi, P., Gambosi, G., Kahn, V., Marchietti-Spaccamela, A. and Protasi, M., 1999, *Complexity and Approximation: Combinatorial Optimization Problems and Their Approximability Properties*, Springer, Berlin.
Berger, J., Barkaoui, M. and Bräysy, O., 2001, A parallel hybrid genetic algorithm for the vehicle routing problem with time windows, *Working Paper*, Defence Research Establishment, Valcartier, Canada.
Bräysy, O., 2003, A reactive variable neighborhood search algorithm for the vehicle routing problem with time windows, *INFORMS J. Comput.* **15**:347–368; available online at http: //www.sintef.no/static/am/ opti/projects/top/.

Bucci, M., Optimization with simulated annealing, 2001, *C/C++ Users J.* November:10–27, simulated annealing source code available from www.cuj.com.

Burke, E. K. and Carter, M., eds, 1998, *Practice and Theory of Automated Timetabling 2: Selected Papers from the 2nd Int. Conf. on the Practice and Theory of Automated Timetabling,* Lecture Notes in Computer Science, Vol. 1408, Springer, Berlin.

Burke, E. and De Causmaecker, P., eds, 2003, *Practice and Theory of Automated Timetabling 4: Selected Papers of the 4th Int. Conf. on Practice and Theory of Automated Timetabling,* Lecture Notes in Computer Science, Vol. 2740, Springer, Berlin.

Burke, E. and Erben, W., eds, 2001, *Practice and Theory of Automated Timetabling 3: Selected Papers of the 3rd Int. Conf. on Practice and Theory of Automated Timetabling,* Lecture Notes in Computer Science, Vol. 2079, Springer, Berlin.

Burke, E. and Ross, P., eds, 1996, *Practice and Theory of Automated Timetabling 1: Selected Papers of the 1st Int. Conf. on the Practice and Theory of Automated Timetabling,* Lecture Notes in Computer Science, Vol. 1153, Springer, Berlin.

Burke, E., Hart, E., Kendall, G., Newall, J., Ross, P. and Schulenburg, S., 2003a, Hyper-heuristics: an emerging direction in modern search technology, in: F. Glover and G. Kochenberger, eds, *Handbook of Meta-Heuristics,* Kluwer, Dordrecht, pp. 457–474.

Burke, E. K., Kendall, G. and Soubeiga, E., 2003b, A tabu search hyperheuristic for timetabling and rostering, *J. Heuristics* **9**:451–470.

Burke, E., MacCarthy, B., Petrovic, S. and Qu, R., 2003c, Knowledge discovery in a hyper-heuristic for course timetabling using case-based reasoning, in: *Practice and Theory of Automated Timetabling IV: Selected Papers from the 4th Int. Conf. on the Practice and Theory of Automated Timetabling (PATAT 2002),* E. K. Burke and P. De Causmaecker, eds, Lecture Notes in Computer Science, Vol. 2740, Springer, Berlin, pp. 276–287.

Burke, E. K., Petrovic, S. and Qu, R., 2005, Case based heuristic selection for timetabling problems, *J. Scheduling,* to appear.

Carter, M., 2004, ftp://ftp.mie.utoronto.ca/pub/carter/testprob (on large exam timetabling problems).

Coffman, E. G., Garey, M. R. and Johnson, D. S., 1996, Approximation algorithms for bin packing: a survey, in: *Approximation algorithms for NP-hard problems,* D. Hochbaum, ed., PWS Publishing, Boston, MA, pp. 46–93.

Crowston, W. B., Glover, F., Thomson, G. L. and Trawick, J. D., 1963, Probabilistic learning combinations of local job shop scheduling rules, *Technical Report ONR Research Memorandum* 117, GSIA, Carnegie Mellon University, Pittsburgh, PA.

Djang, P. A. and Finch, P. R., 1998, Solving one dimensional bin packing problems, http://www.zianet.com/pdjang/binpack/paper.zip.

Fang, H-L., Ross, P. M. and Corne, D., 1994, A promising hybrid GA/heuristic approach for open-shop scheduling problems, in: *Proc. ECAI 94: 11th Eur. Conf. on Artificial Intelligence,* A. Cohn, ed., Wiley, New York, pp. 590–594.

Gambardella, L, 2004, http://www.idsia.ch/~luca/macs-vrptw/problems/welcome.htm (ISDIA vehicle routing data).

Gathercole, C. and Ross, P., 1997, Small populations over many generations can beat large populations over few generations in genetic programming, in: *Genetic Programming 1997: Proc. 2nd Annual Conf.,* J. R. Koza, K. Deb, M. Dorigo, D. B. Fogel, M. Garzon, H. Iba and R. L. Riolo, eds, Morgan Kaufmann, San Mateo, CA, pp. 111–118.

Gehring, H. and Homberger, J., 2004, VRPTW problems at University of Hagen, http://www.fernuni-hagen.de/WINF/touren/menuefrm/probinst.htm

Glover, F. and Kochenberger, G., eds, 2003, *Handbook of Meta-Heuristics,* Kluwer, Dordrecht.

Goldberg, D. E., Deb, K., Kargupta, H. and Harik, G., 1989, Messy genetic algorithms: motivation, analysis and first results, *Complex Syst.* **3**:493–530.

Gratch, J., Chein, S. and de Jong, G., 1993, Learning search control knowledge for deep space network scheduling, in: *Proc. 10th Int. Conf. on Machine Learning,* pp. 135–142.

Hart, E., and Ross, P. M., 1998, A heuristic combination method for solving job-shop scheduling problems, in: *Parallel Problem Solving from Nature V,* A. E. Eiben, T. Back, M. Schoenauer, and H-P. Schwefel, eds, Lecture Notes in Computer Science, Vol. 1498, Springer, Berlin, pp. 845–854.

Hewlett-Packard, 2004, Hewlett-Packard Information Dynamics Lab constraint satisfaction page, http://www.hpl.hp.com/shl/projects/constraints/

Johnson, D. S., 1973, Near-optimal bin-packing algorithms, *Ph.D. Thesis,* MIT Department of Mathematics, Cambridge, MA.

Kendall, G., Cowling, P. and Soubeiga, E., 2002, Choice function and random hyperheuristics, in: *Proc. 4th Asia-Pacific Conf. on Simulated Evolution And Learning SEAL 2002,* pp. 667–671.

Kruskal, J. B., 1956, On the shortest spanning tree of a graph and the travelling salesman problem, *Proc. AMS* **7**:48–50.

KZZI, 2004, Frequency assignment problems. http: //fap.zib.de/

Martello, S. and Toth, P., 1990, *Knapsack Problems. Algorithms and Computer Implementations,* Wiley, New York.

OR, 2004a, OR-library, http://mscmga.ms.ic.ac.uk/jeb/orlib/.

OR, 2004b, http://mscmga.ms.ic.ac.uk/jeb/orlib/tableinfo.html

Petrovic, S., and Qu, R., 2002, Case-based reasoning as a heuristic selector in a hyper-heuristic for course timetabling problems, in: *Proc. 6th Int. Conf. on*

Knowledge-Based Intelligent Information Engineering Systems and Applied Technologies (KES'02), Vol. 82, pp. 336-40.

Ross, P., Marín-Blázquez, J. G., Schulenburg, S. and Hart, E., 2003, Learning a procedure that can solve hard bin-packing problems: A new ga-based approach to hyper-heuristics, in: *Proc. Genetic and Evolutionary Computation Conf.—GECCO 2003,* E. Cantú-Paz et al., ed., Lecture Notes in Computer Science, Vol. 2724, Springer, Berlin, pp. 1295–1306.

Ross, P. M., Márin-Blázquez, J. and Hart, E., 2004, Hyper-heuristics applied to class and exam timetabling problems, in: *Proc. Congress on Evolutionary Computation 2004,* IEEE, Piscataway, NJ.

Ross, P., Schulenburg, S., Marín-Blázquez, J. G., and Hart, E., 2002, Hyper-heuristics: learning to combine simple heuristics in bin packing problems, in: *Proc. Genetic and Evolutionary Computation Conf.—GECCO 2002,* W. B. Langdon et al., eds, pp. 942–948.

Solomon, M. M., 1987, Algorithms for the vehicle routing and scheduling problems with time-window constraints, *Oper. Res.* **35**:254–265.

Terashima-Marín, H., Ross, P. M., and Valenzuela-Rendón, M., 1999, Evolution of constraint satisfaction strategies in examination timetabling, in: *Proc. Genetic and Evolutionary Computation Conf.—GECCO 1999,* W. Banzhaf et al., ed., Morgan Kaufmann, San Mateo, CA, pp. 635–642.

University of Melbourne, http://www.or.ms.unimelb.edu.au/timetabling (timetabling data).

Vázquez, M., and Whitley, D., 2000, A comparison of genetic algorithms for the dynamic job-shop scheduling problem, in: *Proc. Genetic and Evolutionary Computation Conf.—GECCO 2000,* D. Whitley et al., ed., Morgan Kaufmann, San Mateo, CA, pp. 1011–1018.

Voudouris, C., 1997, Guided local search or combinatorial optimisation problems, *Ph.D. Thesis,* Department of Computer Science, University of Essex.

Wilson, S. W., 1995, Classifier systems based on accuracy, *Evol. Comput.* **3**:149–175.

Wolpert, D., and MacReady, W. G., 1995, No free lunch theorems for search, *Technical Report* SFI-TR-92-02-010, Santa Fe Institute, NM.

Chapter 18

APPROXIMATION ALGORITHMS

Carla P. Gomes
Department of Computer Science
Cornell University
Ithaca, NY, USA

Ryan Williams
Computer Science Department
Carnegie Mellon University
Pittsburgh, PA, USA

18.1   INTRODUCTION

Most interesting real-world optimization problems are very challenging from a computational point of view. In fact, quite often, finding an optimal or even a near-optimal solution to a large-scale optimization problem may require computational resources far beyond what is practically available. There is a substantial body of literature exploring the computational properties of optimization problems by considering how the computational demands of a solution method grow with the size of the problem instance to be solved (see e.g. Chapter 11 or Aho et al., 1979). A key distinction is made between problems that require computational resources that grow polynomially with problem size versus those for which the required resources grow exponentially. The former category of problems are called efficiently solvable, whereas problems in the latter category are deemed *intractable* because the exponential growth in required computational resources renders all but the smallest instances of such problems unsolvable.

It has been determined that a large class of common optimization problems are classified as *NP-hard*. See Chapter 11 for more details. It is widely believed—though not yet proven (Clay Mathematics Institute, 2003)—that NP-hard problems are intractable, which means that there does not exist an efficient algorithm (i.e. one that scales polynomially) that is guaranteed to find an optimal solution for such problems. Examples of NP-hard optimization tasks are the minimum traveling salesman problem, the minimum graph coloring problem, and the minimum bin-packing problem. As a result of the nature of NP-hard problems, progress that leads to a better understanding of the structure, computational properties, and ways of solving one of them, *exactly* or *approximately*, also leads to better algorithms for solving hundreds of other different but related NP-hard problems. Several thousand computational problems, in areas as diverse as economics, biology, operations research, computer-aided design and finance, have been shown to be NP-hard. (See Aho et al., 1979, for further description and discussion of these problems.)

A natural question to ask is whether *approximate* (i.e. near-optimal) solutions can possibly be found efficiently for such hard optimization problems. Heuristic local search methods, such as tabu search and simulated annealing (see Chapters 6 and 7), are often quite effective at finding near-optimal solutions. However, these methods do not come with rigorous guarantees concerning the quality of the final solution or the required maximum runtime. In this chapter, we will discuss a more theoretical approach to this issue consisting of so-called "approximation algorithms", which are efficient algorithms that can be proven to produce solutions of a certain quality. We will also discuss classes of problems for which no such efficient approximation algorithms exist, thus leaving an important role for the quite general, heuristic local search methods.

The design of good approximation algorithms is a very active area of research where one continues to find new methods and techniques. It is quite likely that these techniques will become of increasing importance in tackling large real-world optimization problems.

In the late 1960s and early 1970s a precise notion of approximation was proposed in the context of multiprocessor scheduling and bin packing (Graham, 1966; Garey et al., 1972; Johnson, 1974). Approximation algorithms generally have two properties. First, they provide a feasible solution to a problem instance in polynomial time. In most cases, it is not difficult to devise a procedure that finds *some* feasible solution. However, we are interested in having some assured quality of the solution, which is the second aspect characterizing approximation algorithms. The quality of an approximation algorithm is the maximum "distance" between its solutions and the optimal solutions, evaluated over all the possible instances of the problem. Informally, an algorithm approximately solves an optimization problem if it always returns a feasible solution whose measure is close to optimal, for example within a factor bounded by a constant or by a slowly growing function of the input size. Given a constant α, an algorithm $\mathcal{A}$ is an α-approximation algorithm for a given minimization problem Π if its solution is at most α times the optimum, considering all the possible instances of problem Π.

The focus of this chapter is on the design of approximation algorithms for NP-hard optimization problems. We will show how standard algorithm design techniques such as greedy and local search methods have been used to devise good approximation algorithms. We will also show how randomization is a powerful tool for designing approximation algorithms. Randomized algorithms are interesting because, in general, such approaches are easier to analyze and implement, and faster than deterministic algorithms (Motwani and Raghavan, 1995). A randomized algorithm is simply an algorithm that performs some of its choices randomly; it "flips a coin" to decide what to do at some stages. As a consequence of its random component, different executions of a randomized algorithm may result in different solutions and runtime, even when considering the same instance of a problem. We will show how one can combine randomization with approximation techniques in order to efficiently approximate NP-hard optimization problems. In this case, the approximation solution, the approximation ratio, and the runtime of the approximation algorithm may be random variables. Confronted with an optimization problem, the goal is to produce a randomized approximation algorithm with runtime provably bounded by a polynomial and whose feasible solution is close to the optimal solution, *in expectation*. Note that these guarantees hold for every instance of the problem being solved. The only randomness in the performance guarantee of the randomized approximation algorithm comes from the algorithm itself, and not from the instances.

Since we do not know of efficient algorithms to find optimal solutions for NP-hard problems, a central question is whether we can efficiently compute good approximations that are close to optimal. It would be very interesting (and practical) if one could go from exponential to polynomial time complexity by relaxing the constraint on optimality, especially if we guarantee at most a relatively small error.

Good approximation algorithms have been proposed for some key problems in combinatorial optimization. The so-called APX complexity class includes the problems that allow a polynomial-time approximation algorithm with a performance ratio bounded by a constant. For some problems, we can design even better approximation algorithms. More precisely we can consider a family of approximation algorithms that allows us to get as close to the optimum as we like, as long as we are willing to trade quality with time. This special family of algorithms is called an *approximation scheme* (AS) and the so-called PTAS class is the class of optimization problems that allow for a *polynomial time approximation scheme* that scales polynomially in the size of the input. In some cases we can devise approximation schemes that scale polynomially, both in the size of the input and in the magnitude of the approximation error. We refer to the class of problems that allow such *fully polynomial time approximation schemes* as FPTAS.

Nevertheless, for some NP-hard problems, the approximations that have been obtained so far are quite poor, and in some cases no one has ever been able to devise approximation algorithms within a constant factor of the optimum. Initially it was not clear if these weak results were due to our lack of ability in devising good approximation algorithms for such problems or to some inherent structural property of the problems that excludes them from having good approximations. We will see that indeed there are limitations to approximation which are *intrinsic* to some classes of problems. For example, in some cases there is a lower bound on the constant factor of the approximation, and in other cases, we can provably show that there are no approximations within *any* constant factor from the optimum. Essentially, there is a wide range of scenarios going from NP-hard optimization problems that allow approximations to *any* required degree, to problems not allowing approximations at all. We will provide a brief introduction to proof techniques used to derive non-approximability results.

We believe that the best way to understand the ideas behind approximation and randomization is to study instances of algorithms with these properties, through examples. Thus in each section, we will first introduce the intuitive concept, then reinforce its salient points through well-chosen examples of prototypical problems. Our goal is far from trying to provide a comprehensive survey of approximation algorithms or even the best approximation algorithms for the problems introduced. Instead, we describe different design and evaluation techniques for approximation and randomized algorithms, using clear examples that allow for relatively simple and intuitive explanations. For some problems discussed in the chapter there are approximations with better performance guarantees but requiring more sophisticated proof techniques that are beyond the scope of this introductory tutorial. In such cases we will point the reader to the relevant literature results. In summary, our goals for this chapter are as follows:

1 Present the fundamental ideas and concepts underlying the notion of approximation algorithms.

2 Provide clear examples that illustrate different techniques for the design and evaluation of efficient approximation algorithms. The examples include accessible proofs of the approximation bounds.

3 Introduce the reader to the classification of optimization problems according to their polynomial-time approximability, including basic ideas on polynomial-time inapproximability.

4 Show the power of randomization for the design of approximation algorithms that are in general faster and easier to analyze and implement than the deterministic counterparts.

5 Show how we can use a randomized approximation algorithm as a heuristic to guide a complete search method (empirical results).

6 Present promising application areas for approximation and randomized algorithms.

7 Provide additional sources of information on approximation and randomization methods.

In Section 18.2 we introduce precise notions and concepts used in approximation algorithms. In this section we describe key design techniques for approximation algorithms. We use clear prototypical examples to illustrate the main techniques and concepts, such as the minimum vertex cover, the knapsack problem, the maximum satisfiability problem, the traveling salesman problem, and the maximum cut problem. As mentioned earlier, we are not interested in providing the best approximation algorithms for these problems, but rather in illustrating how standard algorithm techniques can be used effectively to design and evaluate approximation algorithms. In Section 18.3 we provide a tour of the main approximation classes, including a brief introduction to techniques to proof lower bounds on approximability. In Section 18.4 we describe some promising areas of application of approximation algorithms. Section 18.6 summarizes the chapter and provides additional sources of information on approximation and randomization methods.

18.2 APPROXIMATION STRATEGIES

18.2.1 Preliminaries

Optimization Problems We will define optimization problems in a traditional way (Aho et al., 1979; Ausiello et al., 1999). Each optimization problem has three defining features: the structure of the input *instance*, the criterion of a feasible *solution* to the problem, and the *measure* function used to determine which feasible solutions are considered to be optimal. It will be evident from the problem name whether we desire a feasible solution with a minimum or maximum measure. To illustrate, the minimum vertex cover problem may be defined in the following way.

Minimum Vertex Cover

Instance: An undirected graph $G = (V, E)$.
Solution: A subset $S \subseteq V$ such that for every $\{u, v\} \in E$, either $u \in S$ or $v \in S$.
Measure: $|S|$.

We use the following notation for items related to an instance I.

- $Sol(I)$ is the set of feasible solutions to I,

- $m_I : Sol(I) \rightarrow \mathbb{R}$ is the measure function associated with I, and

- $Opt(I) \subseteq Sol(I)$ is the feasible solutions with optimal measure (be it minimum or maximum).

Hence, we may completely specify an optimization problem Π by giving a set of tuples $\{(I, Sol(I), m_I, Opt(I))\}$ over all possible instances I. It is important to keep in mind that $Sol(I)$ and I may be over completely different domains. In the above example, the set I is all undirected graphs, while $Sol(I)$ is all possible subsets of vertices in a graph.

Approximation and Performance   Roughly speaking, an algorithm approximately solves an optimization problem if it always returns a feasible solution whose measure is close to optimal. This intuition is made precise below.

Let Π be an optimization problem. We say that an algorithm A *feasibly solves* Π if given an instance $I \in \Pi$, $A(I) \in Sol(I)$; that is, A returns a feasible solution to I.

Let A feasibly solve Π. Then we define the *approximation ratio* $\alpha(A)$ of A to be the minimum possible ratio between the measure of $A(I)$ and the measure of an optimal solution. Formally,

$$\alpha(A) = \min_{I \in \Pi} \frac{m_I(A(I))}{m_I(Opt(I))}$$

For minimization problems, this ratio is always at least 1. Respectively, for maximization problems, it is always at most 1.

Complexity Background   We define a decision problem as an optimization problem in which the measure is 0–1 valued. That is, solving an instance I of a decision problem corresponds to answering a *yes/no* question about I (where *yes* corresponds to a measure of 1, and *no* corresponds to a measure of 0). We may therefore represent a decision problem as a subset S of the set of all possible instances: members of S represent instances with measure 1.

Informally, P (polynomial time) is defined as the class of decision problems Π for which there exists a corresponding algorithm A_Π such that every instance $I \in \Pi$ is solved by A_Π within a polynomial ($|I|^k$ for some constant k) number of steps on any "reasonable" model of computation. Reasonable models include single-tape and multi-tape Turing machines, random access machines, pointer machines, etc.

While P is meant to represent a class of problems that can be efficiently solved, NP (nondeterministic polynomial time) is a class of decision problems

Π that can be efficiently *checked*. More formally, NP is the class of decision problems Π for which there exists a corresponding decision problem Π' in P and constant k satisfying

$$I \in \Pi \text{ if and only if there exists } C \in \{0, 1\}^{|I|^k} \text{ such that } (I, C) \in \Pi'$$

In other words, one can determine if an instance I is in an NP problem efficiently if one is also provided with a certain short string C, which is of length polynomial in I. For example, consider the NP problem of determining if a graph G has a path that travels through all nodes exactly once (this is known as the Hamiltonian path problem). Here, the instances I are graphs, and the proofs C are Hamiltonian paths. If one is given G along with a full description C of a path, it is easy to verify that C describes a Hamiltonian path by checking that

1 the path contains all nodes in G,

2 no node appears more than once in the path, and

3 any two adjacent nodes in the path have an edge between them in G.

However, no polynomial time algorithm is known for finding a Hamiltonian path when one is only given the graph G, and this is the fundamental difference between P and NP. In fact, the Hamiltonian path problem is not only in NP but is also *NP-hard*, see Section 18.1 and Chapter 11.

For $\Pi \in NP$, notice that while a short proof always exists if $I \in \Pi$, it need not be the case that short proofs exist for instances not in Π. Thus, while P problems are considered to be those which are *efficiently decidable*, NP problems are those considered to be *efficiently verifiable* via a short proof.

We will also consider the optimization counterparts to P and NP, which are PO and NPO, respectively. Informally, PO is the class of optimization problems where there exists a polynomial time algorithm that always returns an optimal solution to every instance of the problem, whereas NPO is the class of optimization problems where the measure function is polynomial time computable, and an algorithm can determine whether or not a possible solution is feasible in polynomial time. Our focus here will be on approximating solutions to the "hardest" of NPO problems, those problems where the corresponding decision problem is NP-hard. Interestingly, some NPO problems of this type can be approximated very well, whereas others can hardly be approximated at all.

18.2.2 The Greedy Method

Greedy approximation algorithms are designed with a simple philosophy in mind: repeatedly make choices that get one closer and closer to a feasible solution for the problem. These choices will be optimal according to an imperfect but easily computable heuristic. In particular, this heuristic tries to be as opportunistic as possible in the short run. (This is why such algorithms are called greedy—a better name might be "short-sighted"). For example, suppose my goal is to find the shortest walking path from my house to the theater. If I believed that the walk via Forbes Avenue is about the same length as the walk via Fifth Avenue, then if I am closer to Forbes than Fifth, it would be reasonable to walk towards Forbes and take that route.

Clearly, the success of this strategy depends on the correctness of my belief that the Forbes path is indeed just as good as the Fifth path. We will show that for some problems, choosing a solution according to an opportunistic, imperfect heuristic achieves a non-trivial approximation algorithm.

Greedy Vertex Cover   The minimum vertex cover problem was defined in the preliminaries (Section 18.2.1). Variants on the problem come up in many areas of optimization research. We will describe a simple greedy algorithm that is a 2-approximation to the problem; that is, the cardinality of the vertex cover returned by our algorithm is no more than twice the cardinality of a minimum cover. The algorithm is as follows.

> Greedy-VC: Initially, let S be an empty set. Choose an arbitrary edge $\{u, v\}$. Add u and v to S, and remove u and v from the graph. Repeat until no edges remain in the graph. Return S as the vertex cover.

THEOREM 18.1 *Greedy-VC is a 2-approximation algorithm for Minimum Vertex Cover.*

Proof. First, we claim S as returned by Greedy-VC is indeed a vertex cover. Suppose not; then there exists an edge e which was not covered by any vertex in S. Since we only remove vertices from the graph that are in S, an edge e would remain in the graph after Greedy-VC had completed, which is a contradiction.

Let S^* be a minimum vertex cover. We will now show that S^* contains at least $|S|/2$ vertices. It will follow that $|S^*| \geq |S|/2$, hence our algorithm has a $|S|/|S^*| \leq 2$ approximation ratio.

Since the edges we chose in Greedy-VC do not share any endpoints, it follows that

- $|S|/2$ is the number of edges we chose and

- S^* must have chosen at least one vertex from each edge we chose.

It follows that $|S^*| \geq |S|/2$.   □

Sometimes when one proves that an algorithm has a certain approximation ratio, the analysis is somewhat "loose", and may not reflect the best possible

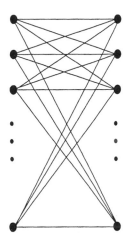

Figure 18.1. A *bipartite* graph is one for which its vertices can be assigned one of two colors (say, *red* or *blue*), in such a way that all edges have endpoints with different colors. Above is a sketch of a *complete* bipartite graph with n nodes colored red and n nodes colored blue. When running Greedy-VC on these instances (for any natural number n), the algorithm will select all $2n$ vertices. This shows the approximation ratio of 2 is tight for Greedy-VC.

ratio that can be derived. It turns out that Greedy-VC is no better than a 2-approximation. In particular, there is an infinite set of Vertex Cover instances where Greedy-VC provably chooses exactly twice the number of vertices necessary to cover the graph, namely in the case of complete bipartite graphs; see Figure 18.1. When such a situation occurs, where the approximation ratio derived for an algorithm is indeed the best possible in the worst case, we say that the bound derived is *tight*.

One final remark should be noted on Vertex Cover. While the above algorithm is indeed quite simple, no better approximation algorithms are known! In fact, it is widely believed that one cannot approximate minimum vertex cover better than $2 - \epsilon$ for any $\epsilon > 0$, unless $P = NP$, see Khot and Regev (2003).

Greedy MAX-SAT The MAX-SAT problem has been very well-studied; variants of it arise in many areas of discrete optimization. To introduce it requires a bit of terminology.

We will deal solely with Boolean variables (that is, those which are either true or false), which we will denote by x_1, x_2, etc. A *literal* is defined as either a variable or the negation of a variable (e.g. x_7, $\neg x_{11}$ are literals). A *clause* is defined as the OR of some literals (*e.g.* $(\neg x_1 \vee x_7 \vee \neg x_{11})$ is a clause). We say that a Boolean formula is in *conjunctive normal form* (CNF) if it is presented as an AND of clauses (*e.g.* $(\neg x_1 \vee x_7 \vee \neg x_{11}) \wedge (x_5 \vee \neg x_2 \vee \neg x_3)$ is in CNF).

Finally, the MAX-SAT problem is to find an assignment to the variables of a Boolean formula in CNF such that the maximum number of clauses are set to true, or are *satisfied*. Formally:

MAX-SAT

Instance: A Boolean formula F in CNF.
Solution: An assignment a, which is a function from each of the variables in F to $\{true, false\}$.
Measure: The number of clauses in F that are set to true (are satisfied) when the variables in F are assigned according to a.

What might be a natural greedy strategy for approximately solving MAX-SAT? One approach is to pick a variable that satisfies many clauses if it is set to a certain value. Intuitively, if a variable occurs negated in several clauses, setting the variable to *false* will satisfy several clauses; hence this strategy should approximately solve the problem well. Let $n(l_i, F)$ denote the number of clauses in F where the literal l_i appears.

> Greedy-MAXSAT: Pick a literal l_i with maximum $n(l_i, F)$ value. Set the corresponding variable of l_i such that all clauses containing l_i are satisfied, yielding a reduced F. Repeat until no variables remain in F.

It is easy to see that Greedy-MAXSAT runs in polynomial time (roughly quadratic time, depending on the computational model chosen for analysis). It is also a "good" approximation for the MAX-SAT problem.

THEOREM 18.2 *Greedy-MAXSAT is a $\frac{1}{2}$-approximation algorithm for MAX-SAT.*

Proof. Proof by induction on the number of variables n in the formula F. Let m be the total number of clauses in F. If $n = 1$, the result is obvious. For $n > 1$, let l_i have maximum $n(l_i, F)$ value, and v_i be its corresponding variable. Let m_{POS} and m_{NEG} be the number of clauses in F that contain l_i and $\neg l_i$, respectively. After v_i is set so that l_i is true (so both l_i and $\neg l_i$ disappear from F), there are at least $m - m_{POS} - m_{NEG}$ clauses left, on $n - 1$ variables.

By induction hypothesis, Greedy-MAXSAT satisfies at least $(m - m_{POS} - m_{NEG})/2$ of these clauses, therefore the total number of clauses satisfied is at least $(m - m_{POS} - m_{NEG})/2 + m_{POS} = m/2 + (m_{POS} - m_{NEG})/2 \geq m/2$, by our greedy choice of picking the l_i that occurred most often.   □

Greedy MAX-CUT   Our next example shows how local search (in particular, *hill climbing*) may be employed in designing approximation algorithms. Hill climbing is inherently a greedy strategy: when one has a feasible solution x, one tries to improve it by choosing some feasible y that is "close" to x, but has a better measure (lower or higher, depending on minimization or maximization). Repeated attempts at improvement often result in "locally" optimal solutions that have a good measure relative to a globally optimal solution (i.e. a member of $Opt(I)$). We illustrate local search by giving an approximation algorithm for the NP-complete MAX-CUT problem which finds a locally optimal satisfying assignment. It is important to note that not all local search strategies try to find a local optimum—for example, simulated annealing attempts to *escape* from local optima in the hopes of finding a global optimum (Kirkpatrick et al., 1983). See Chapter 7 for more details.

MAX-CUT

Instance: An undirected graph $G = (V, E)$.
Solution: A cut of the graph, i.e. a pair (S, T) such that $S \subseteq V$ and $T = V - S$.
Measure: The *cut size*, which is the number of edges crossing the cut, i.e. $|\{\{u, v\} \in E \mid u \in S, v \in T\}|$.

Our local search algorithm repeatedly improves the current feasible solution by changing one vertex's place in the cut, until no more improvement can be made. We will prove that at such a local maximum, the cut size is at least $m/2$.

> Local-Cut: Start with an arbitrary cut of V. For each vertex, determine if moving it to the other side of the partition increases the size of the cut. If so, move it. Repeat until no such movements are possible.

First, observe that this algorithm repeats at most m times, as each movement of a vertex increases the size of the cut by at least 1, and a cut can be at most m in size.

THEOREM 18.3 *Local-Cut is a $\frac{1}{2}$-approximation algorithm for MAX-CUT.*

Proof. Let (S, T) be the cut returned by the algorithm, and consider a vertex v. After the algorithm finishes, observe that the number of edges adjacent to v that cross (S, T) is more than the number of adjacent edges that do not cross, otherwise v would have been moved. Let $\deg(v)$ be the degree of v. Then our observation implies that at least $\deg(v)/2$ edges out of v cross the cut returned by the algorithm.

Let m^* be the total number of edges crossing the cut returned. Each edge has two endpoints, so the sum $\sum_{v \in V}(\deg(v)/2)$ counts each crossing edge at most twice, i.e.

$$\sum_{v \in V}(\deg(v)/2) \leq 2m^*$$

However, observe $\sum_{v \in V} \deg(v) = 2m$: when summing up all degrees of vertices, every edge gets counted exactly twice, once for each endpoint. We conclude that

$$m = \sum_{v \in V} (\deg(v)/2) \leq 2m^*$$

It follows that the approximation ratio of the algorithm is $\frac{m^*}{m} \geq \frac{1}{2}$. $\quad\square$

It turns out that MAX-CUT admits much better approximation ratios than $\frac{1}{2}$; a so-called *relaxation* of the problem to a semi-definite linear program yields a 0.8786 approximation (Goemans and Williamson, 1995). However, like many optimization problems, MAX-CUT cannot be approximated arbitrarily well $(1 - \epsilon$, for all $\epsilon > 0)$ unless P = NP. That is to say, it is unlikely that MAX-CUT is in the *PTAS* complexity class.

Greedy Knapsack   The knapsack problem and its special cases have been extensively studied in operations research. The premise behind it is classic: you have a knapsack of capacity C, and a set of items $1, \ldots, n$. Each item has a particular cost c_i of carrying it, along with a profit p_i that you will gain by carrying it. The problem is then to find a subset of items with cost at most C, having maximum profit.

Maximum Integer Knapsack

Instance: A capacity $C \in \mathbb{N}$, and a number of items $n \in \mathbb{N}$, with corresponding costs and profits c_i, $p_i \in \mathbb{N}$ for all $i = 1, \ldots, n$.
Solution: A subset $S \subseteq \{1, \ldots, n\}$ such that $\sum_{j \in S} c_j \leq C$.
Measure: The *total profit* $\sum_{j \in S} p_j$.

Maximum Integer Knapsack, as formulated above, is NP-hard. There is also a "fractional" version of this problem (we call it Maximum Fraction Knapsack), which can be solved in polynomial time. In this version, rather than having to pick the entire item, one is allowed to choose *fractions* of items, like 1/8 of the first item, 1/2 of the second item, and so on. The corresponding profit and cost incurred from the items will be similarly fractional (1/8 of the profit and cost of the first, 1/2 of the profit and cost of the second, and so on).

One greedy strategy for solving these two problems is to pack items with the largest profit-to-cost ratio first, with the hopes of getting many small-cost high-profit items in the knapsack. It turns out that this algorithm will not give any constant approximation guarantee, but a tiny variant on this approach will give a 2-approximation for Integer Knapsack, and an exact algorithm for Fraction Knapsack. The algorithms for Integer Knapsack and Fraction Knapsack are, respectively:

- Greedy-IKS: Choose items with the largest profit-to-cost ratio first, until the total cost of items chosen is greater than C. Let j be the last item chosen, and S be the set of items chosen before j. Return either $\{j\}$ or S, depending on which one is more profitable.

- Greedy-FKS: Choose items as in Greedy-IKS. When the item j makes the cost of the current solution greater than C, add the *fraction* of j such that the resulting cost of the solution is exactly C.

We omit a proof of the following. A full treatment can be found in Ausiello et al. (1999).

LEMMA 18.4 *Greedy-FKS solves Maximum Fraction Knapsack in polynomial time. That is, Greedy-FKS is a 1-approximation to Maximum Fraction Knapsack.*

We entitled the above as a lemma, because we will use it to analyze the approximation algorithm for Integer Knapsack.

THEOREM 18.5 *Greedy-KS is a $\frac{1}{2}$-approximation for Maximum Integer Knapsack.*

Proof. Fix an instance of the problem. Let $P = \sum_{i \in S} p_i$, the total profit of items in S, and j be the last item chosen (as specified in the algorithm). We will show that $P + p_j$ is greater than or equal to the profit of an optimal Integer Knapsack solution. It follows that one of S or $\{j\}$ has at least half the profit of the optimal solution.

Let S_I^* be an optimal Integer Knapsack solution to the given instance, with total profit P_I^*. Similarly, let S_F^* and P_F^* correspond to an optimal Fraction Knapsack solution. Observe that $P_F^* \leq P_I^*$.

By the analysis of the algorithm for Fraction Knapsack, $P_F^* = P + \epsilon p_j$, where $\epsilon \in (0, 1]$ is the fraction chosen for item j in the algorithm. Therefore

$$P + p_j \geq P + \epsilon p_j = P_F^* \geq P_I^*$$

and we are done. $\square$

In fact, this algorithm can be extended to get a *polynomial time approximation scheme* (PTAS) for Maximum Integer Knapsack, (see (Ausiello et al., 1999)). A PTAS has the property that, for any fixed $\epsilon > 0$ provided, it returns a $(1 + \epsilon)$-approximate solution. Furthermore, the runtime is polynomial in the input size, *provided that ϵ is constant*. This allows us to specify a runtime that has $1/\epsilon$ in the exponent. It is typical to view a PTAS as a *family* of successively better (but also slower) approximation algorithms, each running with a successively smaller $\epsilon > 0$. This is intuitively why they are called an approximation *scheme*, as it is meant to suggest that a variety of algorithms are used. A PTAS is quite powerful; such a scheme can approximately solve a problem with ratios arbitrarily close to 1. However, we will observe that many problems provably do not have a PTAS, unless $P = NP$.

18.2.3   Sequential Algorithms

Sequential algorithms are used for approximations on problems where a feasible solution is a partitioning of the instance into subsets. A sequential algorithm "sorts" the items of the instance in some manner, and selects partitions for the instance based on this ordering.

Sequential Bin Packing   We first consider the problem of Minimum Bin Packing, which is similar in nature to the knapsack problems.

Minimum Bin Packing

Instance: A set of items $S = \{r_1, \ldots, r_n\}$, where $r_i \in (0, 1]$ for all $i = 1, \ldots, n$.
Solution: Partition of S into bins $B_1, \ldots, B_M$ such that $\sum_{r_j \in B_i} r_j \leq 1$ for all $i = 1, \ldots, M$.
Measure: M.

An obvious algorithm for Minimum Bin Packing is an *on-line* strategy. Initially, let $j = 1$ and have a bin B_1 available. As one runs through the input (r_1, r_2, etc), try to pack the new item r_i into the last bin used, B_j. If r_i does not fit in B_j, create another bin B_{j+1} and put a_i in it. This algorithm is "on-line" as it processes the input in a fixed order, and thus adding new items to the instance while the algorithm is running does not change the outcome. Call this heuristic Last-Bin.

THEOREM 18.6 *Last-Bin is a 2-approximation to Minimum Bin Packing.*

Proof. Let R be the sum of all items, so $R = \sum_{r_i \in S} r_i$. Let m be the total number of bins used by the algorithm, and let m^* be the minimum number of bins possible for the given instance. Note that $m^* \geq R$, as the total number of bins needed is at least the total size of all items (each bin holds 1 unit). Now, given any pair of bins B_i and B_{i+1} returned by the algorithm, the sum of items from S in B_i and B_{i+1} is at least 1; otherwise, we would have stored the items of B_{i+1} in B_i instead. This shows that $m \leq 2R$. Hence $m \leq 2R \leq 2m^*$, and the algorithm is a 2-approximation.   □

An interesting exercise for the reader is to construct a series of examples demonstrating that this approximation bound, like the one for Greedy-VC, is tight.

As one might expect, there exist algorithms that give better approximations than the above. For example, we do not even consider the previous bins $B_1, \ldots, B_{j-1}$ when trying to pack an a_i, only the last one is considered.

Motivated by this observation, consider the following modification to Last-Bin. Select each item a_i in decreasing order of size, placing a_i in the *first available* bin out of $B_1, \ldots, B_j$. (So a new bin is only created if a_i cannot fit in any of the previous j bins.) Call this new algorithm First-Bin. An improved approximation bound may be derived, via an intricate analysis of cases.

THEOREM 18.7 *(Johnson, 1974) First-Bin is a $\frac{11}{9}$-approximation to Minimum Bin Packing.*

Sequential Job Scheduling One of the major problems in scheduling theory is how to assign jobs to multiple machines so that all of the jobs are completed efficiently. Here, we will consider job completion in the shortest amount of time possible. For the purposes of abstraction and simplicity, we will assume the machines are identical in processing power for each job.

Minimum Job Scheduling

Instance: An integer k and a multi-set $T = \{t_1, \ldots, t_n\}$ of *times*, $t_i \in \mathbb{Q}$ for all $i = 1, \ldots, n$ (*i.e.* the t_i are fractions).
Solution: An assignment of jobs to machines, i.e. a function a from $\{1, \ldots, n\}$ to $\{1, \ldots, k\}$.
Measure: The completion time for all machines, assuming they run in parallel: $\max \left\{ \sum_{i:a(i)=j} t_i \mid j \in \{1, \ldots, k\} \right\}$.

The algorithm we propose for Job Scheduling is also on-line: when reading a new job with time t_i, assign it to the machine j that currently has the least amount of work; that is, the j with minimum $\sum_{i:a(i)=j} t_i$. Call this algorithm Sequential-Jobs.

THEOREM 18.8 *Sequential Jobs is a 2-approximation for Minimum Job Scheduling.*

Proof. Let j be a machine with maximum completion time, and let i be the index of the last job assigned to j by the algorithm. Let $s_{i,j}$ be the sum of all times for jobs prior to i that are assigned to j. (This may be thought of as the time that job i begins on machine j.) The algorithm assigned i to the machine with the least amount of work, hence all other machines j' at this point have larger $\sum_{i:a(i)=j'} t_i$. Therefore $s_{i,j} \leq \frac{1}{k} \sum_{i=1}^{n} t_i$, i.e. $s_{i,j}$ is less $1/k$ of the total time of all jobs (recall k is the number of machines).

Notice $B = \frac{1}{k} \sum_{i=1}^{n} t_i \leq m^*$, the completion time for an optimal solution, as the sum corresponds to the case where every machine takes exactly the same fraction of time to complete. Thus the completion time for machine j is

$$s_{i,j} + t_i \leq m^* + m^* = 2m^*$$

So the maximum completion time is at most twice that of an optimal solution.

□

This is not the best one can do: Minimum Job Scheduling also has a PTAS (see Vazirani, 1983).

18.2.4   Randomization

Randomness is a powerful resource for algorithmic design. Upon the assumption that one has access to unbiased coins that may be flipped and their values (heads or tails) extracted, a wide array of new mathematics may be employed to aid the analysis of an algorithm. It is often the case that a simple randomized algorithm will have the same performance guarantees as a complicated deterministic (i.e. non-randomized) procedure.

One of the most intriguing discoveries in the area of algorithm design is that the addition of randomness into a computational process can sometimes lead to a significant speedup over purely deterministic methods. This may be intuitively explained by the following set of observations. A randomized algorithm can be viewed as a probability distribution on a set of deterministic algorithms. The behavior of a randomized algorithm can vary on a given input, depending on the random choices made by the algorithm; hence when we consider a randomized algorithm, we are implicitly considering a randomly chosen algorithm from a family of algorithms. If a substantial fraction of these deterministic algorithms perform well on the given input, then a strategy of restarting the randomized algorithm after a certain point in runtime will lead to a speed-up (Gomes et al., 1998).

Some randomized algorithms are able to efficiently solve problems for which no efficient deterministic algorithm is known, such as polynomial identity testing (see Motwani and Raghavan, 1995). Randomization is also a key component in the popular simulated annealing method for solving optimization problems (Kirkpatrick et al., 1983). For a long time, the problem of determining if a given number is prime (a fundamental problem in modern cryptography) was only efficiently solvable using randomization (Goldwasser and Kilian, 1986; Rabin, 1980; Solovay and Strassen, 1977). Very recently, a deterministic algorithm for primality was discovered (Agrawal et al., 2002).

Random MAX-CUT Solution   We saw earlier a greedy strategy for MAX-CUT that yields a 2-approximation. Using randomization, we can give an extremely short approximation algorithm that has the same performance in approximation, and runs in expected polynomial time.

> Random-Cut: Choose a random cut (i.e. a random partition of the vertices into two sets). If there are less than $m/2$ edges crossing this cut, repeat.

THEOREM 18.9 *Random-Cut is a $\frac{1}{2}$-approximation algorithm for MAX-CUT that runs in expected polynomial time.*

Proof. Let X be a random variable denoting the number of edges crossing a cut. For $i = 1, \ldots, m$, X_i will be an indicator variable that is 1 if the ith edge crosses the cut, and 0 otherwise. Then $X = \sum_{i=1}^{m} X_i$, so by linearity of expectation, $E[X] = \sum_{i=1}^{m} E[X_i]$.

Now for any edge $\{u, v\}$, the probability it crosses a randomly chosen cut is $1/2$. (Why? We randomly placed u and v in one of two possible partitions, so u is in the same partition as v with probability $1/2$.) Thus, $E[X_i] = 1/2$ for all i, so $E[X] = m/2$.

This only shows that by choosing a random cut, we expect to get at least $m/2$ edges crossing. We want a randomized algorithm that *always* returns a good cut, and its running time is a random variable whose expectation is polynomial. Let us compute the probability that $X \geq m/2$ when a random cut is chosen. In the worst case, when $X \geq m/2$ all of the probability is weighted on m, and when $X < m/2$ all of the probability is weighted on $m/2 - 1$. This makes the expectation of X as high as possible, while making the likelihood of obtaining an at-least-$m/2$ cut small. Formally,

$$m/2 = E[X] \leq (1 - \Pr[X \geq m/2])(m/2 - 1) + \Pr[X \geq m/2]m$$

Solving for $\Pr[X \geq m/2]$, it is at least $2/(m + 2)$. It follows that the expected number of iterations in the above algorithm is at most $(m + 2)/2$; therefore the algorithm runs in expected polynomial time, and always returns a cut of size at least $m/2$. □

We remark that, had we simply specified our approximation as "pick a random cut and stop", we would say that the algorithm runs in linear time, and has an *expected* approximation ratio of $1/2$.

Random MAX-SAT Solution   Previously, we studied a greedy approach for MAX-SAT that was guaranteed to satisfy half of the clauses. Here we will consider MAX-Ak-SAT, the restriction of MAX-SAT to CNF formulae with *at least k* literals per clause. Our algorithm is analogous to the one for MAX-CUT: *Pick a random assignment to the variables.* It is easy to show, using a similar analysis to the above, that the expected approximation ratio of this procedure is at least $1 - \frac{1}{2^k}$. More precisely, if m is the total number of clauses in a formula, the expected number of clauses satisfied by a random assignment is $m - m/2^k$.

Let c be an *arbitrary* clause of at least k literals. The probability that each of its literals were set to a value that makes them false is at most $1/2^k$, since there is a probability of $1/2$ for each literal and there are at least k of them. Therefore the probability that c is satisfied is at least $1 - 1/2^k$. Using a linearity of expectation argument (as in the MAX-CUT analysis) we infer that at least $m - m/2^k$ clauses are expected to be satisfied.

18.3   A TOUR OF APPROXIMATION CLASSES

We will now take a step back from our algorithmic discussions, and briefly describe a few of the common complexity classes associated with NP optimization problems.

18.3.1   PTAS and FPTAS

Definition   PTAS and FPTAS are classes of optimization problems that some believe are closer to the proper definition of what is efficiently solvable, rather than merely P. This is because problems in these two classes may be approximated with constant ratios *arbitrarily close* to 1. However, with *PTAS*, as the approximation ratio gets closer to 1, the runtime of the corresponding approximation algorithm may grow exponentially with the ratio.

More formally, PTAS is the class of NPO problems Π that have an *approximation scheme*. That is, given $\epsilon > 0$, there exists a polynomial time algorithm A_ϵ such that

- If Π is a maximization problem, A_ϵ is a $(1 + \epsilon)$ approximation, i.e. the ratio approaches 1 from the right.

- If Π is a minimization problem, it is a $(1 - \epsilon)$ approximation (the ratio approaches 1 from the left).

As we mentioned, one drawback of a PTAS is that the $(1 + \epsilon)$ algorithm could be exponential in $1/\epsilon$. The class FPTAS is essentially PTAS but with the additional requirement that the runtime of the approximation algorithm is polynomial in n and $1/\epsilon$.

A Few Known Results   It is known that some NP-hard optimization problems cannot be approximated arbitrarily well unless $P = NP$. One example is a problem we looked at earlier, Minimum Bin Packing. This is a rare case in which there is a simple proof that the problem is not approximable unless $P = NP$.

THEOREM 18.10 *(Aho et al., 1979) Minimum Bin Packing is not in PTAS, unless $P = NP$. In fact, there is no $3/2 - \epsilon$ approximation for any $\epsilon > 0$, unless $P = NP$.*

To prove the result, we use a reduction from the Set Partition decision problem. Set Partitioning asks if a given set of natural numbers can be split into two sets that have equal sum.

Set Partition

Instance: A multi-set $S = \{r_1, \ldots, r_n\}$, where $r_i \in \mathbb{N}$ for all $i = 1, \ldots, n$.

Solution: A partition of S into sets S_1 and S_2; *i.e.* $S_1 \cup S_2 = S$ and $S_1 \cap S_2 = \varnothing$.

Measure: $m(S) = 1$ if $\sum_{r_i \in S_1} r_i = \sum_{r_j \in S_2} r_j$, and $m(S) = 0$ otherwise.

Proof. Let $S = \{r_1, \ldots, r_n\}$ be a Set Partition instance. Reduce it to Minimum Bin Packing by setting $C = \frac{1}{2} \sum_{j=1}^{n} s_j$ (half the total sum of elements in S), and considering a bin packing instance of *items* $S' = \{r_1/C, \ldots, r_n/C\}$.

If S can be partitioned into two sets of equal sum, then the minimum number of bins necessary for the corresponding S' is 2. On the other hand, if S cannot be partitioned in this way, the minimum number of bins needed for S' is at least 3, as every possible partitioning results in a set with sum greater than C. Therefore, if there were a polytime $(3/2 - \epsilon)$-approximation algorithm A, it could be used to solve Set Partition:

- If A (given S and C) returns a solution using at most $(3/2-\epsilon)2 = 3-2\epsilon$ bins, then there exists a Set Partition for S.

- If A returns a solution using at least $(3/2 - \epsilon)3 = 9/2 - 3\epsilon = 4.5 - 3\epsilon$ bins, then there is no Set Partition for S.

But for any $\epsilon \in (0, 3/2)$,

$$3 - 2\epsilon < 4.5 - 3\epsilon$$

Therefore this polynomial time algorithm distinguishes between those S that can be partitioned and those that cannot, so $P = NP$. $\qquad\square$

A similar result holds for problems such as MAX-CUT, MAX-SAT, and Minimum Vertex Cover. However, unlike the result for Bin Packing, the proofs for these appear to require the introduction of *probabilistically checkable proofs*, which will be discussed later.

18.3.2   APX

APX is a (presumably) larger class than PTAS; the approximation guarantees for problems in it are strictly weaker. An NP optimization problem Π is in APX simply if there is a polynomial time algorithm A and constant c such that A is a c-approximation to Π.

A Few Known Results   It is easy to see that $PTAS \subseteq APX \subseteq NPO$. When one sees new complexity classes and their inclusions, one of the first questions to be asked is: how likely is it that these inclusions could be made into equalities? Unfortunately, it is highly unlikely. The following relationship can be shown between the three approximation classes we have seen.

THEOREM 18.11 *(Ausiello et al., 1999)* $PTAS = APX \iff APX = NPO \iff P = NP$.

Therefore, if all NP optimization problems could be approximated within a constant factor, then $P = NP$. Further, if all problems that have constant approximations can be arbitrarily approximated, still $P = NP$. Another way of saying this is: if NP problems are hard to solve, then some of them are hard to approximate as well. Moreover, there exists a "hierarchy" of successively harder-to-approximate problems.

One of the directions stated follows from a theorem of the previous section: earlier, we saw a constant factor approximation to Minimum Bin Packing. However, it does not have a PTAS unless $P = NP$. This shows the direction $PTAS = APX \Rightarrow P = NP$. One example of a problem that cannot be in APX unless $P = NP$ is the well-known Minimum Traveling Salesman problem.

Minimum Traveling Salesman

Instance: A set $C = \{c_1, \ldots, c_n\}$ of *cities*, and a distance function $d : C \times C \to \mathbb{N}$.
Solution: A path through the cities, i.e. a permutation $\pi : \{1, \ldots, n\} \to \{1, \ldots, n\}$.
Measure: The cost of visiting cities with respect to the path, i.e.

$$\sum_{i=1}^{n-1} d(c_{\pi(i)}, c_{\pi(i+1)})$$

It is important to note that when the distances in the problem instances always obey a Euclidean metric, Minimum Traveling Salesperson has a PTAS (Arora, 1998). Thus, we may say that it is the generality of possible distances in the above problem that makes it difficult to approximate. This is often the case with approximability: a small restriction on an inapproximable problem can suddenly turn it into a highly approximable one.

18.3.3   Brief Introduction to PCPs

In the 1990s, the work in probabilistically checkable proofs (PCPs) was *the* major breakthrough in proving hardness results, and arguably in theoretical computer science as a whole. In essence, PCPs only look at a few bits of a proposed proof, using randomness, but manage to capture all of NP. Because the number of bits they check is so small (a constant), when an efficient PCP exists for a given problem, it implies the hardness of *approximately solving* the same problem as well, within some constant factor.

The notion of a PCP arose from a series of meditations on proof-checking using randomness. We know NP represents the class of problems that have "short proofs" we can verify efficiently. As far as NP is concerned, all of the verification done is deterministic. When a proof is correct or incorrect, a polynomial time verifier answers "yes" or "no" with 100% confidence.

However, what happens when we relax the notion of total correctness to include probability? Suppose we permit the proof verifier to toss unbiased coins, and have *one-sided error*. That is, now a randomized verifier only accepts a correct proof with probability at least $1/2$, but still rejects any incorrect proof it reads. (We call such a verifier a *probabilistically checkable proof system*, i.e. a PCP.) This slight tweaking of what it means to verify a proof leads to an amazing characterization of NP: all NP decision problems can be verified by a PCP of the above type, which only flips $O(\log n)$ coins and only checks a *constant* ($O(1)$) number of bits of any given proof! The result involves the construction of highly intricate error-correcting codes. We shall not discuss it on a formal level here, but will cite the above in the notation of a theorem.

THEOREM 18.12 *(Arora et al., 1998)*

$$PCP[O(\log n), O(1)] = NP$$

One corollary of this theorem is that a large class of approximation problems do not admit a PTAS. In particular, we have the following theorem.

THEOREM 18.13 *For* $\Pi \in \{MAX\text{-}Ek\text{-}SAT, MAX\text{-}CUT, Minimum\ Vertex\ Cover\}$ *there exists a c such that* Π *cannot be c-approximated in polynomial time, unless* $P = NP$.

18.4 PROMISING APPLICATION AREAS FOR APPROXIMATION AND RANDOMIZED ALGORITHMS

18.4.1 Randomized Backtracking and Backdoors

Backtracking is one of the oldest and most natural methods used for solving combinatorial problems. In general, backtracking deterministically can take exponential time. Recent work has demonstrated that many real-world problems can be solved quite rapidly, when the choices made in backtracking are randomized. In particular, problems in practice tend to have small substructures within them. These substructures have the property that once they are solved properly, the entire problem may be solved. The existence of these so-called "backdoors" (Williams et al., 2003) to problems make them very tenable to solution using randomization. Roughly speaking, search heuristics will set the backdoor substructure early in the search, with a significant probability.

Therefore, by repeatedly restarting the backtracking mechanism after a certain (polynomial) length of time, the overall runtime that backtracking requires to find a solution is decreased tremendously.

18.4.2   Approximations to Guide Complete Backtrack Search

A promising approach for solving combinatorial problems using complete (exact) methods draws on recent results on some of the best approximation algorithms based on linear programming (LP) relaxations (see Chvatal, 1983; Dantzig, 1998) and so-called randomized rounding techniques, as well as on results that uncovered the extreme variance or "unpredictability" in the runtime of complete search procedures, often explained by so-called heavy-tailed cost distributions (Gomes et al., 2000). Gomes and Shmoys (2002) propose a *complete* randomized backtrack search method that tightly couples constraint satisfaction problem (CSP) propagation techniques with randomized LP-based approximations. They use as a benchmark domain a purely combinatorial problem, the quasigroup (or Latin square) completion problem (QCP). Each instance consists of an n by n matrix with n^2 cells. A complete quasigroup consists of a coloring of each cell with one of n colors in such a way that there is no repeated color in any row or column. Given a partial coloring of the n by n cells, determining whether there is a valid completion into a full quasigroup is an NP-complete problem (Colbourn, 1984). The underlying structure of this benchmark is similar to that found in a series of real-world applications, such as timetabling, experimental design, and fiber optics routing problems (Laywine and Mullen, 1998; Kumar et al., 1999).

Gomes and Shmoys compare their results for the hybrid techniques CSP/LP strategy guided by the LP randomized rounding approximation with a CSP strategy and with a LP strategy. The results show that the hybrid techniques approach significantly improves over the pure strategies on hard instances. This suggest that the LP randomized rounding approximation provides powerful heuristic guidance to the CSP search.

18.4.3   Average Case Complexity and Approximation

While "worst case" complexity has a very rich theory, it often feels too restrictive to be relevant to practice. Perhaps NP-hard problems are hard only for some esoteric sets of instances that will hardly ever arise. To this end, researchers have proposed theories of "average case" complexity, which attempt to probabilistically analyze problems based on randomly chosen instances over distributions; for an introduction to this line of work, see Gurevich (1991). Recently, an intriguing thread of theoretical research has explored the connections between the average-case complexity of problems and their approximation hardness (Feige, 2002). For example, it is shown that if *random 3-SAT* is hard to solve in polynomial time (under reasonable definitions of "random" and "hard"), then NP-hard optimization problems such as Minimum Bisection are hard to approximate in the worst case. Conversely, this implies that improved approximation algorithms for some problems could lead to the average-case tractability of others. A natural research question to ask is: does an PTAS imply average-case tractability, or vice versa? We suspect that some statement of this form might be the case. In our defense, a recent paper (Beier and Vocking, 2003) shows that *Random* Maximum Integer Knapsack is exactly solvable in expected polynomial time! (Recall that there exists an PTAS for Maximum Integer Knapsack.)

18.5 TRICKS OF THE TRADE

One major initial motivation for the study of approximation algorithms was to provide a new theoretical avenue for analyzing and coping with hard problems. Faced with a brand-new interesting optimization problem, how might one apply the techniques discussed here? One possible scheme proceeds as follows:

1. First, try to prove your problem is NP-hard, or find evidence that it is not! Perhaps the problem admits an interesting exact algorithm, without the need for approximation.

2. Often, a very natural and intuitive idea is the basis for an approximation algorithm. How good is a randomly chosen feasible solution for the problem? (What is the expected value of a random solution?) How about a greedy strategy? Can you define a neighborhood such that local search does well?

3. Look for a problem (call it Π) that is akin to yours in some sense, and use an existing approximation algorithm for Π to obtain an approximation for your problem.

4. Try to prove it cannot be approximated well, by reducing some hard-to-approximate problem to your problem.

The first, third, and fourth points essentially hinge on one's resourcefulness: one's tenacity to scour the literature (and colleagues) for problems similar to the one at hand, as well as one's ability to see the relationships and reductions which show that a problem is indeed similar.

This chapter has been mainly concerned with the second point. To answer the questions of that point, it is crucial to prove *bounds* on optimal solutions, with respect to the feasible solutions that one's approaches obtain. For minimization (maximization) problems, one will need to prove *lower bounds* (respectively, upper bounds) on some optimal solution for the problem. Devising lower (or upper) bounds can simplify the proof tremendously: one only needs to show that an algorithm returns a solution with value at most c times the lower bound to show that the algorithm is a c-approximation.

We have proven upper and lower bounds repeatedly (implicitly or explicitly) in our proofs for approximation algorithms throughout this chapter—it may be instructive for the reader to review each approximation proof and find where we have done it. For example, the greedy vertex cover algorithm (of choosing a maximal matching) works because even an optimal vertex cover covers at least one of the vertices in each edge of the matching. The number of edges in the matching is a lower bound on the number of nodes in a optimal vertex cover, and thus the number of nodes in the matching (which is twice the number of edges) is at most twice the number of nodes of an optimal cover.

18.6   CONCLUSIONS

We have seen the power of randomization in finding approximate solutions to hard problems. There are many available approaches for designing such algorithms, from solving a related problem and tweaking its solution (in linear programming relaxations) to constructing feasible solutions in a myopic way (via greedy algorithms). We saw that for some problems, determining an approximate solution is vastly easier than finding an exact solution, while other problems are just as hard to approximate as they are to solve.

In closing, we remark that the study of approximation and randomized algorithms is still a very young (but rapidly developing) field. It is our sincerest hope that the reader is inspired to contribute to the prodigious growth of the subject, and its far-reaching implications for problem solving in general.

SOURCES OF ADDITIONAL INFORMATION

Books on Algorithms

- Data Structures and Algorithms (Aho et al., 1983)

- Introduction to Algorithms (Cormen et al., 2001)

- The Design and Analysis of Algorithms (Kozen, 1992)

- Combinatorial Optimization: Algorithms and Complexity (Papadimitriou and Steiglitz, 1982)

Material on Linear Programming and Duality

- Chapter 3 of this book

- Linear Programming (Chvatal, 1983)

- Linear Programming and Extensions (Dantzig, 1998)

- Integer and Combinatorial Optimization (Nemhauser and Wolsey, 1988)

- Combinatorial Optimization: Algorithms and Complexity (Papadimitriou and Steiglitz, 1982)

- Combinatorial Optimization (Cook et al., 1988)

- Combinatorial Optimization Polyhedra and Efficiency (Schrijver, 2003)

Books on Approximation Algorithms

- Complexity and Approximation (Ausiello et al., 1999)

- Approximation Algorithms for NP-Hard Problems (Hochbaum, 1997)

- Approximation algorithms (Vazirani, 1983)

Books on Probabilistic and Randomized Algorithms

- An Introduction to Probability Theory and Its Applications (Feller, 1971)

- The Probabilistic Method (Alon and Spencer, 2000)

- Randomized Algorithms (Motwani and Raghavan, 1995)

- The Discrepancy Method (Chazelle, 2001)

Surveys

- Computing Near-Optimal Solutions to Combinatorial Optimization Problems (Shmoys, 1995)

- Approximation algorithms via randomized rounding: a survey (Srinivasan)

Courses and Lectures Notes Online

- Approximability of Optimization Problems, MIT, Fall 99 (Madhu Sudan), http://theory.lcs.mit.edu/ madhu/FT99/course.html

- Approximation Algorithms, Fields Institute, Fall 99 (Joseph Cheriyan), http://www.math.uwaterloo.ca/ jcheriya/App-course/course.html

- Approximation Algorithms, Johns Hopkins University, Fall 1998 (Lenore Cowen), http://www.cs.jhu.edu/ cowen/approx.html

- Approximation Algorithms, Technion, Fall 95 (Yuval Rabani), http://www.cs.technion.ac.il/ rabani/236521.95.wi.html

- Approximation Algorithms, Cornell University, Fall 98 (D. Williamson), http://www.almaden.ibm.com/cs/people/dpw/

- Approximation Algorithms, Tel Aviv University, Fall 01 (Uri Zwick), http://www.cs.tau.ac.il/7Ezwick/approx-alg-01.html

- Approximation Algorithms for Network Problems, (Cheriyan and Ravi) http://www.gsia.cmu.edu/afs/andrew/gsia/ravi/WWW/new-lecnotes.html

- Randomized algorithms, CMU, Fall 2000 (Avrim Blum), http://www-2.cs.cmu.edu/afs/cs/usr/avrim/www/Randalgs98/home.html

- Randomization and optimization by Devdatt Dubhashi, http://www.cs.chalmers.se/ dubhashi/ComplexityCourse/info2.html

- Topics in Mathematical Programming: Approximation Algorithms, Cornell University, Spring 99 (David Shmoys), http://www.orie.cornell.edu/ or739/index.html

- Course notes on online algorithms, randomized algorithms, network flows, linear programming, and approximation algorithms (Michel Goemans), http://www-math.mit.edu/ goemans/

Main Conferences Covering the Approximation and Randomization Topics

- IPCO: Integer Programming and Combinatorial Optimization

- ISMP: International Symposium on Mathematical Programming

- FOCS: Annual IEEE Symposium on Foundation of Computer Science

- SODA: Annual ACM-SIAM Symposium on Discrete Algorithms

- STOC: Annual ACM Symposium on Theory of Computing

- RANDOM: International Workshop on Randomization and Approximation Techniques in Computer Science

- APPROX: International Workshop on Approximation Algorithms for Combinatorial Optimization Problems

References

Agrawal, M., Kayal, N. and Saxena, N., 2002, Primes in P, www.cse.iitk.ac.in/news/primality.html

Aho, A. V., Hopcroft, J. E. and Ullman, J. D., 1979, *Computers and intractability: A guide to NP-Completeness,* Freeman, San Francisco.

Aho, A. V., Hopcroft, J. E. and Ullman, J. D., 1983, *Data structures and Algorithms*, Computer Science and Information Processing Series, Addison-Wesley, Reading, MA

Alon, N. and Spencer, J., 2000, *The Probabilistic Method*, Wiley, New York

Arora, S., 1998, Polynomial time approximation schemes for Euclidean traveling salesman and other geometric problems, *J. ACM* **45**:753–782.

Arora, S., Lund, C., Motwani, R., Sudan, M. and Szegedy, M., 1998, Proof verification and the hardness of approximation problems, *J. ACM* **45**:501–555.

Ausiello, G., Crescenzi, P., Gambosi, G., Kann, V., Marchetti-Spaccamela, A. and Protasi, M., 1999, *Complexity and Approximation*, Springer, Berlin.

Beier, R. and Vocking, B., 2003, Random knapsack in expected polynomial time, *Proc. ACM Symp. on Theory of Computing*, pp. 232–241.

Chazelle, B., 2001, *The Discrepancy Method*, Cambridge University Press, Cambridge.

Chvatal, V., 1979, A greedy heuristic for the set-covering, *Math. Oper. Res.* **4**:233–235.

Chvatal, V., 1983, *Linear Programming*, Freeman, San Francisco.

Clay Mathematics Institute, 2003, The millenium prize problems: P vs. NP, http://www.claymath.org/

Colbourn, C., 1984, The complexity of completing partial Latin squares, *Discrete Appl. Math.* **8**:25–30.

Cook, W., Cunningham, W., Pulleyblank, W. and Schrijver, A., 1988, *Combinatorial Optimization*, Wiley, New York.

Cormen, T. H., Leiserson, C. E., Rivest, R. L. and Stein, C, 2001, *Introduction to Algorithms*, MIT Press, Cambridge, MA.

Dantzig, G., 1998, *Linear Programming and Extensions*, Princeton University Press, Princeton, NJ.

Feige, U., 2002, Relations between average case complexity and approximation complexity, in: *Proc. ACM Symp. on Theory of Computing*, pp. 534–543.

Feller, W., 1971, *An Introduction to Probability Theory and Its Applications*, Wiley, New York.

Garey, M. R., Graham, R. L. and Ulman, J. D., 1972, Worst case analysis of memory allocation algorithms, in: *Proc. ACM Symp. on Theory of Computing*, pp. 143–150.

Goemans, M. X. and Williamson, D. P., 1995, Improved approximation algorithms for maximum cut and satisfiability problems using semidefinite programming, *J. ACM* **42**:1115–1145.

Goldwasser, S. and Kilian, J., 1986, Almost all primes can be quickly certified, in: *Proc. Annual IEEE Symp. on Foundations of Computer Science*, pp. 316–329.

Gomes, C., Selman, B., Crato, N. and Kautz, H., 2000, Heavy-tailed phenomena in satisfiability and constraint satisfaction problems, *J. Autom. Reason.* **24**:67–100.

Gomes, C. P., Selman, B. and Kautz, H., 1998, Boosting combinatorial search through randomization, in: *Proc. 15th National Conf. on Artificial Intelligence (AAAI-98),* AAAI Press, Menlo Park, CA.

Gomes, C. P. and Shmoys, D., 2002, The promise of LP to boost CSP techniques for combinatorial problems, in: *Proc. 4th Int. Workshop on Integration of AI and OR Techniques in Constraint Programming for Combinatorial Optimisation Problems (CP-AI-OR'02),* Le Croisic, France, N. Jussien and F. Laburthe, eds, pp. 291–305.

Graham, R. L., 1966, Bounds for certain multiprocessing anomalies, *Bell Syst. Tech. J.* **45**:1563–1581.

Gurevich, Y., 1991, Average Case Complexity, in: *Proc. 18th Int. Colloquium on Automata, Languages, and Programming (ICALP'91),* Lecture Notes in Computer Science, Vol. 510, pp. 615–628, Springer, Berlin.

Hochbaum, D. S., 1997, ed., *Approximation Algorithms for NP-Hard Problems,* PWS Publishing Company, Boston, MA.

Johnson, D. S., 1974, Approximation algorithms for combinatorial problems, *J. Comput. Syst. Sci.* **9**:256–278.

Khot, S. and Regev, O., 2003, Vertex cover might be hard to approximate within 2-e, in: *Proc. IEEE Conf. on Computational Complexity.*

Kirkpatrick, S., Gelatt, C. and Vecchi, M., 1983, Optimization by simulated annealing, *Science* **220**:671–680.

Kozen, D., 1992, *The design and analysis of algorithms,* Springer, New York.

Kumar, S. R., Russell, A. and Sundaram, R., 1999, Approximating Latin square extensions, *Algorithmica* **24**:128–138.

Laywine, C. and Mullen, G., 1998, *Discrete Mathematics using Latin Squares,* Discrete Mathematics and Optimization Series, Wiley-Interscience, New York.

Motwani, R. and Raghavan, P., 1995, *Randomized Algorithm,* Cambridge University Press, Cambridge.

Nemhauser, G. and Wolsey, L., 1988, *Integer and Combinatorial Optimization,* Wiley, New York.

Papadimitriou, C. and Steiglitz, K., 1982, *Combinatorial Optimization: Algorithms and Complexity,* Prentice-Hall, Englewood Cliffs, NJ.

Rabin, M. (1980). Probabilistic algorithm for testing primality, *J. Number Theory* **12**:128–138.

Schrijver, A., 2003, *Combinatorial Optimization Polyhedra and Efficiency,* Springer, Berlin.

Shmoys, D., 1995, Computing near-optimal solutions to combinatorial optimization problems, in: *Combinatorial Optimization,* Discrete Mathematics and Theoretical Computer Science Series, W. Cook, L. Lovasz and P. Seymour, eds, American Mathematical Society, Providence, RI.

Solovay, R. and Strassen, V., 1977, A fast Monte Carlo test for primality, *SIAM J. Comput.* **6**:84–86.

Srinivasan, A., Approximation algorithms via randomized rounding: a survey. Available from: citeseer.nj.nec.com/493290.html

Vazirani, V., 1983, *Approximation Algorithms,* Springer, Berlin.

Williams, R., Gomes, C. P. and Selman, B., 2003, Backdoors to typical case complexity, In *Proc. Int. Joint Conf. on Artificial Intelligence (IJCAI).*

Chapter 19

FITNESS LANDSCAPES

Colin R. Reeves

School of Mathematical and Information Sciences
Coventry University, UK

19.1   HISTORICAL INTRODUCTION

One of the most commonly-used metaphors to describe the process of heuristic methods such as local search in solving a combinatorial optimization problem is that of a "fitness landscape". However, describing exactly what we mean by such a term is not as easy as might be assumed. Indeed, many cases of its usage in both the biological and optimization literature reveal a rather serious lack of understanding.

The landscape metaphor appears most commonly in work related to evolutionary algorithms, where it is customary to trace the usage of the term back to a paper by the population geneticist Sewall Wright (Wright, 1932), although Haldane had already introduced a similar notion (Haldane, 1931). The metaphor has become pervasive, being cited in many biological texts that discuss evolution.

Wright's original idea of a fitness landscape was somewhat ambiguous. It appears that what he initially had in mind concerned within-species variation where the "axes" of a search space represented unspecified gene combinations, but Dobzhansky's subsequent enthusiastic use of the metaphor (Dobzhansky, 1951) seems to have established the consensus view of the axes of the search landscape as the frequency of a particular allele of a particular gene in a particular population. This can be seen in many textbooks on the subject of evolution, such as Ridley (1993). In the hands of Simpson, who seems to have thought primarily in terms of phenotypic characters (Simpson, 1953), the story became even more highly developed, although even more divorced from empirical reality. Despite Wright's later attempts to clarify the situation (Wright, 1967; Wright, 1988), the ambiguity remains. There is thus an interesting paradox in evolutionary biology: according to Futuyma (1998, p. 403),

> [The] adaptive landscape is probably the most common metaphor in evolutionary genetic[s]

yet nobody seems sure what exactly is the reality to which the metaphor is supposed to relate! However, it remains extremely popular: the book of Dawkins (1996), for example, makes considerable use of the notion, as its title *Climbing Mount Improbable* suggests.

Although we may have a vague idea of what the search space is, it is rather harder to define any axes for such a search space, as we have seen. Fitness in evolutionary biology is also a rather slippery concept. It is discussed as if there is some objective a priori measure, yet as usually defined, "fitness" concerns an organism's reproductive success, which can only be measured a posteriori.[1] Add to this the confusion over what the search space axes represent, and it becomes almost impossible to relate them to some quantifiable measure of fitness. It is thus generally dealt with by prestidigitation, and so, for all its popularity, the popular idea of a fitness landscape in biology is a mirage, displaying what is to a mathematician a distressing lack of rigor. (Happily some biologists agree, as in the cogent arguments against the hand-waving approach in Eldredge and Cracraft, 1980.)

A more serious approach was foreshadowed by Eigen (Eigen et al., 1989; Eigen, 1983). In his work on viruses, he introduced the concept of a quasi-species: a group of similar sequences. Each sequence S_k is a string of symbols drawn from some alphabet, the natural one to consider for viruses being the RNA bases adenine, cytosine, guanine and uracil: $\{A, C, G, U\}$. Differences in members of the quasi-species correspond to point mutations—replacement of one symbol by another one.

This interpretation falls somewhat short of the grand ideas in the popular biology textbooks, but it does make a formal mathematical development of the concept of a fitness landscape much more feasible, and following pioneering work by Weinberger (1990) in particular, a fairly complete formal statement of landscape theory was proposed by Stadler (1995). Recent work has developed this idea further (Reidys and Stadler, 2002), but some quite extensive mathematical knowledge is needed in order to appreciate it fully. In the expectation that the mathematical background of the readers of this volume will be somewhat variable, this tutorial will try to survey some of the themes most relevant to combinatorial optimization, without using advanced mathematical ideas. Some basic ideas of set theory, matrix algebra and functional analysis will be required, but the more complex ideas found in Reidys and Stadler (2002) will not be covered. Illustrative numerical examples will also be used at key points in an attempt to aid understanding.

---

[1] The *Oxford Dictionary of Biology*, for example, defines fitness as "The condition of an organism that is well adapted to its environment, as measured by its ability to reproduce itself."

19.2   COMBINATORIAL OPTIMIZATION

We can define combinatorial optimization problems as follows: we have a discrete search space $\mathcal{X}$, and a function

$$f : \mathcal{X} \mapsto I\!R$$

The general problem is to find

$$x^* = \arg \max_{x \in \mathcal{X}} f$$

where x is a vector of *decision variables* and f is the *objective function*. (Of course, minimization can also be the aim, but the modifications are always obvious). In the field of evolutionary algorithms, the function f is often called the *fitness*; hence the associated landscape is a *fitness landscape*. The vector x^* is a global optimum: that vector which is the "fittest" of all. (In some problems, there may be several global optima—different vectors of equal fitness.)

With the idea of a fitness landscape comes the idea that there are also many local optima or false peaks, in which a search algorithm may become trapped without finding the global optimum. In continuous optimization, notions of continuity and concepts associated with the differential calculus enable us to characterize quite precisely what we mean by a landscape, and to define the idea of an optimum. It is also convenient that our own experiences of hill climbing in a three-dimensional world gives us analogies to ridges, valleys, basins, watersheds, etc, which help us to build an intuitive picture of what is needed for a successful search, even though the search spaces that are of interest often have dimensions many orders of magnitude higher than 3.

However, in the continuous case, the landscape is determined only by the fitness function, and the ingenuity needed to find a global optimum consists in trying to match a technique to this single landscape. There is a major difference when we come to discrete optimization. Indeed, we really should not even use the term "landscape" unless we can define the topological relationships of the points in the search space $\mathcal{X}$. Unlike the continuous case, we have some freedom to specify these relationships, and in fact, that is precisely what we do when we decide to use a particular technique.

19.2.1   An Example

In practice, one of the most commonly used search methods for a combinatorial optimization problem is *neighborhood search*. This idea is at the root of modern "metaheuristics" such as simulated annealing (see Chapter 7) and tabu search (see Chapter 6)—as well as being much more involved in the methodology of genetic algorithms than is sometimes realized.

A *neighborhood structure* is generated by using an operator that transforms a given vector x into a new vector x'. For example, if the solution is represented by a binary vector (as is often the case for genetic algorithms (see Chapter 4), for instance), a simple neighborhood might consist of all vectors obtainable by "flipping" one of the bits. The "bit flip" neighbors of (00000), for example, would be

$$\{(10000), (01000), (00100), (00010), (00001)\}$$

Consider the problem of maximizing a simple function

$$f(z) = z^3 - 60z^2 + 900z + 100$$

where the solution z is required to be an integer in the range $[0, 31]$. Regarding z for a moment as a continuous variable, we have a smooth unimodal function with a single maximum at $z = 10$—as is easily found by calculus. Since this solution is already an integer, this is undoubtedly the most efficient way of solving the problem.

However, suppose we chose instead to represent z by a binary vector x of length 5. By decoding this binary vector as an integer it is possible to evaluate f, and we could then use neighborhood search, for example, to search over the binary hypercube for the global optimum using some form of hill-climbing strategy.

This discrete optimization problem turns out to have four optima (three of them local) when the bit flip operator is used. If a "steepest ascent" strategy is used (i.e. the *best* neighbor of a given vector is identified before a move is made) the local optima are as shown in Table 19.1. Also shown are the "basins of attraction"—the set of initial points from which the search leads to a specified optimum. For example, if we start the search at any of the points in the first column, and follow a strict best improvement strategy, the search will finish up at the global optimum. However, if a "first improvement" strategy is used (where the first change that leads uphill is accepted without ascertaining if a still better one exists), the basins of attraction are rather different, as shown in Table 19.2.

In fact, there are even more complications: in Table 19.2, the order of searching the components of the vector is "forward" (left to right). If the search is made in the reverse direction (right to left) the basins of attraction are different, as shown in Table 19.3.

Thus, by using flipping with this binary representation, we have created local optima that did not exist in the integer version of the problem. Further, although the optima are still the same, the chances of reaching a *particular* optimum can be seriously affected by a change in hill-climbing strategy.

However, the bit flip operator is not the only mechanism for generating neighbors. An alternative neighborhood could be defined as follows: for $k = 1, \ldots, 5$, flip bits $\{k, \ldots, 5\}$. Thus, the neighbors of (00000), for example, would now be

$$\{(11111), (01111), (00111), (00011), (00001)\}$$

Table 19.1. Local optima and basins of attraction for steepest ascent with the bit flip operator in the case of a simple cubic function. The bracketed figures are the fitnesses of each local optimum.

Local optimum	0 1 0 1 0	0 1 1 0 0	0 0 1 1 1	1 0 0 0 0
	(4100)	(3988)	(3803)	(3236)
Basin	0 0 0 0 0	0 0 1 0 0	0 0 1 1 0	1 0 0 0 0
	0 0 0 0 1	0 1 1 0 0	0 0 1 1 1	1 0 0 0 1
	0 0 0 1 0	1 1 1 0 0	1 0 1 1 0	1 0 0 1 0
	0 0 0 1 1		1 0 1 1 1	1 0 0 1 1
	0 0 1 0 1			1 0 1 0 0
	0 1 0 0 0			
	0 1 0 0 1			
	0 1 0 1 0			
	0 1 0 1 1			
	0 1 1 0 1			
	0 1 1 1 0			
	0 1 1 1 1			
	1 0 1 0 1			
	1 1 0 0 0			
	1 1 0 0 1			
	1 1 0 1 0			
	1 1 0 1 1			
	1 1 1 0 1			
	1 1 1 1 0			
	1 1 1 1 1			

We shall call this the "CX operator", and it creates a very different landscape. In fact, there is now only a single global optimum (01010); *every* vector is in its basin of attraction. This illustrates the point that it is not merely the choice of a binary representation that generates the landscape—the search operator needs to be specified as well.

Incidentally, there are two interesting facts about the CX operator. Firstly, it is closely related to the one-point crossover operator frequently used in genetic algorithms. (For that reason, it has been termed the complementary crossover or CX operator). Secondly, if the 32 vectors in the search space are re-coded using a *Gray* code, it is easy to show that the bit-flip neighbors of a point in Gray-coded space are identical to those in the original binary-coded space under CX. This is an example of an *isomorphism* of landscapes. (An isomorphism in mathematics refers to mappings between mathematical objects that preserve structure. It comes from the Greek *iso* (equal) and *morphe* (shape). For example, two graphs are isomorphic if there is a one-to-one mapping σ

Table 19.2. Local optima and basins of attraction for first improvement (forward search) using the bit flip operator.

Local optimum	0 1 0 1 0	0 1 1 0 0	0 0 1 1 1	1 0 0 0 0
	(4100)	(3988)	(3803)	(3236)
Basin	0 0 1 0 1	0 0 1 0 0	0 0 1 1 1	0 0 0 0 0
	0 0 1 1 0	0 1 0 0 0	0 1 1 1 1	0 0 0 0 1
	0 1 0 0 1	0 1 1 0 0	1 0 1 1 1	0 0 0 1 0
	0 1 0 1 0	1 0 1 0 0	1 1 1 1 1	0 0 0 1 1
	0 1 0 1 1	1 1 0 0 0		1 0 0 0 0
	0 1 1 0 1	1 1 1 0 0		1 0 0 0 1
	0 1 1 1 0			1 0 0 1 0
	1 0 1 0 1			1 0 0 1 1
	1 0 1 1 0			
	1 1 0 0 1			
	1 1 0 1 0			
	1 1 0 1 1			
	1 1 1 0 1			
	1 1 1 1 0			

Table 19.3. Local optima and basins of attraction for first improvement (reverse search) using the bit flip operator.

Local optimum	0 1 0 1 0	0 1 1 0 0	0 0 1 1 1	1 0 0 0 0
	(4100)	(3988)	(3803)	(3236)
Basin	0 1 0 0 0	0 1 1 0 0	0 0 0 0 0	1 0 0 0 0
	0 1 0 0 1	0 1 1 0 1	0 0 0 0 1	1 0 0 0 1
	0 1 0 1 0	0 1 1 1 0	0 0 0 1 0	1 0 0 1 0
	0 1 0 1 1	0 1 1 1 1	0 0 0 1 1	1 0 0 1 1
			0 0 1 0 0	1 0 1 0 0
			0 0 1 0 1	1 0 1 0 1
			0 0 1 1 0	1 0 1 1 0
			0 0 1 1 1	1 0 1 1 1
				1 1 0 0 0
				1 1 0 0 1
				1 1 0 1 0
				1 1 0 1 1
				1 1 1 0 0
				1 1 1 0 1
				1 1 1 1 0
				1 1 1 1 1

between their respective sets of vertices such that for every edge (x, y) of one graph, $(\sigma(x), \sigma(y))$ is an edge of the other.)

19.3 MATHEMATICAL CHARACTERIZATION

Now that some of the typical features of a landscape have been illustrated, we can provide a mathematical characterization. We define a landscape $\mathcal{L}$ for the function f as a triple $\mathcal{L} = (\mathcal{X}, f, d)$ where d denotes a distance measure $d : \mathcal{X} \times \mathcal{X} \to I\!\!R^+ \cup \{\infty\}$ for which it is required that, $\forall s, t, u \in \mathcal{X}$,

$$d(s, t) \geq 0; \quad d(s, t) = 0 \Leftrightarrow s = t; \quad d(s, u) \leq d(s, t) + d(t, u)$$

Note that we do not need to specify the representation explicitly, since this is assumed to be implied in the description of $\mathcal{X}$. We have also decided, for the sake of simplicity, to ignore questions of search strategy and other matters in the definition of a landscape, unlike the more comprehensive definition of, for example, Jones (1995).

This definition says nothing about how the distance measure arises. In fact, for many cases a "canonical" distance measure can be defined. Often, this is symmetric, i.e. $d(s, t) = d(t, s) \ \forall s, t \in \mathcal{X}$, so that d also defines a *metric* on $\mathcal{X}$. This is clearly a nice property, although it is not essential.

19.3.1 Neighborhood Structure

The distance measure is typically related to the neighborhood structure. Every solution $x \in \mathcal{X}$ has an associated set of *neighbors*, $N_\omega(x) \subset \mathcal{X}$, called the neighborhood of x. This neighborhood is generated by applying an operator ω to a vector s in order to transform it into a vector t. What we may call a canonical distance measure d_ω is that induced by ω whereby

$$t \in N_\omega(s) \Leftrightarrow d_\omega(s, t) = 1$$

The distance between non-neighbors is defined as the length of the shortest path between them (if one exists). The operator ω generally takes a parameter, which means that it is technically a one-to-many function, able to generate many neighbors from one initial vector. The size of the neighborhood will be denoted by n.

For example, if $\mathcal{X}$ is the binary hypercube $\{0, 1\}^\ell$, the bit flip operator can be defined as

$$\phi(i) : \{0, 1\}^\ell \times \mathbb{Z} \to \{0, 1\}^\ell \qquad \begin{cases} z_i' = 1 - z_i \\ z_k' = z_k & \text{if } i \neq k \end{cases}$$

where z is a binary vector of length ℓ, and i is the parameter specifying the bit to be flipped. It is clear that the distance metric induced by ϕ is the well-known

Hamming distance

$$d_H(x, y) = \sum_{i=1}^{\ell} [x_i \neq y_i]$$

where the square brackets [*expr*] denote an *indicator* function, which takes the value 1 if the logical expression *expr* is true and 0 otherwise. Thus we could describe this landscape as a Hamming landscape (with reference to its distance measure), or as the bit-flip landscape (with reference to the operator). Similarly, we can define the CX operator as

$$\gamma(i) : \{0, 1\}^{\ell} \times \mathbb{Z} \to \{0, 1\}^{\ell} \qquad \begin{cases} z'_k = 1 - z_k & \text{for } k \geq i \\ z'_k = z_k & \text{otherwise} \end{cases}$$

The distance measure induced here is clearly more complicated than the Hamming landscape, and cannot be described by a simple function. In both of these cases the size of the neighborhood is $n = \ell$.

As an example of an asymmetric distance measure, consider the case where $\mathcal{X}$ is Π_m, the space of permutations of length m, which is often relevant for scheduling problems. A familiar neighborhood is defined by the "forward shift" operator, taking two parameters in this case,

$$\mathcal{FSH}(i, j) : \Pi_m \times \mathbb{Z} \times \mathbb{Z} \to \Pi_m \qquad \begin{cases} \pi'_{k-1} = \pi_k & \text{for } j < k \leq i \\ \pi'_i = \pi_j \\ \pi'_k = \pi_k & \text{otherwise} \end{cases}$$

The neighbors of (1234), for example, would be

$$\{(2134), (2314), (2341), (1324), (1342), (1243)\}$$

(note that the size of this neighborhood is $n = \binom{m}{2}$). It is easily seen that (1234), however, is not a neighbor of (2314), (2341) or (1342), so $\mathcal{FSH}$ is not symmetric. Other neighborhood operators (for example, "exchange", where two items in the sequence are swapped) induce different distance measures, so there may be advantages in choosing operator-independent distance measures (Reeves, 1999) for practical comparisons.

Distance measures may become even more complicated: for instance, in the problem of biological sequence comparison (RNA, DNA and protein sequences: see Waterman, 1995), it is common to compare sequences in terms of the minimal number of genetic operations necessary to convert one string into another (the "string edit" distance). Thus, even finding the distance measure becomes an optimization problem.

19.3.2   Local Optima

We can now give a formal statement of a fundamental property of fitness landscapes: for a landscape $\mathcal{L} = (\mathcal{X}, f, d)$, a vector $x^{\circ} \in \mathcal{X}$ is *locally optimal* if

$$f(x^\circ) > f(t) \quad \forall\, t \in N(x^\circ)$$

We shall denote the set of such optima as $\mathcal{X}^\circ$, and the set of *global optima* (recall that we allow the possibility of more than one) as $\mathcal{X}^*$ where the vector $x^* \in \mathcal{X}^\circ$ is a global optimum if

$$f(x^*) \geq f(x^\circ) \quad \forall\, x^\circ \in \mathcal{X}^\circ$$

Landscapes that have only one local (and thus also global) optimum are commonly called *unimodal*, while landscapes with more than one local optimum are said to be *multimodal*.

The number of local optima in a landscape clearly has some bearing on the difficulty of finding the global optimum. In our previous example, it is clearly more difficult to find the global optimum using bit-flipping than if we used CX. However, it is not the only indicator: in our example the steepest ascent strategy increased the chance of finding the global optimum, since there were more initial solutions that led to the global optimum than under first-improvement.

19.3.3 Basins of Attraction

We can also now define more precisely the idea of a *basin of attraction*. Neighborhood search can be interpreted as a function

$$\mu : \mathcal{X} \mapsto \mathcal{X}^\circ$$

where if x is the initial point, $\mu(x)$ is the optimum that it reaches. With this in mind, we can define the basin of attraction of x° as the set

$$B(x^\circ) = \{x : \mu(x) = x^\circ\}$$

The problem is that $B(x^\circ)$ is not independent of the search strategy, as the example of Section 19.2.1 demonstrated. In fact, it is only well defined for the case of steepest ascent. For other search strategies, such as first improvement, the order of searching may be highly influential. Our example showed that the basin of attraction of the global optimum was much larger for steepest ascent than for the other strategies, but it is possible to find examples where the opposite is the case.

19.3.4 Graph Representation

Neighborhood structures are clearly just another way of defining a graph Γ, which can be described by its $(n \times n)$ *adjacency matrix* A. The elements of A are given by $a_{ij} = 1$ if the indices i and j represent neighboring vectors, and

$a_{ij} = 0$ otherwise. For example, the graph induced by the bit flip ϕ on binary vectors of length 3 has adjacency matrix

$$A_\phi = \begin{bmatrix} 0 & 1 & 1 & 0 & 1 & 0 & 0 & 0 \\ 1 & 0 & 0 & 1 & 0 & 1 & 0 & 0 \\ 1 & 0 & 0 & 1 & 0 & 0 & 1 & 0 \\ 0 & 1 & 1 & 0 & 0 & 0 & 0 & 1 \\ 1 & 0 & 0 & 0 & 0 & 1 & 1 & 0 \\ 0 & 1 & 0 & 0 & 1 & 0 & 0 & 1 \\ 0 & 0 & 1 & 0 & 1 & 0 & 0 & 1 \\ 0 & 0 & 0 & 1 & 0 & 1 & 1 & 0 \end{bmatrix}$$

where the vectors are indexed from 0 to 7 in the usual binary-coded integer order (i.e. (000), (001), etc). By way of contrast, the adjacency matrix for the CX operator is

$$A_\gamma = \begin{bmatrix} 0 & 1 & 0 & 1 & 0 & 0 & 0 & 1 \\ 1 & 0 & 1 & 0 & 0 & 0 & 1 & 0 \\ 0 & 1 & 0 & 1 & 0 & 1 & 0 & 0 \\ 1 & 0 & 1 & 0 & 1 & 0 & 0 & 0 \\ 0 & 0 & 0 & 1 & 0 & 1 & 0 & 1 \\ 0 & 0 & 1 & 0 & 1 & 0 & 1 & 0 \\ 0 & 1 & 0 & 0 & 0 & 1 & 0 & 1 \\ 1 & 0 & 0 & 0 & 1 & 0 & 1 & 0 \end{bmatrix}$$

It is simply demonstrated that permuting the rows and columns so that they are in the order 0, 1, 3, 2, 7, 6, 4, 5 reproduces the adjacency matrix A_ϕ—another way of demonstrating the isomorphism mentioned earlier. In other words,

$$P^{-1}A_\phi P = A_\gamma$$

where P is the associated permutation matrix of the binary-to-Gray transformation. It is also clear that the eigenvalues and eigenvectors are the same (up to a permutation).

As a final example, we may consider the adjacency matrix for $\mathcal{FSH}$ in the space Π_3:

$$A_{\mathcal{FSH}} = \begin{bmatrix} 0 & 1 & 1 & 1 & 0 & 0 \\ 1 & 0 & 0 & 0 & 1 & 1 \\ 1 & 1 & 0 & 1 & 0 & 0 \\ 0 & 0 & 1 & 0 & 1 & 1 \\ 1 & 1 & 0 & 0 & 0 & 1 \\ 0 & 0 & 1 & 1 & 1 & 0 \end{bmatrix}$$

where the permutations are indexed in lexicographic order $(123), \ldots, (321)$. The lack of symmetry in the distance measure is of course reflected in an asymmetric matrix.

19.3.5   Laplacian Matrix

The *graph Laplacian* Δ is defined as

$$\Delta = A - D$$

where D is a diagonal matrix such that d_{ii} is the degree of vertex i. Usually, these matrices are vertex-regular and $d_{ii} = k\ \forall i$, so that

$$\Delta = A - kI$$

This notion recalls that of a Laplacian operator in the continuous domain; the effect of this matrix, applied as an operator at the point s to the fitness function f is

$$\Delta f(s) = \sum_{t \in N(s)} (f(t) - f(s))$$

so it functions as a kind of differencing operator. In particular, $\Delta f(s)/n$ is the average difference in fitness between the vector s and its neighbors. Grover has shown (Grover, 1992) that the landscapes of several combinatorial optimization problems satisfy an equation of the form

$$\Delta f + \frac{Cf}{n} = 0$$

where C is a problem-specific constant and n (in Grover's notation) is the size of the problem instance. From this it can be deduced that *all* local optima are better than the mean ($\bar{f}$) over all points on the landscape. Furthermore, it can also be shown that under mild conditions on the nature of the fitness function, the time taken by neighborhood search to find a local optimum in a maximization problem is $\mathcal{O}(n \log_2[f_{max}/\bar{f}])$ where f_{max} is the fitness of a global maximum. (A similar result can be obtained, *mutatis mutandis*, for minimization problems.)

19.3.6   Graph Eigensystem

In the usual way, we can define eigenvalues and eigenvectors of the matrices associated with the graph Γ. The set of eigenvalues is called the *spectrum* of the graph. For an $n \times n$ matrix A the spectrum is

$$\begin{pmatrix} \lambda_0 & \lambda_1 & \cdots & \lambda_{n-1} \end{pmatrix}$$

where λ_i is the ith eigenvalue, ranked in (weakly) descending order. Similarly, the spectrum of the Laplacian is

$$\begin{pmatrix} \mu_0 & \mu_1 & \cdots & \mu_{n-1} \end{pmatrix}$$

where, again, μ_i is the ith eigenvalue, ranked this time in (weakly) ascending order. For a regular connected graph it can be shown that

$$\mu_i = k - \lambda_i \ \forall i$$

Further, from the corresponding eigenvectors $\{\varphi_i\}$, f can be expanded as

$$f(s) = \sum_i a_i \varphi_i(s)$$

Stadler and Wagner (1998) call this a "Fourier expansion".

Unfortunately, the size of these graphs rapidly becomes very large. However, graphs can often be partitioned in a way that makes it possible to reduce the scale of the problem. This enables the formation of a *collapsed matrix* $\tilde{A}$ whose eigenvalues are the *distinct* eigenvalues of A, with multiplicities given by the cardinalities of the partitions. (Relevant mathematical details may be found in the books of Biggs, 1993, and Godsil, 1993.) If the diameter of such a graph is δ, the number of distinct eigenvalues is only $\delta + 1$, so a considerable reduction in size is possible—at least in principle.

Similarly, the Fourier expansion can be partitioned into a sum

$$f(s) = \sum_p \beta_p \tilde{\varphi}_p(s)$$

over the distinct eigenvalues of Δ. The corresponding values

$$|\beta_p|^2 = \sum |a_k|^2$$

(where the sum is over the coefficients that correspond to the pth distinct eigenvalue) form the *amplitude spectrum*, which expresses the relative importance of different components of the landscape.

Ideally, such mathematical characterizations could be used to aid our understanding of the important features of a landscape, and so help us to exploit them in designing search strategies. But beyond Grover's rather general results above, it is possible to carry out further analytical studies only for small graphs or graphs with a special structure, as illustrated for example, by Stadler (Stadler, 1995). In the important case of the Hamming landscape of a binary search space analytical results for the graph spectrum show that the eigenvectors are thinly disguised versions of the familiar Walsh functions.[2] For the case of recombinative operators the problem is considerably more complicated, and

---

[2]For readers who are unfamiliar with Walsh functions, they are digital analogs of trigonometrical functions, forming an orthonormal set of rectangular waveforms. An introduction to their uses in the analysis of optimization methods can be found in Reeves and Rowe (2002).

necessitates the use of "P-structures" (Stadler and Wagner, 1998). The latter are essentially generalizations of graphs in which the mapping is from pairs of "parents" (x, y) to the set of possible strings that can be generated by their recombination. However, it can be shown that for some "recombination landscapes" (such as that arising from the use of uniform crossover) the eigenvectors are once more the Walsh functions. Whether this is also true in the case of one- or two-point crossover, for example, is not known, but Stadler and Wagner conjecture that it is. In view of the close relationship between the bit-flip and CX landscapes as demonstrated above, it would not be surprising if this is a general phenomenon. However, to obtain these results, some assumptions have to be made—such as a uniform distribution of parents—that are unlikely to be true in a specific finite realization of a genetic search.

In the case of the bit-flip landscape, the distinct eigenvalues correspond to sets of Walsh coefficients of different orders, and the amplitude spectrum is exactly the set of components of the "epistasis variance" associated with other attempts to measure problem difficulty (for a review, see Reeves and Rowe, 2002). For the cubic function of Section 19.2.1 above, the components of variance for the different orders of Walsh coefficients can be shown to be (0.387, 0.512, 0.101, 0, 0) respectively; i.e. 61.3% of the variation in the landscape is due to interactions of order 2 and 3. This is consistent with the relatively poor performance of the bit flip hill-climber.

Of course, the eigenvalues and eigenvectors are exactly the same (up to a permutation) for the CX landscape of this function, and the set of values for the Walsh coefficients in the Fourier decomposition is also the same. However, the effect of the permutation inherent in the mapping from the bit flip landscape to the CX landscape is to re-label some of the vertices of the graph, and hence some of the Walsh coefficients. Thus some coefficients that previously referred to linear effects now refer to interactions, and vice versa. Taking the cubic function as an example again, the components of variance or amplitude spectrum becomes (0.771, 0.174, 0.044, 0.011, 0.000). We see that the linear effects now predominate (77.1%), and this is consistent with the fact that the hill-climber in the CX landscape always finds the optimum.

19.3.7   Recombination Landscapes

If we look at the "recombination landscapes" derived from the P-structures of Stadler and Wagner (1998), we find that once again the Walsh coefficients are obtained, but labeled in yet another way. The coefficients in the bit flip and CX landscapes are grouped according to the number of 1s in their binary- and Gray-coded index representations respectively. However, in a recombination landscape—such as that generated by one-point crossover—it is the *separation*

Table 19.4. Illustration of the different groupings of the Walsh coefficients associated with the bit flip, CX and recombination landscapes.

Index	Binary coding	Bit flip	CX	Recom	Index	Binary coding	Bit flip	CX	Recom
0	0000	0	0	0	8	1000	1	2	1
1	0001	1	1	1	9	1001	2	3	4
2	0010	1	2	1	10	1010	2	4	3
3	0011	2	1	2	11	1011	3	3	4
4	0100	1	2	1	12	1100	2	2	2
5	0101	2	3	3	13	1101	3	3	4
6	0110	2	2	2	14	1110	3	2	3
7	0111	3	1	3	15	1111	4	1	4

between the outermost 1-bits that defines the groupings. Table 19.4 shows the groupings for a 4-bit problem.

Several things can be seen from this table: firstly, the linear Walsh coefficients (and hence the linear component of epistasis variance) are the same in both the bit flip and the recombination landscapes. Secondly (as already explained), the coefficients in the CX landscape are simply a re-labeling of those in the bit flip landscape. Thirdly, the coefficients in the recombination landscape do not form a natural grouping in terms of interactions, and consequently the different components of variance for the recombination landscape do not have a simple interpretation as due to interactions of a particular order.

19.3.8   Summary

This section has set out some of the basic mathematics necessary for the analysis of landscapes. As has probably become obvious, the details can require an extensive mathematical knowledge. Furthermore, the full analysis of a particular landscape (i.e. its eigensystem) may need the gathering of a large amount of empirical information, perhaps equivalent to a complete knowledge of the fitness function at all points of the search space! Landscape analysis in such cases can be no more than an a posteriori justification (or lack of it!) for the choice of a particular neighborhood. Further discussion on some of these points may be found in Reeves and Rowe (2002).

While it is undeniably useful that we can construct mathematical techniques to help us neatly summarize certain facts about a landscape, we must also recognize that there are other features—possibly very important ones—that are not captured by these methods. In the simple example of the cubic function we have seen that the search strategy adopted can make a big difference to the likelihood of a hill-climber finding the global optimum.

Mathematical analysis holds out some tantalizing prospects of future progress, but for the moment we turn to a consideration of the results of experimental work on landscapes.

19.4 STATISTICAL MEASURES

If mathematical analysis of a landscape is a difficult task, then it is natural to ask if there is something we can learn about the nature of a landscape, simply from the process of searching it. Several ideas have been suggested.

19.4.1 Autocorrelation

One of the earliest attempts to obtain some statistical measure of a landscape was by Weinberger, who showed that certain quantities obtained in the course of a random walk can be useful indicators (Weinberger, 1990). If the fitness of the point visited at time t is denoted by f_t, we can estimate the *autocorrelation function* (usually abbreviated to *acf*) of the landscape during a random walk of length T as

$$r_j = \frac{\sum_{t=1}^{T-j} (f_t - \bar{f})(f_{t+j} - \bar{f})}{\sum_{t=1}^{T}(f_t - \bar{f})^2}$$

Here $\bar{f}$ is of course the mean fitness of the T points visited, and j is known as the *lag*. The concept of autocorrelation is of course an important one in time series analysis, but its interpretation in the context of landscapes is interesting.

For "smooth" landscapes, and at small lags (i.e. for points that are close together), the *acf* is likely to be close to 1 since neighbors are likely to have similar fitness values. However, as the lag increases the correlations will diminish. "Rugged" landscapes are informally those where close points can nevertheless have completely unrelated fitnesses, and so the *acf* will be close to zero at all lags. Landscapes for which the *acf* has significant negative values are conceptually possible, but they would have to be rather odd.

A related quantity is the *correlation length* of the landscape, usually denoted by τ. Classical time series analysis (Box and Jenkins, 1970) can be used to show that the standard error of the estimate r_j is approximately $1/\sqrt{T}$, so that there is only approximately 5% probability that $|r_j|$ could exceed $2/\sqrt{T}$ by chance. Values of r_j less than this value can be assumed to be zero. The correlation length τ is then the last j for which r_j is non-zero:

$$\tau = j : |r_{j+1}| < 2/\sqrt{T} \wedge \left\{ |r_k| > 2/\sqrt{T} \quad \forall\, k \leq j \right\}$$

The *acf* and the correlation length are useful indicative measures of the rugged-ness of a landscape, but they are rather crude statistics, and it is difficult to attach a great deal of meaning to their values for particular instances.

19.4.2   Number of Optima

Although it is not the full story, the number of local optima is widely ac-knowledged as a very important factor in how easy or difficult it is to find a global optimum of a landscape, and is clearly much more directly relevant for a particular instance than the correlation measures. Recently, some attempts have been made (Reeves, 2001; Eremeev and Reeves, 2002, 2003; Reeves and Eremeev, 2004) to obtain direct estimates of the number of optima using sta-tistical principles.

It is assumed that a heuristic search method can be restarted many times using different initial solutions. Given the landscape framework we have dis-cussed above, by randomly generating initial solutions, we will sample many basins of attraction. Of course, this will be evident by the number of *different* final solutions $\{x^o\}$ that are found. Suppose this number is k, and the number of restarts is r ($\geq k$). Various statistical models may be used in order to estimate the number of optima v.

Waiting-Time Model   We can ask for the distribution of the waiting-time to find all optima. If r exceeds k substantially, we can use this fact to estimate the probability that all optima have been found. This would also imply, a fortiori, that the global optimum had been found, and thus provides us with an objective confidence level concerning the quality of the best solution obtained.

Counting Model   In the event—unfortunately, a common one—that k is not much smaller than r, it is unlikely that we have seen many of the optima. However, a counting model can be used to estimate the value of v, in a similar way to those used by ecologists to estimate the size of an unknown animal population. This can be quite illuminating in showing the differences between landscapes generated by different neighborhood operators.

Non-parametric Estimates   Fairly restrictive assumptions are needed in or-der to obtain tractable statistical models of landscapes. Where these estimates can be checked against actuality (by enumerating all points on a landscape), it appears that the effect of these assumptions is to produce negatively biased estimates—i.e. the estimate of v is consistently smaller than the true value. Removing the assumptions by creating more powerful models would probably be impossible, so some *non-parametric* approaches have been explored, and found to provide useful estimates of v, although the problem of negative bias remains. All these models are summarized in Reeves and Eremeev (2004).

19.5   EMPIRICAL STUDIES

Besides explicit statistical models of landscape features, several empirical studies have been aimed at providing some idea of the "big picture". Although multi-dimensional fitness landscapes have few similarities with "real" 3D landscapes, certain empirical findings can be interpreted sensibly in terms of characteristics of real landscapes, which provides us with some insights into ways we can approach hard optimization problems.

One of the most interesting observed properties of fitness landscapes has been seen in many different studies: it is a feature of Kauffman's well-known "NK-landscapes" (Kauffman, 1993),[3] and it also appears in many examples of combinatorial optimization problems, such as the traveling salesman problem (Lin, 1965; Boese et al., 1994), graph partitioning (Merz and Freisleben, 1998), and flowshop scheduling (Reeves, 1999).

In the first place, such studies have repeatedly found that, on average, local optima are very much closer to the global optimum than are randomly chosen points, and closer to each other than random points would be. That is, the distribution of local optima is not "isotropic"; rather, they tend to be clustered in a "central massif" (or—if we are minimizing—a "big valley"). This can be demonstrated graphically by plotting a scatter graph of fitness against distance to the global optimum. Secondly, if the basins of attraction of each local optimum are explored, size is quite highly correlated with quality: the better the local optimum, the larger is its basin of attraction. (If true, this also impinges on the estimation problem we discussed in the previous section: although there is a negative bias in the estimate of ν, the "big valley" phenomenon implies that it is only the small basins and low-quality optima that we are missing.)

Of course, there is no guarantee that this property holds in any particular case, but it provides an explanation for the success of "perturbation" methods (Johnson, 1990; Martin et al., 1992; Zweig, 1995) which currently appear to be the best available for the traveling salesman problem. It is also tacitly assumed by such methods as simulated annealing and tabu search, which would lose a great deal of their potency if local optima were isotropically distributed.

19.5.1   Practical Applications

These studies also suggest a starting point for the development of new heuristic search algorithms, such as the "adaptive multi-start" algorithm of Boese et al. (1994). As a more recent example, we shall consider the "path

---

[3] In Kauffman's notation, N is the length of a binary string, and K is the maximum number of genes that are allowed to interact with any other; e.g., if $K = 1$, each gene can interact with just one other. There are several different ways in which the sets of interacting genes can be chosen, but essentially they turn out to make little difference.

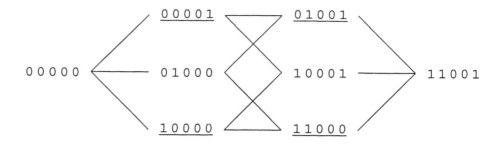

Figure 19.1. The diagram shows the set of paths that could be traced between the parents 00000 and 11001. Only those intermediate vectors indicated by underlines can be generated by one-point crossover, but all can be generated by uniform crossover.

tracing" algorithms introduced in Reeves and Yamada (1998) and Yamada and Reeves (1998), which can be motivated either as a use of the idea of a landscape, or in terms of extending the boundaries of evolutionary algorithms.

If we consider the case of crossover of vectors in $\{0, 1\}^{\ell}$, it is easily seen that any "child" produced from two "parents" will lie on a path that leads from one parent to another. Figure 19.1 demonstrates this fact.

In an earlier paper (Reeves, 1994), such points were described as "intermediate vectors". In other search spaces, the distance measure may be more complicated, but the principle is still relevant. Crossover is re-interpreted as finding a point lying "between" two parents in some landscape in which we hope the big valley conjecture is true. This "path-tracing crossover" was implemented for both the makespan and the flowsum versions of the flowshop sequencing problem; Figure 19.2 shows in a two-dimensional diagram the idea behind it, while full details can be found in Reeves and Yamada (1998, 1999).

In this way, the concept of recombination can be fully integrated with traditional neighborhood search methods, and the results obtained for flowshop instances (see Reeves and Yamada (1998) and Yamada and Reeves (1998) for details) were gratifyingly good. For the makespan problem, embedded path tracing helped the GA to achieve results of outstandingly high quality: several new best solutions were discovered for well-known benchmarks. For the flowsum version, optimal solutions are not known, but the path-tracing GA consistently produced better solutions than other proposed techniques.

This idea has also recently been applied to multi-constrained knapsack problems (Levenhagen et al., 2001), where the need was confirmed for a "big valley" structure in order to benefit from this approach.

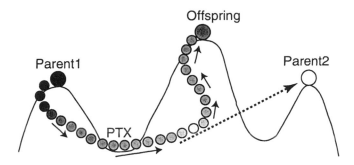

Figure 19.2. Path tracing crossover combined with local search: a path is traced from one parent in the direction of the other. In the "middle" of the path, solutions may be found that are not in the basins of attractions of the parents. A local search can then exploit this new starting point by climbing to the top of a hill (or the bottom of a valley, if it is a minimization problem)—a new local optimum. The acronym PTX signifies "path-tracing crossover".

19.6 SOME PROMISING AREAS FOR FUTURE APPLICATION

Finally, we should remark that several interesting future research questions are suggested. On the theoretical side, a deeper knowledge of the connections between algebra and graph theory may provide further useful analytical results. For example, it would be useful to have analytical results for all the common operators in permutation spaces analogous to those derived for the simpler case of binary strings. Real fitness landscapes for a number of combinatorial optimization problems also have to cope with extensive areas where there is no change in the fitness for many steps. Measuring the extent and effects of such "plateaux" formations also needs further study, as does the characterization of basins if attraction. (Some promising ideas based on the notion of a "barrier tree" have already been put forward by Stadler and colleagues in Flamm et al., 2002.)

Building on such notions, it would be helpful if we could provide a formal definition of what it means for a "big valley" structure to exist, and how it could be related to mathematical constructs associated with neighborhood structures. Does the big valley exist almost everywhere? If not, can we define classes of problems and neighborhood structures for which it does not occur? Further empirical analysis, such as that described by Levenhagen et al. (2001) and Watson et al. (2002), should be of considerable assistance in suggesting fruitful avenues to explore.

More generally, it is clear that crude correlation measures can only be a general guide to the nature of a landscape instance, and we need to find better ways of characterizing landscapes from empirical measurements. Some suggestions have been made in Reeves (2004) for further work in this direction.

In the area of implementation, it is important to see if we can further refine the path tracing methodology and its integration into heuristic search methods such as evolutionary algorithms. Also, the methodological developments pioneered in Reeves (2001), Eremeev and Reeves (2002, 2003) and Reeves and Eremeev (2004) for deducing properties of an instance of a landscape from the results of heuristic search offer the possibility of making principled probability statements about the quality of solutions obtained.

19.7   TRICKS OF THE TRADE

Mathematical analysis of landscapes is generally possible only for small problems, and then can only really be useful as an a posteriori validation (or questioning) of the decisions already made. However, empirical analysis is relatively easy and may provide some useful insights.

Correlation analysis can be a helpful indicator of the type of landscape with which we are dealing. Typically this proceeds by making a random walk on the landscape for several thousand steps and collecting data on fitness. The resultant "time series" can be analyzed with standard statistical tools. The drawback of this approach is that even when it is complete, knowing how smooth or rugged the landscape is from the perspective of a random walk does not help very much in deciding which heuristic search method to adopt. Further, much computation has been carried out yet the search for an optimum has not even started!

For those wishing to make use of empirical landscape analysis as part of a general research program, it should be realized that much of the necessary information is inherently generated in the course of applying a heuristic search method to a combinatorial optimization problem. Of course, if a single run is all that is used, nothing much can be gleaned, but if independent restarts or a Metropolis-type search are used, it becomes possible to collect statistics and make use of them.

The existence of a "big valley" is usually an encouraging feature, and requires little checking. Assuming the global optimum is unknown it will not be possible to do a complete analysis, but useful information can be gained by computing the distance of each local optimum from the best local optimum, and plotting this against their corresponding differences in fitness. A strong correlation is indicative of a "big valley", and motivates the application of metaheuristics that perform intensive searches in the region of "good" local optima.

If every local optimum ever found is distinct, not much more can be done, but if it is noticed that specific local optima are being detected multiple times, it becomes possible to provide indications of solution quality, using statistical estimation tools based on the waiting-time or counting models mentioned above. For low values of the ratio k/r (see above), it may even be possible to provide a (probabilistic) guarantee that the global optimum has indeed been found.

19.8   CONCLUSIONS

This chapter has reviewed and discussed in some detail the basic mathematical theory and methods associated with the concept of a fitness landscape. While these methods can be very useful in enhancing our understanding of evolutionary algorithms, it has been emphasized that they cannot provide a complete explanation for the performance of a specific algorithm on their own—even in the case of very simple functions. Secondly, and more briefly, some empirically determined properties of many search landscapes have been described, and one approach whereby such properties can be exploited has been outlined.

As our understanding of the nature of fitness landscapes and how to exploit them develops, this promises to become an important area of research into the theory and application of heuristic search.

SOURCES OF ADDITIONAL INFORMATION

- For technical and theoretical analysis, there are many papers associated with Peter Stadler and his co-workers. The paper of Reidys and Stadler (2002) is perhaps the most readily accessible and recent treatment of theoretical properties of landscapes, although the seminal work is still Stadler (1995). Many of these papers can be found on the University of Vienna website:
  http://www.tbi.univie.ac.at/~studla/publications.html
  and also at the Santa Fe Institute:
  http://www.santafe.edu/sfi/publications/working-papers.html.

- Several papers give a general low-tech introduction to landscapes, (for example, Reeves, 1999, 2000), as does the chapter in the book by Reeves and Rowe (2002).

- For correlation analysis, Weinberger (1990) is still a major source of information, supplemented by more recent work in papers by Stadler and co-workers (see the Vienna website); another useful reference is Hordijk (1996).

- For work relating to the "big valley" and its exploitation, there are several important papers: Boese et al. (1994), Reeves and Yamada (1998), Merz and Freiselben (1998), and Reeves (1999); a chapter by Reeves and Yamada in Corne et al. (1999) is also an accessible introduction.

▪ The statistical approach to estimation of landscape properties is described in a series of papers (Eremeev and Reeves, 2002, 2003; Reeves and Eremeev, 2003). Its extension to the use of the Metropolis algorithm is considered in Reeves and Aupetit-Bélaidouni (2004).

References

Biggs, N. L., 1993, *Algebraic Graph Theory*, Cambridge University Press, Cambridge.

Boese, K. D., Kahng, A. B. and Muddu, S., 1994, A new adaptive multi-start technique for combinatorial global optimizations, *Oper. Res. Lett.* **16**:101–113.

Box, G. E. P. and Jenkins, G. M., 1970, *Time Series Analysis, Forecasting and Control*, Holden Day, San Francisco.

Corne, D. A., Dorigo, M. and Glover, F., eds, 1999, *New Methods in Optimization*, McGraw-Hill, London.

Dawkins, R. (1996) *Climbing Mount Improbable*, Viking, London.

Dobzhansky, T., 1951, *Genetics and the Origin of Species*. Columbia University Press, New York.

Eigen, M., 1993, Viral quasispecies, *Sci. Am.* **269**:32–39.

Eigen, M., McCaskill, J. and Schuster, P., 1989, The molecular quasi-species, *Adv. Chem. Phys.* **75**:149–263.

Eldredge, N. and Cracraft, J., 1980, *Phylogenetic Patterns and the Evolutionary Process*, Columbia University Press, New York.

Eremeev, A. V. and Reeves, C. R., 2002, Non-parametric estimation of properties of combinatorial landscapes, in: *Applications of Evolutionary Computing*, Lecture Notes in Computer Science, Vol. 2279, J. Gottlieb and G. Raidl, ed., Springer, Berlin, pp. 31–40.

Eremeev, A. V. and Reeves, C. R., 2003, On confidence intervals for the number of local optima, in: *Applications of Evolutionary Computing*, Lecture Notes in Computer Science, Vol. 2611, G. Raidl et al., ed., Springer, Berlin, pp. 224–235.

Flamm, C., Hofacker, I. L., Stadler, P. F. and Wolfinger, M. T., 2002, Barrier trees of degenerate landscapes, *Z. Phys. Chem.* **216**:155–173.

Futuyma, D. J., 1998, *Evolutionary Biology*, Sinauer Associates, Sunderland, MA.

Godsil, C. D., 1993, *Algebraic Combinatorics*, Chapman and Hall, London.

Grover, L. K., 1992, Local search and the local structure of NP-complete problems, *Oper. Res. Lett.* **12**:235–243.

Haldane, J. B. S., 1931, A mathematical theory of natural selection, Part VI: Metastable populations, *Proc. Camb. Phil. Soc.* **27**:137–142.

Hordijk W., 1996, A measure of landscapes, *Evol. Comput.* **4**:335–360.

Johnson, D. S., 1990, Local optimization and the traveling salesman problem, in: *Automata, Languages and Programming*, Lecture Notes in Computer Science, Vol. 443, G. Goos and J. Hartmanis, eds, Springer, Berlin, pp. 446–461.

Jones, T. C., 1995, *Evolutionary Algorithms, Fitness Landscapes and Search*, Doctoral dissertation, University of New Mexico, Albuquerque, NM.

Kauffman, S., 1993, *The Origins of Order: Self-Organization and Selection in Evolution*, Oxford University Press, Oxford.

Levenhagen, J., Bortfeldt, A. and Gehring, H., 2001, Path tracing in genetic algorithms applied to the multiconstrained knapsack problem, in: *Applications of Evolutionary Computing*, E. J. W. Boers et al., eds, Springer, Berlin, pp. 40–49.

Lin, S., 1965, Computer solutions of the traveling salesman problem, *Bell Syst. Tech. J.* **44**:2245–2269.

Martin, O., Otto, S. W. and Felten, E. W., 1992, Large step Markov chains for the TSP incorporating local search heuristics. *Oper. Res. Lett.* **11**:219–224.

Merz, P. and Freisleben, B., 1998, Memetic algorithms and the fitness landscape of the graph bi-partitioning problem, in: *Parallel Problem-Solving from Nature—PPSN V*, A. E. Eiben, T. Bäck, M. Schoenauer and H-P. Schwefel, eds, Springer, Berlin, pp. 765–774.

Reeves, C. R., 1994, Genetic algorithms and neighbourhood search, in: *Evolutionary Computing: AISB Workshop, Leeds, UK, April 1994; Selected Papers*, T. C. Fogarty, ed., Springer, Berlin, pp. 115-130.

Reeves, C. R. and Yamada, T., 1998, Genetic algorithms, path relinking and the flowshop sequencing problem, *Evol. Comput.*, **6**:45–60.

Reeves, C. R., 1999, Landscapes, operators and heuristic search. *Ann. Oper. Res.* **86**:473–490.

Reeves, C. R. and Yamada, T., 1999, *Goal-Oriented Path Tracing Methods*, in: *New Methods in Optimization*, D. A. Corne, M. Dorigo and F. Glover, eds, McGraw-Hill, London.

Reeves, C. R., 2000, Fitness landscapes and evolutionary algorithms, in: *Artificial Evolution: 4th Eur. Conf., AE99*, Lecture Notes in Computer Science, Vol. 1829, C. Fonlupt, J-K. Hao, E. Lutton, E. Ronald and M. Schoenauer, eds, Springer, Berlin, pp. 3–20.

Reeves, C. R., 2001, Direct statistical estimation of GA landscape features, in: *Foundations of Genetic Algorithms 6*, W. N. Martin and W. M. Spears, eds, Morgan Kaufmann, San Mateo, CA, pp. 91–107.

Reeves, C. R. and Rowe, J. E., 2002, *Genetic Algorithms—Principles and Perspectives*, Kluwer, Norwell, MA.

Reeves, C. R. and Eremeev, A. V., 2004, Statistical analysis of local search landscapes, *J. Oper. Res. Soc.* **55**:687–693.

Reeves, C. R., 2004, Partitioning landscapes. Available online at http://www.dagstuhl.de/04081/Talks/

Reeves, C. R. and Aupetit-Bélaidouni, M., 2004, Estimating the number of solutions for SAT problems, in: *Parallel Problem-Solving from Nature—PPSN VIII*, X. Yao et al., eds, Springer, Berlin, pp. 101–110.

Reidys, C. M. and Stadler, P. F., 2002, Combinatorial landscapes, *SIAM Rev.* **44**:3–54.

Ridley, M., 1993, *Evolution*, Blackwell, Oxford.

Simpson, G. G., 1953, *The Major Features of Evolution*, Columbia University Press, New York.

Stadler, P. F., 1995, *Towards a Theory of Landscapes*, in: *Complex Systems and Binary Networks*, R. Lopéz-Peña, R. Capovilla, R. García-Pelayo, H. Waelbroeck and F. Zertuche, eds, Springer, Berlin, pp. 77–163.

Stadler, P. F. and Wagner, G. P., 1998, Algebraic theory of recombination spaces, *Evol. Comput.* **5**:241–275.

Waterman, M. S., 1995, *Introduction to Computational Biology*, Chapman and Hall, London.

Watson, J-P, Barbalescu, L., Whitley, L. D. and Howe, A. E., 2002, Contrasting structured and random permutation flow-shop scheduling problems: Search-space topology and algorithm performance, *INFORMS J. Comput.* **14**:98–123.

Weinberger, E. D., 1990, Correlated and uncorrelated landscapes and how to tell the difference, *Biol. Cybernet.* **63**:325–336.

Wright, S., 1932, The roles of mutation, inbreeding, crossbreeding and selection in evolution, in: *Proc. 6th Int. Congress on Genetics*, D. Jones, ed., **1**:356–366.

Wright, S., 1967, Surfaces of selective value, *Proc. Natl Acad. Sci. USA* **102**:81–84.

Wright, S., 1988, Surfaces of selective value revisited, *Am. Nat.*, **131**:115–123.

Yamada, T. and Reeves, C. R., 1998, Solving the C_{sum} permutation flowshop scheduling problem by genetic local search, in: *Proc. 1998 Int. Conf. on Evolutionary Computation*, IEEE, Piscataway, NJ, pp. 230–234.

Zweig, G., 1995, An effective tour construction and improvement procedure for the traveling salesman problem, *Oper. Res.* **43**:1049–1057.

Index